OPTICAL METHODS
IN CELL PHYSIOLOGY

VOLUME 40

IN THE SOCIETY OF GENERAL
PHYSIOLOGISTS SERIES

OPTICAL METHODS IN CELL PHYSIOLOGY

Editors:

PAUL DE WEER
Department of Cell Biology and Physiology
Washington University
St. Louis, Missouri

BRIAN M. SALZBERG
Department of Physiology
University of Pennsylvania
Philadelphia, Pennsylvania

Society of General Physiologists and

WILEY-INTERSCIENCE

New York • Chichester • Brisbane • Toronto • Singapore

Library of Congress Cataloging in Publication Data:
Optical methods in cell physiology.

(Society of General Physiologists series: v. 40)
Based on a symposium organized by the Society of
General Physiologists, held in July 1985 at Woods Hole, Mass.
Includes index.
1. Cell physiology—Congresses. 2. Imaging systems
—Congresses. 3. Indicators (Biology)—Congresses.
I. De Weer, Paul, 1938-   . II. Salzberg, Brian M.
(Brian Matthew), 1942-   . III. Society of General
Physiologists. IV. Series. [DNLM: 1. Cells—physiology—
congresses. 2. Image Enhancement—
congresses. 3. Optics—congresses. Wi SO872G v.40/

QH 631 062 1985]
QH631.068 1986   574.87'6'028   85-22493
ISBN 0-471-82215-9

CONTRIBUTORS

ROBERT DAY ALLEN   Dartmouth College, Hanover, New Hampshire

L. ANGLISTER   Department of Neurobiology, Weizmann Institute of Science, Rehovot, Israel

S. M. BAYLOR   Department of Physiology, University of Pennsylvania, Philadelphia, Pennsylvania

JOHN R. BLINKS   Department of Pharmacology, Mayo Graduate School of Medicine, Rochester, Minnesota

S. R. BOLSOVER   Department of Ophthalmology, Washington University School of Medicine, St. Louis, Missouri

J. E. BROWN   Department of Ophthalmology, Washington University School of Medicine, St. Louis, Missouri

LEE D. CHABALA[+]   Division of Biology, California Institute of Technology, Pasadena, California

JAMES M. COGGINS   Department of Computer Science, Worcester Polytechnic Institute, Worcester, Massachusetts

C. S. COHAN   Department of Biology, University of Iowa, Iowa City, Iowa

LAWRENCE B. COHEN   Department of Physiology, Yale University School of Medicine, New Haven, Connecticut

STEPHEN DILLON   Department of Pharmacology, College of Physicians and Surgeons, Columbia University, New York, New York

GORDON W. ELLIS   University of Pennsylvania, Philadelphia, Pennsylvania

ELLIOT L. ELSON   Department of Biological Chemistry, Division of Biology and Biomedical Sciences, Washington University School of Medicine, St. Louis, Missouri

[+]*Present address*: Department of Physiology and Biophysics, Cornell Medical College, New York, New York.

FREDRIC S. FAY   Department of Physiology, University of Massachusetts Medical School, Worcester, Massachusetts

NED FEDER   National Institute of Arthritis, Diabetes and Digestive and Kidney Diseases, National Institutes of Health, Bethesda, Maryland

KEVIN E. FOGARTY   Department of Physiology, University of Massachusetts Medical School, Worcester, Massachusetts

J. A. FREEMAN   Department of Anatomy, Vanderbilt University, Nashville, Tennessee

HAROLD GAINER   National Institutes of Health, Bethesda, Maryland

Y. E. GOLDMAN   Department of Physiology, University of Pennsylvania School of Medicine, Philadelphia, Pennsylvania

T. H. GOLDSMITH   Department of Biology, Yale University, New Haven, Connecticut

A. GRINVALD   Department of Neurobiology, Weizmann Institute of Science, Rehovot, Israel

ALISON M. GURNEY   Division of Biology, California Institute of Technology, Pasadena, California

R. HILDESHEIM   Department of Neurobiology, Weizmann Institute of Science, Rehovot, Israel

JOSEPH F. HOFFMAN   Department of Physiology, Yale University School of Medicine, New Haven, Connecticut

S. HOLLINGWORTH   Department of Physiology, University of Pennsylvania, Philadelphia, Pennsylvania

SHINYA INOUÉ   Marine Biological Laboratory, Woods Hole, Massachusetts

THEODORE INOUÉ   Universal Imaging Corporation, Falmouth, Massachusetts

G. A. JACOBS   Department of Zoology, University of California, Berkeley, California

JACK H. KAPLAN   Department of Physiology, University of Pennsylvania School of Medicine, Philadelphia, Pennsylvania

S. B. KATER   Department of Biology, University of Iowa, Iowa City, Iowa

J. I. KORENBROT   Department of Physiology, University of California Medical School, San Francisco, California

U. KUHNT   Max-Planck Institut für Biophysik und Chemie, Göttingen, West Germany

PHILIP C. LARIS   Department of Biological Sciences, University of California, Santa Barbara, California

SARAH LESHER   Department of Physiology, Yale University School of Medicine, New Haven, Connecticut

HENRY A. LESTER   Division of Biology, California Institute of Technology, Pasadena, California

J. A. LONDON   Department of Physiology, Yale University School of Medicine, New Haven, Connecticut

A. MANKER   Department of Neurobiology, Weizmann Institute of Science, Rehovot, Israel

D. L. MILLER   Department of Physiology, University of California Medical School, San Francisco, California

J. P. MILLER   Department of Zoology, University of California, Berkeley, California

EDWIN D. W. MOORE   Department of Pharmacology, Mayo Graduate School of Medicine, Rochester, Minnesota

MARTIN MORAD   Department of Physiology, University of Pennsylvania, Philadelphia, Pennsylvania

L. J. MULLINS   Department of Biophysics, University of Maryland School of Medicine, Baltimore, Maryland

JEANNE M. NERBONNE*   Division of Biology, California Institute of Technology, Pasadena, California

ANA LIA OBAID   University of Pennsylvania, Philadelphia, Pennsylvania

D. L. OCHS   Department of Physiology, University of California Medical School, San Francisco, California

BRIAN M. SALZBERG   University of Pennsylvania, Philadelphia, Pennsylvania

M. SEGAL   Department of Neurobiology, Weizmann Institute of Science, Rehovot, Israel

ROBERT E. SHERIDAN‡   Division of Biology, California Institute of Technology, Pasadena, California

STEPHEN J. SMITH   Section of Molecular Neurobiology, Yale University School of Medicine, New Haven, Connecticut

*Present address: Department of Pharmacology, Washington University School of Medicine, St. Louis, Missouri.
‡Present address: Department of Pharmacology, Georgetown University School of Medicine and Dentistry, Washington, D.C.

WALTER W. STEWART   National Institute of Arthritis, Diabetes and Digestive and Kidney Diseases, National Institutes of Health, Bethesda, Maryland

JOHN A. THOMAS   Department of Biochemistry, University of South Dakota School of Medicine, Vermillion, South Dakota

ROGER Y. TSIEN   Department of Physiology–Anatomy, University of California, Berkeley, California

ALAN S. WAGGONER   Department of Biological Sciences, Carnegie-Mellon University, Pittsburgh, Pennsylvania

JAMES WEISS   Division of Cardiology, University of California-Los Angeles Health Center, Los Angeles, California

J. A. WILLIAMS   Department of Medicine, University of California Medical School, San Francisco, California

BRIAN E. WOLF   Chemistry Department, Amherst College, Amherst, Massachusetts

D. ZEĆEVIĆ   Department of Physiology, Yale University School of Medicine, New Haven, Connecticut

PREFACE

The editors' aim was to assemble under a single cover a series of authoritative chapters concerned with the use of advanced optical imaging, recording, measurement, and photochemical techniques in the study of cell function and of functional interaction within ensembles of cells. Throughout, emphasis was placed not only on the physiological insight that can be gained from such studies, but also on the theory behind the techniques, as well as on the instrumentation requirements, so that both established researchers and students may become more cognizant of the applicability, technical limitations, and specific advantages and disadvantages of the various approaches. Four broad areas are represented.

Part 1 deals with computer-enhanced video techniques which have greatly advanced the study of intact cell motility, cytoplasmic streaming, reconstituted motile systems, flagellar motion, spindle formation, and so forth. Video enhancement of low-level fluorescent marker molecules, including labelled antibodies, now permits long-term observations on living cells and the morphological changes they undergo. A chapter on the optical reconstruction of the three-dimensional distribution of fluorescent molecules in single smooth muscle cells, after deconvolution of a point-spread function, is representative of several recent developments in this area.

Part 2 examines the use of molecular optical probes (potentiometric dyes) for monitoring transmembrane electrical events in excitable cells and in regions of cells which are otherwise inaccessible. More sensitive and less obtrusive probes are still being developed and rational dye design based on a physicalchemical understanding of their mechanism now appears possible. Powerful techniques for multiple site optical recording of transmembrane potential (MSORTV) are presented, and their application to systems as diverse as vertebrate nerve terminals and an intact mammalian heart are described.

Part 3 of this book is devoted to methods using intracellular indicators whose optical properties (absorbance, fluorescence, luminescence) change in response to changes in this environment. These indicators are introduced into cells either by microinjection or by exploiting intracellular esterases which hydrolyze a relatively permeant ester, trapping the impermeant indicator inside the cell. Optical monitoring of the binding state of intracellular indicators allows fine spatial and temporal resolution of signals reflecting free concentrations of important ions, as described in this volume. This field is continuing to advance on a variety of fronts: hitherto unreported interference of magnesium in the response of the photoprotein aequorin to calcium is described; the stoichiometry and reaction rate constants of calcium indicators such as Arsenazo III are being refined; binding of indicators to cell protein is better understood; new and more powerful indicators such as fura-2 and others are being developed.

Part 4 of this volume includes chapters on the photobleaching recovery technique in the study of lateral mobility of membrane proteins, and the manufacture and use of photo-labile ''caged'' compounds that can be used to deliver extra- and intracellular pulses of H^+, Ca^{2+}, ATP, cAMP, neurotransmitters, and so forth.

For a number of years the Society of General Physiologists has organized and published timely symposia on important topics of interest to cell physiologists. This volume is the Society's reaction to a perceived need for a single volume that could serve as a reference source for a variety of optical techniques in the study of cell physiology. If the response of those who attended the symposium is any indication, this volume could satisfy that need.

PAUL DE WEER
BRIAN M. SALZBERG

Woods Hole, Massachusetts
July 1985

ACKNOWLEDGMENTS

We thank the Director and Staff of the Marine Biological Laboratory, Woods Hole, Massachusetts, and the officers of the Society of General Physiologists for their contribution to the success of this Symposium. The obvious enthusiasm of the speakers, and the very high caliber of their presentations, made this gathering an extremely enjoyable one for all who attended. We trust that our readers will also sense the excitement that pervaded this meeting. This Symposium was supported in part by grants from the National Science Foundation and the Office of Naval Research.

CONTENTS

PART 1   IMAGE ENHANCEMENT TECHNIQUES

1. *Video-Enhanced Microscopy*   3
    Robert Day Allen

2. *Computer-Aided Light Microscopy*   15
    Gordon W. Ellis, Shinya Inoué, and Theodore Inoué

3. *Image Intensification of Stained, Functioning,
    and Growing Neurons*   31
    S. B. Kater, C. S. Cohan, G. A. Jacobs, and J. P. Miller

4. *Analysis of Molecular Distribution in Single Cells Using
    a Digital Imaging Microscope*   51
    Fredric S. Fay, Kevin E. Fogarty, and James M. Coggins

5. *Attempts to Synthesize a Red-Fluorescing Dye
    for Intracellular Injection*   65
    Walter W. Stewart and Ned Feder

PART 2   OPTICAL PROBES OF MEMBRANE POTENTIAL

6. *Optical Monitoring of Membrane Potential; Methods
    of Multisite Optical Measurement*   71
    Lawrence B. Cohen and Sarah Lesher

7. *Optical Studies of the Mechanism of Membrane Potential
    Sensitivity of Merocyanine 540*   101
    Brian E. Wolf and Alan S. Waggoner

8. *Simultaneous Monitoring of Activity of Many Neurons from
    Invertebrate Ganglia Using a Multielement Detecting System*   115
    J. A. London, D. Zécvič, and L. B. Cohen

9. *Optical Studies of Excitation and Secretion at Vertebrate Nerve
    Terminals*   133
    Brian M. Salzberg, Ana Lia Obaid, and Harold Gainer

10. *Real-Time Optical Mapping of Neuronal Activity
    in Vertebrate CNS in Vitro and in Vivo*   165
    **A. Grinvald, M. Segal, U. Kuhnt, A. Manker, L. Anglister,
    J. A. Freeman, and R. Hildesheim**

11. *Optical Determination of Electrical Properties of Red Blood
    Cell and Ehrlich Ascites Tumor Cell Membranes with
    Fluorescent Dyes*   199
    Philip C. Laris and Joseph F. Hoffman

12. *An Acousto-Optically Steered Laser Scanning System for
    Measurement of Action Potential Spread in Intact Heart*   211
    Martin Morad, Stephen Dillon, and James Weiss

PART 3   INTRACELLULAR INDICATORS

13. *Practical Aspects of the Use of Photoproteins as Biological
    Calcium Indicators*   229
    John R. Blinks and Edwin D. W. Moore

14. *Strategies for the Selective Measurement of Calcium
    in Various Regions of an Axon*   239
    L. J. Mullins

15. *Polychromator for Recording Optical Absorbance Changes
    from Single Cells*   255
    Stephen J. Smith

16. *Calcium Transients in Frog Skeletal Muscle Fibers Injected
    with Azo1, a Tetracarboxylate Ca^{2+} Indicator*   261
    S. Hollingworth and S. M. Baylor

17. *Intracellular pH of* Limulus *Ventral Photoreceptor Cells:
    Measurement with Phenol Red*   285
    S. R. Bolsover, J. E. Brown, and T. H. Goldsmith

18. *Intracellularly Trapped pH Indicators*   311
    John A. Thomas

19. *New Tetracarboxylate Chelators for Fluorescence Measurement and Photochemical Manipulation of Cytosolic Free Calcium Concentrations*   327
   Roger Y. Tsien

20. *The Use of Tetracarboxylate Fluorescent Indicators in the Measurement and Control of Intracellular Free Calcium Ions*   347
   J. I. Korenbrot, J. E. Brown, D. L. Ochs, J. A. Williams, and D. L. Miller

PART 4   PHOTOBLEACHING AND PHOTOACTIVATION

21. *Membrane Dynamics Studied by Fluorescence Correlation Spectroscopy and Photobleaching Recovery*   367
   Elliot L. Elson

22. *Caged ATP as a Tool in Active Transport Research*   385
   Jack H. Kaplan

23. *Laser Pulsed Release of ATP and Other Optical Methods in the Study of Muscle Contraction*   397
   Y. E. Goldman

24. *Design and Application of Photolabile Intracellular Probes*   417
   Jeanne M. Nerbonne

25. *Experiments with Photoisomerizable Molecules at Nicotinic Acetylcholine Receptors in Cells and Membrane Patches from Rat Muscle*   447
   Henry A. Lester, Lee D. Chabala, Alison M. Gurney, and Robert E. Sheridan

   Index   463

PART **1**

IMAGE ENHANCEMENT TECHNIQUES

CHAPTER 1

VIDEO-ENHANCED MICROSCOPY

ROBERT DAY ALLEN

Dartmouth College
Hanover, New Hampshire

	INTRODUCTION	4
1.	THE MICROSCOPE IMAGE	5
2.	THE TELEVISION CAMERA	5
3.	ANALOGUE ENHANCEMENT	6
4.	DIGITAL SUBTRACTION OF MOTTLE	7
5.	REDUCTION OF PIXEL NOISE BY AVERAGING	7
6.	ENHANCEMENT BY DIGITAL IMAGE PROCESSING	7
7.	MOTION DETECTION AND OBLITERATION	8
8.	SOME APPLICATIONS	9
9.	FUTURE PROSPECTS	10
	9.1 VIDEO-ENHANCED MICROSCOPY IN THE ULTRAVIOLET	10
	9.2 PHASE MODULATION VIDEO-ENHANCED MICROSCOPY	10
	9.3 QUANTITATIVE MICROSCOPY	11
	CONCLUSION	12
	REFERENCES	12

INTRODUCTION

It has only recently become known that optical microscopes transmit vast amounts of information that is not detectable or comprehensible to human vision without the aid of electronic imaging and some form of image processing.

It was accidentaly discovered during a short course I have offered annually on optical microscopy at the Marine Biological Laboratory that under certain conditions a television camera can make visible and enhance fine details in images that the unaided eye simply cannot detect. This finding marked the beginning of a revolution in optical microscopy through the development of a collection of powerful techniques of video enhancement applicable to the various modes of optical microscopy used in biomedical and materials science (1, 3, 6).

Enhancement of optical images is not new. Each improvement in the design of microscope lenses that has led to present internally corrected planapochromatic objectives in the Zeiss Axiomat (23) can be considered to be a step in the process of optical enhancement. Equally important have been the inventions of spatial filtration techniques, such as dark-field, the various forms of anaxial (oblique) illumination, phase contrast, double-beam and especially differential interference contrast, that generate enough image contrast to allow us to see details in phase objects that are not visible with bright-field microscopy. We are indebted to many pioneers in optical microscopy for their important contributions to optical image enhancement of phase objects, including Ernst Abbe, Frits Zernike, F. H. Smith, Horst Piller, Andrew F. Huxley, Shinya Inoué, and Georges Nomarski.

Polarized light microscopy of birefringent and dichroic objects has been a specialized but particularly fertile field for optical image enhancement through successful efforts to reduce stray light resulting mainly from strain birefringence in lens elements and depolarization due to reflection. The invention of polarization rectifiers independently by Inoué and Hyde (19) and A. F. Huxley (18) laid the foundations for the refined technique of high extinction polarized light microscopy, which until recently has been the method of choice for studies of such cellular processes as mitosis (20).

A third form of image enhancement has been photographic techniques that utilize either "photographic inertia" or high gamma development and printing to generate image contrast as an aid to vision and documentation.

The fourth and by far the most powerful form of image enhancement is techniques of electronic imaging, both analogue and digital image processing. It has been in electronic imaging that greatest progress has been achieved in terms of technical improvements, instrument development, and new scientific information.

I should therefore like to describe the theory and technique of video-enhanced microscopy with particular reference to a commercial instrument that I helped to develop: the Hamamatsu C-1966 Photonic **Microscope** System (distributed in the United States by Photonic Microscopy, Inc., Oak Brook, IL 60521). This system was conceived in January 1981, and its development was undertaken in October 1982. A first prototype was tested in May 1983; a second was demonstrated in August 1983 in

Woods Hole, Massachusetts. The final instrument was exhibited in San Antonio, Texas, in November 1983 at the Annual Meeting of the American Society for Cell Biology, at MICRO '83 in London, and at the International Congress of Cell Biology, Tokyo, in August 1984.

1. THE MICROSCOPE IMAGE

Any microscope can be used for video-enhanced microscopy, and the images it produces will be indeed enhanced. However, each microscope has its own set of limitations, some of which can limit severely the quality of images in the case of inexpensive microscopes at magnifications of $\sim 10,000\times$. We have selected the inverted Zeiss Axiomat because of its mechanical stability, the superior image quality of its internally corrected planapochromatic lenses, its zoom magnifier, and the convenient auxilliary aperture plane for the placement of spatial filtration devices (23).

Images obtained by any of the normal contrasting modes of the light microscope can be enhanced by the electronic procedures to be described (1). However, the microscope is usually adjusted in a somewhat different manner from that for visual microscopy. For example, the iris diaphragm is usually kept open. This naturally floods the field of the microscope with stray light, making the image useless for visual microscopy. However, doing so provides optimal conditions for high-resolution video-enhanced microscopy.

The microscope is set up according to the principles of Köhler illumination (21) in order to make the brightness as uniform as possible over the field. Considerable care has to be taken in centering the light source, as uneven illumination can limit the extent of analogue enhancement that can be achieved. In addition, an even slightly off axis source can produce a shadow-cast image when it is not desired.

It is possible to enhance any of the many modes of microscopy applied to phase objects, including phase contrast, and various forms of anaxial illumination. However, the greatest gains are achieved using methods of polarization or interference microscopy because the setting of the compensator plays a large role in generating contrast in video enhancement (3, 6).

2. THE TELEVISION CAMERA

Many types of vidicons are available, and almost any can be used to obtain some degree of enhancement. The most important characteristics appear to be high dynamic range, low lag, and sensitivity, in that order. For this reason the current camera of choice for the methods I shall describe is the Chalnicon, with the Newvicon second best. Intensification cameras available so far are not recommended for enhancement because of their relatively small dynamic range. Solid-state cameras are expected to become available in the future with highly desirable characteristics.

3.   ANALOGUE ENHANCEMENT

Before 1981, analogue enhancement of television images was not practiced to any significant degree, and cameras, even those used for scientific image recording, had very limited ranges of black-level ("pedestal") adjustment and gain, if any. We discovered that when the optical bias retardation was $\lambda/9$–$\lambda/4$, the video image could be greatly enhanced by the use of a wide-range gain (up to 16 ×) and offset. Offset is a negative voltage (up to -3 V) that makes it possible to establish an arbitrary black level in the television image, regardless of the conditions of illumination in the optical image or amount of gain applied to the video output signal. This feature of the AVEC methods embodied in the C-1966 Photonic Microscope System is patented (U.S. Patent No. 4412246) and provides the entire raison d'être for the digital image processing equipment designed for image enhancement in optical microscopy. Neither digital subtraction nor computer enhancement or other processing applied alone or in combination with the visually optimal image results in significant gains in image quality per se, without prior analogue enhancement by the AVEC methods.

The explanation is as follows: Contrast (C) for the human visual system may be expressed as

$$C = \frac{I_B - I_S}{I_B}$$

where I_B and I_S are the brightnesses ($\approx$ intensities) of the background and specimen, respectively. Increasing the bias retardation in any polarizing or interference microscope from the traditional range of $\lambda/100$–$\lambda/50$ to the range of $\lambda/9$–$\lambda/4$ sharply increases the *potential* for a contrasty image by increasing the numerator of the above expression manyfold; applying gain increases it still further. However, the even greater increase in the denominator at high bias retardations makes the image appear washed out (in fact, usually invisible) to the eye. When the offset voltage, which is analogous to negative brightness, I_V, is subtracted from the denominator to create video contrast, (C_V), the image quality is at first rapidly restored, then enormously improved by the increase in contrast:

$$C_V = \frac{I_B - I_S}{I_B - I_V}$$

The arbitrary establishment of a black level with the offset voltage gives the operator the power to select a contrast comparable to that in an instrument having almost any high value of extinction factor desired. It is only in this way that structures in the 10- to 100-nm size range, such as single microtubules and intermediate filaments, can be detected in the AVEC-POL, AVEC-DIC, AVEC-DBI, and other optical modes. These methods do not require polarizing rectifiers to limit stray light and can give satisfactory results even if the optical extinction factor is spoiled by deliberately offsetting one of the crossed polarizing filters.

4. DIGITAL SUBTRACTION OF MOTTLE

A serious but correctable disadvantage of high-gain analogue enhancement is the contamination of the image by a pattern of mottle contributed by dirt and surface imperfections on the many glass elements in the microscope. This mottle is never seen in the visual image, but it can be photographed at small viewing apertures.

When real-time digital subtraction using an arithmetic logic unit and a single frame memory was introduced to solve this problem (2), the resulting images were remarkable for their clarity and revealed structures as small as microtubules (2) and synaptic vesicles moving along microtubules in axons (4, 5).

Specimens often introduce stray light from out-of-focus diffraction or birefringence. An important feature of the AVEC methods is the capacity to suppress stray light yet amplify the in-focus contrast to detect structures an order of magnitude smaller than the resolution limit. Hayden and Allen (17) have shown that it is possible not only to detect single microtubules (confirmed by electron microscopy) but to record the bidirectional motion of particles along these microtubules. Visual observation of these same cells at a similar or lower magnification with an extinction factor of about 1000 could not detect these structures or motions. It is perhaps still more remarkable that synaptic vesicles can be observed in motion in one optical section of a squid axon (5) or in extruded axoplasm (14) despite considerable light scattered by vast numbers of large and small particles throughout the 500-μm-thick axon.

5. REDUCTION OF PIXEL NOISE BY AVERAGING

At highest levels of analogue enhancement, pixel noise may be so noticeable that small details in the image may be lost. It was found that pixel noise could be dramatically reduced by either a rolling or jumping average over several video frames. The pixel noise averaged over n frames is reduced by a factor of $\sqrt{n}$. Since there are 30 frames per second, averaging over 16 frames—for example, ~ 0.5 sec—reduces pixel noise by a factor of 4.

6. ENHANCEMENT BY DIGITAL IMAGE PROCESSING

During the development of image processing as an art, microscopists were conspicuously absent from the scene, apparently secure in their collective opinion that microscope images seemed "good enough" not to require further processing. However, a number of textbooks cover the spectrum of digital image processing operations that are available (12, 16, 24).

Walter and Berns (28) awakened microscopists to some of the possibilities of off-line digital image processing by showing, as an example, how phase contrast images could be derivatized to produce shadow-cast differential images resembling those from differential interference contrast (DIC). In these differential images, there was an apparent superiority in image quality, but since the limiting factor was the image quality of the phase contrast image, it probably was attributable to the fact that the human visual system is more accustomed to differential images.

If the microscope is adjusted for optimal visual contrast, there is a limited amount of information available to a television and image processing system. Computer enhancement alone of such an image can improve it only to a limited degree. It is far better to use AVEC analogue video enhancement first to take full advantage of full aperture illumination and sufficient contrast to achieve resolution according to the Sparrow criterion (2). The only disadvantage in this approach is that some extra pixel noise will be introduced near the top of the dynamic range of the television camera tube. However, it is rare that the noise exceeds a level that can be removed by averaging over 2–16 frames.

Digital enhancement of analogue-enhanced images produces modest improvements in image quality by pixel histogram stretching or manipulation of the gamma of the display. These are processes similar to darkroom procedures for the control of photographic contrast, except that they are virtually instantaneous in a hard-wired arithmetic logic unit, such as the Hamamatsu C-1966 Photonic Microscope System. One can, for example, arrange the gamma correction in the display to reduce the brightness of structures that would otherwise dominate the image because of excessive brightness in the uncorrected image.

Another effective means of generating contrast intelligible to the eye is pseudo-color conversion. The eye is well known to be more sensitive to different hues than to different monochrome brightness levels. Therefore, the intelligent use of look-up tables specifying a color density wedge for a given range of brightness levels provides a certain measure of gain in image quality.

7.   MOTION DETECTION AND OBLITERATION

The real-time *rolling average* operation used to reduce pixel noise also serves to blur or obliterate any movements that occur during the averaging interval. On the other hand, the *jumping average* operation emphasizes any movements by presenting them as periodic, discrete motions. Another way of detecting or emphasizing motions is by *autosubtraction*. The real-time image is subtracted from a stored, in-focus image at time 0. As motions occur, they are the only events seen in an otherwise featureless autosubtraction image. This is a remarkably sensitive method of motion detection that works properly only with a highly stable microscope kept in a temperature-controlled, draft-free room. If the initial frame is refreshed at regular intervals, the operation becomes *sequential subtraction*, which has the advantage that the image does not become excessively complex with time, yet motions are clearly and selectively observed. If sequentially subtracted images of a moving bacterium or ciliate, for example, are presented simultaneously, as in *trace* mode, one can easly record the trajectory and analyze velocities from stored images.

The digital image processing portion of the Photonic Microscope system is capable of a number of additional arithmetic functions that either enhance an image, such as *differential* mode or several schemes for computer enhancement. It also performs matrix filtration and summing of light-limited images, functions that are of particular importance in low-light imaging.

8. SOME APPLICATIONS

Our laboratory has been interested since 1964 in a dramatic motility process known as reticulopodial movement in marine Foraminifera and Radiolaria. The reticulopodial networks of these organisms are used for locomotion, feeding, and excretion (7). We were intrigued by the bidirectional transport of particles along these thin cellular extensions.

In 1965, a graduate student, William Burdwood, who was familiar with reticulopodial movement in Foraminifera, reported similar movements of particles being transported bidirectionally in cultured chick dorsal root neurons in culture (15). For several years we made parallel studies of the movements in the foraminiferan, *Allogromia* (22), and in neurons (13). The former were easier to grow and study, so we were eager to find out whether their motility and particle transport processes were similar in mechanism.

Microtubules were found to be the structural members in the vicinity of which the particle transport occurred, both in *Allogromia* (22) and in chick neurons (13). At that time, the light microscope could not detect single microtubules, although in mitotic spindles it was clear that a bundle of a few of them could be detected (11).

The discovery of the AVEC methods of video enhancement in 1980 led us to have yet another look—first at *Allogromia* and then at neurons. In *Allogromia*, the microtubular bundles that carried particles in both directions could be clearly seen, and microtubules were seen for the first time to be undergoing movements, both lateral and axial (6, 27). These findings suggested the possibility that microtubules themselves might be the motive-force-producing elements, although it was equally possible that the microtubules were being moved by particles with force transducers on their surfaces or by a second system of nonactin, 5-nm filaments (26).

In 1981 we began to examine giant neurons of the squid, lobster, and the marine worm, *Myxicola*. In all, we observed particles associated with microtubules, apparently moving in both directions along them. At that time it could not be determined that one microtubule could support bidirectional particle motion. This was demonstrated first by Hayden and Allen (17) and confirmed by others using the AVEC method (25).

The squid axon proved to be particularly favorable material for exploring the function of microtubules in particle transport. We found that axoplasm could be extruded and then dissociated without losing the motion of particles along microtubules (14). Sometimes fine "filaments," which we suspected were single microtubules because they had the same contrast in polarized light and DIC as reassembled microtubles, would work their way out of the axoplasm–buffer interface, and they would have particles moving in both directions along them.

More recently, we have completely dissociated these "filaments" from extruded squid axoplasm and have shown that they are native microtubules that have independent motility. When sheared into lengths of 1–30 μm and provided with >1.0 mM of ATP, they glide at uniform velocity over glass surfaces treated in a number of different ways. These same microtubule segments transport particles in one or both directions *while gliding* (10). [A video tape was presented showing, as examples of the techniques discussed, fast axonal transport in the intact squid giant axon and recent experiments, to be published elsewhere (10), on the isolation of native microtubules that transport particles. Fragments of these microtubules glide on glass surfaces if ATP is available.]

9.   FUTURE PROSPECTS

9.1   Video-Enhanced Microscopy in the Ultraviolet

We have begun to explore the possibilities of video-enhanced contrast microscopy in the UV region of the spectrum. Since proteins, nucleic acids, and many other cell constituents absorb in the ultraviolet, UV microscopy has already been used extensively in the bright-field mode for high-resolution microspectrophotometry.

The resolution attainable at the shorter wavelength limit of the transmittance of microscope optics for the ultraviolet should be 2–3 times better than the ~ 100- to 200-nm resolution attained in green light. The detectability of small phase objects is predicted to benefit not only from the improved resolution but from the fact that such objects will generate higher contrast because of the larger phase shifts they induce in shorter wavelengths of light.

The prospects for observing cellular processes in UV light are dimmed somewhat by the well-known thermal and photochemical damage to living cells caused by high-energy photons. However, the use of highly sensitive television cameras now under development may make it possible to observe living processes over some tens of minutes using an attenuated source and integration of low-dose images as part of the image processing strategy.

Even if living cells should prove too sensitive to the damage caused by observing them in UV light, there are excellent prospects for observing model systems simulating living processes, such as microtubular motility and actin–myosin interactions, and for seeing details of chromosome and mitotic spindle structure, for example, at a resolution of 30–50 nm.

9.2   Phase Modulation Video-Enhanced Microscopy

In the 1960s the method of phase modulation microscopy was introduced as a means of measuring small changes in phase retardations due to birefringence or refraction

(8). It was a microbeam method in which the sampling area could be as small as an Airy disk. The incident polarized light beam was phase-modulated by an electrooptical light modulator consisting of a Pockels cell, and phase shifts were detected by an AC synchronous detection system and measured by a servo loop. Phase retardations as small as 0.1 Å could be detected and measured, and larger, rapid changes in retardation could be detected in 1 ms. The method served the purpose for which it was intended (9) but had a limited application because of its lack of spatial resolution unless the specimen was scanned over the measuring beam: a slow process at best.

Phase modulation techniques can be adapted easily to the present methods of video-enhanced microscopy and have much to offer. By phase-modulating the incoming polarized light in synchrony with the video framing rate, it is possible, by subtracting adjacent video frames or matched sequences of averaged frames, to display subtraction images that convey contrast due to a single optical property, such as birefringence. In principle, the contributions of absorption, scattering, and refraction to image contrast are identical at complementary states of elliptically polarized light, so these stray sources of contrast would be subtractively removed from the image, leaving specimen birefringence as the single optical property observed, and with doubled contrast. We have "walked through" this type of arithmetic operation by manual operation of both the compensator and digital subtraction. The images obtained are remarkably clear and crisp and free of bright-field contrast as predicted (2). There is, however, a small, spurious contrast contribution in the form of background mottle due to the reinforcement (by subtraction) of any residual lens birefringence. This, too, can be removed arithmetically by storing it in memory from blank-field images and subtracting blank-field contrast to remove this special form of mottle due to lens birefringence.

The prospects for further improvements in video enhancement methods by phase modulation techniques are excellent. With polarized light, we can look forward to highly sensitive imaging modes for the specific detection of optical rotation, birefringence, and linear and circular dichroism. The sensitivity of double-beam and differential interference contrast microscopy can similarly be doubled along with a similar removal of extraneous contrast.

9.3 Quantitative Microscopy

Although light microscopists have striven for quantitation of the various optical properties of specimens, especially those that convey molecular or at least chemical information, the goal has been elusive. It has been rare that a microscope image could be rigorously interpreted quantitatively in terms of a single optical property. With phase modulation techniques we can use the same principle used in classical quantitative polarized light and interference microscopy and, by time-varying changes in contrast, isolate the information we wish to record and analyze. To realize this goal, it will be necessary to use one plane of memory for shading correction to ensure that each pixel has the same sensitivity.

CONCLUSION

Video enhancement has literally revolutionized light microscopy. Structures and processes that have long lain far beyond the limits of the light microscope have suddenly become accessible. The technology for video enhancement has been at hand for a decade or more, waiting for its utility to be discovered. We now have commercial instruments and methods for their use that open many avenues in both biomedical and materials research.

REFERENCES

1. Allen, R. D. (1985) New observations on cell architecture and dynamics by video-enhanced contrast optical microscopy. *Annu. Rev. Biophys. Biochem.*, **14**, 265–290.

2. Allen, R. D., and N. S. Allen (1983) Video-enhanced microscopy with a computer frame memory. *J. Microsc.*, **129**, 3–17.

3. Allen, R. D., N. S. Allen, and J. L. Travis (1981) Video-enhanced contrast, differential interference contrast (AVEC-DIC) microscopy: A new method capable of analyzing microtubule related motility in the reticulopodial network of *Allogromia laticollaris. Cell Motil.* **1**, 291–302.

4. Allen, R. D., D. T. Brown, S. P. Gilbert, and H. Fujiwake (1983) Transport of vesicles along filaments dissociated from squid axoplasm. *Biol. Bull.*, **165**, 523.

5. Allen, R. D., J. Metuzals, I. Tasaki, S. T. Brady, and S. P. Gilbert (1982) Fast axonal transport in squid giant axon. *Science*, **218**, 1127–1129.

6. Allen, R. D., J. L. Travis, N. S. Allen, and H. Yilmaz (1981) Video-enhanced contrast polarization (AVEC-POL) microscopy: A new method applied to the detection of birefringence in the motile reticulopodial network of *Allogromia laticollaris. Cell Motil.*, **1**, 275–289.

7. Allen, R. D. (1964) "Cytoplasmic streaming and locomotion in marine foraminifera," in R. D. Allen and N. Kamiya, Eds., *Primitive Motile Systems in Cell Biology*, Academic, New York, pp. 407–431.

8. Allen, R. D., J. W. Brault, and R. Zeh (1966) "Image contrast and phase modulated light methods in polarization and interference microscopy," in R. Bauer and V. Cosslett, Eds., *Recent Advances in Optical and Electron Microscopy*, Academic, New York. pp. 77–114.

9. Allen, R. D., D. W. Francis, and R. Zeh (1971) Direct test of the positive pressure gradient theory of pseudopod extension and retraction in amoeba. *Science*, **174**, 1237–1240.

10. Allen, R. D., D. G. Weiss, J. H. Hayden, D. T. Brown, H. Fujiwake, and M. Simpson (1985) Gliding movement of and bidirectional organelle transport along single native microtubules from squid axoplasm: Evidence for an active role of microtubules in cytoplasmic transport. *J. Cell Biol.* 100, 1736–1752.

11. Bajer, A., and R. D. Allen (1966) Structure and organization of the living mitotic spindle of *Haemanthus katherinae. Science*, **151**, 572–574.

12. Baxes, G. A. (1984) *Digital Image Processing: A Practical Primer*. Prentice-Hall, Englewood Cliffs, NJ.

13. Berlinrood, M., S. M. McGee-Russel, and R. D. Allen (1972) Patterns of particle movement in nerve fibers in vitro: An analysis of photokymography and microscopy. *J. Cell Biol.*, **11**, 875–886.

14. Brady, S. T., R. J. Lasek, and R. D. Allen (1982) Fast axonal transport on extruded axoplasm from squid giant axon. *Science*, **218**, 1129–1131.

15. Burdwood, W. O. (1965) Rapid bidirectional particle movement in neurons. *J. Cell Biol.*, **27**, 115a.

16. Castleman, K. R. (1979) *Digital Image Processing*. Prentice-Hall, Englewood Cliffs, NJ.

17. Hayden, J. H., and R. D. Allen (1984) Detection of single microtubules in living cells: Particle transport can occur in both directions along the same microtubule. *J. Cell Biol.*, **99**, 1785–1793.

18. Huxley, A. F. British patent 856,621. London, England.

19. Inoué, S., and W. L. Hyde (1957) Studies on depolarization of light on microscope lens surfaces. II. The simultaneous realization of high resolution and high sensitivity with the polarizing microscope. *J. Biophys. Biochem. Cytol.*, **3**, 831–838.

20. Inoué, S. (1964) "Organization and function of the mitotic spindle," in R. D. Allen and N. Kamiya, Eds., *Primitive Motile Systems in Cell Biology*, Academic, New York, pp. 549–594.

21. Köhler, A. (1893) Ein neues Beleuchtungsverfahren für mikrophotographische Zwecke. *Z. Wiss Mikrosk.*, **10**, 433.

22. McGee-Russell, S. M., and R. D. Allen (1971) "Reversible stabilization of labile microtubules in the reticulopodial network of *Allogromia*," in E. J. Duprow, Ed., *Advances in Cell and Molecular Biology*, Vol. I, Academic, New York, pp. 153–184.

23. Michel, K. (1974) Zeiss Axiomat. A new concept in microscope design. *Zeiss Inform.*, **21**, 1–32.

24. Pratt, W. K. (1978) *Digital Image Processing*. Wiley, New York.

25. Schnapp, B., R. Vale, M. Sheetz, and T. B. Reese (1984) "Directed movements of organelles along filaments dissociated from squid giant axon," in S. Seno and Y. Okado, Eds., *International Cell Biology*, Japan Society for Cell Biology, Tokyo, p. 132.

26. Travis, J. L., and R. D. Allen (1981) Studies on the motility of the Foraminifera. I. Ultrastructure of the reticulopodial network of *Allogromia laticollaris* (Arnold). *J. Cell Biol.*, **90**, 211–221.

27. Travis, J. L., J. F. Kenealy, and R. D. Allen (1983) Studies on the motility of Foraminifera. II. The dynamic microtubular cytoskeleton of the reticulopodial network of *Allogromia laticollaris*. *J. Cell Biol.*, **97**, 1668–1676.

28. Walter, R. J., and M. W. Berns (1981) Computer-enhanced video microscopy: Digitally procesed microscope images can be produced in real time. *Proc. Natl. Acad. Sci. U.S.A.*, **78**, 6927–6931.

CHAPTER 2

COMPUTER-AIDED LIGHT MICROSCOPY

GORDON W. ELLIS

University of Pennsylvania
Philadelphia, Pennsylvania

SHINYA INOUÉ

Marine Biological Laboratory
Woods Hole, Massachusetts

THEODORE INOUÉ

Universal Imaging Corporation
Falmouth, Massachusetts

1.	INTRODUCTION	16
2.	DIGITAL IMAGE PROCESSING	16
3.	HARDWARE	18
4.	SOFTWARE	22
5.	RESULTS AND DISCUSSION	24
6.	CONCLUSIONS	29
	REFERENCES	29

1. INTRODUCTION

Despite its limited spatial resolution, the persisting value of light microscopy is that it enables the nondestructive examination of living cells under relatively normal physiological conditions. During the four decades since Zernike (1) introduced the phase contrast microscope, several other advances in optical contrast generation with the light microscope were made. These include double-beam (2–6) and multiple-beam (7) interference microscopes, differential interference contrast (DIC) microscopes (8, 9), the rectified polarizing microscope (10), reflection contrast microscopes (11–13), and single–sideband microscopes (14, 15). Each of these enhances the visibility or quantifiability of some feature of living cells.

Over the past two decades low light level microscope images, made visible by image intensifiers (16), have become accessible to improved video cameras (17, 18). More recently, video image enhancement (19–22) and digital image processing and analysis (23–26) have become of practical interest to microscopists, substantially owing to the increasingly economical availability of instrumentation cameras, recorders, processors, and computers.

2. DIGITAL IMAGE PROCESSING

Digital image processing has proceeded along two relatively independent lines which can be called image enhancement and image analysis. Image enhancement results in the improvement in visibility or discernibility of the image itself; image analysis concentrates on the extraction of numeric or geometric information about the image or the object it represents.

Image enhancement operations can be divided into two categories: contrast enhancement operations and spatial enhancement operations (26). Contrast enhancements are alterations in the relative brightness of the components of an image, and spatial enhancements result in modification of the structural details of the image. Both are helpful in improving the intelligibility of an image. Generally, contrast enhancements require point processing in which each pixel is processed independently from its neighbors, whereas spatial enhancements require group processing, as in convolution.

In any case, processing has the same beginning. The image to be processed must be converted into a digital representation of the two-dimensional intensity distribution of the original. For real-time image enhancement, if one is willing to settle for a maximum of 512 video lines of resolution in the horizontal direction and 480 video lines vertically, the hardware must be able to convert 245,760 picture elements (pixels) from a continuous analog function into discrete digital values 30 times each second (7,372,800 pixels per second—actually, the rate required is somewhat greater, because this number does not account for the nondisplayed portion of the video signal), modify each of these according to the processing function in use, and simultaneously reconvert them back to a continuous analog function complete with the necessary synchronization pulses.

When digital image processing began, about 30 years ago, real-time video image processing was a task well beyond the capability of the largest, most advanced computers available. Today, image processing hardware that easily meets these requirements is available for operation by microcomputers. Table 1 is a list of suppliers of image processing hardware of varying capabilities and priced from

TABLE 1.   IMAGE PROCESSING EQUIPMENT SUPPLIERS

Arlunya
The Dindima Group Pty. Ltd.
P.O. Box 106, Vermont
Victoria, 3133 Australia

Colorado Video
Box 928
Boulder, CO 80306

Comtal/3M
505 W. Woodbury Rd.
Altadena, CA 91001

Datacube, Inc.
4 Dearborn Rd.
Peabody, MA 01960

Gould DeAnza
1870 Lundy Ave.
San Jose, CA 95131

Grinnel Systems Corp.
6410 Via del Oro
San Jose, CA 95119

Hughes Aircraft Co.
Industrial Products Division
Carlsbad, CA 92008

Imaging Technology, Inc.
600 West Cummings Park
Woburn, MA 01801

Interactive Video Systems
358 Baker Ave.
Concord, MA 01742

International Imaging Systems
1500 Buckeye Dr.
Milpitas, CA 95035

Interpretation Systems, Inc.
6322 College Blvd.
Overland Park, Kansas 66211

Joyce-Loebl/Nikon
Nikon Inc.
Instrument Division
623 Stewart Ave.
Garden City, NY 11530

E. Leitz Inc.
Rockleigh, NJ 07647

Micro Consultants Ltd.
Princeton Electronics Products, Inc.
P.O. Box 101
North Brunswick, NJ 08902

Octek, Inc.
7 Corporate Pl., S. Bedford St.
Burlington, MA 01803

Optronics International Inc.
7 Stuart Road
Chelmsford, MA 01824

Perceptive Systems, Inc.
5231 Whittier Oaks
Friendswood, TX 77546

Photonic Microscopy Inc.
Box 648
Waltham, MA 02254

Quantex Corporation
252 N. Wolfe Rd.
Sunnyvale, CA 94086

Spatial Data Systems, Inc.
P.O. Box 249
508 S. Fairview Ave.
Goleta, CA 93017

Vicom
Scientific Systems Sales Corp.
36 Commerce Way
Woburn, MA 01801

Carl Zeiss, Inc.
One Zeiss Drive
Thornwood, NY 10594

around $10,000 to over $100,000. In this field, capability is not a simple function of price.

In 1980, we concluded that the development of microprocessor-based computers had reached the state where they should be capable of serious use for digital image processing. Accordingly we began a search for the equipment that would satisfy our needs at subastronomical costs. Disappointingly, commercial development of image processing hardware lagged behind our optimistic expectations. In 1982, a new company, Imaging Technology, Inc., announced a line of board-level image processing peripherals for microcomputer operation. This hardware, which included frame buffers, analog processors, and arithmetic logic units, incorporates major computational capabilities into these dedicated peripheral units, thereby relieving the host computer of the need to perform the relatively simple, but very repetitious, numerical operations required for a variety of real-time image processing procedures.

Thus the opportunity was realized for developing a compact system capable of real-time, or very nearly real-time, manipulation, enhancement, and quantitation of microscope images. Furthermore, it appeared that the system could be made highly flexible to meet the varying needs of microscopists to allow direct and interactive manipulations of the image under observation with simple keyboard or mouse inputs.

We have pretty much completed the testing of the hardware and development of the software that fulfill these conditions. In this paper we report on the characteristics of the hardware and capabilities of the software that are now available at costs no greater than the cost of many research microscopes.

3.   HARDWARE

Digital image processing hardware is rapidly becoming both less expensive and more effective as a result of the explosive development of microcomputers and the large-scale digital integrated circuits that make them possible. Several manufacturers are now offering image processors suitable for use with instrumentation cameras with EIA RS-170 or RS-330 format. These are available as complete, stand-alone image processors or as component units for operation in a host computer environment.

For reasons of both price and flexibility, we have chosen to use the IP-512 family of image processing circuit boards manufactured by Imaging Technology, Inc. These are available both in the intel Multibus (IEEE 796 Standard) and in the Digital Equipment Company (DEC) Q-Bus formats. The establishment of the IEEE standard for the Multibus and the availability of a variety of microprocessor-based single-board computers in this format influenced our choice of the Multibus version. Those who have existing PDP-11 series minicomputers or LSI-11 microcomputers might choose the Q-Bus versions.

The IP-512 family's principal members are the AP-512 analog processor, the FB-512 frame buffer, and the ALU-512 arithmetic processor. Regardless of the host bus format, these boards also communicate directly with each other through a separate high-speed video bus (HSVB) independently of the host. Other boards newly available for this family include the RBG-512 three-color analog processor, the HF-512 histogram and feature extraction board, and the MFB-512 memory-mapped frame buffer. We have not used these latter boards in the system of this report.

Our system consists of an AP-512, three FB-512's, and an ALU-512 all operating in a Multibus host system using a Z-80-based single board computer; the Heurikon MLZ-91A, two single-sided, double-density, 8-inch floppy disks; an ADDS Viewpoint terminal; and a Random Access three-button mouse. Figure 1a shows a block diagram of this system.

The AP-512 accepts RS-170 or -330 video input from either of two sources under software control. Before digitization, the input gain can be adjusted to any one of 256 values between zero and 2, and the black level can be set at any one of 256 levels between zero and 1 volt. The conditioned signal is then passed through an 8-bit flash digitizer at either 10 million picture element samples (10 megasamples) per second or five megasamples per second, depending on the horizontal resolution setting. The digitized data stream is passed through a "look-up table" (LUT) (see below) and through the HSVB to the ALU-512, where it is either arithmetically processed and stored in a frame buffer, passed directly to a frame buffer, or returned directly to the output section of the AP.

Digital data entering the output section of the AP-512 are passed in parallel through three LUTs to three independent D/A converters. There it becomes analog video at red, green, and blue output connections for operation of a RGB monitor.

The AP-512's LUT are 256 byte read/write memories arranged so that the pixel data from the digitizer, in the case of the input LUT, or from the frame buffer, in the case of the output LUTs, act as an address to the LUT memory, which then passes on the contents of the LUT memory at that address as the modified pixel value. The AP-512 has 4 input LUTs and 12 output LUTs. Thus, one can have four different sets of LUTs preprogrammed with different contrast or pseudo-color mappings that can be switched into use for rapid changes in display mode.

An AP-512 can also be used with a single frame-buffer, in the absence of an ALU (Fig. 1b), for limited real-time contrast enhancement processes. Here the data stream from the input LUT passes directly (via the HSVB) to the frame buffer. At each location in the frame buffer, before writing the new pixel byte, the previous contents can be read into the output section of the AP-512. Thus, as the video data stream is digitized and stored, the modified data from the previous frame are read out and displayed, essentially synchronously with the input video. With a minimal system of this sort, other types of processing must be done off line by the host processor.

The FB-512 frame buffers each have 262 kbytes of high-speed read/write memory. This memory does not reside in the memory space of the host computer and is not directly addressable as part of its memory. Instead, host computer access to each frame buffer memory is through x and y position registers and a data register, which are addressable either as input/output (I/O) ports or as local areas of host memory. In this way an 8-bit microprocessor with a memory addressing capacity of only 64 kbytes can control and access several frame buffers, each 4 times the size of its own address space.

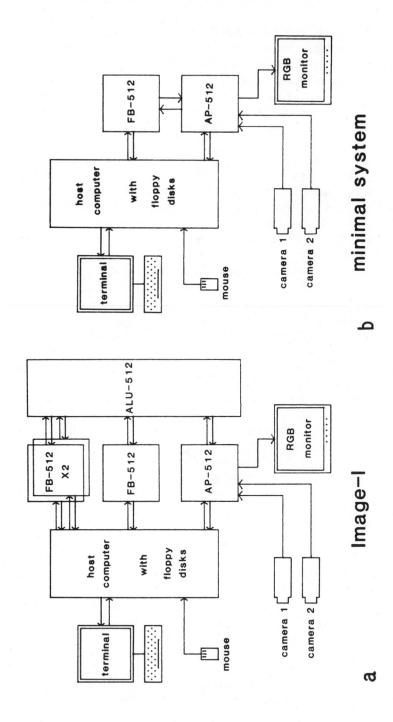

Figure 1. (*a*) Block diagram of the Image-I image processing system. (*b*) Block diagram of minimal image processing system.

Normally, the frame buffer memory addresses directly map the location of each pixel in the video image, with the upper left corner defined as x = zero, y = zero. The incoming video from a standard RS-170 or -330 camera supplies two interlaced fields per frame. The odd field lines are stored at the odd-numbered y addresses, and the even field lines at the even-numbered y addresses. In this way the two fields of each frame are interlaced in memory just as they are on the screen. For simultaneous image aquisition and display from the same frame buffer, the addressing must obviously be the same for input and output, but when the display is from a frozen image, its position on the screen can be panned and/or scrolled by changing the addressing sequence for reading out the image. The displayed image can also be changed in size by changing the addressing and timing sequences for frame buffer readout.

For example, if every second pixel is read from a horizontal scan, but at the standard pixel rate (10 megapixels/s), and only every second line is scanned, the resulting image is reduced to one-half size (occupying one-fourth of the original area). On the other hand, if the scanning rate is cut in half (5 megapixels/s) and the pixels are scanned in their regular sequence (for half of each line) and each line is scanned in its normal sequence (for half the lines of the frame), the result is that the area scanned in the memory (one-fourth of the stored frame) now occupies the full display and appears magnified 2 to 1. The process can be repeated for greater magnification or reduction.

When more than one frame buffer is used, each can have its own address registers, or several can use the same address registers but may be assigned different address spaces. In this way, four frame buffers using the same address registers can be assigned address spaces such that they act as a single frame buffer of 1024 × 1024 8-bit pixels. The FB-512 address registers have the capacity to address frame buffer arrays of up to 4096 × 4096 16-bit pixels. In the system we are demonstrating at the meetings, we use three frame buffers: one by itself as the normal 512 × 512 8-bit pixel buffer, and the other two as a 512 × 512 16-bit pixel buffer. The two buffers that are linked as a 16-bit buffer can still be accessed independently as the high byte or as the low byte at each pixel address. We have not yet had the occasion to use larger than 512 × 512 arrays, but one could find larger arrays useful for work with scanning electron microscope images.

The AP-512 can be used with a single frame buffer for real-time contrast enhancement or pseudo-color generation, but more complex operations at the live rate require the use of the ALU-512 along with extra frame buffers. In such systems, the ALU-512, in addition to arithmetic and logical processing, also serves to convey the data streams to and from the frame buffers and the analog processor by means of the high-speed video data buses.

The ALU-512 is a high-speed, pipelined processor that is capable of operating on data from up to four frame buffer arrays and one analog processor in parallel. At its processing rate of 10 megapixels/s, it can perform, in real time, addition or subtraction to 16-bit accuracy and exponentially weighted averaging. It can perform convolutions with $m \times n$ kernels in $m \times n$ thirtieths of a second.

The ALU-512 has three independent video bus connectors which carry data streams to and from two 16-bit frame buffer arrays and the analog processor for a total of five 8-bit channels (high and low byte for each frame buffer array and one for the AP). Three additional inputs are from registers which can be assigned any 8-bit values for use as constants. Five data selectors choose which bytewide data stream will flow to each of five internal data buses in the ALU. One of these buses carries its data stream directly to the HSVB, which takes it to the display section of the analog processor; the others act in pairs: two pass their data streams to an 8×8 multiplier (or bypass it via delay registers) with the 16-bit product of the multiplier (or the delay registers) going to one of the two 16-bit inputs to the arithmetic and logic processing chip; the other two pass their data (via delay registers that match the time spent by the other data streams in the multiplier) to the other 16-bit input of the arithmetic and logic chip. This chip can add the two 16-bit inputs, subtract either from the other, or combine them using the logic operations *and*, *or*, and *exclusive or*. From the arithmetic processor, the 16-bit resultant is passed through a barrel shifter that can perform binary multiplication and division. Finally, either byte of the 16-bit output of the shifter can be sent to either bytewide output path of each of the two frame buffer HSVBs.

The modes of operation of the ALU-512, as well as of the FB-512s and the AP-512, are set by the host processor by depositing instructions in the boards' control and address registers. For all real-time video processing, with this system, the processor's role is limited to this configuration process and to writing the new LUTs as needed. At the maximum, it need only pass any desired changes to the boards within the interval of one frame time, 1/30 of a second. Most of the time it merely sets the operating conditions, and the video processing continues in that mode without further intervention by the host computer. The host is then free for other activities.

For operations beyond the real-time repertoire of the system, the host computer plays a more active role, and, given time, virtually any kind of processing is possible. The terminal and the mouse provide the means for interactive control of the system by the operator.

4. SOFTWARE

The hardware provides the system with the capability for a variety of real-time image enhancement and image analysis procedures. Actually performing any of these procedures requires that the computer be programmed to accomplish the needed functions. While the board manufacturer does supply illustrative examples of simple procedures in 8080 assembler code, what was needed to make this system a convenient tool for microscopy required extensive programming. Fortunately, one of us (T.I.) turned out to be really adept at this task. Under the auspices of the Universal Imaging Corporation, he has produced a menu-driven program that provides an effective means of control of this image processor for video microscope use. Functions available through the menu include the following:

Set image aquisition mode (1) to display live video, (2) to freeze the current frame, (3) to sum incoming frames into frame buffer, (4) to perform exponentially weighted running average on incoming video, (5) to perform running averages of image with background continuously subtracted, and (6) to perform jumping averages (automatically displays sequentially the results of 2^n-frame sums).

Set display mode (1) to display image in monochrome with 256 gray levels, (2) to display image in pseudo color, (3) to display image with selectable number of brightness contours, (4) to display absolute values (for cases where processing produces negative pixel values), and (5) to display image through sigmoid LUTs.

Manipulate contrast (1) to set analog gain and black level, (2) to enhance contrast via alteration of input or output LUT, (3) to stretch contrast to cover linear range between specified lower and upper limits, (4) to fix the current LUT pattern on a selected area of the image, (5) to invert the contrast, (6) to quantize current LUT into 16 discrete levels, and (7) to enhance contrast interactively based on gray-level histogram.

Spatial enhancement operations (1) perform selected arithmetic and logic functions between frame memories, (2) spatially filter image through convolution to sharpen, soften, or detect edges in stored image, (3) differentially enhance edges through image subtraction with interactive control of image shear, and (4) detect motion by display of difference between consecutive frames or by accumulating differences to display motion paths.

Analyze the image (1) display X and Y position and pixel brightness for any pixel in an image (brightness data optionally labeled on image with a single key press), (2) display horizontal or vertical line scans of image brightness, (3) map out 3D display of intensity profile of entire image (profile is shaded, and hidden lines removed to aid interpretation), (4) plot graph of brightness versus time for up to 100 points at sampling intervals from 0.1 to 25 s, (5) display image gray-level histogram and interactively read selected data on terminal, and (6) measure area using image histogram with measured area displayed in red.

Edit and annotate image and generate graphics (1) alphanumerically label image in any color or brightness, (2) copy selected areas of image with selectable rotation and zoom, (3) display paired half images side by side for comparison or for making stereo pairs, and (4) draw or paint in any color or brightness using mouse controlled "brush" of selectable shape and area.

General utilities (1) one keystroke return to unenhanced live video for comparison with stored image, (2) copy images between the three frame buffers, (3) toggle between the user-programmable LUTs, (4) draw a gray wedge on the image to visualize the effect of output LUTs, (5) convert stored image to values displayed through LUT (allows multiple processing of image), (6) zoom or shrink image by factors of 2, (7) move image around on screen, and (8) display on-line help files for use of functions.

A functionally equivalent set of processing hardware, combined with the software package just described, is now available as a complete "turnkey" package, designated as the Image-I (image one) image processor, from Interactive Video Systems (Table 1).

5.   RESULTS AND DISCUSSION

The image-processing system we are describing (henceforth referred to as Image-I) does not require a special camera for its input and will work with any camera that satisfies the standard RS-170 or RS-330 formats, but, when working at high-contrast enhancement levels, camera defects can become painfully obvious. Fortunately, these defects can usually be minimized by background subtraction (*note:* for best results, background subtraction procedures require that any automatic camera gain and black-level controls be turned off). Nevertheless, the old computer adage, GIGO (garbage in, garbage out), applies to image processing as well, so efforts to provide improved optical images to the camera, and an improved analog signal from the camera to Image-I, are rewarded.

Figures 2–6 show some, but not all, of the capabilities of Image-I. Figure 2 shows the improvement in image contrast that can be realized when working with specimens whose visual contrast, when looking at the direct optical image, is too low to discern the object, much less its details. It also demonstrates the use of a band-pass convolution kernel that smoothes out pixel-level noise while sharpening detail at the next larger dimensional scale.

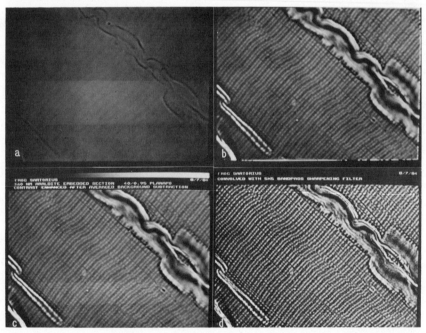

Figure 2.   Frog sartorius muscle embedded in Araldite and sectioned at 360 nm. Nikon 40 × 0.95NA Plan Apo. Polarized light image with the polarizer offset from crossed position to reduce initial contrast. (*a*) Unenhanced image from monitor. (*b*) Image has been contrast-enhanced, and an averaged out-of-focus image has been subtracted to level the illumination. A gray wedge showing the current LUT gray scale is displayed at the top of the picture. (*c*) Image as in (*b*), label added from keyboard. (*d*) Image has been sharpened by convolution with a 5 × 5 band-pass kernel. The lighter band across the screen in some frames is the result of camera shutter timing error.

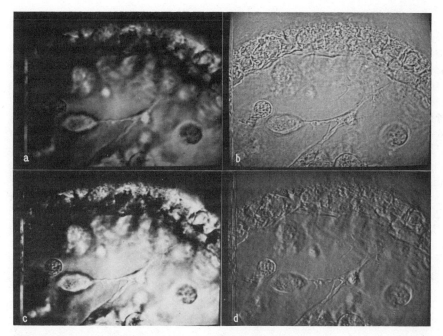

Figure 3.   Portion of an actively swimming *Lytechinus variegatus* gastrula. Nikon Plan Apo 40 ×
0.95NA. Polarizer and analyzer parallel. Input gain at 2, black level at 0.3. (*a*) Two-frame average with
output LUT expanded 2 × . (*b*) Result of convolution of image of (*a*) with 3 × 3 Laplacian kernel for edge
detection. Output LUT modified to lighten background. (*c*) Result of convolution of image of (*a*) with 3 ×
3 sharpening kernel. (*d*) Differential edge enhancement of image of (*c*). Image is duplicated into second
frame buffer, and result is obtained by subtracting one image from the other—in this case, offset by 1 pixel
width horizontally. The amount and the direction of the offset are interactively selectable under mouse
control.

Figure 3 demonstrates three other spatial enhancement features. In Fig. 3*b* we
show the result of edge detection with a 3 × 3 Laplacian convolution kernel which
yields a pixel value of zero unless a neighboring pixel, in any direction, has a value
different from the sampled pixel (26). The raw result is white outlines on a black
background. Here we have modified the output LUT to improve the intelligibility of
the result. Figure 3*c* shows the use of a 3 × 3 high-pass kernel, and Fig. 3*d* shows the
use of differential edge enhancement (24, 26) of the result from Fig. 3*c*. Differential
edge enhancement results in an image that resembles that of DIC microscopy,
because both are the result of subtraction of slightly displaced but otherwise identical
images. However, the interpretation of the images is not necessarily the same, since
differential edge enhancement can be applied to images whose initial contrast results
from other optical phenomena (e.g., from specimen birefringence), whereas DIC is
limited to the display of optical path difference gradients. One could imagine the use
of an image-duplicating interference microscope to record images that could yield
quantitative optical path difference data but that could also be displayed, through

differential edge enhancement, as true DIC images. In principle, one might also accomplish the converse, integrating a DIC image to return optical path differences.

Figure 4 shows the noise cancellation features of Image-I. The noisy, low-light-level image of Fig. 4a, in which the microspheres are barely discernible, is strengthened by summation and leveled by background subtraction in Fig. 4b and cleaned up by thresholding in Fig. 4c to reveal a crisp image of the fluorescent microspheres against a black background. The use of running (or jumping) averages with continual background subtraction would allow time-lapse recording of fluorescent events with minimal light exposure to the cells. Fig. 4d displays a three-dimensional plot (with hidden line removal) of the pixel brightness values from Fig. 4c

Figure 5 illustrates some of Image-I's image analysis features. In Fig. 5a and b we show a superimposed horizontal line scan which in Fig. 5a plots the pixel values of the gray wedge displayed at the top of the frame, showing the sigmoid shape of the

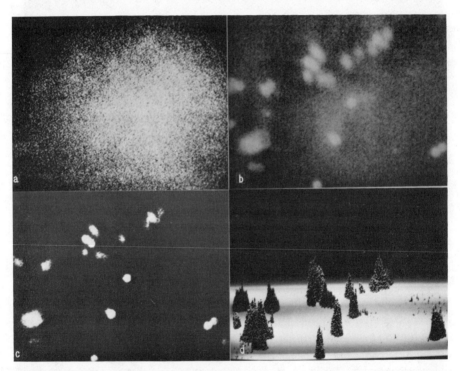

Figure 4.   SIT image of rhodamine-stained 1.25-μm microspheres. Zeiss Neofluar 25 × 0.85NA. (a) Single frame unenhanced image. The microspheres are hard to sort out from the noise. (b) This image is the result of subtracting a 256-frame sum of the background from a 256-frame sum of the focused image. (c) Image of (b) with threshold raised to eliminate background. (d) This is an Image-I–generated 3D plot of pixel gray values as seen in image (c). Pixel intensity is plotted as height above the perspective representation of the image plane. The shading of the image aids the interpretation of the image by relating the vertically displaced bright points to their appropriate place in the image plane. This is a monochrome photograph of a pseudo-color display. A variety of other shadings, in grays or pseudo color, are available under keyboard/mouse control.

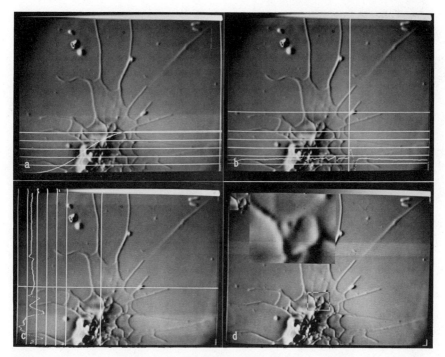

Figure 5. Petalocytes from *L. variegatus* viewed with Nikon 40× 0.95NA Plan Apo DIC. (*a*) Superimposed on the image is a pixel intensity plot of a horizontal scan through the gray wedge at the top of the frame, showing the sigmoid pattern of the LUT being used. In the original, the scan line and the scale lines are displayed in red. In this monochrome print the scan line cannot be followed past the dark part of the wedge. (*b*) Here the scan line is through a portion of the petalocyte. (*c*) Now we show a vertical scan. (*d*) Image extraction and selective zoom features. The portion of the image marked by the rectangular outline has been picked up and placed in the upper left corner and then zoomed to 4× its original size. Obviously, we cannot get detail that is not recorded in the frame buffer. Consequently, the expansion of the zoomed image represents an expansion of the size of the pixels making up the zoomed image, as can be seen.

LUT employed for that picture. Figure 5*b* shows the line scan through the petalocyte. Figure 5*c* demonstrates the vertical line scan capability. Figure 5*d* shows the ability of Image-I to extract arbitrarily selected rectangular portions of the image for selective processing of that portion—in this case zooming. A similar area designation procedure can be used to apply different LUT values to selected areas of a given image.

Figure 6 demonstrates two of the utility features of Image-I. Figures 6*a* and *b* show the use of the mouse to designate, and label with their pixel value, selected points on the image. Figures 6*c* and *d* show the split-screen comparison feature, showing in this case, before-and-after comparisons of images sharpened by 3 × 3 high-pass filters. The split-screen feature is also useful for the display of stereo pairs.

Notable among the Image-I functions not illustrated in this report are the histogram plotting and manipulating functions and the ability to measure and display plots of intensity versus time for up to 100 cursor designated image points at sampling

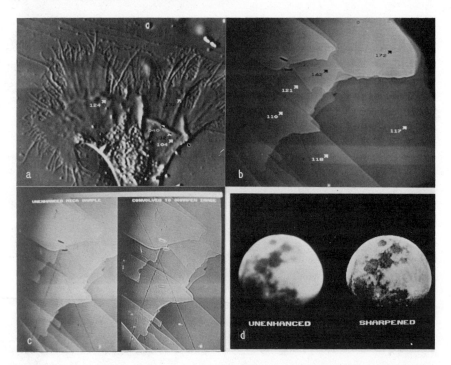

Figure 6. Parts (a) and (b) show the pixel labeling feature, and parts (c) and (d) show the split screen display feature. (a) Petalocyte DIC image with selected points labeled to show their pixel brightness. (b) Mica chip viewed by polarized light with selected pixels labeled. To use this feature, the mouse is moved to guide a cursor over the image. At each location of the cursor, the current pixel location and gray value are displayed on the screen of the terminal. At any cursor position, if you wish to label the image, a single key press will deposit on the monitor image an arrow indicating the location of the measured pixel and a label stating its gray value. (c) This view of the mica chip shows a split-screen comparison of the image in an unmodified form, on the left, and the same view, after enhancing the edges by convolution with a 3 × 3 high-pass kernel, on the right. (d) The unenhanced view of the moon shown on the left was obtained by hand-holding a 7 × 50 binocular in front of a tripod-mounted video camera. The image on the right shows the effect of the 3 × 3 high-pass filter.

intervals from 0.1 to 25 s/point. These were, however, shown at the demonstration sessions of the meeting.

While the individual features incorporated into Image-I are not necessarily original, digital image processing being some 30 years old now, the simplicity and speed of operation of Image-I and the comprehensive character of its software represent a new evolutionary step in microscopy. At the same time, and no less important, the improvement in the real-time performance of the Image-I (which can digitize and perform many of its processes on a full 512 × 512 eight-bit frame in real-time) relative to the performance of the earlier system described by Berns and his associates (24, 25), which was limited to 6-bit real-time performance, is accomplished at a net reduction in cost. The latter is a feat for which the authors can take no credit, but which illustrates the continuing progress in hardware development.

6. CONCLUSIONS

With the interactive and flexible video image processing capabilities built into processors such as Image-I, we can now explore the physiology, behavior, and structural organization of living cells with vastly improved sensitivity for detecting low levels of contrast and fine image details using very low levels of light. This provides not only new power to the light microscope but also an opportunity to explore living cells with less perturbation. Supravital indicator dyes can be used at lower concentrations, and fluorescence can be detected with substantially reduced levels of excitation as also can small phase differences, birefringence, dichroism, and other optical parameters that measure the molecular organization in dynamic cellular constituents. In addition, the quantitative functions of the processors provide immediate display of photometered functions over many image points.

Applied to electron and other forms of microscopy, Image-I can again be used to raise contrast, enhance the higher spatial frequency details, reveal image features selectively, and remove unwanted image noise. Perhaps more importantly, reduction of specimen damage through electron beam minimization becomes readily feasible, and substantial image contrast could be achieved with less reliance on differential extraction or massive exposure of the frail specimen to heavy metal staining.

In summary, computer-aided image enhancement and photometry, using video systems coupled to the microscope, maximize the contrast-yielding, high spatial frequency-enhancing, and optical data-gathering capabilities of the microscope, thus promising optimum utilization of the optical innovations that have accrued over the past several decades. With these advanced capabilities, not only can one explore new realms of the microscopic and submicroscopic world, but, equally significantly, studies of the physiological behavior and dynamic structural changes of living cells can now be carried out with diminishing interference with their vital, responsive state. These are indeed exciting developments.

ACKNOWLEDGMENTS

We thank Robert Wang of Imaging Technology, Inc., for his help in selecting and trouble-shooting the system components and for his insight and encouragement in the development of the software. This study was supported by National Institutes of Health grant GM 31617-04 and National Science Foundation grant PCM 8216301.

REFERENCES

1. Zernike, F. (1955) How I discovered phase contrast. *Science,* **121**, 345–349 (Nobel address).
2. Dyson, J. (1950) An interference microscope. *Proc. R. Soc.,* **A204**, 170.
3. Smith, F. H. (1956) "Microscopic interferometry," in A. E. J. Vickers, Ed., *Modern Methods of Microscopy,* Butterworths, London, p. 76.
4. Horn, W. (1958) "Mikro-Interferenz II" in *Jahrbuch für Optik und Feinmechanik 3,* Pegasus-Verlag, Wetzlar, W. Germany.

5. Huxley, A. F. (1954) A high power interference microscope. *J. Physiol.*, **125**, 11–13 p.

6. Krug, W., J. Rienitz, and G. Schulz (1964) *Contributions to Interference Microscopy*. J. H. Dickson, trans. Hilger and Watts Ltd., London.

7. Tolansky, S. (1948) *Multiple Beam Interferometry*. Clarendon Press, Oxford.

8. Smith, F. H. (1950) Br. Pat. Spec. #639014.

9. Nomarsky, G. (1955) Microinterférométrie différentielle à ondes polarisées. *J. Phys. Radium*, **16**, 9.

10. Inoué, S., and W. L. Hyde (1957) Studies on depolarization of light at lens surfaces. II. The simultaneous realization of high resolution and high sensitivity with the polarizing microscope. *J. Biophys. Biochem. Cytol.*, **3**, 831–838.

11. Ploem, J. S. (1975) Reflection-contrast microscopy as a tool for investigation of the attachment of living cells to a glass surface, in R. von Furth, Ed., *Mononuclear Phagocytes in Immunity, Infection and Pathology*, Blackwell, Melbourne, London, pp. 405–421.

12. Izzard, C. S., and L. R. Lochner (1976) Cell-to-substrate contacts in living fibroblasts: An interference reflexion study with an evaluation of the technique. *J. Cell Sci.*, **21**, 129–159.

13. Bereiter-Hahn, J., C. H. Fox, and B. Thorell (1979) Quantitative reflection contrast microscopy of living cells. *J. Cell Biol.*, **82**, 767–779.

14. Hoffman, R., and L. Gross (1975) Modulation contrast microscopy. *Appl. Optics*, **14**, 1169–1176.

15. Ellis, G. W. (1978) "Advances in visualization of mitosis in vivo," in Dirksen, Prescott, and Fox, Eds., *Cell Reproduction: In Honor of Daniel Mazia*, Academic, New York, pp. 465–476.

16. Reynolds, G. T. (1972) Image intensification applied to biological problems. *Quant. Rev. Biophys.*, **5**, 295.

17. Willingham, M., and I. Pastan (1978) The visualization of fluorescent proteins in living cells by video intensification microscopy. *Cell*, **13**, 501.

18. Reynolds, G. T., and D. L. Taylor (1980) Image intensification applied to light microscopy. *Bioscience*, **30**, 586.

19. Dvorak, J. A., L. H. Miller, W. C. Whitehouse, and T. Shiroishi (1975) Invasion of erythrocytes by malaria merozoites. *Science*, **181**, 748–750.

20. Inoué, S. (1981) Video image processing greatly enhances contrast, quality, and speed in polarization-based microscopy. *J. Cell Biol.*, **89**, 346–356.

21. Allen, R. D., J. L. Travis, N. S. Allen, and H. Yilmaz (1981) Video-enhanced contrast polarization (AVEC-POL) microscopy. *Cell Motil.*, **1**, 275–289.

22. Allen, R. D., N. S. Allen, and J. L. Travis (1981) Video-enhanced contrast differential interference contrast (AVEC-DIC) microscopy. *Cell Motil.*, **1**, 291–302.

23. Castleman, K. R. (1979) *Digital Image Processing*. Prentice-Hall, Englewood Cliffs, NJ.

24. Walter, R. J., and M. W. Burns (1981) Computer-enhanced video microscopy: Digitally processed microscope images can be produced in real time. *Proc. Natl. Acad. Sci. U.S.A.*, **78**, 6927–6931.

25. Burns, G. S., and M. W. Berns (1982) Computer-based tracking of living cells. *Exp. Cell Res.*, **142**, 103–109.

26. Baxes, G. A. (1984) *Digital Image Processing: A Practical Primer*. Prentice-Hall, Englewood Cliffs, NJ.

CHAPTER 3

IMAGE INTENSIFICATION OF STAINED, FUNCTIONING, AND GROWING NEURONS

S. B. KATER
C. S. COHAN

Department of Biology
University of Iowa
Iowa City, Iowa

G. A. JACOBS
J. P. MILLER

Department of Zoology
University of California
Berkeley, California

1. INTRODUCTION   32
2. VIDEO MICROSCOPY OF LIVING NEURONS   34
    2.1 INSTRUMENTATION FOR SIT CAMERA
         MICROSCOPY   34
    2.2 OBSERVATION OF CHANGES IN NEURONAL
         MORPHOLOGY   35
    2.3 OBSERVATION OF DYE-FILLED NEURONS   37
    2.4 QUANTITATIVE ANALYSIS OF FLUORESCENT
         IMAGES   39

Acknowledgments: We thank Denise Dehnboistel for preparation of figures and Drs. P. G. Haydon and R. D. Hadley for allowing publication of previously unreported data. This work was supported in part by NSF grant BNS-8202416, a Sloan Foundation Fellowship to J.P.M., and PHS grants NS 21217, NS 15350, NS 18819, and HD 18577.

	2.5	DYE EJECTION FROM MICROELECTRODES	42
3.		ZAP AXOTOMY	43
4.		PHOTOINACTIVATION OF SINGLE IDENTIFIED DENDRITES IN SITU	44
	4.1	EPIFLUORESCENCE ATTACHMENT	45
	4.2	LASER	47
	4.3	TRINOCULAR ATTACHMENT AND IMAGE INTENSIFIER	47
	4.4	MICROSCOPE MOUNT	47
	4.5	OPERATION OF THE APPARATUS	47
	4.6	DENDRITIC ARCHITECTURE OF WIND-SENSITIVE INTERNEURONS	48
5.		CONCLUSIONS	50
		REFERENCES	50

1. INTRODUCTION

One of the primary goals of the neurosciences is to interrelate the structure and electrical activity of individual neurons within the context of larger neuronal ensembles. In its earliest stages, the neurosciences benefited enormously from two independently derived techniques—the Golgi technique on the one hand, and microelectrode recording on the other. The Golgi technique, by virtue of its capricious staining, allowed the resolution of individual neurons and their complete morphology without the confusion of the other elements surrounding them. This technique has provided invaluable information on the anatomy of neurons. Many years later, the development of the microelectrode provided the physiological counterpart by sorting out the activity of one neuron from the context of the electrical activity of the thousands of neurons that surround it. For approximately 25 years these two technologies developed on separate paths: Morphologists knew little of the electrophysiological activities of the neurons they could see, and electrophysiologists knew little or nothing about the morphology of the neurons from which they could obtain high-quality electrical recordings. In 1968 the neurosciences were given a tremendous technical boost that provided a wealth of new data. The work of Stretton and Kravitz (15) with the dye procion yellow allowed the incorporation of morphological data with electrophysiological recordings. These investigators discovered a dye and devised a method of injection that allowed the microelectrode to become not only a recorder of electrical information but also a carrier of a sensitive nontoxic fluorescent dye that could be injected into living neurons. Such neurons when processed histologically could be seen with good detail and without interference from the numerous processes that surround them. Thus, the major schism that had existed for nearly a quarter of a century between structure and function in the nervous system was breached by these investigators.

Subsequent to these investigations it became clear that many compounds could serve as dyes for injection through microelectrodes. Most noteworthy, however, is the analytical approach of Walter Stewart (14) with his synthesis of a compound which has proved a tremendous advance over procion yellow—namely, the fluorescent dye lucifer yellow. The increased fluorescence and ease of injection of this dye further facilitated the routine recording and staining of neurons anywhere in nervous systems throughout the animal kingdom. The ease of processing these fluorescent dyes (as compared with substances such as horseradish peroxidase) has made them the choice for many investigators examining such structure–function relationships.

Even with the technological advances offered by fluorescent dyes, anatomical studies still required the fixation and histological preparation of tissue. Thus, while electrophysiological techniques could be used to study dynamic processes in the nervous system, morphological correlates of such events were unobtainable. Observation of morphological changes often required the viewing of dye-filled preparations for extended periods of time. However, extended viewing causes photobleaching, which occurs when fluorescent dyes are irradiated with normal viewing levels of light. Such photobleaching is particularly troublesome when fine structures such as dendrites, neurites, axons, and terminals are viewed. To circumvent this problem we have routinely used low-light microscopy as facilitated by instruments such as the silicon intensifier target camera and second-generation microchannel plate image intensifiers.

Photobleaching is not the only problem associated with fluorescent dyes. Miller and Selverston (13) demonstrated that individual dye-filled neurons could be killed selectively by irradiation with normal viewing intensities of blue light. These investigators used this method to great advantage to remove neurons selectively from a particular neural circuit. It became clear from such studies that prolonged viewing of neurons, for purposes other than killing them, was not possible with normal light levels. To this end, Kater and Hadley (9, 10) turned to low-light microscopy for viewing growing neurons.

In addition to simply viewing living neurons, it would be of enormous interest to be able to modify the morphology of a neuron while recording electrophysiological events, with the aim of determining how these precise modifications might alter the electrophysiological behavior of a given neuron. To this end, we have combined the technique of intracellular injection of fluorescent dye with the localized irradiation of parts of neurons. The present communication describes the methods and approaches used on two separate preparations with two separate goals in mind. Both of these goals were oriented around modifying neuronal architecture selectively without damage or alteration to surrounding tissue. On the one hand, the work on the snail *Helisoma*, which we will describe, is oriented around viewing growing neurons, both in cell culture and within ganglia. We have not only been able to selectively view growing neurons, but, by taking advantage of the irradiation effects described by

Miller and Selvertson, we have also been able to selectively evoke growth by specific neuronal axotomy. The second preparation to be discussed in this communication utilizes the well-defined sensory interneuronal systems of the cricket sixth abdominal ganglion. In this system, where precise details of dendritic morphology of second-order sensory interneurons can be defined, it is possible to determine, by concomitant electrophysiological recordings, the precise projection of sensory input onto specific regions of the dendritic tree.

2.   VIDEO MICROSCOPY OF LIVING NEURONS

The work described in this chapter makes use of two quite different systems: one, an assemblage of straightforward off-the-shelf components for use on the compound microscope, and the other, described later in this chapter, a custom-designed, special-purpose microirradiation device used on a dissecting microscope. Our plan in presenting this material is not only to show how experiments were performed, but also to suggest, through a series of exemplary uses, the additional directions that could readily be taken in any neurosciences-oriented laboratory by simply applying these methods.

2.1.   Instrumentation for SIT Camera Microscopy

A stable, inverted microscope forms the backbone of our original instrumentation and permits viewing of living neurons both in situ and in cell culture. We have interfaced a silicon intensifier target camera (SIT: RCA No. TC 1030/H) to the microscope as the primary viewing device (Fig. 1). This camera, because of its low light requirement, is, as will be discussed below, the critical component of this system. Specimen illumination levels (whether for bright-field, phase, interference contrast, or epifluorescence) are kept low by insertion of neutral density filters into the appropriate light path. In this way, specimens are viewed on a video monitor at illumination levels inadequate for distinguishing them directly through the oculars of the microscope. The output of the camera can be handled in various ways. For highest-quality image reproduction, we connected the output directly to a high-resolution monitor from which 35-mm photographs could be taken. For other experiments it proved useful to record ongoing events on video tape (though on our ½-in. time-lapse system there was considerable loss of detail). Most recently we have interfaced the camera output to an Apple II microcomputer through an inexpensive, commercially available card (DS-65, Microworks). In this way it was possible to obtain quantitative data on light intensities for each pixel (picture element) in a given field. With this simple and easily implemented apparatus, a broad spectrum of new quantitative experimental approaches is possible, recognizing, of course, that all of the normal constraints of light microscopy still apply.

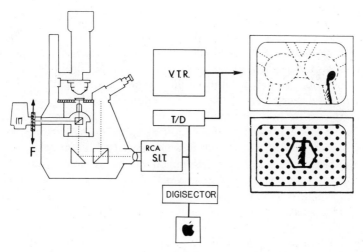

Figure 1. System for low-light viewing of dye-filled neurons under the compound fluorescence microscope. The preparation is placed on an inverted microscope (Zeiss ICM) equipped for both trans- and epiillumination. A halogen light source above the preparation provides white light for transillumination. Light from a fluorescence source (50-W mercury) at the rear of the microscope passes through a neutral density filter (F) and is reduced by 99% before epiilluminating the specimen. The intensity of fluorescence induced by this light is too low to provide a visible image through the microscope oculars, so the image must be viewed on a video monitor. The microscope image is transmitted to a silicon intensifier target (SIT) camera which is connected to a time–date generator (T/D) and video tape recorder (RCA). An iris diaphragm field stop in the light path of the microscope is used to limit the area of illumination (lower monitor), and the neutral density filter can be removed for selective irradiation (see text). For quantitative measurements, the analog signal from the camera is digitized (digisector) and stored in a computer (Apple).

2.2.   Observation of Changes in Neuronal Morphology

One approach to relating events within specific parts of a neuron to the neuron as a whole is to explant a neuron from its normal environment to cell culture. Identified neurons from the snail *Helisoma* can be dissected from ganglia and plated as individual somata in culture. Within hours, such neurons grow neurites and generate distinctive morphologies specific to each identified neuron (7). We have been interested in this growth process for two reasons: On the one hand, growth status has been shown to influence the formation of electrical synapses (5), and, on the other, we have recently found that the growth state of neurons can be regulated in a highly specific manner by the neurotransmitter serotonin (6). To understand these events more fully, it was necessary to examine the behavior of growing neurites in detail and over long periods of time. To this end, we focused on methods for direct microscopic observation of living, growing neurons in cell culture.

Observation of Cultured, Unfilled Neurons

It has been suggested that conventional microscopic observation of even unfilled, cultured neurons can deleteriously affect their growth properties, presumably owing to the intensity of white-light illumination (Letourneau, personal communication). We tested the effect of normal levels of illumination (i.e., that sufficient to view cultured *Helisoma* neurons under phase contrast through the microscope oculars) on neurite outgrowth. This was accomplished by time-lapse observations of growth cone motility during low-light and subsequent normal-light viewing. Neurite elongation proceeded normally when growth cones were viewed with low-light microscopy using the SIT camera (Fig. 2). Under these conditions, growth cone movements could be analyzed for several hours at a given time, and observations could be repeated over several days with no deleterious effects. However, as little as 20 min of illumination at normal light intensities produced total inhibition of growth cone advance (in confirmation of the findings of Letourneau). This occurred even though care was taken to limit the wavelength of the incident light to 520 nm with an interference filter in the light path. Thus, viewing of neurite outgrowth under phase contrast optics requires low levels of illumination.

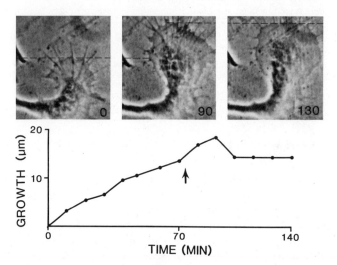

Figure 2. Normal levels of bright-field illumination of growing neurons suppress outgrowth. Left photograph shows an actively growing growth cone from an isolated identified neuron of *Helisoma* as viewed with low-light illumination with the SIT camera. Middle and right photographs show the same growth cone at successive times after illumination is increased to normal viewing levels. Note that growth cone elongates until 20 min after light level is increased, and then growth stops. Graph shows measurements of growth cone advance in greater detail at 10 min intervals. Arrow indicates point at which light intensity was increased to normal levels.

2.3. Observation of Dye-Filled Neurons

Frequently, when two or more neurons are plated together in culture it is desirable to discern from the maze of intertwined outgrowth which neurites belong to which particular cell body. To this end, we turned to the use of intracellularly injected fluorescent dyes. A cultured neuron can be readily filled with fluorescent dye in less than 5 min by passing 1 or 2 nA of negative current through a dye-containing microelectrode which penetrates its cell body. However, when viewed even for as little as 1 min with normal light levels, such neurons rapidly show distinct signs of damage in the form of blebs developing along neurites (Fig. 3). Not only are such conditions unacceptable for studying the morphology of growing neurons, but they completely obviate the use of these preparations for short-term electrophysiological experiments, since electrogenic properties also rapidly degrade. Such problems do not arise when neurons are viewed with levels of fluorescence illumination that are reduced by 2 log units (i.e., 99%) with neutral density filters. Figure 4 shows the finely detailed morphology of a dye-filled neuron from which normal electrophysiological recordings were obtained.

Although we have not observed deleterious effects at this reduced level of illumination, we still remain dubious about the complete lack of damage, since even at these low light levels growth cones of dye-filled neurons are notably retarded in their growth. We have avoided such problems by combining low-light microscopy with intermittent (rather than continuous) viewing of limited duration. In this way,

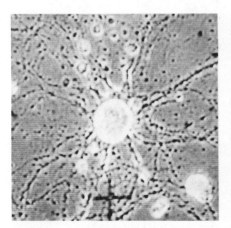 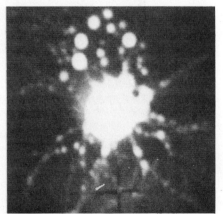

Figure 3. Typical cellular damage caused by viewing cultured, fluorescent dye-filled neurons with levels of fluorescence illumination normally used to produce an image through oculars. Photographs from TV monitor with normal viewing levels of light. Left panel shows a phase contrast image of the neuron. Right panel shows the neuron after a short time of viewing at normal intensity of fluorescent light. Note blebs above cell body.

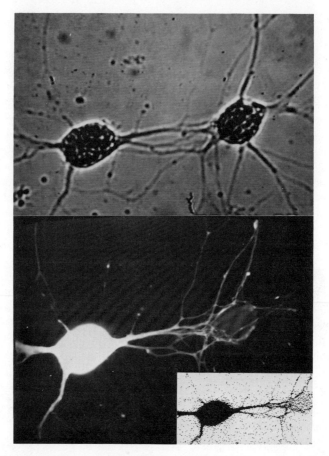

Figure 4. Normal morphology is preserved, and overlapping neurites can be distinguished using fluorescent dye injection and SIT camera viewing at low exciting light levels. Upper photo shows phase contrast image of neurite outgrowth from two isolated identified neurons in culture. Lower photo shows SIT camera image after neuron on left was injected with lucifer yellow and illuminated at low intensity. Notice that neurites of injected neuron can be distinguished from neurites of other cells in this photo. Inset shows computer reconstruction of digitized video image.

changes in neuronal morphology that occur over extended periods of time can be observed. This approach has been used in situ (Fig. 5) to observe successive stages of outgrowth and changes in connectivity over a period of days (10). Thus, the growth of new interconnecting neurites could be followed during successive time intervals in much the same way that repenetration of neurons with microelectrodes provided a picture of changes in physiological activity (4).

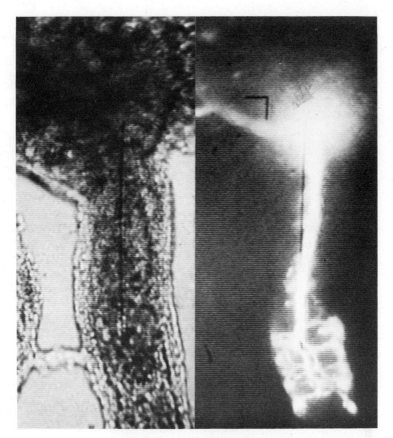

Figure 5.   Neurite outgrowth of lucifer yellow–filled neuron observed in situ under low exciting illumination with SIT camera. Left photo shows transilluminated, white image of buccal ganglion. Right photo (same field) shows the identified neuron 5, which has been injected with fluorescent dye and viewed with SIT camera. Growth of new neurites is seen at distal end of crushed axon.

2.4.   Quantitative Analysis of Fluorescent Images

One of the more recent additions to video microscopy is the ability to digitize the analogue video signals produced by the camera. A variety of systems ranging from tabletop microcomputers such as the Apple II to small mainframes such as the VAX (Digital Equipment Corp.) are now available to handle video signals. While there are obvious differences in the power, resolution, and flexibility of these systems, the smaller, inexpensive microcomputers can provide a reasonable degree of quantification of variables such as intensity. One popular video interface for the

Apple II(DS-65 by Microworks) can resolve 256×256 pixels, has a conversion time of 12 μs, and provides 64 levels of gray scale. On the other hand, a video digitizer for the IBM-PC (PC-Eye by Chorus Data Systems) can resolve 640×480 pixels, has a conversion time of 80 ns, and provides up to 256 levels of grey scale. These inexpensive systems can even "frame grab" individual images (Fig. 4, inset), though more formidable expense may be involved in manipulating such frames. The resolving power and individual requirements of each experimental situation will determine the appropriate computer for use.

As one example of the type of quantitative data that can be obtained, Fig. 6 shows measurements of dye movement along the axon of an injected neuron (in this case the neuron is within the buccal ganglion of *Helisoma*). Depending on the degree of precision a given experiment requires, it is possible to use this relatively simple system to obtain highly reproducible data on spatially distributed points. An additional advantage of computer implementation is that intensity sampling at different time points can be automatically accomplished on an interrupt basis. Here an

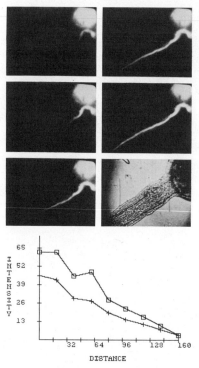

Figure 6. Quantitative measurements of fluorescence intensity obtained with an Apple II + microcomputer. Photographs from video monitor show movement of dye at successive times (from upper left to lower right) after injection into neuron soma. Video image was digitized, and intensities were obtained for nine spatially distributed points (20 μm each) along axon. Two curves are plotted for measurements made at two different times (□ first, then +). Digitized intensity values are based on 64 levels of gray scale.

addressable timer (such as the Rockwell 6522) can be used to generate a wide range of sampling periods. With our system it was necessary to establish a linear range of camera gain over which intensity measurements could be made. On-line analysis proved superior to analysis of taped data owing to image jitter, which was associated with our video system. It should also be realized that quantitative analysis may require attention to other details such as constancy and uniformity of exciting

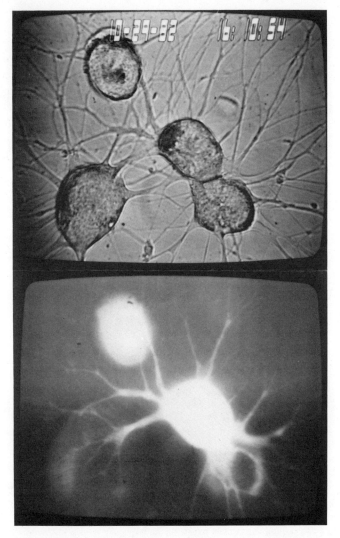

Figure 7.   Transfer of fluorescent dye between two electrically coupled neurons. Upper photo shows low light image of four isolated neurons that were plated in cell culture. Dye injected into one neuron selectively spreads to another electrically coupled neuron, with no dye transfer to neurons in contact but not coupled (lower photo, SIT image).

illumination and changes in dye concentration with time. Once these requirements were met, however, our system provided easily accessible and reliable data on fluorescence intensity over a range of spatially and temporally addressable points. One of our initial motivations for implementing quantitative techniques stemmed from the success of J. Sherridan (personal communication) with experiments on dye transfer between electrically coupled cells. An example of selective dye transfer between two highly coupled neurons of a plated network of four *Helisoma* neurons is shown in Fig. 7.

2.5. Dye Ejection from Microelectrodes

One of the most interesting by-products of our experiments on dye coupling came from control observations on the iontophoretic movement of fluorescent dye in microelectrodes. It was necessary in our dye-fill experiments to assure defined conditions under which no dye would flow from the micropipette. Conventional wisdom suggested that a bucking potential, reversed in sign from that used for passing dye through the microelectrode, was the method of choice. Figure 8 shows direct observations of dye ejection from a microelectrode. As expected, increasing

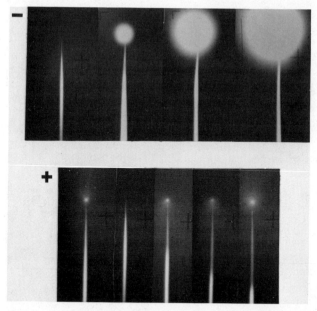

Figure 8. Iontophoretic movement of the negatively charged dye lucifer yellow at tip of microelectrode as viewed with SIT camera. Upper panel shows that negative currents of increasing magnitude (left to right) eject increasing amounts of the dye from the pipette tip. Initial photo was taken just before dye had reached the pipette tip. Lower panel shows effects of positive bucking currents of increasing magnitude (left to right). Dye leaking from microelectrode tip (left) first stopped but then began to leak again as amplitude of positive current increased. Note depletion of dye and its rearward movement from pipette tip in last three images. Microelectrode tip was placed in *Helisoma* saline for these experiments.

magnitudes of hyperpolarizing current resulted in increasing amounts of dye emitted from the electrode. Also as expected, dye leakage could be retarded by a bucking current of low magnitude. Quite unexpectedly, however, increases in magnitude of the bucking current resulted in dye ejection from the microelectrode tip. Depletion of the dye from the distal tip of the electrode was also apparent in this case. While several explanations (e.g., electroosmosis) may be mustered to account for this phenomenon, one point is clear. One cannot rely with absolute certainty on *not* injecting dye into a neuron unless one has, in fact, optically monitored the process throughout the period of cell penetration.

3.   ZAP AXOTOMY

One of the major questions arising from our work on neuronal regeneration and the formation of new connections concerns the influence of tissues other than the neurons that were experimentally treated. Growth from individual neurons is routinely evoked by crushing entire nerve trunks. This obviously damaged many, if not all, axons in the nerve as well as other nonneuronal cells. The precision of experimental work on regeneration could be markedly increased with a method for selectively evoking growth from individual neurons without disturbing others. To this end, we developed a technique for selectively severing a particular axon without perturbing any of the neighboring elements, neuronal or nonneuronal. This technique derived from the findings of Miller and Selverston (13), who showed that whole neurons could be killed by photoinactivation if they were irradiated with normal levels of blue light after being filled with dye. We found that further selectivity could be achieved by restricting the area of blue-light irradiation.

Selective (zap) axotomy can reliably be obtained by local irradiation of specific regions of dye-filled neurons (2). Dye-filled neurons are first viewed under low light with the SIT camera, and the region to be irradiated is centered in the field. An iris diaphragm in the light path is then closed to form a small spot of appropriate diameter for irradiation. The size of this spot and the precision of localization depend, of course, on the magnification of the microscope objective (in most of our experiments a spot of about 25 μm was used with a 40 $\times$ water immersion objective, NA = 0.75). The duration of irradiation for effective axotomy is largely dependent upon fluorescence intensity, which in turn depends on several variables (e.g., amount of dye in neurons, amount of overlying tissue, intensity of irradiating source, NA of objective lens). For our experiments on neuron 5 of the *Helisoma* buccal ganglion, axotomy was produced routinely with exposures of about 20 s. It is important to point out, however, that for reasons as yet undefined, dye-filled axons of other neurons (even when located in the same nerve trunk) could have quite different zap axotomy requirements. Neuron 4, for example, often proved highly refractory to zap axotomy even when the duration of light exposure was more than 3 min. This was quite similar to the results obtained in the laser irradiation experiments described below (see legend, Fig. 10).

Irradiation levels that are sufficient for axotomy produce a characteristic sequence of events. When the dye-filled neuron is first viewed under the SIT camera, its axon has a quite homogeneous fluorescence. During irradiation, the SIT camera is turned off (to avoid light-saturation damage), and the neutral density filters are removed from the light path. As soon as irradiation is complete, filters are repositioned, and the SIT camera is turned on. Immediately thereafter, one detects the photobleaching of the irradiated portion of axon as compared to the bright glow of adjacent axon regions. Within a few hours, however, this situation often reverses such that the irradiated spot is fluorescent (Fig. 9) whereas the adjacent nonirradiated areas of axon are quite faint. We presumed this was the result of a photoactivated modification in the binding properties of irradiated area while dye elsewhere in the cell diffused away (note that the fluorescent dye used here was 6-carboxyfluorescein; see ref. 2).

The technique of zap axotomy has important physiological implications as well. Conduction of impulses is blocked shortly after irradiation. This is accomplished without physiological damage to adjacent irradiated but not dye-filled, control axons (2). Additionally, and most important for our requirements, selective and profuse neurite outgrowth is obtained after 1 or 2 days. Outgrowth is confined to the dye-filled neuron, whereas adjacent neurons show no morphological defects, demonstrating the selective nature of the method. It is still possible, however, that some tissue damage has occurred but escaped observation. Nonetheless, when compared to the gross and massive damage resulting from nerve crush, zap axotomy must be regarded as a far preferable alternative.

The technique described above is a refinement of the whole-cell photoinactivation technique of Miller and Selverston (13). Further refinement has been accomplished (12) to allow ablation of selected regions of neurons at the same time behavioral studies are performed in conjunction with intracellular microelectrode recording. Although analyses and explanations of neuronal function often incorporate arguments based on observations of neuronal morphology, it has not been possible before to test such hypotheses directly by modifying neuronal structure in situ during intracellular recording. We have developed a technique (12) by which single dendrites of a neuron can be isolated from the rest of the neuron, allowing a direct analysis of the functions of separate regions of a single cell.

4. PHOTOINACTIVATION OF SINGLE IDENTIFIED DENDRITES IN SITU

The apparatus for selective ablation of individual dendrites is based on a commercially available stereodissecting microscope, which is sufficiently unique to warrant more detailed description here. The five other essential components include an epifluorescence illuminator, a laser, a trinocular attachment, an image intensifier, and a movable optical bench (Fig. 10). The necessary specifications for these components are presented below.

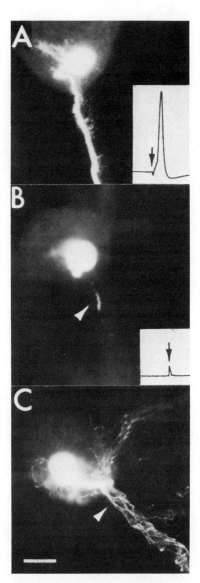

Figure 9. Selective (zap) axotomy of individual neurons produced by irradiating restricted areas of a dye-filled neuron with normal viewing intensities of fluorescent light. (A) Normal morphology and antidromic action potential recorded in cell body (inset) of the identified neuron 5 filled with dye. (B) Twelve hours after restricted irradiation of a portion of its axon (at arrowhead), dye is seen fixed at the point of irradiation, and antidromic action potentials fail to propagate past this site (inset). Arrows in insets show stimulus artifacts. (C) Several days after zap axotomy, refilling with lucifer yellow reveals neurite outgrowth for considerable distance beyond point of irradiation. Other neurons in nerve trunk were not affected.

4.1. Epifluorescence Attachment

An essential feature of this apparatus is the projection of the illuminating light beams down through the objective optics of the stereodissecting microscope. The excellent optics of the objective stage allow very high-quality, small-diameter beams to be projected onto the specimen. For most of our experiments, the laser beam diameter at

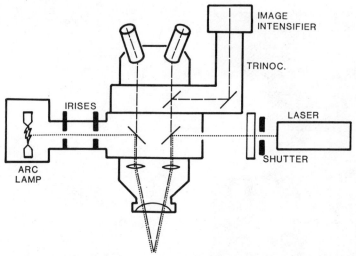

Figure 10.   Schematic diagram of the stereomicroscope epiilluminator and image intensifier system. The preparation is viewed with a stereodissecting microscope (Wild M5-A) that is specially equipped for epiillumination. One fluorescence light source (arc lamp) is used to illuminate the specimen. The diameter and intensity of its beam are controlled by iris diaphragms. A second light source, a laser, provides the intense levels of illumination necessary for rapid photoinactivation. Its beam is controlled by a shutter and variable attenuator. An image intensifier (second-generation microchannel plate intensifier), connected to a trinocular head, is used to view the specimen under low-light illumination. Nerve cells in different preparations varied in their susceptibility to photoinactivation. Effective durations of laser irradiation varied from a few seconds in Arthropoda and Crustacea to several minutes in some neurons of Aplysia.

the target is adjusted to 30 μm, though a diameter of 10 μm can readily be obtained. Since the microscope objective serves as the condenser for the illuminating beams, the target locations for photoinactivation are selected by moving the microscope with respect to the tissue. The light beams always maintain the same locations in the microscope's field of view.

Epiillumination is achieved by the addition of a separate optical section between the ocular and objective stages of the microscope. Just as in a conventional epiilluminator of a compound microscope, a dichroic "filter cube" is inserted into each of the two optical paths of the stereoscope. One element in each filter set reflects blue light down through the objective onto the specimen but does not allow any of this "excitation" beam to pass up toward the eyepieces. Only the spectral components at longer wavelengths (including the yellow fluorescence of the lucifer or fluorescein dyes) are transmitted through these filters along the optic axes. A narrow band-pass barrier filter is incorporated into each filter set between the eyepieces and the dichroic mirror to selectively transmit the fluorescence of the irradiated dye back up to the oculars.

We use either of two different light sources for epiillumination of the tissue: a mercury arc lamp or a laser. The 100-W mercury arc lamp is rigidly attached to the dichroic filter housing. The diameter of its beam can be narrowed to about 500 μm using an iris diaphragm at the image plane of the optical path. Intensity of the beam is controlled by a second iris at a nonimage plane in the condenser. The ability to decrease the intensity of the light from this source is essential for low-level illumination of the sample. An epifluorescence attachment that meets all of these specifications has been designed by one of us (J.P.M.) and is available from a commercial manufacturer as an option to one of their microscopes (Wild M5-A).

4.2. Laser

The second light source used in our studies is a helium-cadmium laser (Liconix 4210), which emits in the blue region of the spectrum (442 nm). The laser beam is directed into a second beam path of the stereomicroscope through a small port directly opposite the mercury arc lamp condenser arm, utilizing the second dichroic filter set. Since the laser and mercury arc light travel through separate beam paths, they can be used independently or simultaneously. The laser beam characteristics are controlled by a variable attenuator, a beam expander, and an electronic shutter. The beam is directed into the microscope using a series of front-surfaced mirrors.

4.3. Trinocular Attachment and Image Intensifier

As discussed earlier, illumination of dye-filled neurons with high-intensity light results in immediate cellular damage. Therefore, the intensity of illumination must be kept below this damaging threshold while the microscope apparatus is being oriented to target the laser on a particular location. This necessitates the use of an image intensifier. In contrast to SIT camera technology presented earlier in this chapter, a second-generation microchannel plate image intensifier tube (ITT #MX9644/UV) was attached to the trinocular head of the stereomicroscope (Fig. 10). This compact, solid-state intensifier forms an image of the fluorescent sample on a 25-mm fiber optic screen, yielding a luminance gain of about 80,000. This screen is viewed directly, or it can be photographed with a 35-mm video camera. Also visible on the intensifier screen is an image of a cross hair reticule, indicating the location at which the laser beam will be targeted when the shutter is opened.

4.4. Microscope Mount

All the above equipment, except the laser, must be mounted together on a mechanical stage that can be moved in three dimensions. As well as serving as a mechanical support, this stage must serve as an optical bench for the mirrors that direct the laser beam into the microscope. Our design uses three heavy-duty linear translation stages driven by DC stepper motors (Daedal #44045, with the appropriate drivers).

4.5. Operation of the Apparatus

System operation can best be described by a narration of the procedures during a typical experiment. The goal of the experiment described below was to determine the properties of synaptic inputs onto a single identified dendrite of a primary sensory interneuron in the cricket cercal afferent system.

4.6.   Dendritic Architecture of Wind-Sensitive Interneurons

The sensory interneuron called ''10–3'' is one of a class of wind-sensitive interneurons that encode the orientations of wind stimuli directed at the animal (11). The primary afferent input to this interneuron is from sensory neurons associated with filiform hairs of the cerci (3), located at the rear end of the cricket. There are about 1000 filiform hairs arranged over each of the two cerci like the bristles on a bottle brush. Each of these hairs has an optimal response to wind stimuli from one particular direction. The anatomy of four main classes of cercal afferents was examined in detail by Bacon and Murphey (1). All the filiform hair afferents project to a region within the terminal abdominal ganglion called the cercal glomerulus. Each class of sensory afferents terminates in a different region, so that the cercal glomerulus can be represented as being divided into functionally distinct regions corresponding to sensory inputs from four different wind directions.

This spatial mapping of afferent terminals into distinct regions of the ganglion has important functional consequences for the wind-sensitive interneurons. An interneuron with dendrites located in one or more of these regions could receive synaptic input from afferents responding to the respective wind directions. The interneuron's overall ''directional sensitivity'' would thus be determined in part by the location of its dendritic branches within these different ganglionic regions. By careful examination of the morphology of 10–3, predictions were made about the

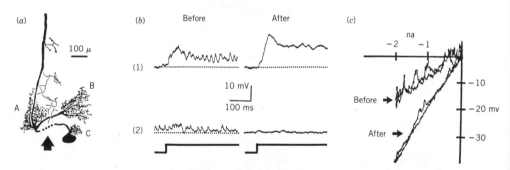

Figure 11.   Isolation of a single dendrite by microbeam photoinactivation. (*a*) Camera lucida drawing of a sensory interneuron 10–3 from the terminal abdominal ganglion of the cricket. Each of the three different dendrites has been labeled with a letter A, B, or C. The target for photoinactivation in this experiment is indicated with an arrow. Beam diameter was 30 μm; the dotted line delineates the total region observed to degenerate after the irradiation. (*b*) Responses recorded in the cell body to wind puffs from two different directions, before and 15 min after the laser illumination. Responses labeled (1) were due to wind directed at the medial face of the right cercus, and responses labeled (2) were due to wind directed at the lateral face. The onset and duration of the wind stimuli are indicated by the step functions below the responses. (*c*) Input resistance measured at the cell body, before and 15 min after the laser illumination. These are the voltage responses to a 0.25-Hz sine wave of hyperpolarizing current, plotted as voltage versus current. Therefore, the slope of the line is proportional to the input resistance.

directional sensitivity of the whole cell and about the directional sensitivity of individual dendrites projecting into the different neurophil regions. Specifically, we predicted that the dendrite of 10–3 labeled C in Fig. 11 would show a maximal response to a wind stimulus directed at the medial face of the right cercus, and little or no response to wind directed at the lateral face of the right cercus. To test this hypothetical relationship between structure and function, we needed to observe the response of 10–3 with *only* dendrite C connected to the cell body. To do so, the neuron soma was first penetrated with an intracellular microelectrode and then filled iontophoretically with 6-carboxyfluorescein. This dye had several advantages over lucifer yellow. Electrodes containing 6-carboxyfluorescein clogged much less frequently, had lower resistances, and showed less polarization than lucifer yellow electrodes. The laser was then aimed at the appropriate spot on the neurite. This was accomplished by (a) illuminating the ganglion with the mercury arc lamp at an extremely low intensity level, (b) viewing the fluorescence of the dye on the fiber optic output screen of the image intensifier, and (c) moving the microscope apparatus until the reticule cross hairs on the intensifier screen were directly over the intended target. The electronic shutter in the laser beam path was then opened for 3 s, allowing the full intensity of the beam to illuminate the neurite at the target point.

Immediately after the illumination, the section of dendrite receiving direct illumination was bleached, as also noted for zap axotomy (see above). Over the next 10 min, an additional neurite on each side of the illuminated segment was observed to disappear, presumably owing to secondary degeneration. Over this same time period, there was an initial decrease in input resistance, followed by a steady increase in input resistance. By 15 min after the irradiation, the neurite appeared to have sealed off at a point about 30 μm to either side of the targeted area, and the input resistance had stabilized at approximately twice the initial value, as shown in Fig. 11. This increase in input resistance is the expected result of sealing off the neurite at the lesion point, which effectively removes a substantial portion of the cell surface area. Such increases in input resistance should be considered a necessary observation for effective ablation in all such experiments.

The responses of the cell to wind jets from several different directions were recorded and compared to the response obtained before the laser ablation. The responses at the two selected test portions are shown in Fig. 11. Several features are of particular interest. As predicted, a maximal response was obtained at position 1, which corresponds to wind directed at the medial face of the cercus. The response was actually larger *after* the irradiation. This increase in response amplitude is due to the increase in input resistance as discussed above. Also as predicted, the response at wind position 2 was totally eliminated. This corresponds to wind directed at the lateral face of the cercus, which was predicted to activate only dendrite B. Note that no action potentials are visible in the responses recorded after the laser ablation. This is because the spike initiating zone in 10–3 is located distal to the targeted region of the neurite (8).

By performing several experiments similar to the one described here, we have greatly refined our knowledge about the relationships between neuronal form and function in this system. The response properties of these sensory interneurons can now be interpreted in terms of the synaptic interactions at single dendritic branches. In this particular cell, the overall directional sensitivity of the neuron can be characterized as a nearly linear sum of the directional sensitivities of the three different dendrites.

5. CONCLUSIONS

The goal of a knowledge of structure–function relationships in individual neurons is now significantly closer than ever before. With the milestone contribution of intracellular staining of individual neurons by Stretton and Kravitz (15), the neurosciences took a major step forward. By adapting quite standard optical equipment for use in conjunction with neurophysiological experiments, *living* neurons now can be readily examined and, more importantly, routinely *altered* to assess more precisely the functional roles of particular structures as well as to determine the degree of plasticity present in individual elements of the nervous system.

REFERENCES

1. Bacon, J. P., and R. K. Murphey (19) *J. Physiol. (Lond.)*, **84**, 352, 601.
2. Cohan, C. S., R. D. Hadley, and S. B. Kater (1983) *Brain Res.*, **270**, 93.
3. Edwards J. S., and J. Palka (1974) *Proc. R. Soc. Lond., Ser. B* **185**, 83.
4. Hadley, R. D., R. G. Wong, S. B. Kater, D. L. Barker, and A. G. M. Bulloch (1982) *J. Neurobiol.*, **13**, 217.
5. Hadley, R. D., S. B. Kater, and C.S. Cohan (1983) *Science*, **221**, 466.
6. Haydon, P. G., D. P. McCobb, and S. B. Kater (19) *Science*, **84**, 226, 561.
7. Haydon, P. G., C. S. Cohan, D. P. McCobb, H. R. Miller, and S. B. Kater (1985) *J. Neurosci. Res.*, **1**, 13, 135.
8. Jacobs, G. A. (1984) Thesis, Department of Biology, State University of New York at Albany.
9. Kater, S. B., and R. D. Hadley (1982) *Trends Neurosci.*, **5**, 80.
10. Kater, S. B., and R. D. Hadley (1982) "Intracellular staining combined with video fluorescence microscopy for viewing living identified neurons," in V. Chan-Palay and S. L. Palay, Eds., *Cytochemical Methods in Neuroanatomy*, Liss, New York, p. 441.
11. Levine, R. B., and R. K. Murphey (1980) *J. Comp. Physiol.*, **135**, 269.
12. Miller, J. P., and G. A. Jacobs (19) *J. Exp. Biol.*, **84**, 112, 129.
13. Miller, J. P., and A. I. Selverston (1979) *Science*, **185**, 181.
14. Stewart, W. W. (1978) *Cell*, **14**, 741.
15. Stretton, A. O. W., and E. A. Kravitz (1968) *Science*, **162**, 132.

CHAPTER 4

ANALYSIS OF MOLECULAR DISTRIBUTION IN SINGLE CELLS USING A DIGITAL IMAGING MICROSCOPE

FREDRIC S. FAY
KEVIN E. FOGARTY

Department of Physiology
University of Massachusetts Medical School
Worcester, Massachusetts

JAMES M. COGGINS
Department of Computer Science
Worcester Polytechnic Institute
Worcester, Massachusetts

1.	INTRODUCTION	52
2.	DIGITAL IMAGING MICROSCOPE	53
3.	SOFTWARE	55
4.	IMAGE DECONVOLUTION	55
5.	IMAGE SIMPLIFICATION	59
6.	DISPLAY SOFTWARE	60
7.	CONCLUSION	61
	REFERENCES	62

1. INTRODUCTION

Many cell functions appear to involve changes in the number or distribution of molecules within cells. Perhaps the best-studied example is the contraction of skeletal muscle which appears to result from changes in the cellular distribution of aggregates of proteins such as actin, myosin, and α-actinin (Huxley, 1971). The signal that initiates contraction itself involves changes in the distribution of an ion, Ca^{2+}, which moves from the sarcoplasmic reticulum to binding sites associated with the contractile apparatus (Blinks et al., 1979; Winegrad, 1968). Although we know far less about the changes in molecular distribution that underlie the response to hormones and neurotransmitters, in many cells translocation of both hormone and receptor from a domain in the plasma membrane to other sites either within the plasma membrane or to specific regions within the cell is believed to be essential for the complex response often seen to hormones or transmitters (Stiles et al., 1984; Szego, 1984). Finally, because virtually all differentiated cells are asymmetric and this asymmetry appears to be essential to cell function, the development of specialized cell function must be the consequence of changes in the distribution of molecules within a cell as it develops. An understanding of the events responsible for the development of specialized cell function must therefore require an understanding of changes in molecular distribution that underlie cellular differentiation.

To date, insight into the changes in distribution of molecules within cells under virtually all of these conditions has been largely indirect and quite often has required methods that preclude on-line continuous analysis of molecular distribution simultaneous with the analysis of cell function. We believe that the approach we have developed for analyzing molecular distribution in single cells promises to obviate many of these difficulties. The approach was developed originally in order to obtain insights into the mechanism of contraction of smooth muscle.

We have been interested in this question for the past few years and to this end have devoted considerable effort to studying the cellular physiology of single enzymatically isolated smooth muscle cells (Fay et al., 1979, 1982; Scheid et al., 1979). Although we have been able to measure active shortening or changes in isometric force in response to a variety of stimuli (Fay, 1977; Warshaw and Fay, 1983; Fay and Singer, 1977), the subcellular events underlying these changes were not well understood. Insight into these events must hinge critically on knowledge of the organization of the contractile proteins into a contractile apparatus. Such insights had been lacking at least in part because images at the level of the light microscope using standard contrasting techniques did not yield resolution of cellular substructure (Fay and Delise, 1973), whereas images at the level of the electron microscope, although allowing for a clear differentiation of various structures presumed to be involved in contraction (Fay et al., 1976), sample such a small region of the cell that longer range order cannot be discerned. Furthermore, dynamic analyses of structural changes underlying contraction were difficult to pursue because of the destructive nature of the preparative techniques for electron microscopy. In an attempt to circumvent these difficulties, we began some time ago to explore the use of fluorescent probes in conjunction with a device we were developing to obtain three-dimensional images of both living and fixed single cells.

Our efforts in this direction were spurred at least in part by recent developments in molecular biology, immunology, and organic synthesis which provided the means for producing probes with a high degree of specificity for proteins, nucleic acids, and even ions. When these probes are in turn coupled to fluorescent chromophores, a reagent is produced that upon introduction into a cell allows for the analysis of the numbers and distribution of a particular molecular species. We have begun the analysis of the organization of the contractile apparatus in smooth muscle by studying the distribution of the protein α-actinin, which is known to be present in smooth muscle principally in cytoplasmic and plasma membrane dense bodies (Shollmeyer et al., 1976). These cytoplasmic dense bodies appear to play an analogous role to the z-disks in skeletal muscle acting to anchor oppositely directed thin actin-containing filaments (Bond and Somlyo, 1982); the dense bodies along the plasma membrane also appear to anchor thin filaments. We chose to begin this study with α-actinin because we reasoned that if contractile proteins are indeed organized in some repeating pattern within the cell, then these α-actinin-containing bodies ought to demarcate the boundaries between adjacent contractile units. Furthermore, because oppositely directed thin filaments appear to emerge from the ends of these cytoplasmic dense bodies, the direction of these bodies should indicate the lines of force generated by the contractile apparatus within the smooth-muscle cell. We thus set out to investigate the 3D pattern of distribution of α-actinin in single isolated smooth-muscle cells utilizing tetramethylrhodamine-labeled anti-α-actinin. In order to accomplish this goal, an image of fluorescence amenable to quantitative analysis at the subcellular level must be produced. To this end we have developed a device, the digital imaging microscope, which consists of several parts: a computer-controlled light microscope, a system for recording a digital image of fluorescence, and hardware for processing and then redisplaying that image for subsequent analysis. The objective in designing this system was to obtain images that recorded light within a cell with extremely high accuracy and sensitivity so that we could detect very small numbers of molecules in a limited region of a cell. In this paper we present an overview of the general approach, and details of our methods will be published shortly (Fay et al., in preparation, a, b, c).

2. DIGITAL IMAGING MICROSCOPE

The digital imaging microscope is shown schematically in Fig. 1. It consists of a microscope equipped for epifluorescence which has been modified in two important ways. First, an electronic shutter under computer control has been placed in the fluorescence excitation path. Illumination of the specimen can thus be carefully controlled to minimize bleaching of the fluorescent probes and possible light-induced changes in cell function due to prolonged exposure to a strong light source. Second, the focus mechanism has been motorized and is under computer control using feedback from an eddy current sensor. As a result the focus of the microscope can be

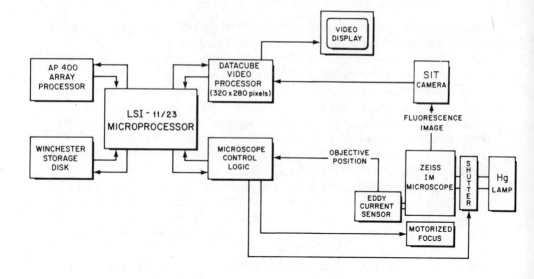

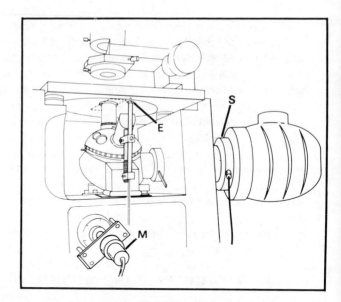

Figure 1. Schematic illustration of components of the digital imaging microscope. Images of fluorescence are obtained with the SIT camera, whose output is digitized to 8-bit resolution by the Datacube Video Processor, which samples a video frame at 320 × 240 pixels; the camera is operated in a noninterlaced sequential mode. Multiple video frames are summed in the AP-400 array processor, which acquires each digitized video frame via its direct memory access port. Multiple summed images are stored in the 11/23 microprocessor. Corrections for dark current in the image are applied by subtracting an equal number of averaged images with the shutter (S) in the excitation path closed. Images at multiple focal planes are obtained under computer control utilizing a stepping motor (M) to adjust the focus and feedback from an eddy current sensor (E), which detects the position of the objective relative to the stage to an accuracy of 0.025 μm. The array processor facilitates high-speed repetitive calculations on an image. Images are stored on the high-capacity hard disk.

changed in a programmed manner in order to obtain a series of optical sections of fluorescence for 3D analysis of molecular distributions within a cell. The image of fluorescence formed by this microscope is recorded by an ultrasensitive SIT camera whose output is digitized to a resolution of 320 × 240 picture elements with each digitized point in the image assigned a value from 0 to 255 depending on the intensity. The digitized image is then transferred to the PDP-11/23 microcomputer memory for subsequent processing. Long-term storage is accomplished by subsequent transfer to a high-capacity hard disk.

3. SOFTWARE

In addition to the hardware required to obtain a digitized image of fluorescence, the digital imaging microscope consists of software that serves two general functions. First, there are programs to carry out the various functions of the microscope described above. Second, there are programs that carry out strategies we have developed (1) to optimize acquisition of individual images at any one focal plane, (2) to minimize distortions inherent to acquisition of a 3D image composed of multiple optical sections, (3) to extract features of interest from the images, and (4) to display two- and three-dimensional images for viewing as well as interactive quantitative analysis.

The strategies we have developed to obtain an optimum image at each plane within a cell are illustrated by the sequence of images shown in Fig. 2. To reduce the contribution of noise to the image, we average multiple video frames. While this strategy is quite effective in reducing the contamination of the image by time-variant noise, it is ineffective at reducing the contribution of noise that is spatially variant but constant in time such as that due to spatial variations in dark current in the camera. To reduce this source of noise we subtract from the averaged image an equal number of video frames obtained with the excitation source blocked. A final source of distortion we correct for results from spatial inhomogeneities in ''system gain'' due to inhomogeneities in both illumination of the cell and gain of the SIT camera. To correct for this distortion we obtain an image of a uniform fluorescent specimen (a solution of the fluorochrome used in the experiment) and from that image assess spatial variations in system gain and then correct for these variations in gain in the image of the cell. In this manner we obtain an image of fluorescence distribution in a single cell with high sensitivity and minimal distortion. To assess the 3D distribution of a fluorescent probe we obtain a series of images of this kind at equally spaced intervals (usually 0.25 μm) along the depth of the cell. The set of 3D data must be further processed before it can be interpreted.

4. IMAGE DECONVOLUTION

The need for further processing results from the fact that any fluorescent point within the cell emits light that is clearly detectable over a wide focal range, as can be seen in Fig. 2. Thus any point is represented redundantly in the series of images, and there is

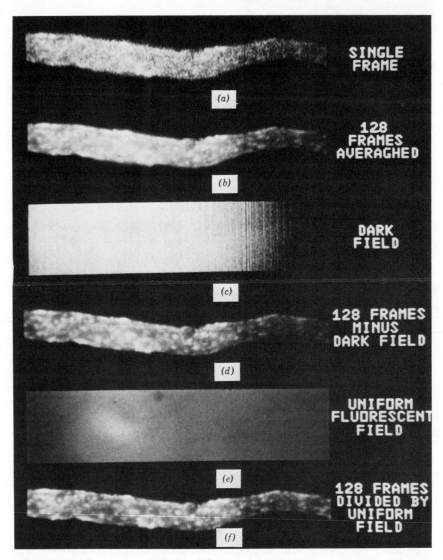

Figure 2. Illustration of the procedure for acquisition of a digitized fluorescence image. From top to bottom: (*a*) a single digitized video frame of fluorescence in an isolated smooth muscle cell stained with rhodamine-labeled anti-α-actinin; (*b*) the average of 128 frames of the same cell as in (*a*); (*c*) the average of 128 frames obtained with the excitation source blocked; (*d*) average of 128 frames of cellular fluorescence in (*b*) following subtraction of 128 frames with excitation source blocked; (*e*) image of "uniform fluorescent field" consisting of solution of rhodamine-labeled IgG, 128 averaged frames minus dark current as in (*d*); (*f*) image in (*d*) following correction for spatial variations in system gain by multiplication on a pixel-by-pixel basis with a gain normalization factor; the gain normalization factor was defined from image in (*e*) as the ratio of the intensity at each pixel divided by the mean gray level in the image. Note (1) the improvement in signal-to-noise ratio in (*b*) by averaging multiple frames resulting in improved resolution; (2) the pattern of dark current in the image in (*b*) which is removed by subtraction in (*d*); (3) the artifactual variations in intensity of stained elements within the image that are eliminated by multiplication by a gain normalization "map." All images are displayed so that the lowest-intensity pixel was displayed as black (intensity = 0) and the brightest pixel was displayed as white (intensity = 255). Magnification, 2000×.

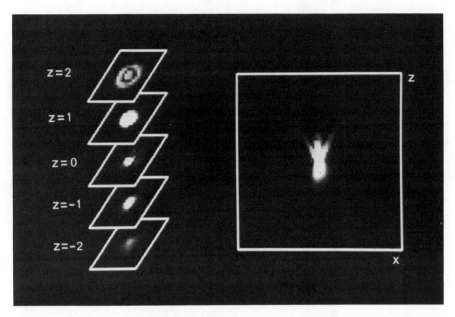

Figure 3.   Selected views of the point spread function of the digital imaging microscope for a fluorescent point source. Images of a 0.2-μm-diameter polystyrene sphere, to which was conjugated tetramethyl rhodamine, were obtained as in Figure 2 at 0.25-μm planes. The image at $z = 0$ is the most in focus view of the bead, and the distance above (+) or below (−) focus for the other views is indicated in microns. The image at the left represents a sample along the y axis through the center of the 3D data set comprised of 64 planes at ± 16 μm of optimum focus. It is thus a side view of the 3D point spread function of the system. Images all displayed as in Figure 2. Magnification, 1440 × .

uncertainty in any one view as to which visible structures are in fact in the same plane within the specimen. These problems might be solved if the distortion imposed by the imaging system, especially along the z axis, could be reversed. We have explored several methods to accomplish this and have obtained the best results using an iterative restoration technique based loosely on a method proposed by Jansson et al. (1970) for restoration of chemical spectra and most recently applied to images by Agard and Sedat (1983). To reverse distortion in the image this method requires only that we know the manner in which a point source is imaged three-dimensionally by our system (i.e., the empirical 3D point spread function [PSF] of the system). This algorithm has as its central premise that what we see in the image reflects what is in the cell convolved with the PSF of the imaging system. Now if we take a guess as to what the true object looks like and convolve it with the PSF, we will obtain an image of that object guess. If the guess is right, this predicted image and the observed image will be identical. If not, the difference between the observed and predicted images is added back to the original guess as described in Fig. 4, bringing the guess closer to the true object that formed the image. This process of updating the object guess on the basis of the degree of matching of observed and predicted images is repeated about 30 times until much of the image distortion is reversed (see Fig. 3). Application of the

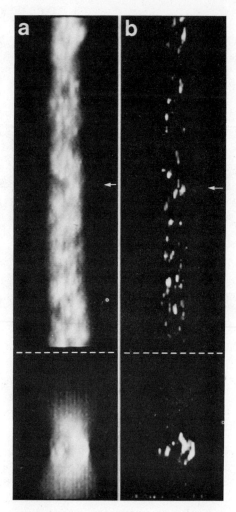

Figure 4. Restoration of a 3D fluorescence image of a single smooth-muscle cell stained with rhodamine-labeled anti-α-actinin. The restoration procedure was applied to a 3D data set comprised of 64 image planes obtained at 0.25-μm intervals. The restoration was accomplished in an iterative manner in accordance with the following scheme. Given that the observed 3D image is produced by convolution (denoted by the symbol *) of the object (the cell in this case) with the 3D point spread function (PSF) of the optical system, then we can obtain a guess of the object itself in an iterative manner (i.e., the distribution of fluorescent molecules within the cell) because a perfect guess would, upon convolution with the PSF, yield the observed image. We take as our first guess of the object (object^k) the observed image, which obviously is not accurate but is a good starting point. The next guess of the real object (object^{k+1}) is calculated as follows: object^{k+1} = object^k + gain factor × [image − object^k * PSF]. The gain factor function is designed to slow corrections to the estimate of the object as the resulting predicted image approaches some upper and lower bounds as detailed in Fay et al. (in preparation). Furthermore, the object guess is constrained from ever becoming negative, as negative fluorescence intensities have no physical meaning. The restoration shown here utilized the empirically determined point spread function and involved 30 progressive estimates of the object. Shown here are the same x–y image planes from the original and restored 3D image, as well as a computer-generated x–z plane obtained from the 3D data set at the position denoted by the arrow. Note that the background light from objects above and below this plane of focus is largely eliminated and the individual bodies are more clearly defined. Both restored and original images are displayed as indicated in Figure 2. Magnification, 2900×.

iterative restoration to a 3D image of a smooth-muscle cell stained with rhodamine-labeled anti-α-actinin resulted in suppression of much of the background haze due to light emitted from points above or below the particular plane of focus shown. In addition, the profile of individual bodies is more sharply defined.

5.   IMAGE SIMPLIFICATION

While 3D images restored in this manner may be viewed directly by creating stereo pairs from the 3D data set, such images are still too complex to effectively analyze spatial relations that may exist within these cells. We have thus developed strategies to simplify the image so that only features of interest are represented. In the case of these single smooth-muscle cells stained with anti-α-actinin, we want to know where within the 3D space occupied by the cell the bodies rich in α-actinin are located and what direction they are pointing. To extract these important features from our image we have developed an artificial 3D visual system that utilizes the responses to three sets of filters to recognize certain properties of the 3D image. The first filter in this series is used to determine where these fusiform bodies are located within the image. It is shown in a schematic manner in Fig. 5a (center) and consists of two cones each having sides that make an angle of 30° with respect to the vertical axis and has an overall length of 1.5 μm at the level of the cell. This filter optimally passes the energy contained within a body when situated over the body center and does so best for bodies having the shape and orientation of the fusiform-stained bodies (see Fig. 5b). Points in the original image where body centers are located are thus characterized in the image following convolution with this filter as energy maxima in 3D which are particularly intense. Thus by applying a threshold to the 3D extrema in this filtered image of the cell, we can identify those points in the restored 3D image corresponding to centers of these fusiform-stained bodies. The final step in the simplification of our image of α-actinin distribution involves determination of the direction that each of these bodies is pointing—that is, the 3D orientation of the long axis of these bodies. This is accomplished using the two sets of filters shown schematically in Fig. 5a (top and bottom). Each set of filters occupies the same space in toto as the original filter used to locate the body centers. One series of filters (the theta series—Fig. 5a, top) is used to determine the orientation with respect to the vertical of each body, and the second series of filters (the phi series—Fig. 5a, bottom) is used to determine the azimuthal orientation of each body. To determine the orientation of each body we compare the relative amount of energy captured by each of these filters when centered over each body with the relative energies captured when this same series of filters is centered over model bodies having various known orientations. The orientation of each body within the cell is assigned that of the model body whose response pattern to these filters most closely matches it. At the end of this process, as can be seen in Plate 1a, each fusiform-stained element within the cell is replaced by a single vector running through the point defined as the body center and oriented in 3D according to the methods just described. The larger plaques along the periphery of the cell are identified by methods generally similar to those described above and are replaced by a sphere.

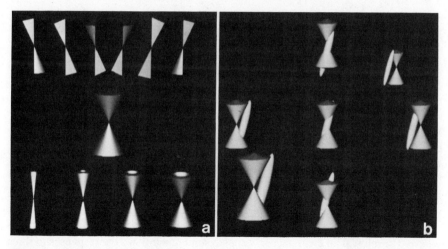

Figure 5. Illustration of the filtering method utilized to identify and characterize fusiform bodies in single cells. (*a*) Illustration of filters utilized to identify and characterize fusiform bodies rich in α-actinin in 3D images of single smooth-muscle cells. The double cone in the center is used to locate fusiform bodies within the cell. These bodies are located by analysis of a 3D cell image following convolution with this filter. The approach is illustrated in panel (*b*), where six discrete stages in the convolution of a simple image containing only one fusiform body are illustrated. Convolution of this image with the filter involves progressively moving the filter through the image and at each position multiplying on a pixel-by-pixel basis the filter and image, summing these products, and then storing the result in a new "filtered image" at a location equivalent to the current position of the center of the filter. As illustrated here, the filtered image will be most intense when the center of the filter is situated directly over the center of the fusiform body because of maximum overlap of filter and fusiform body at this position. Thus the positions of body centers in the original image are characterized in the filtered image as 3D intensity extrema. The length of the filter in real space is 1.5 μm, and its sides are at 30° to vertical, thus optimizing the capture of energy (overlap) from fusiform bodies within the smooth-muscle cells which are on average 1.25 ± 0.10 (SD) μm long and at ± 30° to the cells long axis in relaxed and moderately contracted cells. Because of this design, positions in the filtered image corresponding to the centers of fusiform-stained elements are 3D intensity maxima which are more intense than 3D intensity maxima resulting from convolution of the filter with bodies of other shapes and thus can be identified by appropriately thresholding the 3D intensity extrema in the filtered image. The two series of filters along the top and bottom of Figure 5*a* are cut from the central double cone and are used to determine the orientation of the long axis of each body. The set of filters in the top row (theta series) is a series of hollow cones subtending a narrow range of angles with respect to the vertical, and are used to determine the orientation of the long axis of each body with respect to the vertical as described in the text. The set of filters in the bottom row are a series of double wedges cut from the original double cone filter at six different azimuthal angles. This set of filters is used to determine the azimuthal angle of the long axis of each body as described in the text.

6.   DISPLAY SOFTWARE

This binary image may in turn be interactively analyzed utilizing software we have developed for displaying and interactively analyzing both 3D and 2D images. Though the display of 2D images is quite straightforward on a standard video monitor, 3D data must be conveyed either by use of a "rotating display," by simultaneously viewing two separate static 2D images that reflect two projections of the 3D image (usually $\pm$ 6° of the desired view), or by using reflectance cues from solid model representations of the bodies. Because the 3D data are stored in the computer, the cell image may easily be presented from different perspectives, often greatly facilitating interactive analysis of the images and providing new views of cellular order that have not been possible heretofore (see Plate 1). The image in Plate 1b shows the distribution of α-actinin-rich bodies as viewed sighting along one of these bodies from inside the cell. These computer-generated images can be interactively analyzed utilizing a 3D cursor with appropriate software and specific portions of the image marked by increasing the intensity or changing the color of specific bodies and in this way patterns of organization analyzed. Application of this approach to the pattern of organization of α-actinin in single isolated smooth-muscle cells has revealed (see Plate 1b and c) that the fusiform bodies in the cytoplasm are organized into strings with the elements comprising these strings having a quite regular spacing between sequential elements (mean center to center spacing = 2.2 μm). Furthermore, regions within the cell where such strings appear to terminate on larger plaques along the periphery of the cell may be identified. If these strings we have identified by interactive analysis of our images of α-actinin distribution indeed reflect the basic pattern of organization of the contractile apparatus in smooth muscle, one would expect that other proteins such as actin and myosin would show similar patterns of organization. Studies of the distribution of these proteins has only just begun, but recent restored 3D images of the distribution of rhodamine–phalloidin, a fluorescent probe for actin (Wieland, 1977), have revealed discrete cables that can be tracked for over 10 μm in any one image plane—a finding in strong support of the organization of the contractile proteins into such stringlike arrays. That these stringlike arrays of α-actinin–rich bodies indeed function in this manner is also supported by the observation that the center-to-center spacing between these bodies is diminished in cells following active cell shortening (Fay et al., 1983). It appears from these studies of molecular distribution using the digital imaging microscope that the contractile proteins in smooth muscle are in fact much more highly organized than would be expected from earlier observations with the light or electron microscope. Furthermore these studies suggest that changes in contractile state of the smooth muscle cell result from changes in the interaction between elements comprising these stringlike arrays. Clearly, continued analysis of the distribution of various molecular species during changes in contractile state will further enhance our understanding of the events underlying contraction of smooth muscle. As indicated at the beginning of this brief discussion, the approach we have developed on the basis of the digital imaging microscope should also prove quite helpful in increasing our understanding of a wide range of other biological questions.

7. CONCLUSION

We believe that the digital imaging microscope represents a powerful new tool, especially when combined with molecularly specific optical probes, for investigating the changes in molecular distribution underlying many changes in cell function. The digital imaging microscope itself is a name given to a microscope coupled to modern optoelectronic and computer hardware in combination with strategies for data processing and analysis. The device at present is capable of generating a molecularly specific CAT-scan-like image of a cell which may be analyzed by man in cooperation with a computer having some pattern analysis capabilities itself. This general approach for studying cell function is very much in evolution, and development of new molecular probes, electromagnetic energy sensors, and image analysis strategies and hardware promises to enhance capabilities even beyond those conceivable at present.

ACKNOWLEDGMENTS

This work was supported by grants from the NIH (HL14523) and the Muscular Dystrophy Association of America. We gratefully acknowledge gifts of equipment from Digital Equipment Corporation through the P.E.E.R. Program and the Lexidata Corp., which have made this work possible. We thank Ms. Shirley Borsuk for her skilled assistance in the preparation of this manuscript.

REFERENCES

Agard, D. A., and J. W. Sedat (1983) Three-dimensional architecture of a polytene nucleus. *Nature* **302**, 676–681.

Blinks, J. R., R. Rudel, and S. R. Taylor (1978) Calcium transients in isolated amphibian skeletal muscle fibers: Detection with aequorin. *J. Physiol.*, **277**, 291–323.

Bond, M., and A. V. Somlyo (1982) Dense bodies and actin polarity in vertebrate smooth muscle. *J. Cell Biol.*, **95**, 403–413.

Coggins, J. M. (1983) *A Framework for Texture Analysis Based on Spatial Filtering*. Ph.D. Dissertation, Michigan State University, University Microfilms, Ann Arbor, MI.

Fay, F. S. (1977) Isometric contractile properties of single isolated smooth muscle cells. *Nature*, **265**, 553–556.

Fay, F. S., and C. M. Delise (1973) Contraction of isolated smooth muscle cells—structural changes. *Proc. Natl. Acad. Sci. USA*, **70**, 641–645.

Fay, F. S., P. H. Cooke, and P. G. Canaday (1976) "Contractile properties of isolated smooth muscle cells," in E. Bulbring and M. F. Shuba, Eds., *Physiology of Smooth Muscle*, Raven Press, New York, pp. 249–264.

Fay, F. S., H. H. Shlevin, W. C. Granger, and S. R. Taylor (1979) Aequorin luminescence during activation of single isolated smooth muscle cells. *Nature*, **280**, 506–508.

Fay, F. S., K. Fujiwara, D. D. Rees, and K. E. Fogarty (1983) Distribution of α-actinin in single isolated smooth muscle cells. *J. Cell Biol.*, **96**, 783–795.

Fay, F. S., K. E. Fogarty, and J. M. Coggins. 3-Dimensional molecular distribution in single cells revealed by digital imaging microscopy—analytical framework. *Biophys. J.*, in preparation, a.

Fay, F. S., K. E. Fogarty, and J. M. Coggins. An artificial visual system for analysis and graphic modeling of 3-dimensional cellular images. *Pattern Anal. Machine Intelligence*, in preparation, b.

Fay, F. S., K. E. Fogarty, and J. M. Coggins. 3-Dimensional molecular distribution in single cells revealed by digital imaging microscopy—α-actinin in smooth muscle cells. *Biophy. J.*, in preparation, c.

Huxley, H. E. (1971) The structural basis of muscular contraction. *Proc. R. Soc. Lond.*, **B160**, 442–448.

Jansson, A., R. H. Hunt, and E. K. Plyler (1970) Resolution enhancement of spectra. *J. Opt. Soc. Am.*, **60**, 596–599.

Scheid, C. R., T. W. Honeyman, and F. S. Fay (1979) Mechanism of β-adrenergic relaxation of smooth muscle. *Nature*, **277**, 32–36.

Schollmeyer, J. E., L. T. Furcht, D. E. Goll, R. M. Robson, and M. H. Stromer (1976) "Localization of contractile proteins in smooth muscle cells and in normal and transformed fibroblasts," in R. Goldman, T. Pollard, and J. Rosenbaum, Eds., *Cell Motility*, Cold Spring Harbor Laboratory, Cold Spring Harbor, NY, pp. 361–388.

Stiles, G. L., M. G. Caron, and R. J. Lefkowitz (1984) β-Adrenergic receptors: Biochemical mechanisms of physiological regulation. *Physiol. Rev.*, **64**, 661–743.

Szego, C. M. (1984) Mechanism of hormone action: Parallels in receptor-mediated signal propagation for steroid and peptide effectors. *Life Sci.*, **35**, 2383–2396.

Warshaw, D. M., and F. S. Fay (1983) Cross-bridge elasticity in single smooth muscle cells. *J. Gen. Physiol.*, **82**, 157–199.

Wieland, T. (1977) Modification of actins by phallotoxins. *Naturwissenschaften*, **64**, 303–309.

Winegrad, S. (1968) Intracellular calcium movements during recovery from tetanus. *J. Gen. Physiol.*, **51**, 65–83.

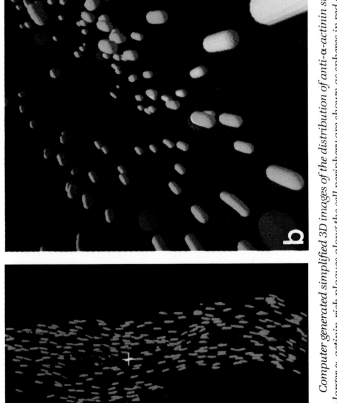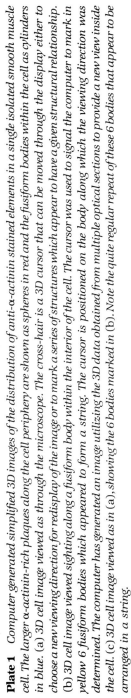

Plate 1 Computer generated simplified 3D images of the distribution of anti-α-actinin stained elements in a single isolated smooth muscle cell. The larger α-actinin-rich plaques along the cell periphery are shown as spheres in red and the fusiform bodies within the cell as cylinders in blue. (a) 3D cell image viewed as through the microscope. The cross-hair is a 3D cursor that can be moved through the display either to choose a new viewing direction for redisplay of the image or to mark a series of structures which appear to have a given structural relationship. (b) 3D cell image viewed sighting along a fusiform body within the interior of the cell. The cursor was used to signal the computer to mark in yellow 6 fusiform bodies which appeared to form a string. The cursor is positioned on the body along which the viewing direction was determined. The computer has generated an image utilizing the 3D data obtained from multiple optical sections to provide a new view inside the cell. (c) 3D cell image viewed as in (a), showing the 6 bodies marked in (b). Note the quite regular repeat of these 6 bodies that appear to be arranged in a string.

CHAPTER 5

ATTEMPTS TO SYNTHESIZE A RED-FLUORESCING DYE FOR INTRACELLULAR INJECTION

WALTER W. STEWART
NED FEDER

National Institute of Arthritis, Diabetes, and Digestive and Kidney Diseases
National Institutes of Health
Bethesda, Maryland

1. DISCUSSION   65
    REFERENCES   68

1. DISCUSSION

Over the past few years we have attempted to synthesize red-fluorescing dyes that would be suitable as intracellular tracers. Such dyes could be used, in conjunction with a yellow-fluorescing dye like Lucifer Yellow, to study pairs of neighboring neurons. (For studies involving intracellular injection and pairs of dyes with contrasting colors, see Refs. 1, 2, 4, 5.)

There are surprisingly few groups of red-fluorescing dyes with high quantum yields. We chose rhodamines because their quantum yields are high and they are said to be resistant to photobleaching. A disadvantage of the rhodamines is the scarcity of published information about their synthesis and properties.

One interesting property of rhodamines is the strong dependence of their absorption and emission maxima on the degree of alkyl substitution of the two

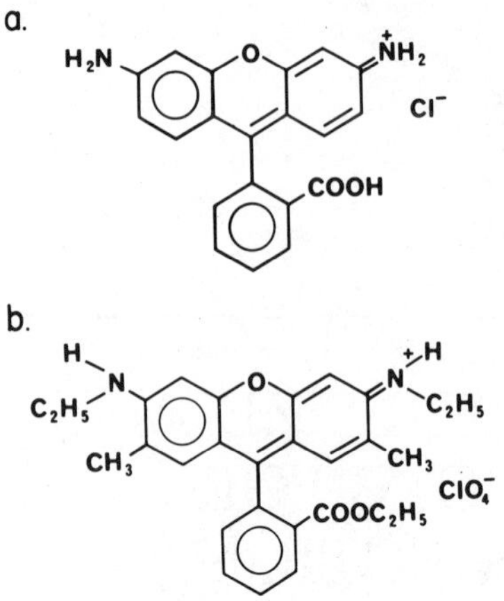

Figure 1. (*a*) A nonalkylated rhodamine and (*b*) a tetraalkylated rhodamine.

aromatic amino groups. The compound in Fig. 1*a* (Rhodamine 110, Eastman #11927), for example, is reported by the manufacturer to absorb maximally at 510 nm and to emit maximally at 532 nm. These maxima are only slightly to the red of fluorescein's (about 490 nm and 515 nm, respectively, for absorption and emission); the two dyes are not reliably distinguishable by eye. The emission maximum in particular is actually to the blue of that for Lucifer dyes (around 540 nm). Thus nonalkylated rhodamines are unsuitable as red-fluorescent intracellular tracers. Similarly, bisalkylated rhodamines do not emit sufficiently far in the red. An example is the compound in Fig. 1*b* (Rhodamine 6G perchlorate, Eastman #11954), which absorbs maximally at 530 nm and emits maximally at 552 nm (compared with about 540 nm for Lucifer dyes).

For our purposes, then, it seemed that a tetraalkylated rhodamine would be required. These dyes have an emission maximum around 580 nm and have a distinctly orange-red fluorescence color.

As discussed elsewhere (3), a tracer suitable for intracellular injection has to possess several properties. It should be highly soluble in water and should have an intense fluorescence not easily quenched. Preferably the fluorescence should be resistant to photobleaching. Another desired property is the ability to bind covalently to tissue, either through an aldehyde fixative or directly by means of a chemically reactive group on the dye molecule itself. (The former property was achieved in the case of Lucifer Yellow CH by the inclusion of a hydrazido group, which is known to react with aldehydes.) The dye should also have a fixed charge (i.e., negligible proportion of uncharged species at physiological pH), a property associated with impermeability of the cell membrane to the dye and thus important if dye is to be retained within the injected cell. Finally, it has been our impression that compounds with a net negative charge are easier to inject iontophoretically through a micropipette.

Initially we tried to start with commercially available rhodamines and modify their chemical structure in order to achieve the desired properties. Unfortunately some commercial rhodamines, perhaps the majority, prove impure when they are examined by thin-layer chromatography. We have generally found that it is difficult to purify rhodamines by crystallization, and for this reason we preferred to work only with dyes that were initially pure and to look for chemical reactions that gave products that were directly pure. Our attempts to base a synthesis on commercially available rhodamine dyes did not succeed.

We then turned to the possibility of actually synthesizing the rhodamine skeleton, using appropriate starting materials. A key step in the synthesis of tetraalkylated rhodamines is the condensation of 2 moles of a dialkyl -*m*-aminophenol with 1 mole of an aromatic aldehyde or anhydride. For ease of synthesis we concentrated on symmetrical rhodamines (i.e., those incorporating 2 moles of the same dialkyl -*m*-aminophenol). We are currently working on a rhodamine of the type shown in Fig.

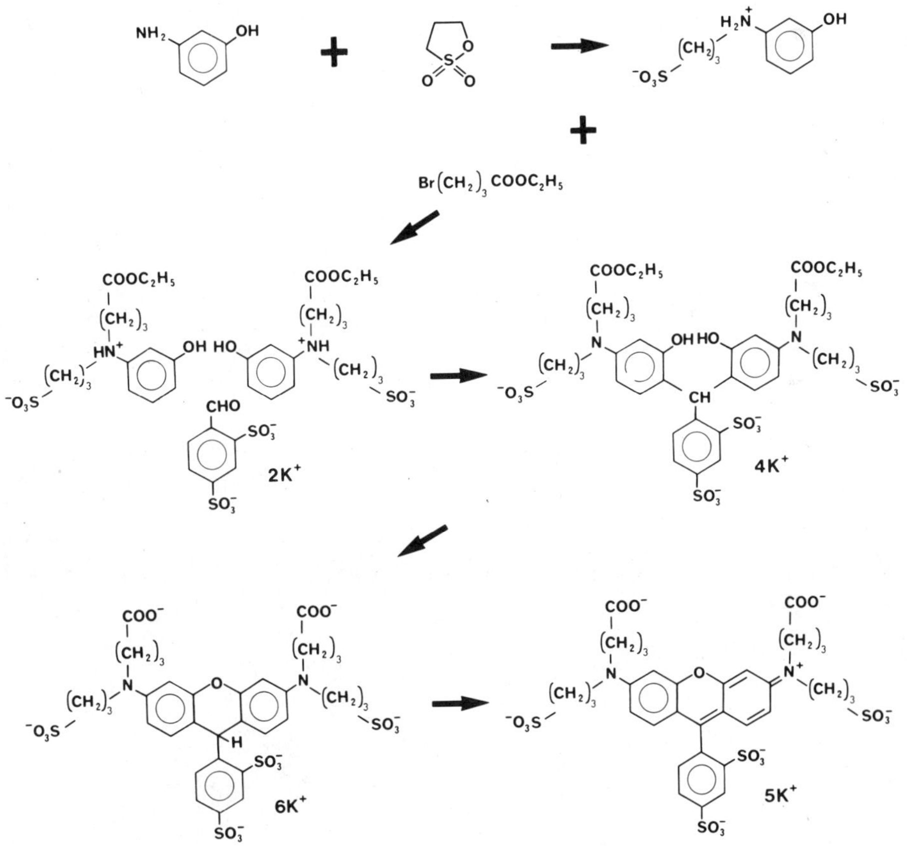

Figure 2. Proposed synthesis of a rhodamine.

2. A rhodamine such as this might serve as a starting material for the synthesis of a dye with the desired properties.

2.   REFERENCES

1.   Gilbert, P., H. Kettenmann, R. K. Orkand, and M. Schachner (1982) Immunocytochemical cell identification in nervous system culture combined with intracellular injection of a blue fluorescing dye (SITS). *Neurosci. Lett.*, **34**, 123–128.

2.   Kaneko, A. (1971) Electrical connexions between horizontal cells in the dogfish retina. *J. Physiol.*, **213**, 95–105.

3.   Stewart, W. W. (1978) Functional connections between cells as revealed by dye-coupling with a highly fluorescent naphthalimide tracer. *Cell*, **14**, 741–759.

4.   Vaney, D. I. (1984) ''Coronate'' amacrine cells in the rabbit retina have the ''starburst'' dendritic morphology. *Proc. R. Soc. Lond. B*, **220**, 501–508.

5.   Westbrook, G. L., and P. G. Nelson (1983) Electrophysiological techniques in dissociated tissue culture. *Methods Enzymol.*, **103**, 111–132.

PART 2

OPTICAL PROBES OF MEMBRANE POTENTIAL

CHAPTER **6**

OPTICAL MONITORING OF MEMBRANE POTENTIAL: METHODS OF MULTISITE OPTICAL MEASUREMENT

LAWRENCE B. COHEN
SARAH LESHER
Department of Physiology
Yale University School of Medicine
New Haven, Connecticut

1. INTRODUCTION 72
2. SOME OPTICAL SIGNALS ARE POTENTIAL-DEPENDENT 73
3. CHOICE OF ABSORPTION, BIREFRINGENCE, OR FLUORESCENCE 77
4. DYES 79
   4.1 SCREENING 79
   4.2 PHARMACOLOGIC EFFECTS 79
   4.3 PHOTODYNAMIC DAMAGE AND DYE BLEACHING 80
5. MEASURING TECHNOLOGY 80
   5.1 PHOTODETECTORS 82
   5.2 EXTRANEOUS NOISE 83
   5.3 LIGHT SOURCES 84
   5.4 IMAGE-RECORDING DEVICES 87
   5.5 OPTICS 89
   5.6 AMPLIFIERS 92
   5.7 COMPUTER 93

6. FUTURE DEVELOPMENTS   95
    REFERENCES   97

1. INTRODUCTION

An optical measurement of membrane potential using a molecular probe might be beneficial in a variety of circumstances. "Such a probe could, we believe, provide a powerful new technique for measuring membrane potential in systems where, for reasons of scale, topology, or complexity, the use of electrodes is inconvenient or impossible" (B. M. Salzberg, personal sentence). The possibility of using optical methods was first suggested in 1968 by the discovery of potential-dependent changes in intrinsic optical properties of squid giant axons (Cohen et al., 1968). Shortly thereafter, Tasaki et al. (1968) found stimulus-dependent changes in fluorescence of stained axons, and in 1971 we (Cohen et al., 1971) began a search for dyes that would give signals large enough to be useful for monitoring membrane potential. By now more than 1000 dyes have been tested for their ability to act as molecular transducers of changes in membrane potential into changes in three types of optical signals: absorption, birefringence, and fluorescence. This screening effort has resulted in the

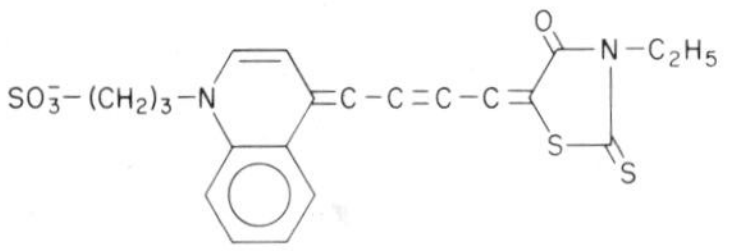

XVII, Merocyanine, Absorption, Birefringence

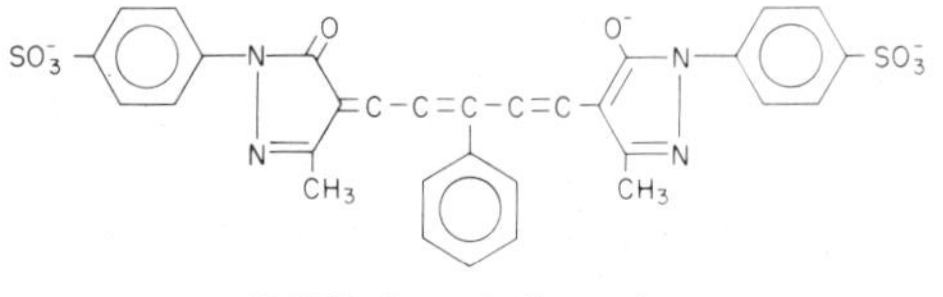

RH155, Oxonol, Absorption

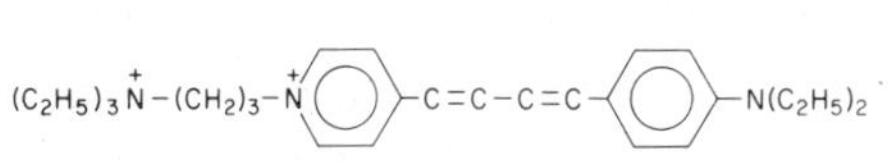

RH414, Styryl, Fluorescence

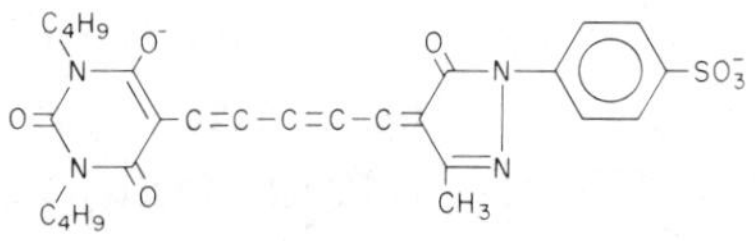

XXV, Oxonol, Fluorescence, Absorption

Figure 1. Structures of several dyes that have been used to monitor membrane potential. The merocyanine (XVII) was the dye used in the experiments illustrated in Figs. 3, 4, and 5. Dye XVII and the oxonal XXV are available from Dr. A. S. Waggoner, Center for Fluorescence, Carnegie Mellon University, 4400 Fifth Ave., Pittsburgh, PA, as WW375 and WW781. Dye XVII is available commercially as NK 2495 from Nippon Kanko-Shikiso Kenkyusho Co. Ltd. The oxonol, RH155, and styryl, RH414, are available from Amiram Grinvald, Department of Neurobiology, Weizmann Institute, Rehovot, Israel. RH414 is available commercially as dye 1112 from Molecular Probes, Junction City, OR. RH155 is available as NK3041.

discovery of dyes with a signal-to-noise ratio 100 times larger than was available from any signal in 1971. Several of these dyes (see, e.g., Fig. 1) have been used to monitor changes in potential in a variety of preparations. For reviews, see Cohen and Salzberg (1978), Waggoner (1979), Freedman and Laris (1981), Cohen and Hoffman (1982), Salzberg (1983), Grinvald (1985), and the chapters that follow.

In this chapter we begin with the evidence that has been used to show that optical signals are potential-dependent. Then we discuss the selection of signal type, dye, light source, photodetectors, optics, and computer hardware and software. This concern about apparatus arises because the signal-to-noise ratios in optical measurements are often smaller than one would like; attention to detail is required in order to maximize signal size. Some of the discussion of method is very specific and applies only to our own apparatus; other aspects of the paper are more general and would apply to any multisite optical measurement.

All of the optical signals described in this paper are fast signals as defined in Cohen and Salzberg (1978). These signals are presumed to arise from membrane-bound dye and follow changes in membrane potential with time courses that are rapid compared to the rise time of an action potential. The use of a second, slower type of potential-dependent signal, redistribution signals, is described in the chapter by Laris and Hoffman (See chapter 11).

2. SOME OPTICAL SIGNALS ARE POTENTIAL-DEPENDENT

The squid giant axon has provided a useful preparation for distinguishing among possible origins for optical signals and for testing new dyes. Figure 2 is a schematic drawing of the apparatus used to measure absorption and fluorescence using a giant axon: a very simple filter spectrophotometer and spectrofluorimeter. The axon is

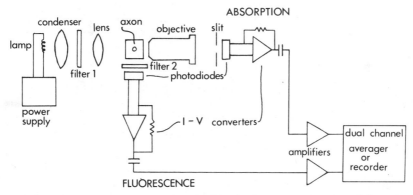

Figure 2. Schematic diagram of the apparatus used for simultaneous measurements of absorption and fluorescence of a squid axon. Filter 1 was an interference filter whose width at half-height was usually 30 nm. The barrier filters (filter 2) were obtained from Schott Optical Glass, Inc. For the absorption measurements an image of the axon was formed with a $10\times$ objective lens, and slits in the image plane were used to block the light that did not pass through the axon. From Ross et al. (1977).

perpendicular to the page and is suspended in a cuvette filled with seawater. Birefringence is measured in the absorption path by inserting two polarizers—one between filter 1, the interference filter, and the axon, and the second between the objective and the photodetector. The polarizers are crossed (oriented perpendicular to each other) and at an angle of 45° to the long axis of the axon. The light intensity reaching the photodiode in an absorption measurement is about 10^3 times larger than in fluorescence. In birefringence measurements the intensity is intermediate between fluorescence and absorption.

Figure 3 shows the results of a measurement of light absorption during an action potential in a squid axon stained with a merocyanine dye (XVII; Roman numerals refer to dyes in Fig. 1 or in Table I of Gupta et al. (1981). The dotted trace is the light intensity transmitted through the axon at 750 nm; the smooth curve is the potential measured between internal and external electrodes. The axon membrane was space-clamped, and the internal potential-sensing electrode was in the illuminated region of the axon, so the optical and electrical measurements were made at the same time and place. Because the two measurements had very similar time courses, it seemed likely that the absorption signal was related to the changes in membrane potential of the action potential and not to the ionic currents or the membrane permeability increases that occur during the action potential. More direct evidence for potential dependence can be obtained from voltage–clamp experiments such as the one illustrated in Fig. 4. The top trace is the absorption signal; clearly, it has a time course similar to that of the membrane potential (middle trace) and distinctly different from that of the permeability changes or the ionic currents (bottom trace). This kind of result has been obtained for many dyes, including the four illustrated in Fig. 1.

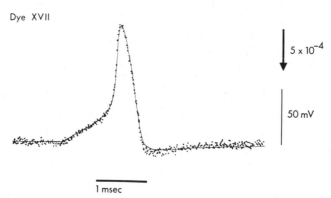

Dye XVII

5×10^{-4}

50 mV

1 msec

Figure 3.   The change in absorption (dots) of a giant axon stained with dye XVII during a membrane action potential (smooth trace) simultaneously recorded with an internal electrode. The change in absorption and the action potential had the same time course. In this figure and in Fig. 4 the direction of the arrow adjacent to the optical trace indicates the direction of an increase in absorption; the size of the arrow represents the stated value of a change in absorption, ΔA, in a single sweep divided by the resting absorption due to the dye, A_r. Incident light of 750 nm was used; 32 sweeps were averaged. The response time constant of the light measuring system was 5 µs. From Ross et al. (1977).

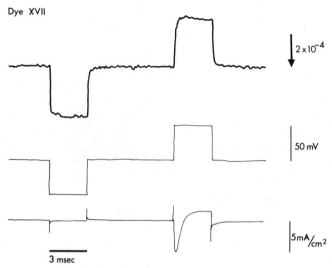

Figure 4.   Changes in absorption of a giant axon stained with dye XVII (top trace) during hyperpolarizing and depolarizing steps (middle trace). The bottom trace is the current density. The absorption changes had the same shape as the potential changes and were insensitive to the large currents and conductance changes that occurred during the depolarizing step. The holding potential was the resting potential, and hyperpolarization is represented downward; inward currents are downward. Incident light of 750 nm was used; 128 sweeps were averaged; the time constant of the light-measuring system was 20 μs. From Ross et al. (1977).

Inspection of Fig. 4 might suggest that signals with a time course similar to the currents or permeability are smaller than 5% of the total signal. In fact, a conclusion this strong is unwarranted because the result in Fig. 4 was obtained with a rather arbitrary amount of compensation for the resistance in series with the axon membrane. Although the compensation used implied a series resistance within the range of previously reported values, the series resistance was not measured independently in this experiment. This ambiguity has been resolved by Salzberg and Bezanilla (1983).

If a voltage clamp experiment is performed using four potential steps, and the size of the absorption changes versus the size of the potential step is plotted, then the result shown in Fig. 5A is obtained. The absorption change in squid axons was linearly related to membrane potential over the range ± 100 mV from the resting potential. In subsequent experiments with larger steps carried out with dyes XVII and XXII, the absorption signals were found to be linearly related to membrane potential over the range − 130 to + 200 mV from the resting potential (Gupta et al., 1981). As Fig. 5B shows, the birefringence signal obtained with dye XVII was also linear. A linear relationship between optical signal and membrane potential has been obtained with many dyes. Thus, in many instances there is strong evidence that the signals obtained with millisecond potential steps in squid axons depended in some manner on changes in the transmembrane potential (Cohen et al., 1970, 1974; Conti and Tasaki, 1970;

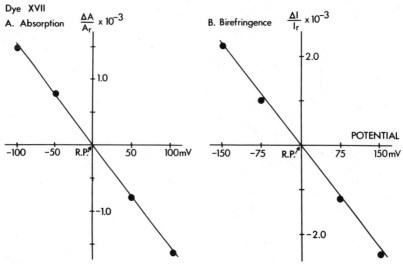

Figure 5. Change in dye XVII absorption and birefringence as a function of membrane potential in a squid axon. Both the absorption and the birefringence changes of dye XVII were linearly related to membrane potential. The duration of the voltage clamp steps was 1 ms. From Ross et al. (1977).

Patrick et al., 1971; Davila et al., 1974; Conti, 1975; Ross et al., 1977; Gupta et al., 1981), although there was some earlier disagreement about this conclusion (Conti et al., 1971; Tasaki et al., 1972).

These dye signals can be rapid. Our initial attempts to measure differences between the potential change and the optical signal showed that some of the signals were too fast to measure and thus lagged behind the change in membrane potential by less than 10μs (Ross et al., 1977). Measurements with a faster apparatus by Salzberg, Bezanilla, and Obaid showed that the signals obtained with dyes XXVI (styryl), XVII, and XXII (merocyanines) were still too rapid to measure and thus lagged behind the change in membrane potential by less than 2 μs (Loew et al., 1985; B. M. Salzberg, F. Bezanilla, and A. L. Obaid, unpublished results). This rapid tracking of potential change by optical signal was, however, sometimes not obtained either when relatively low or relatively high concentrations of dye are used; with low concentrations, time constants as slow as 70 μs were obtained with dye I, and with high concentrations, very slow components may appear (Ross et al., 1977; Ross and Krauthamer, 1984).

The results in Fig. 3–5 suggest that when these dye signals are used to measure rapid events, the signals represent changes in membrane potential. However, in a number of preparations, much slower signals 100–1000 ms) are obtained (e.g., Orbach and Cohen, 1983; Orbach et al., 1985; Kauer et al., 1984; Lev-Ram and Grinvald, 1984). Clearly, it would be useful to have independent evidence that such slow signals represent a change in membrane potential. But evidence has not been obtained, and thus the origins of slow optical signals must be interpreted with caution.

A number of studies have been made to determine the molecular mechanisms that result in potential-dependent optical properties. This subject is discussed by Waggoner in a later chapter (See chapter 7).

3. CHOICE OF ABSORPTION, BIREFRINGENCE, OR FLUORESCENCE

Sometimes it is possible to decide in advance which kind of optical signal will give the best signal-to-noise ratio, but in other situations a careful experimental comparison is necessary.

The choice of signal type may depend on the optical characteristics of the preparation. Birefringence signals are relatively large in preparations that, like giant axons, have a cylindrical shape and radial optic axis (Gupta et al., 1981). However, in preparations with spherical symmetry (e.g., molluscan cell soma), the birefringence signals in adjacent quadrants will cancel, and thus in these preparations, birefringence can be measured only with a detection system with high spatial resolution. Achieving this spatial resolution leads to degradation of the signal-to-noise ratio (Boyle and Cohen, 1980). Because birefringence can be measured at wavelengths outside the absorption band of the dye (Ross et al., 1977), eliminating photodynamic effects and dye bleaching, birefringence would be preferable to absorption or fluorescence in measurements of propagation along axons. In one case where it should have been tested, it was not (Shrager et al., 1985).

A second instance where the preparation dictated the choice of signal was in measurements from mammalian cortex. Here transmitted light measurements are not feasible (without subcortical implantation of a light guide), and the small size of absorption signals that can be detected in reflected light (Ross et al., 1977; Orbach and Cohen, 1983) meant that fluorescence had to be used (Orbach et al., 1985).

An additional factor that affects the choice of absorption or fluorescence is that the signal-to-noise ratio in fluorescence is relatively sensitive to the amount of dye bound to extraneous material. Figure 6 illustrates a spherical cell surrounded by extraneous material. In Fig. 6A we assume that dye binds only to the cell; in Fig. 6B we assume that there is 10 times as much dye bound to extraneous material. To calculate the transmitted intensity we assume that there is one dye molecule for every 2.5 phospholipid molecules (a large concentration if one would also expect to maintain physiological function) and an extinction coefficient of 10^5 (some of the best available dyes have extinction coefficients this large). The amount of light transmitted by the cell is still 0.99 of the incident light. Thus, even if this dye were to completely disappear as a result of a change in potential, the fractional change ($\Delta I/I_o$) in transmission would be only 1% (10^{-2}). The amount of light reaching the photodetector in fluorescence will be much lower, say $0.0001 I_o$. Several factors account for the lower intensity. First only $0.01 I_o$ is absorbed by the dye. Second, we assumed a fluorescence efficiency (photons emitted/photon absorbed) of 0.1. And third, if we assume a light-collecting system of 0.8 NA (numerical aperture), only 0.1 of the emitted light reaches the photodetector. But even though the light reaching the

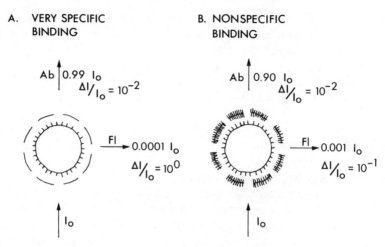

Figure 6. (A) The light transmission and fluorescence intensity when only a neuron binds dye and (B) when both the neuron and extraneous material binds dye. In (A), assuming that one dye molecule is bound per 2.5 phospholipid molecules, 0.99 of the incident light is transmitted. If a change in membrane potential causes the dye to disappear, the fractional change in transmission is 1%, but in fluorescence it is 100%. In (B), nine times as much dye is bound to extraneous material. Now the transmitted intensity is reduced to 0.9, but the fractional change is still 1%. The fluorescence intensity is increased 10-fold, and therefore the fractional change is reduced by the same factor. Thus extraneously bound dye degrades fractional changes and signal-to-noise ratios much more rapidly in fluorescence than in absorption.

fluorescence detector is small, disappearance of dye would result in a 100% decrease in fluorescence—a fractional change of 1×10^0. Thus the fractional change in fluorescence can be much larger than the fractional change in transmission in situations where dye is bound only to the cell membrane and there is only one cell in the light path. However, the relative advantage of fluorescence is reduced if dye binds to extraneous material. If 10 times as much dye is bound to the extraneous material as was bound to the cell membrane (Fig. 6B), the transmitted intensity is reduced to approximately 0.9. If a potential change again causes the cell-bound dye to disappear, the fractional change in transmission is nearly unaffected. By contrast, the resting fluorescence intensity is now higher by a factor of 10, so the fractional change is reduced by the same factor. The fluorescence fractional change is more severely affected than absorption by dye binding to extraneous material. It does not matter whether the extraneous material happens to be connective tissue, glial membrane, or neighboring neuronal membranes. In Fig. 6B, the fractional change in fluorescence was still larger than in transmission. However, the light intensity in fluorescence was about 10^3 smaller, and this reduces the signal-to-noise ratio in fluorescence (see below). Partly because of the signal degradation due to extraneous dye, fluorescence signals have been found most useful in monitoring activity from tissue-cultured neurons whereas absorption has been preferred in measurements from ganglia and brain slices. In ganglia and brain slices the fractional changes in both transmission and fluorescence are small; they range between 10^{-4} and 10^{-2} for a 100-mV potential change.

4. DYES

The choice of dye is a very important factor in maximizing the signal-to-noise ratio in an optical measurement. In only a few instances, where there is a large density of synchronously active membrane, has it been possible to obtain large signals without testing a fair number of dyes. In addition, in some preparations, photodynamic damage due to illumination of the dye in the presence of oxygen may also affect the choice of dye. Pharmacologic effects and dye bleaching are also considered in this section.

4.1   Screening

Using squid giant axons, more than 1000 dyes have been tested for signal size in response to changes in membrane potential. This screening was made possible by the heroic synthetic efforts of two laboratories. Alan Waggoner, Jeff Wang, and Ravender Gupta of Amherst College and Rina Hildesheim and Amiram Grinvald at the Weizmann Institute have each synthesized more than 300 dyes. In addition, Leslie Loew has rekindled interest in and synthesized a number of styryl dyes (Loew and Simpson, 1981; Hassner et al., 1984), and several merocyanine dyes have been made by Nippon Kankoh-Shikiso Kenkyusho Co. Ltd. Included in these syntheses were about 100 analogues of each of the four dyes illustrated in Fig. 1. In each of these four groups there are 10 or 20 dyes that gave similarly large signals (within a factor of 2 of the dye illustrated) on squid axons.

   However dyes that gave nearly identical signals on squid axons gave very different responses on other preparations, and thus many dyes had to be tested to maximize the signal. Examples of preparations where a number of dyes had to be screened are the *Navanax* and *Aplysia* buccal ganglia (Cohen et al., 1986; London et al., this volume), rat cortex (Orbach et al., 1985), and tissue-cultured neurons (Ross and Reichardt, 1979; Grinvald et al., 1981a). Some of the dyes were unable to penetrate through connective tissue or along intercellular spaces to the membrane of interest. Others appeared to have a relatively low affinity for neuronal versus nonneuronal (connective) tissue. However, in some instances, the dye penetrated well and the staining appeared to be specific, but nonetheless the signals were small. Ross and Krauthamer (1984) have reported a case where supraesophageal ganglia from different barnacle species of the same genus had qualitatively different signals. In only a few instances was it not necessary to screen to obtain relatively large signals. These included nudibranch buccal ganglia (Boyle et al., 1983), salamander olfactory bulb (Orbach and Cohen, 1983) frog neurohypophysis (Salzberg et al., 1983), and muscle (Salama and Morad, 1976; Baylor et al., 1981; Nakajima and Gilai, 1981). (See chapter 8).

4.2   Pharmacologic Effects

In many preparations high concentrations of dye had pharmacologic effects. However, in most instances the dye concentration needed to obtain the maximum signal size was lower than the concentration at which pharmacologic effects were detected. These include the squid giant axon (Cohen et al., 1974; Ross et al., 1977; Gupta et al., 1981), neuroblastoma cells in tissue culture (Grinvald et al., 1982), heart muscle (Morad and Salama, 1979), skeletal muscle (Nakajima and Gilai, 1980; Baylor et al., 1981), the barnacle supraesophageal ganglion (Salzberg et al., 1977; Grinvald et al., 1981b), embryonic chick heart (Fujii et al., 1981), embryonic semilunar ganglion (Saki et al., 1985), and the *Navanax* buccal ganglion (London et al., this volume). Thus, pharmacologic effects are not known to be a major difficulty, even in the *Navanax* experiments where the buccal ganglion was stained in a minimally dissected preparation and feeding behavior was measured with and without staining. Here complex synaptic interactions are probably required to generate the correct behavior. However, in the optical experiments on salamander olfactory bulb, frog optic tectum, and rat cortex, the ability to detect pharmacologic effects was very limited. There, one can only say that pharmacologic effects were not disastrous (See chapter 8).

4.3   Photodynamic Damage and Dye Bleaching

In certain experiments—for example, on neuroblastoma neurons using the styryl dye RH414 (Fig. 1)—photodynamic damage (due to the interaction of light, dye, and oxygen) limited the duration of the experiments (Grinvald et al., 1982). In others—for example, on *Navanax* buccal ganglia using the oxonol dye RH155 (Fig. 1)—it was difficult to detect photodynamic damage (London et al., this volume). Similarly, dye bleaching has caused difficulties in some preparations but not in others. Grinvald et al. (1982) reported a 3% bleaching from 350 ms of illumination using a styryl dye, whereas we found bleaching difficult to detect after 10 min of illumination using the oxonol (London et al., this volume). This difference in severity is, in part, due to the difference in dyes that were used; however, in addition, higher light intensities were used in the experiments where damage and bleaching were severe. Thus, advantages of increased intensities in terms of signal-to-noise ratio (see below) may be partially counterbalanced by increased damage and bleaching. Since both effects are dye-dependent (Cohen et al., 1974; Ross et al., 1977; Gupta et al., 1981), additional dye screening may be necessary in preparations where they cause difficulty. Although bleaching and photodynamic damage are sometimes correlated (Gupta et al., 1981), in other instances bleaching can occur without detectable damage (Ross and Krauthamer, 1984).

5.   MEASURING TECHNOLOGY

The limit of accuracy with which light can be measured is set by the shot noise that arises from the statistical nature of photon emission and detection. Fluctuations in the number of photons emitted per unit time will occur, and if an ideal light source (tungsten filament) emits an average of 10^{14} photons/ms, the root-mean-square (RMS) deviation in the number emitted is the square root of this number or 10^7 photons/ms. In the shot noise limited (ideal) case, the signal-to-noise ratio is

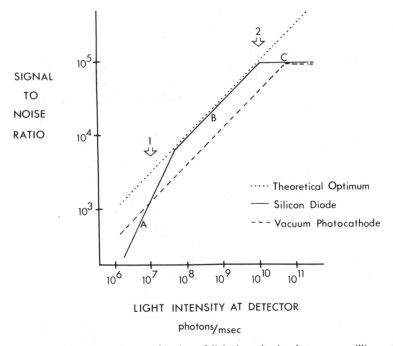

Figure 7. Signal-to-noise ratio as a function of light intensity in photons per millisecond. The approximate light intensity per detector in fluorescence measurements from ganglia or vertebrate cortex using a tungsten filament bulb is indicated by arrow 1. The approximate intensity in absorption measurements in ganglia or brain slices is indicated by arrow 2. The theoretical optimum signal-to-noise ratio (dotted line) is the shot-noise limit. The signal-to-noise ratio expected with a silicon diode detector is indicated by the solid line. The silicon diode signal-to-noise ratio approaches the theoretical maximum at intermediate light intensities (segment B) but falls off at low intensities (segment A) because of dark noise and falls off at high intensities (segment C) because of extraneous noise. The expected signal-to-noise ratio for a vacuum photocathode detector is indicated by the dashed line. At low intensities the vacuum photocathode is better than a silicon diode because it has less dark noise. At intermediate intensities it is not as good because of its lower quantum efficiency.

proportional to the square root of the number of measured photons and the square root of the bandwidth of the photodetection system (Braddick, 1960; Malmstadt et al., 1974). This ideal result is indicated by the dotted line in Fig. 7, which is a plot of signal-to-noise ratio versus light intensity reaching the photodetector. In a shot-noise limited measurement, improvements in the signal-to-noise ratio can be obtained only by increasing the illumination intensity or the light-gathering efficiency of the measuring system or by reducing the amplifier bandwidth. Only a small fraction of the 10^{14} photons/ms emitted by a 3300°F tungsten filament will reach the photodetector. An ideal 0.7-NA lamp collector lens would collect 0.06 of the emitted light. For small objects the luminous-field diaphragm (Piller, 1977) would reduce the solid angle even further. Only 0.2 of the emitted and collected photons (at 3300°F) are in the visible wavelength range; the remainder are in the infrared (heat). An interference filter of 30 nm width at half-height might transmit 0.05 of the visible light. Additional losses will occur in the condenser lens and at all air–glass interfaces. Thus, the light reaching the preparation might typically be reduced to 10^{10} photons/ms. If the light-collecting system has high efficiency (e.g., in an absorption measurement), then $\sim 10^{10}$ photons/ms will reach the photodetector, and if the photodetector has a quantum efficiency (photoelectrons/photon) of 1.0, then 10^{10} electrons/ms will be measured. The RMS shot noise will be 10^5 electrons/ms; thus the relative noise is 10^{-5}.

5.1 Photodetectors

Since the signal-to-noise ratio in a shot noise limited measurement is proportional to the square root of the number of photons measured (converted into photoelectrons), the quantum efficiency (photoelectrons/photon) of this conversion is an important figure of merit. As is indicated in Table 1, silicon photodiodes have quantum efficiencies approaching the ideal at the wavelengths where most dyes absorb or emit light (500–900 nm). In contrast, only specially chosen vacuum photocathode devices (phototubes, photomultipliers, or image intensifiers) have a quantum efficiency as high as 0.15. Thus, in a shot-noise limited situation a silicon diode will have a signal-to-noise ratio that is at least 2.5 times larger. This advantage of silicon diode over vacuum photocathode is indicated in Fig. 7 by the fact that the diode curve (solid line) is higher than the vacuum photocathode curve (dashed line) over much of the intensity range (segment B).

There are three types of noise that can degrade the signal-to-noise ratio from the theoretical limit. The first is dark noise, the system's noise in the absence of light. The dark noise is generally far larger in a silicon diode system than in a vacuum photocathode system (Table 1). Thus at low light levels ($<10^7$ photons/ms), a vacuum photocathode device will provide a larger signal-to-noise ratio. When the light level is reduced so that the shot noise is less than the dark noise of a silicon diode ($\sim 10^8$ photons/s), the signal-to-noise ratio of the diode decreases linearly with light intensity (segment A, Fig. 7). The crossover in signal-to-noise ratio between the silicon diode and vacuum photocathode device occurs at about 10^7 photons/ms (arrow 1, Fig. 7). This crossover occurs near the intensities obtained in fluorescence measurements from ganglia and intact cortex (P. Saggau and L. B. Cohen, unpublished results), and thus the optimal choice of photodetector at these intensities requires testing carefully selected examples of both kinds of devices. Cooling a silicon diode reduces the dark noise, improving its performance at low intensities and

TABLE 1 DETECTOR COMPARISON

	Silicon Diode	Vacuum Photocathode
Quantum efficiency	0.9	0.15
Dark noise equivalent power	$\sim 10^7$ photons/s	$< 10^5$ photons/s
$1/f$ noise	Some diodes	No

shifting the crossover in signal-to-noise ratio to the left. On the other hand, future improvements in the quantum efficiency of vacuum photocathode devices would shift the crossover to the right.

Some silicon diodes may have an additional light-dependent noise. David Kleinfeld (personal communication) found this second type of noise, called excess noise or $1/f$ noise, in PIN 6D and PIN 10D diodes (United Detector Technology) and in a Hewlett-Packard 5082-4203 diode. One indication of the presence of excess noise was that the noise current was directly proportional to the photocurrent rather than having the square root proportionality of shot noise. We measured the relationship between noise and intensity in three diodes—a United Detector Technology PIN 5D, an EG&G PV 444, and one element of a 12 × 12 Centronic array (Fig. 9). For both the PV 444 and the Centronic diode the noise was proportional to intensity to the 0.55 power (W. N. Ross, J. A. London, D. Zečević, and L. B. Cohen, unpublished results), close to the expected relationship for shot noise. This result suggests that $1/f$ noise does not make a large contribution to the light noise at the bandwidth we tested (10–100 Hz). However, with the PIN 5D the noise was proportional to intensity to the 0.65 power. This deviation from 0.5 suggests the presence of $1/f$ noise in this kind of diode, in agreement with the results obtained by Kleinfeld. It should be noted that optical measurements on vertebrate preparations are made using frequencies lower than those we tested. Since by definition, measurements at low frequencies are more likely to suffer from interference from $1/f$ noise, additional testing would be informative.

5.2 Extraneous Noise

While dark noise is a concern at low intensity, and $1/f$ noise may be avoided by photodiode selection, a third type of noise, termed extraneous noise, is more apparent at higher light intensities where the sensitivity of the measurement is high because the fractional shot noise is low. One type of extraneous noise, due to fluctuations in the output of the light source, will be discussed in the next section. Two other sources of extraneous noise are vibrations in the light path and movement of the preparation. A number of precautions for reducing vibrational noise are described in Salzberg et al. (1977). We found that embedding ganglia in 1–3% agar would further reduce vibrations (London et al., 1985). Nevertheless, it has been difficult to reduce vibrational noise to much less than 10^{-5} of the total light in experiments on invertebrate ganglia. With this amount of vibrational noise, increases in measured intensity past 10^{10} photons/s would reduce the shot noise, but, because both vibrational noise and signal will increase linearly with intensity, there would be no improvement in the signal-to-noise ratio (segment C of Fig. 7).

We recently found that the pneumatic isolation mounts on two vibration isolation tables which we used were providing only minimal isolation in the frequency range 20–60 Hz. By replacing the pneumatic mounts with air-filled soft rubber tubes we have reduced the vibration noise to an undetectable level in absorption measurements on *Aplysia* ganglia (D. Zečević, J. A. London, and L. B. Cohen, unpublished

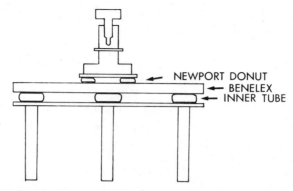

Figure 8.   Schematic drawing of the vibration isolation system we currently use. There are two stages of isolation provided by air-filled soft-rubber tubes. This system reduces noise in the 20–60-Hz range by at least a factor of 10. The lower stage uses three inner tubes (6-in.-diameter inner hole) purchased from a tire distributor. The second stage uses four donuts purchased from Newport Research Corporation, Fountain Valley, CA. The microscope and lamp housing are mounted on a half-inch aluminum plate, which is isolated from a Benelex table top by the Newport donuts. The table top also has ancillary equipment and lead bricks on it weighing 200 kg. The top itself weighs about 75 kg. The top is isolated from a support base with the inner tubes. (D. Zećević, J. A. London, and L. B. Cohen, unpublished results).

results). Figure 8 is a schematic diagram of a system incorporating two stages of isolation. This system reduces noise in the 20–60-Hz range by at least a factor of 10. Thus the argument given at the end of the preceding paragraph is probably no longer valid.

Noise due to movement is a major problem in measurement on in vivo preparations. Some methods for reducing the movements or the resulting artifacts in molluscan and mammalian experiments have been described (London et al., this volume; Orbach et al., 1985) (See chapter 8).

5.3   Light Sources

Three kinds of sources have been used. Tungsten filament lamps are a stable source, but their intensity is relatively low, particularly at wavelengths less than 550 nm. Arc lamps and lasers are less stable but can provide more intense illumination.

Tungsten Filament Lamps.   It is not difficult to provide a power supply (constant voltage or constant current source) that is stable enough that the output of the bulb fluctuates by less than 1 part in 10^5. In absorption measurements, where the fractional changes in intensity are relatively small, only tungsten filament sources have been used. On the other hand, fluorescence measurements often have large fractional changes that will better tolerate light sources with systematic noise, and the measured intensities are low, which makes possible improvements in signal-to-noise ratio from more intense sources attractive. Hence, arc lamps or laser sources have sometimes been used in fluorescence measurements.

Arc Lamps. Both mercury and mercury–xenon arc lamps have been used to obtain higher intensities. However, comparison of the excitation intensity obtainable from tungsten filament and arc sources is not simple. Grinvald et al. (1982) reported that the intensity from a mercury arc lamp was 50–100 times higher than that from a tungsten filament lamp using a 540-nm filter with a width at half-height of 18 nm. However, the advantage implied by such a comparison may be misleading. Because the excitation "action" spectrum of some dyes (i.e., the styryl, RH414) is quite broad, it is possible to use a filter with a width at half-height of 90 nm. This will substantially increase the incident intensity from the tungsten filament source. However, because the mercury lamp has a distinct emission line at 546 nm, using a wider filter adds little intensity with this source. Furthermore, the tungsten filament bulb can be overrun to increase its color temperature, increasing its output intensity by about 75%. Finally, the intensity will depend on the area of the object that is illuminated. Using critical illumination, the arc lamp, which approximates a point source, will be relatively preferred for smaller objects. Thus the increase in intensity obtained by using an arc lamp in lieu of tungsten will often not be as great as that indicated by Grinvald et al. (1982) and will depend on the size of the preparation (P. Saggau, L. B. Cohen, and A. Grinvald, unpublished results).

The main difficulty with arc lamps is output intensity fluctuations (in part resulting from arc wander). Using a high-speed constant current power supply (KEPCO JQE 75-15(M) HS), the peak output fluctuations were reduced to 4×10^{-4} of the total intensity over the bandwidth of 0.5 Hz–1 kHz (Davila et al., 1974). J. Pine (personal communication) has utilized the circuit illustrated in Fig. 9 to further reduce output fluctuations. With this circuit the output noise was 10^{-4} over the bandwidth of 10 Hz–10 KHz. The arc lamp requires a very high voltage for ignition, and the use of some starters has led to damage of ancillary electronic equipment. However, the Zeiss starter for the HBO 100 W/2 lamp has not had this problem (W. N. Ross, and P. Saggau, personal communications). At present it is difficult to make an a priori decision between a tungsten filament and an arc source for fluorescence measurements. The decision will depend on the fractional change, the predominant source of noise, the level of difficulty associated with dye bleaching or photodynamic damage, and future improvements in quieting an arc lamp.

Lasers. It has been possible to take advantage of two useful characteristics of laser sources. First, the laser output can be focused onto a small spot in preparations with no scattering, allowing measurement of membrane potential from small processes in tissue-cultured neurons (Grinvald and Farber, 1981). Second, the laser beam can be positioned flexibly and rapidly using acoustooptical deflectors (Dillon and Morad, 1981; Morad et al., this volume; Hill and Courtney, 1985) (See chapter 12). However, lasers have thus far only been used in situations where the fractional change in intensity was large. There have been two difficulties: first, the amplitude of the laser light output is not stable, and second, there appears to be excess noise that may be due to laser speckle (Dainty, 1984). Commercially available (Uniphase) helium–neon lasers (the 632.8-nm line can be used to excite certain oxonol dyes;

ARC–LAMP SHUNT REGULATOR (J. PINE)

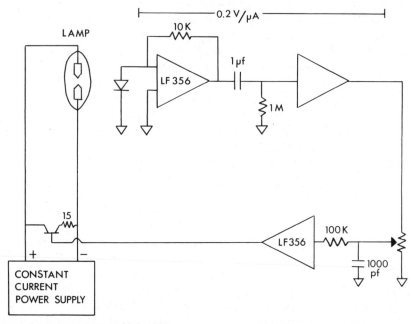

Figure 9. Feedback circuit for reducing noise in the output of an arc lamp. A small fraction of the lamp output is directed at the photodetector. The detector current is converted to a voltage, amplified, and then fed to a power amplifier whose output is fed back to the power supply such that an increase in intensity at the diode leads to a decrease in power to the lamp. The converse occurs when the light intensity decreases. From J. Pine (personal communication).

XXV, Fig. 1) with modified power supplies have intensity fluctuations of less than 2×10^{-5} of the total intensity (B. M. Salzberg, personal communication). Argon or krypton lasers (the 514.5-nm line can be used to excite certain styryl dyes; RH414, Fig. 1) have intensity fluctuations of 5×10^{-3} of the total intensity. Two kinds of schemes for reducing the effects of amplitude fluctuations have been used. Both involve beam splitting to deflect a portion of the light to a reference photodetector. The output of the reference detector can be used in a negative-feedback loop with a Pockels cell (Cohen et al., 1972), or the reference signal can be subtracted from (or divided into) the experimental signal. Each of these procedures can reduce the effect of amplitude fluctuations by a factor of about 10.

However, the noise at the photodector can be surprisingly large when the noise in the laser source appears to be small. In fluorescence measurements on invertebrate ganglia, we found that the fractional noise on each detector was substantially larger than the fractional noise present in the incident light (B. M. Salzberg, D. Senseman, L. B. Cohen, and A. Grinvald, unpublished results). This excess noise may be due to laser speckle, and it has been possible to partially eliminate this noise by introducing high-frequency mode scrambling into the laser beam (G. Ellis and B. M. Salzberg, personal communication).

Amplitude fluctuations in currently available solid-state lasers (Liconix, Sunnyvale, CA) are apparently very small, less than 10^{-4} of the total intensity (B. M. Salzberg, personal communication). They have not yet been tested in optical measurements.

5.4   Image-Recording Devices

The major motivation for developing optical methods for monitoring membrane potential was the possibility of making simultaneous multisite measurements of activity. Only one of the many available types of imaging devices has been used thus far. The need to avoid reductions in signal-to-noise ratio has often restricted this choice. One type of imager that has outstanding spatial and temporal resolution is movie film. But because it is difficult to obtain quantum efficiencies of even 1% with film (Shaw, 1979), there would be an automatic factor of 10 degradation in signal-to-noise ratio in comparison with a silicon diode. This and other possible difficulties, including frame-to-frame and within-frame emulsion nonuniformity, has discouraged attempts to use film. The lower quantum efficiencies of vacuum photocathodes has discouraged their use. Hence only silicon diode imaging devices have been used for multisite optical recording.

In all of the systems now in use, the image-recording device has been placed in the objective image plane of a microscope. However, Tank and Ahmed (1985) are implementing a scheme by which a hexagonal close-packed array of 256 optical fibers is positioned in the image plane and individual photodiodes are connected at the opposite ends of these fibers.

Parallel versus Serial Recording.   A number of imaging devices lend themselves to electronic readout. This recording can be either parallel or serial. An example of a parallel readout device that has been used in several laboratories is the diode array illustrated in Fig. 10. The dark squares (1.4 × 1.4 mm) in this 10 × 10 array (Centronics, Ltd.) are the individual detectors (pixels). Each detector has a wire that runs to an amplifier (see below). In the 12 × 12 array currently employed, 124 elements are followed by 124 amplifiers. With this parallel recording, the high-frequency cutoff of each amplifier is determined by the highest frequencies that are present in the signal of interest. If, for example, molluscan action potentials have frequency components that go up to 200 Hz, and one wants to know the time course accurately, then the amplifiers should have high frequency cutoffs of 200 Hz and the output of each amplifier should be recorded every millisecond.

An alternative, serial readout mechanism, which would save the construction and maintenance cost of the array of amplifiers, would be to follow the diode array with a multiplexer and then to use a single amplifier. However, this single amplifier must operate on 100 signals every millisecond, and thus its high frequency cutoff would have to be in the 100 kHz range—many times higher than in the case of a parallel

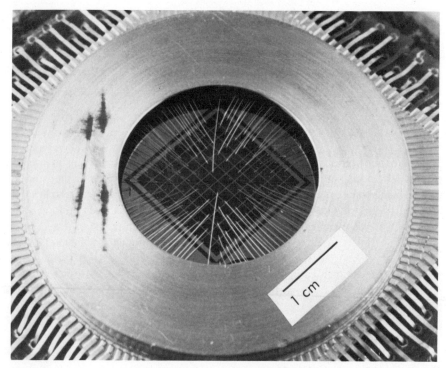

Figure 10. A 10 × 10 array of silicon photodiodes manufactured by Centronic (New Addington, Croydon, England). The dark squares (1.4 × 1.4 mm) are the photodiodes. The insulating regions between the diodes are 0.1 mm thick. The output of each diode is carried from the array to a current to voltage converter (Fig. 11) via a wire. At the present time a number of laboratories are using a 12 × 12 array that is identical to the one shown except for the additional 44 detectors.

readout. Since the signal-to-noise ratio is proportional to the square root of bandwidth, a serial recording will have a relatively large dark noise. Thus a serial readout can be used only in situations where optimizing the signal-to-noise ratio is not a primary consideration or where other noise sources dominate. One example of the use of a serial recording is the laser scanning apparatus developed by Dillon and Morad (1981; Morad et al., this volume) for monitoring the propagation of cardiac action potentials (See chapter 12).

A second advantage of a parallel readout scheme is that lower recording accuracy is required. In a serial readout, determining the changes in intensity, which are a very small fraction of the total intensity, can be done only by subtracting two relatively large numbers: the total intensity at time t minus the total intensity at time t-1. To measure the total intensity with the accuracy of one part in 10^5 that would be necessary for some absorption signals, an analog-to-digital (A–D) conversion accurate to 16 or 17 bits would be needed. This kind of accuracy is not inexpensively achieved at the required data rates. However, in a parallel readout, either capacity coupling or DC subtraction (see below) can be used so that the A–D converter needs to have only enough resolution to measure the change in intensity. In this situation 8-bit resolution has usually been adequate.

These two advantages of parallel recording schemes are probably the major factors that account for the use of the Centronic arrays. Even though the number of pixels is very small when compared to a charge-coupled device (CCD) array, the CCD can only be read out serially, and there are no commercially available CCD's that will read out the change in intensity rather than the total intensity. Other limitations of CCD devices—larger dark noise, saturation, and a smaller fraction of the device area covered by photosensor—might also cause difficulty. Furthermore, increasing the number of pixels could be difficult with the computer hardware currently in use (see below).

If improved spatial resolution (many more pixels) were implemented, in fluorescence measurements, the number of photons per detector would be lower and probably in the intensity region where vacuum–photocathode devices would have the best signal-to-noise ratio (Fig. 7). One of these, a Radechon (Kazan and Knoll, 1968), has an output proportional to the changes in the input.

5.5   Optics

The need to maximize the number of measured photons has also dominated the choice of optical components. The number of photons that are collected by an objective lens in forming the image is proportional to the square of the numerical aperture (NA) of the lens. In epifluorescence both the excitation light and the emitted light pass through the objective, and the intensity reaching the photodiodes is proportional to the fourth power of its numerical aperture. Accordingly, objectives (and condensers) with high numerical apertures have been employed. With a Leitz Ortholux microscope, use of a lamp collector with 0.7 NA (Aspherab, Oriel Corp.) increased the intensity reaching the object by a factor of 1.5–2 in comparison to a 0.4-NA lens (Shoeffel Instrument Corp.). In addition, in epifluorescence, the illumination was more homogeneous with the 0.7-NA collector (L. B. Cohen, J. A. London, and D. Zećević, unpublished results). Knox Chandler (personal communication) pointed out that the increase we measured appears to depend on the position of the luminous-field diaphragm (Piller, 1977) in the Leitz microscope. When the field diaphragm was moved closer to the lamp collector lens, the differences in intensity between the two collector lenses disappeared. (See Piller (1977) for a further discussion of the subject.)

Because the Centronic diode arrays that are currently available have a pixel size of 1.4×1.4 mm, using the 12×12 array is most convenient on a microscope where the diameter of the primary image plane formed by the objective is about 20 mm. While the Leitz Ortholux and Zeiss UEM microscopes do have an image of this size, a Zeiss Universal microscope has an image size that is significantly smaller (A. Grinvald, personal communication). In addition, it is convenient if the primary image plane is easily accessible. In an Olympus-Vanox AHB LBI microscope, the objective image plane is inside the trinocular head (K. Kamino, personal communication), and thus a secondary image must be formed for the array.

Depth of Focus.   Salzberg et al. (1977) determined the effective depth of focus for a 0.4-NA objective lens on an ordinary microscope by recording an optical signal from a neuron when it was in focus and then moving the neuron out of focus by various distances. They found that the neuron had to be moved 300 μm out of focus to cause a 50% reduction in signal size. (This result will be obtained only when the diameter of the neuron image and the diameter of the detector are similar.) Using 0.6-NA optics, the effective depth of focus was reduced by about 50% (A. Grinvald and L. B. Cohen, unpublished results). This large effective depth of focus can be advantageous in some circumstances. If, for instance, one would like to record from all the neurons in a 500-μm-thick invertebrate ganglion, then using 0.4-NA optics one could focus at the middle of the ganglion and the signals from neurons on the top and bottom of the ganglion would be reduced in size by less than 50%. (This situation will, of course, result in the superposition of signals from 2 or more neurons on some detectors; subsequent sorting out will be required (London et al., this volume). On the other hand, the large effective depth of field makes it difficult to determine the depth in a preparation at which a signal originates (See chapter 8).

Effects of Light Scattering.   Light scattering can limit the spatial resolution of an optical measurement (See chapter 8). London et al. (this volume) measured the scattering of 705 nm light in *Navanax* buccal ganglia. They found that insertion of a ganglion in the light path caused light from a 30-μm spot to spread such that the diameter of the circle of light that included intensities greater than 50% of the total was now about 50 μm. The spread was greater, to about 100 μm, with light of 510 nm. Since the blurring is not large compared to the average cell diameter at the wavelengths we used (705 nm), it does not lead to a large overestimate of cell size or to severe degradation of the signal-to-noise ratio. Figure 11 illustrates the results of similar experiments that were carried out on the salamander olfactory bulb. The top section indicates that when no tissue is present, essentially all of the light (750 nm) from the small spot falls on one detector. The bottom section illustrates the result when a 500-μm-thick slice of olfactory bulb is present. The light from the small spot is spread to about 200 μm (Orbach and Cohen, 1983). Certainly mammalian cortex does not appear to scatter less than the olfactory bulb. Thus, light scattering will cause considerable blurring of signals in intact vertebrate preparations.

   A second possible source of blurring is signal from regions that are out of focus. For example, if the active region is a cylinder (a column) that is perpendicular to the plane of focus, and the objective is focused at the middle of the cylinder, then the light from the middle will have the correct diameter at the image plane, but the light from the regions above and below are out of focus and will have a diameter that is too large. The middle section of Fig. 11 illustrates the effect of moving the small spot of light 500 μm out of focus. The light from the small spot is spread to about 200 μm.

Confocal Microscope.   Petran and Hadravsky (1966) patented a modification of the microscope that would substantially reduce both the scattered light and the out-of-focus light that contributes to the image. Egger and Petran (1967) demonstrated the improvement that could be achieved with this microscope by

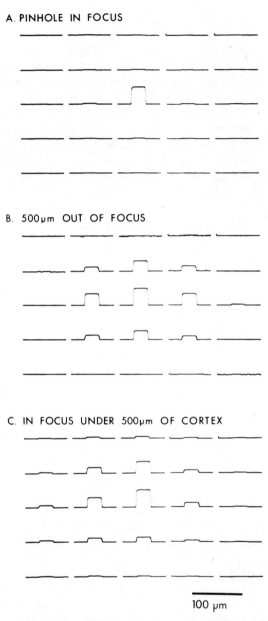

Figure 11. Effects of focus and scattering on the distribution of light from a point source onto the array. (A) A 40-μm pinhole in aluminum foil covered with saline was illuminated with light at 750 nm. The pinhole was in focus. More than 90% of the light fell on one detector. (B) The stage was moved downward by 500 μm. Light from the out-of-focus pinhole was now seen on several detectors. (C) The pinhole was in focus but covered by a 500-μm slice of salamander cortex. Again the light from the pinhole was spread over several detectors. A 10 × 0.4 NA objective was used. Kohler illumination was used before the pinhole was placed in the object plane. The recording gains were adjusted so that the largest signal in each of the three trials would be approximately the same size in the figure. From Orbach and Cohen (1983).

showing that it provided much clearer images of dorsal root ganglia neurons, and later Sheppard and Wilson (1981) named this kind of microscope a confocal microscope.

The Petran invention is based on the following idea. A small spot of the object is illuminated. If a pinhole is positioned at the corresponding spot in the image plane and only light passing through the pinhole is used to form the image, then most of the scattered light and the light from out-of-focus regions will be rejected. As is illustrated in Fig. 11, both the scattered and out-of-focus light reach a relatively large area of the image plane, and most of this light is rejected in the confocal microscope. To form a complete image, the spot and pinhole are moved in parallel to each pixel in the field of view. The implementation of this idea by Petran and Hadravsky involved symmetrical incident light and imaging pathways with a rotating disk of very carefully positioned pinholes. The pinholes in the image path corresponded exactly to the pinholes in the incident path.

One of the difficulties of Petran's design was that only 10^{-7} of the illuminating light is available for forming the image (Boyde et al., 1983), and thus very bright light sources were required; Egger and Petran (1967) used sunlight. However, if a scanning laser spot (Davidovitz and Egger, 1969) were used to illuminate the preparation, then the light loss in the illuminating path would be eliminated. Thus, by using a confocal microscope with a laser light source, one might be able to obtain signals from intact vertebrate preparations with much better spatial resolution than was achieved with ordinary microscopy. P. Saggau and G. Gerstein (personal communications) are attempting to construct laser-based confocal microscopes with imaging rates in the millisecond range. Because they will have to contend with the "speckle" noise and the possible degradation of signal-to-noise due to serial readout (see above), this may be a difficult task.

5.6   Amplifiers

In our apparatus, amplifier noise does not make a substantial contribution to either the dark noise or the total noise. The dark noise usually did not decrease when a National Semiconductor, LF 356N, (Fig. 12) was replaced with a Burr-Brown OPA 111AM, an amplifier with substantially lower noise specifications (suggested by J. Meyer). However, certain relatively noisy LF356N amplifiers did add to the dark noise, and thus this amplifier appears to define the limit at which amplifier noise will add to the dark noise.

There are two difficulties with the use of AC coupled amplifiers. First, the AC coupling will restrict the ability to measure slow signals (low frequencies), and second, the optical recording cannot begin the instant the light is turned on because one must wait about 10 time constants for the input to the second amplifier to settle. This can be disadvantageous if there is significant photodynamic damage and the actual recording period is short compared to the settling time. The long settling time can be eliminated if the amplifiers have two coupling time constants with rapid switching between the two (Fig. 12; B. M. Salzberg, personal communication). A fast time constant is used initially to provide a short settling period, and the second, slower time constant is chosen to be long compared to the signal.

A useful feature of the AC recording was that the second amplifier could have a high gain so that a relatively large, ± 10-V signal could be fed to the multiplexer and A–D converter. This allowed less expensive multiplexers and A–D converters to be used. The advantage of using a high-gain, second-stage amplifier can also be obtained along with a DC recording if the DC light level is measured when the light is first turned on and then this value is subtracted from all subsequent measurements. This can be implemented either with sample-and-hold amplifiers (Nakajima and Gilai, 1980) or digitally (Senseman and Salzberg, 1980; J. Meyer, personal communication).

5.7 Computer

The discussion in this section is limited to the hardware and software currently in use in our laboratory.

Hardware.   Recording from 124 pixels (plus 4 electrical inputs) where each output is sampled every 1.0 ms generates $1.28 \cdot 10^5$ data points per second. This recording

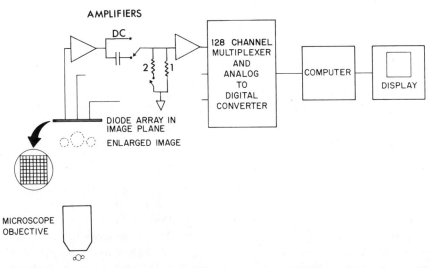

Figure 12.   Schematic drawing of the amplifiers and computer hardware used to make a parallel recording from a 12 × 12 photodiode array. This is the system currently used in our laboratory. The output of each diode, used in a photovoltaic mode, is converted into a voltage signal by the first LF 356 amplifier. There are two fixed high-frequency cutoff RC filters at this stage (not shown). The output of the first amplifier is fed to the second via one of three switch-selectable pathways. The DC coupling mode allows the resting light intensity to be recorded. A short time-constant AC coupling (both resistors in) is used to allow quick settling of the input to the second amplifier just after the light is turned on. The long time-constant AC coupling (only one resistor) is used during the signal measurement. The second amplifier is used to provide the gain (1000–5000 ×), necessary for the A–D conventors which are used as 8-bit converters with a ± 10-V input range. Amplifiers of this type can be purchased from Vic Pantani, Department of Physiology, Yale University School of Medicine, New Haven, CT. The A–D multiplexer cards (ADAC Corp 600-11) are inserted into the Unibus backplane of the DEC PDP 11/34 minicomputer.

rate can be maintained for periods of several minutes at 7.68M bytes/min using a 16-bit minicomputer (Digital Equipment Corp., PDP 11/34) by simultaneously inputting data into a buffer in main memory and writing data from a previously filled buffer to a Winchester disk. In the system we are now using (Fig. 12), two multiplexer converter cards (ADAC, Inc.) are run in parallel. Each multiplexes 64 channels with a minimum conversion time of about 13 μs. The most significant eight bits from each ADAC card are combined into a 16-bit word in the data register of a direct-memory-access (DMA) interface and deposited via the Unibus into a 64-K byte section of memory. During the filling of a second 64-K byte buffer the data from the first buffer are written to a large file on an RLO1 disk. The data input rate might be increased by a factor of 5 by using additional or higher-speed A–D–DMA hardware. However, we calculated that the maximum throughput rate to the RLO1 disk was only $2.2 \cdot 10^5$ bytes/s, and thus a substantial increase in either the number of pixels, or the time resolution, or the number of bits of A–D resolution would require faster hardware. R. Lombardi and G. Salama (personal communication) found that the RLO1 interface for a Q-bus took control of the bus during the entire time it was writing one track. This eliminated the possibility of simultaneous data input and writing to the disk with the RLO1–Q-bus system. Thus, the capabilities of the present apparatus is limited by the hardware currently in use.

Programs. Our programs are written in assembly language (Macro 11) called by Fortran under the RT 11 operating system (version 5.0) and are available from the authors. R. Lombardi and G. Salama and A. Grinvald have added additional functions to these programs. N. Lasser-Ross and W. Ross have written similar programs using only assembly language.

We have three programs for recording and displaying data from the array. The first two use only the computer memory (136K bytes of extended memory) for real-time storage and are thus limited to about 1000 bytes (data points) per diode. The first program, SUNTIM, records and displays the results of a single trial: 1088 points per detector. The second program, FILTSA, signal-averages. One-third of extended memory is used for storing the 8-bit data from the most recent trial (357 points per detector). The remaining two-thirds of memory are used to store the sum. Using two bytes for each time point for the sum allows at least 256 trials to be added. Three types of triggering can be used with this program. In the first, the program itself puts out a trigger pulse and the data are added to the sum, time-locked to this trigger. In the second, the computer puts out a pulse to start the trial, but it automatically determines the time of occurrence of an action potential on an electrode recording and then adds the data to the sum, time-locked to this action potential. This mode allows averaging off the time of events when the timing of the event relative to the trigger is not precisely fixed. In the third type of triggering, the start of the sweep is determined by an external event, such as an electrocardiogram signal. In this mode, two recordings are made—one with a stimulus and one without. Then the recording without the stimulus is subtracted from the recording with the stimulus. This procedure reduces the effect of noise that is time-locked to the external event (e.g., movement of a rat cortex that is locked to the heart beat).

The data from SUNTIM and FILTSA can be stored in files on a disk or tape which are examined by a third program, STUDY. STUDY provides a number of functions including additional high-frequency filtering of the data, a display of the derivative of the signal, or a display of signals divided by the resting-light intensity. In this case the data are displayed according to the relative fractional change in intensity. The outputs of selected detectors can be summed, scaled, and superimposed to allow comparison of time courses.

The third program, SVLTPR, uses the double-buffering scheme described above to record for longer periods, usually 20 s (2.56M bytes of data). We tested an assembly language subroutine to write to the disk, but it was not noticeably faster than using the programmed request, .WRITEW. The study program, LSTUDY, is designed to optimize the detection of action potentials in a situation where the signal-to-noise ratio may be small. Both high- and low-frequency filtering are provided. To avoid rounding errors and still use integer arithmetic, the byte data must be moved into a word buffer and increased in size. The filtered data from each detector (20K points) are displayed on an expanded scale, 1K/sweep, to aid in spike identification. Once the detectors whose outputs have action potentials are known, the 20K recordings are displayed on long (2-m) sheets of paper (Tektronix 4633A) in groups of 20 neighboring detectors. This kind of display facilitates the identification of duplicate signals on adjacent detectors that arise when light from relatively large neurons reaches several detectors. After a trace containing the activity of each active neuron is specified, a spike identification program is used to enter the time of occurrence of the action potentials into the computer. The next section of the program checks for duplicates, and finally a raster diagram of action potentials in individual cells is generated. To obtain recording times of up to 20 min, Hirota et al. (1985) have interposed a tape recorder between the analog-to-digital converters and the computer.

Programming a 16-bit computer, which directly addresses only 64K bytes of memory, is difficult with the large amounts of data from these experiments. About half of the programming time is devoted to extended memory programming and to implementing program overlays. This difficulty may be relieved by using a 32-bit computer. A more modern computer would also relieve the hardware restrictions on the data input.

6. FUTURE DEVELOPMENTS

It is clear that improvements in the signal-to-noise ratios of optical measurements of membrane potential would be useful. A number of avenues remain partially or completely unexplored.

Only three optical properties of stained membranes have been examined for signals in response to changes in membrane potential. The possibilities of finding large changes in other optical properties—for example, energy transfer, circular dichroism (optical rotation), or absorption-enhanced Raman scattering—have been largely neglected. Ehrenberg and Berezin (1984) have used resonance Raman to study surface potential, and Ehrenberg and Loew (personal communication) are planning to investigate its use in measuring transmembrane potential. There was a report of holographic signals in leech neurons (Sharnoff et al., 1978a), but signals were not found in subsequent experiments on squid axons (Sharnoff et al., 1978b). There were also reports of changes in intrinsic infrared absorption (Sherebrin, 1972; Sherebrin et al., 1972), but these have not been pursued further. Thus, one approach to looking for larger signals would be to investigate new types of optical phenomena.

A second approach involves improvement of the apparatus. In situations where a very coarse recording of the image is done, as with a 12 × 12 array, the spatial resolution of the microscope lens is not fully utilized. It might be possible to design lenses with increased numerical aperature if resolution is sacrificed. A second useful apparatus improvement would be further quieting of arc and laser light sources and investigation of new kinds of light sources. The successful implementation of a confocal microscope for preparations with substantial scattering or thickness would greatly improve spatial resolution.

The third approach is finding or designing better dyes. All of the dyes in Fig. 1 and the vast majority of those synthesized in recent years are of the general class named cyanines (Hamer, 1964), a class that was first used in photographic processes. It is certainly possible that improvements in signal size can be obtained with new cyanine dyes (see Waggoner and Grinvald (1977) for a discussion of maximum possible fractional changes in absorption and fluorescence). On the other hand, the fractional change on squid axons has not increased much in recent years (Gupta et al., 1981; L. B. Cohen, A. Grinvald, K. Kamino and B. M. Salzberg, unpublished results), and most improvements (Grinvald et al., 1982) have involved synthesizing analogues that work well on new preparations. Radically different synthetic approaches or a modicum of cooperation from corporations like Eastman Kodak might prove to be very useful.

The results in the following chapters show that optical methods can already provide new and previously unobtainable information about cell and organ function. Clearly there has been dramatic progress since the first recording of an action potential in a leech neuron (Salzberg et al., 1973). We hope that additional improvements will further increase the utility of these methods.

ACKNOWLEDGMENTS

The authors are indebted to their collaborators Vicencio Davila, Amiram Grinvald, Kohtaro Kamino, Jill London, Bill Ross, Brian Salzberg, Alan Waggoner, and Dejan Zećević for numerous discussions about optical methods. We also thank Pancho Bezanilla, David Kleinfeld, Richard Lombardi, Ana Lia Obaid, Jerome Pine, Guy Salama, Peter Saggau, and David Senseman, who have allowed us to cite unpublished results. The experiments carried out in our laboratory were supported by NIH grant NS-08437.

REFERENCES

Baylor, S. M., W. K. Chandler, and M. W. Marshall (1981) "Studies in skeletel muscle using optical probes of membrane potential," in A. D. Grinnel and M. A. B. Brazler (Eds.), *The Regulation of Muscle Contraction: Excitation–Contraction Coupling*, Academic, New York, pp. 97–130.

Boyde, A., M. Petran, and M. Hadravsky (1983) Tandem scanning reflected light microscopy of internal features in whole bone and tooth samples. *J. Microsc.*, **132**, 1–7.

Boyle, M. B., and L. B. Cohen (1980) Birefringence signals that monitor membrane potential in cell bodies of molluscan neurons. *Fed. Proc.*, **39**, 2130.

Boyle, M. B., L. B. Cohen, E. R. Macagno, and H. S. Orbach (1983) The number and size of neurons in the CNS of gastropod mollusks and their suitability for optical recording of activity. *Brain Res.*, **266**, 305–317.

Braddick, H. J. J. (1960) Photoelectric photometry. *Rep. Prog. Physics*, **23**, 154–175.

Cohen, L. B., and J. F. Hoffman (1982) "Optical monitoring of membrane potential," in P. F. Baker (Ed.), *Techniques in Cellular Physiology*, Elsevier/North Holland, Amsterdam, pp. 118, 1–13.

Cohen, L. B., and B. M. Salzberg (1978) Optical measurement of membrane potential. *Rev. Physiol. Biochem. Pharmacol.*, **83**, 35–88.

Cohen, L. B., B. Hille, and R. D. Keynes (1968) Light scattering and birefringence changes during nerve activity. *Nature*, **218**, 438–441.

Cohen, L. B., D. Landowne, B. B. Shrivastav, and J. M. Ritchie (1970) Changes in fluorescence of squid axons during activity. *Biol. Bull.*, **139**, 418–419.

Cohen, L. B., H. V. Davila, and A. S. Waggoner (1971) Changes in axon fluorescence. *Biol. Bull.*, **141**, 382.

Cohen, L. B., R. D. Keynes, and D. Landowne (1972) Changes in axon light-scattering that accompany the action potential: current dependent components. *J. Physiol.*, **244**, 727–752.

Cohen, L. B., B. M. Salzberg, H. V. Davila, W. N. Ross, D. Landowne, A. S. Waggoner, and C. H. Wang (1974) Changes in axon fluorescence during activity: Molecular probes of membrane potential. *J. Membr. Biol.*, **19**, 1–36.

Cohen, L. B., J. A. London, and D. Zećević (1986) Simultaneous optical recording of activity from many neurons during feeding in *Navanax*. (Submitted.)

Conti, F. (1975) Fluorescent probes in nerve membranes. *Annu. Rev. Biophys. Bioeng.*, **4**, 287–310.

Conti, F., I. Tasaki, and E. Wanke (1971) Fluorescence signals in ANS-stained squid axons during voltage clamp. *Biophys. J.*, **8**, 58–70.

Dainty, J. C. (Ed.) (1984) *Laser Speckle and Related Phenomena*, 2nd ed., Springer-Verlag, New York.

Davidovitz, P. and M. D. Egger (1969) Scanning laser microscope. *Nature*, **223**, 831.

Davila, H. V., L. B. Cohen, B. M. Salzberg, and B. B. Shrivastav (1974) Changes in ANS and TNS fluorescence in giant axons from *Loligo*. *J. Membr. Biol.*, **15**, 29–46.

Dillon, S., and M. Morad (1981) Scanning of the electrical activity of the heart using a laser beam with acousto-optics modulators. *Science*, **214**, 453–456.

Egger, M. D., and M. Petran (1967) New reflected light miroscope for viewing unstained brain and ganglion cells. *Science*, **157**, 305–307.

Ehrenberg, B., and Y. Berezin (1984) Surface potential on purple membranes and its sidedness studied by resonance Raman dye probe. *Biophys. J.*, **45**, 663–670.

Freedman, J. C., and P. C. Laris (1981) Electrophysiology of cells and organelles: Studies with optical potentiometric indicators. *Int. Rev. Cytol.* (Suppl.) **12**, 177–246.

Fugii, S., A. Hirota, and K. Kamino (1981) Action potential synchrony in embryonic precontractile chick heart: Optical monitoring with potentiometric dyes. *J. Physiol.* **312**, 253–263.

Grinvald, A. (1985) Real-time optical mapping of neuronal activity: From growth cones to the intact mammalian brain. *Ann. Rev. Neurosci.*, **8**, 263–305.

Grinvald, A., and I. Farber (1981) Optical recording of Ca^{2+} action potentials from growth cones of cultured neurons using a laser microbeam. *Science*, **212**, 1164–1169.

Grinvald, A., W. N. Ross, and I. C. Farber (1981a) Simultaneous optical measurements of electrical activity from multiple sites on processes of cultured neurons. *Proc. Natl. Acad. Sci. USA*, **78**, 3245–3249.

Grinvald, A., L. B. Cohen, S. Lesher, and M. B. Boyle (1981b) Simultaneous optical monitoring of activity of many neurons in invertebrate ganglia using a 124 element photodiode array. *J. Neurophysiol.*, **45**, 829–840.

Grinvald, A., R. Hildesheim, I. C. Farber, and L. Anglister (1982) Improved fluorescent probes for the measurement of rapid changes in membrane potential. *Biophys. J.*, **39**, 301–308.

Gupta, R. K., B. M. Salzberg, A. Grinvald, L. B. Cohen, K. Kamino, S. Lesher, M. B. Boyle, A. S. Waggoner, and C. H. Wang (1981) Improvements in optical methods for measuring rapid changes in membrane potential. *J. Membr. Biol.*, **58**, 123–137.

Hamer, F. M. (1964) *The Cyanine Dyes and Related Compounds.* Wiley, New York.

Hassner, A., D. Birnbaum, and L. M. Loew (1984) Charge shift probes of membrane potential: synthesis. *J. Organic Chem.*, **49**, 2546–2550.

Hill, B. C., and K. R. Courtney (1985) Optical monitoring of myocardial conduction: Observations of reentry and examples of lidocaine action. *Biophys. J.*, **47**, 496a.

Hirota, A., K. Kamino, H. Komuro, T. Saki, and T. Yada (1985) Early events of development of electrical activity and contraction in embryonic rat heart monitored using voltage-sensitive dyes. *J. Physiol.*, in press.

Kauer, J. S., D. Senseman, and L. B. Cohen (1984) Voltage-sensitive dye recording from the olfactory system of the tiger salamander. *Soc. Neurosci. Abstr.*, **10**, 846.

Kazan, B., and M. Knoll (1968) *Electronic Image Storage.* Academic, New York, pp. 134–141.

Lev-Ram, V., and A. Grinvald (1984) Is there a potassium dependent depolarization of the paranodal region of myelin sheath? Optical studies of rat optic nerve. *Soc. Neurosci. Abstr.*, **10**, 948.

Loew, L. M., and L. L. Simpson (1981) Charge shift probes of membrane potential: A probable electrochromic mechanism for p-aminostyrylpyridinium probes on a hemispherical bilayer. *Biophys. J.*, **34**, 353–365.

Loew, L. M., L. B. Cohen, B. M. Salzberg, A. L. Obaid, and F. Bezanilla (1985) Charge shift probes of membrane potential. Characterization of aminostyrylpyridinium dyes on the squid giant axon. *Biophys. J.*, **47**, 71–77.

Malmstadt, H. V., C. G. Enke, S. R. Crouch, and G. Harlick (1974) *Electronic Measurements for Scientists.* Benjamin, Menlo Park, CA, pp. 738–749.

Morad, M., and G. Salama (1979) Optical probes of membrane potential in heart muscle. *J. Physiol.*, **292**, 267–295.

Nakajima, S., and A. Gilai (1980) Action potentials of isolated single muscle fibers recorded by potential sensitive dyes. *J. Gen. Physiol.*, **76**, 729–750.

Orbach, H. S., and L. B. Cohen (1983) Optical monitoring of activity from many areas of the in vivo and in vitro salamander olfactory bulb: A new method for studying functional organization in the vertebrate central nervous system. *J. Neurosci.*, **3**, 2251–2262.

Orbach, H. S., L. B. Cohen, and A. Grinvald (1985) Optical mapping of electrical activity in rat somatosensory and visual cortex. *J. Neurosci.*, **5**, 1886-1895.

Patrick, J., B. Valeur, L. Monnerie, and J.-P. Changeux (1971) Changes in extrinsic fluorescence intensity of the electroplax membrane during electrical excitation. *J. Membr. Biol.*, **5**, 102–120.

Petran, M., and M. Hadravsky (1966) Czechoslovakian patent appl. 7720.

Piller, H. (1977) *Microscope Photometry*. Springer-Verlag, Berlin.

Ross, W. N., and V. Krauthamer (1984) Optical measurements of potential changes in axons and processes of neurons of a barnacle ganglion. *J. Neurosci.*, **4**, 659–672.

Ross, W. N., and L. F. Reichardt (1979) Species-specific effects on the optical signals of voltage sensitive dyes. *J. Membr. Biol.*, **48**, 343–356.

Ross, W. N., B. M. Salzberg, L. B. Cohen, A. Grinvald, H. V. Davila, A. S. Waggoner, and C. H. Wang (1977) Changes in absorption, fluorescence, dichroism, and birefringence in stained giant axons: Optical measurement of membrane potential. *J. Membr. Biol.*, **33**, 141–183.

Saki, T., A. Hirota, and H. Komuro, S. Fujii, and K. Kamino (1985) Optical recording of membrane potential responses from early embryonic chick ganglia using voltage sensitive dyes. *Develop. Brain. Res. 17*:39-51.

Salama, G., and M. Morad (1976) Merocyanine 540 as an optical probe of transmembrane electrical activity in the heart. *Science*, **191**, 485–487.

Salzberg, B. M. (1983) "Optical recording of electrical activity in neurons using molecular probes," in J. L. Barker and J. F. McKelvy, Eds., *Current Methods in Cellular Neurobiology*, Wiley, New York, pp. 139–187.

Salzberg, B. M., and F. Bezanilla (1983) An optical determination of the series resistance in *Loligo*. *J. Gen. Physiol.*, **82**, 807–817.

Salzberg, B. M., H. V. Davila, and L. B. Cohen (1973) Optical recording of impulses in individual neurons of an invertebrate central nervous system. *Nature*, **246**, 508–509.

Salzberg, B. M., A. Grinvald, L. B. Cohen, H. V. Davila, and W. N. Ross (1977) Optical recording of neuronal activity in an invertebrate central nervous system: Simultaneous monitoring of several neurons. *J. Neurophysiol.*, **40**, 1281–1291.

Salzberg, B. M., A. L. Obaid, D. H. Senseman, and H. Gainer (1983) Optical recording of action potentials from vertebrate nerve terminals using potentiometric probes provides evidence for sodium and calcium components. *Nature*, **306**, 36–40.

Senseman, D. M., and B. M. Salzberg (1980) Electrical activity in an exocrine gland: Optical recording with a potentiometric dye. *Science*, **208**, 1269–1271.

Sharnoff, M., R. W. Henry, and D. M. J. Bellezza (1978a) Holographic visualization of the nerve impulse. *Biophys. J.*, **21**, 109a.

Sharnoff, M., N. J. Romer, L. B. Cohen, B. M. Salzberg, M. B. Boyle, and S. Lesher (1978b) Differential holography of squid giant axons during excitation and rest. *Biol. Bull.*, **155**, 465–466.

Shaw, R. (1979) Photographic detectors. *Appl. Optics Optical Eng.*, **7**, 121–154.

Sheppard, C. J. R., and T. Wilson (1981) The theory of the direct-view confocal microscope. *J. Microsc.*, **124**, 107–117.

Sherebrin, M. H. (1972) Changes in infrared spectrum of nerve during excitation. *Nature*, **235**, 122–124.

Sherebrin, M. H., B. A. E. MacClement, and A. J. Franko (1972) Electric-field-induced shifts in the infrared spectrum of conducting nerve axons. *Biophys. J.*, **12**, 977–989.

Shrager, P., S. Y. Chiu, J. M. Ritchie, D. Zečević, and L. B. Cohen (1985) Optical measurement of propagation in normal and demyelinated frog nerve. *Soc. Neuroscience Abstracts*, **11**, 147.

Tank, D. W., and Z. Ahmed (1985) Multiple-site monitoring of activity in cultured neurons. *Biophys. J.*, **47**, 476a.

Tasaki, I., A. Watanabe, R. Sandlin, and L. Carnay (1968) Changes in fluorescence, turbidity, and birefringence associated with nerve excitation. *Proc. Natl. Acad. Sci. USA*, **61**, 883–888.

Tasaki, I., A. Watanabe, and A. Hallett (1972) Fluorescence of squid axon membrane labeled with hydrophobic probes. *J. Membr. Biol.*, **8**, 109–132.

Waggoner, A. S. (1979) Dye indicators of membrane potential. *Annu. Rev. Biophys. Bioeng.*, **8**, 47–68.

Waggoner, A. S., and A. Grinvald (1977) Mechanisms of rapid optical changes of potential sensitive dyes. *Ann. N.Y. Acad. Sci.*, **303**, 217–241.

CHAPTER 7

OPTICAL STUDIES OF THE MECHANISM OF MEMBRANE POTENTIAL SENSITIVITY OF MEROCYANINE 540

Brian E. Wolf
Chemistry Department,
Amherst College, Amherst, Massachusetts

Alan S. Waggoner
Department of Biological Sciences,
Carnegie-Mellon University, Pittsburgh, Pennsylvania

1.	INTRODUCTION	102
2.	MATERIALS AND METHODS	103
3.	RESULTS	105
	3.1 DIFFERENCE SPECTRA	105
	3.2 SIGNAL SIZE VERSUS DYE STRUCTURE	107
4.	DISCUSSION	111
	REFERENCES	112

1. INTRODUCTION

Merocyanine 540 (structure in Fig. 1) applied to membranes responds to electrical potential changes with corresponding changes in light absorption and fluorescence (Ross et al., 1974). Although M540 has been replaced by more sensitive and less toxic dyes in physiological studies (Ross et al., 1977; Gupta et al., 1981), it is important to understand the molecular mechanism of its response, because similar mechanisms may be in effect for some of the newer dyes. Furthermore, knowledge of mechanisms is useful for design and synthesis of better probes, helps prevent misuse of probes, and increases our understanding of how small molecules, such as metabolites and drugs, interact with excitable membranes.

Ross et al. (1974) were the first to suggest that the fluorescence and absorption signals of M540 in the squid axon were different expressions of a single mechanism and that membrane-bound dye exists in a potential-dependent equilibrium between monomers and dimers having different absorption maxima and different quantum yields. The membrane-associated monomer absorbs at 570 nm and is highly fluorescent, whereas the dimer absorbs near 520 nm and is only slightly fluorescent. Additional support for the membrane-localized monomer–dimer mechanism was published by Tasaki and Warashina (1976). Dragsten and Webb (1978) studied fluorescence changes of M540-stained hemispherical bilayer membranes illuminated by polarized laser light. Their results indicate that at low dye concentrations the optical signal results mainly from a reorientation of the membrane-associated chromophores in the electric field. At higher dye concentrations (in the relevant range for squid axon experiments), effects from the formation of nonfluorescent aggregates were seen. Waggoner and Grinvald (1977) provided further support for the involvement of dimers in the mechanism by titrating phosphatidyl choline vesicles with M540 and resolving the absorption spectrum of dimers that are formed at higher concentrations.

This report describes studies of potential-dependent light absorption changes of M540 associated with spherical oxidized cholesterol membranes and is therefore

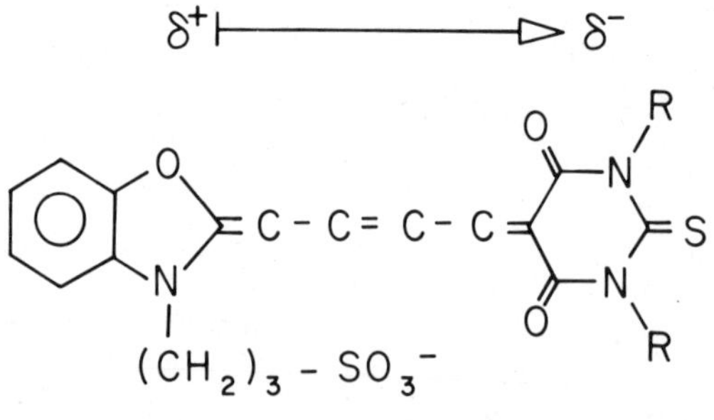

Figure 1. Structure of M540. The R groups are —CH$_2$)$_N$—CH$_3$. Arrow indicates the probable direction of the absorption transition moment. The ground-state permanent dipole moment is probably also in the same general direction.

complementary to the fluorescence studies of Dragsten and Webb (1978), who used the same model membrane system. Absorption difference spectra that result from membrane potential changes were obtained using polarized illumination of the bilayer membranes in the presence of dye. Comparison of the difference spectra with the spectra of membrane-bound M540 monomers and dimers (Waggoner and Grinvald, 1977) support the mechanism in which potential-dependent reorientation of highly dipolar (μ = 9.7 Debye; Dragsten and Webb, 1978; Kushner and Smyth, 1949) M540 molecules results in shifts between three dye populations. Population I consists of monomers with optical transition moments oriented relatively perpendicular to the plane of the membrane, population II is made up of monomers oriented more in the plane of the membrane, and population III contains dimers oriented in the plane of the membrane. We also propose a quantitative explanation for the strong dependence of the absorption signal on the amount of aliphatic hydrocarbon attached to M540 analogues we have added to bilayer membranes.

2.   MATERIALS AND METHODS

Merocyanine 540 was obtained from Eastman Kodak. Analogues of M540 were synthesized by procedures described in Ross et al. (1977). The membrane-forming solution consisted of 4% (w/v) cholesterol (Eastman Kodak) in n-octane (Eastman practical grade) that had been refluxed 6 h with oxygen bubbling. Older solutions tended to form more stable membranes.

The apparatus used to measure optical changes in M540-stained bilayers is shown schematically in Fig. 2. Spherical bilayer membranes were formed at the end of a Becton Dickinson 1-cc tuberculin syringe with the top and the needle removed. The top of the syringe was attached to a Manostat screw pipette so that small volumes of salt solution could be drawn into or extruded from the tip of the syringe which was immersed in a 100-mM salt solution contained by a Pyrex 1 × 1 cm cuvette. Membrane-forming solution was wiped across the syringe tip with a brush or Teflon spatula, and a spherical bilayer membrane about 2–3 mm in diameter was formed by exerting hydrostatic pressure with the pipette. Membranes usually thinned within a few minutes. Thinning could be monitored by observing capacitance spikes in the membrane current signal. Dye in ethanol (usually 15 μL) was added with a Hamilton syringe to the salt solution while the membrane was thinning. A stir bar in the bottom of the cuvette aided mixing of the dye, but stirring had to be turned off later for stable optical measurements.

Trains of square step voltage changes across the membrane were generated by a Hewlett-Packard 8011A pulse generator. The voltage steps were symmetrical about ground potential, which was defined to be zero in the salt solution outside the spherical membrane. Ag–AgCl electrodes were placed within the syringe and in the cuvette in positions as close to the membrane as possible without interfering with the optical path. Both the voltage and the current applied to the membrane were monitored on an oscilloscope.

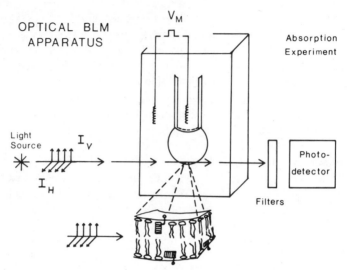

Figure 2. Apparatus for optical studies. The directions of the electric vector of polarized light are indicated by I_H and I_V.

The illumination source for the optical measurements consisted of a 100-W FCR tungsten halogen lamp in an Oriel 6325 lamp housing powered by a stable Kepco JQE15-12M DC power supply. Before reaching the membrane, the light passed through an aperature, a heat filter, one of twenty 10-nm band-pass interference filters (Ditric, Inc.) mounted on a wheel for rapid change, a polarizer, and a $10 \times$-long working distance objective. On the opposite side of the cuvette, light that had passed through the bilayer was collected with another $10 \times$-long working length objective and passed to a PIN-10DP planar diffused silicon photodiode (United Detector Technology Inc.). The syringe was attached to a micromanipulator so that different parts of the spherical bilayer could be placed in the focal region of the illumination beam.

The photocurrent was converted to a photovoltage with an Analog Devices 40J operational amplifier with a selectable gain. This signal was monitored with a 3A9 amplifier of a Tektronix oscilloscope, and the signal amplified by this unit was passed to an EG&G-PAR 4202 digital signal averager and also to an EG&G-PAR model 128A phase-sensitive detector supplied by the same company. Thus the amplitude of the optical signal could be read directly off the CRT of the signal averager (which gave the wave form of the signal as well) or could be recorded with a chart recorder attached to the lock-in amplifier. The magnitude of this signal is defined as I. The DC voltage reaching the signal averager or the lock-in amplifier was also measured. The DC voltage, I, is proportional to the average light level reaching the photodetector, and ΔI is proportional to the change in the light level that occurs during membrane voltage pulses.

3. RESULTS

Voltage changes across M540-stained oxidized cholesterol bilayers (Fig. 2) caused changes in the intensity of light passed through the bilayer. Changes in the light intensity, ΔI, normalized by the average light intensity hitting the photodetector, I, are proportional to absorbance changes ΔA of light passing through the membranes according to the relation (Waggoner and Grinvald, 1977):

$$\Delta A = - \Delta I/2.3I = 1000\Delta\epsilon \, \Delta n$$

Positive values of $\Delta I/I$ mean that an increase in light transmittance (decrease in absorbance) occurs during the phase of a voltage step that makes the salt solution inside of the bilayer more positive.

3.1. Difference Spectra

A typical signal ($\Delta I/I$) for nonpolarized light directed at the bottom of the spherical bilayer is shown in Fig. 3a. The signal plotted as a function of wavelength is a "difference spectrum." The difference spectrum in Fig. 3a is similar to that seen with the squid axon (Ross et al., 1977). The polarity of the signal is reversed if dye is added to the salt solution inside the bilayer rather than to the outside solution. The magnitude of the signal is linear over a range of pulse sizes to 300 mV (voltage steps between -150 and $+150$ mV). The signal magnitude increases with dye concentration, half-saturating at 2 μM and saturating near 6 μM. Difference spectra are more informative if polarized light is used. The absorption difference spectra of M540 on squid axons have been studied by Tasaki and Warashina (1976), Ross et al. (1977), and Waggoner and Grinvald (1977). The use of spherical bilayers, where polarized light can be focused either perpendicular or tangential to a planar region of membrane at the bottom of the sphere, provides an ideal way to look at light absorption by dye molecules with their absorption transition moments oriented parallel or perpendicular to the plane of the membrane (Dragsten and Webb, 1978). Since the orientation of the absorption transition moment of M540 is more or less directed along the long axis of the chromophore (Fig. 1), the term orientation of the dye will be synonymous with orientation of the transition moment. Light will be maximally absorbed when it is polarized in the same direction in which the dye is oriented.

Difference spectra for vertically and horizontally polarized light focused on the bottom of the bilayer at three different M540 concentrations are shown in Figs. 3d, 3e, and 3f. Vertically polarized light produces a large decrease in transmittance at 570 nm at all three dye concentrations during positive potential changes. Monomeric M540 molecules in hydrophobic solvents and on phosphatidylcholine vesicles (Waggoner and Grinvald, 1977) absorb strongly at this wavelength. Thus positive potential changes produce an increase in the amount of dye with a component of transition moment perpendicular to the plane of the membrane. This is what is

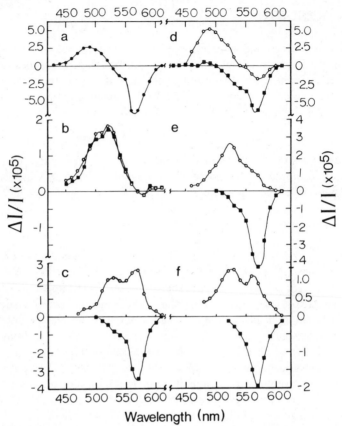

Figure 3. Dichroic transmittance difference spectra for M540. The direction of the polarized illumination is indicated by ●, unpolarized; ○, horizonally polarized (I_H); ■, vertically polarized (I_v). (*a*) Signal for illumination tangential to the bottom of the membrane with unpolarized light; 5 μg/mL dye. (*b*) Face of the spherical bilayer was illuminated. Both I_H and I_v are perpendicular to the plane of the bilayer in this configuration; 1 μg/mL dye. (*c*) M540 analogue with —N(CH$_3$)$_3$$^+$ instead of —SO$_3$$^-$. Tangential illumination with 0.5 μg/mL dye. (*d*)–(*f*) Obtained with tangential illumination but different M540 concentrations: (*d*) 5 μg/mL, (*e*) 1 μg/mL, (*f*) 0.5 μg/mL.

expected, because a positive potential change favors an orientation of dye that places the more negative end (thiobarbituric acid nucleus) of the dipolar molecule closer to the inside of the bilayer (Ross et al., 1974; Tasaki and Warashina, 1976; Waggoner and Grinvald, 1977; Dragsten and Webb, 1978). Without further study it is impossible to give the exact orientation of this new population of dye, which is labeled I in the model shown in Fig. 6. The next question is what is the source of dye molecules that appear in population I.

Difference spectra for experiments at intermediate dye concentrations using horizontally polarized light (Fig. 3*f*) disclose a reduction (during positive potential steps) in two distinct populations of dye that are oriented more in the plane of the membrane. One of the two populations has a difference absorption peak in the vicinity of 560 nm. This peak represents the disappearence of a monomeric species in the plane of the membrane—that is, disappearance of population II (Fig. 6). The other species that is reduced during positive potential steps absorbs light at 520 nm and corresponds to a dimeric species first described by Ross et al. (1974). The dimeric species is relatively nonfluorescent and is oriented more in the plane of the bilayer (population III in Fig. 6). Waggoner and Grinvald (1977) resolved the absorption spectra of membrane-bound monomers and dimers by titrating phosphatidylcholine vesicles with M540. In that membrane system the dimer absorbed maximally at 520 nm (see also data of Lelkes and Miller, 1980). We found it impossible to form oxidized cholesterol vesicles that could be titrated with dye in a similar manner. Nevertheless, it is likely that the position and shape of the merocyanine dimer spectrum on oxidized cholesterol membranes are similar to those of the corresponding spectrum from phosphatidylcholine membranes.

With increasing dye concentration the positive peak at 560 nm observed with horizontally polarized light disappears eventually, leaving a negative peak at this wavelength when the concentration reaches 5 μM (Fig. 3d–f). Thus, at higher dye concentrations, populations I and III have grown considerably and are responsible for the large optical signals in Fig. 3d, but population II, which absorbs at 560 nm, does not seem to increase proportionally. At intermediate concentrations all three populations contribute to potential-dependent absorption changes (Fig. 3f). Dragsten and Webb (1978) used very low dye concentrations in some experiments and were able to see fluorescence signals involving only populations I and II but no dimers.

Further analysis of the difference spectra of Fig. 3 is warranted. The negative peak near 570 nm (Fig. 3d) can be explained if population I is tilted from perpendicular to the membrane. Potential-dependent movement of dimers, absorbing at 520 nm to tilted monomers, absorbing at 570 nm, will produce transmittance decreases with both vertically and horizontally polarized 570-nm light. The relative size of these two negative peaks at 570 nm in spectrum Fig. 3d depends on the average angle of the dye molecules in population I. It depends also on the amount of dye in population II, which, if large, will tend to increase $\Delta I/I$ at this wavelength for horizontally polarized light.

Difference spectra obtained with illumination of the front face of the spherical bilayer (incoming light normal to the plane of the bilayer; Fig. 3b) are also consistent with the mechanism discussed above. The large positive peak near 520 nm is due to reduction of the dimer population. However, compared with the phosphatidylcholine dimer peak in Fig. 4, difference spectrum (Fig. 3b) appears somewhat truncated at longer wavelengths. A simultaneous increase in tilted monomers that absorb at 570 nm would explain the truncation, because they would absorb some of the light that is polarized parallel to the plane of the membrane.

3.2. Signal Size versus Dye Structure

The magnitude of the absorption signal of M540 analogues is strongly dependent on the length of the alkyl groups attached to the thiobarbituric acid nucleus. This dependence is shown in Fig. 5, which depicts the spherical bilayer absorption signal

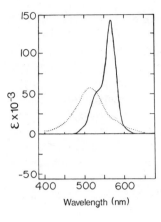

Figure 4. Absorption spectra of M540. Solid spectrum, 2.7 μM dye plus 620 μM phosphatidylcholine in distilled water; dotted spectrum, 21 μM dye in 0.5 M NaCl plus 620 μM phosphatidylcholine.

size versus N, the number of —CH_2— groups between the nitrogen atoms and the terminal methyl groups of the analogues. Also included are data from Ross et al. (1974) indicating the signal to noise ratio for M540 analogues. Notice that M540, for which $N = 6$, gives the strongest signal for both the squid axon and the model systems and that as —CH_2— groups are either added to or subtracted from the thiobarbituric acid nucleus of the M540 structure, the signal size decreases. We will now address quantitatively the dependence of absorption signal size on the number of —CH_2— groups attached to the dye structure. We will begin with the following expression for the signal size adapted from Eq. (5) of Waggoner and Grinvald (1977). This expression is similar to the one used by Dragsten and Webb (1978):

$$\Delta A = A \cdot B(N) \cdot C(N) \tag{1}$$

In this equation, A is a constant that we assume to be independent of the number of —CH_2— groups on the dye. Instead, A depends only on the size of the potential change across the membrane, the dipole moment of the dye, the extent of dye rotation between populations I and II, the thickness of the membrane, and the change in extinction coefficient of the dye when it moves between two populations (Waggoner and Grinvald, 1977). Each of these contributions to A we assume remains constant in the experiments that provided the data in Fig. 5. The other two factors, $B(N)$ and $C(N)$, depend on the number of —CH_2— groups on the dye. The factor $B(N)$ is the total amount of dye bound to the membrane. We expect the partitioning of dye between the bathing solution and the membrane to depend on N and also to show saturation as the concentration of dye in the bathing solution becomes large. On the other hand, $C(N)$ accounts for the effect on signal size of the relative distribution of dye between populations I, II, and III of the membrane-bound dye but not on the total concentration of dye on the membrane. The expressions for $B(N)$ and $C(N)$ described below will contain two free-energy terms that can be adjusted so that the product of $B(N)$ and $C(N)$, which is proportional to signal size, will optimize fit for the experimental data in Fig. 5.

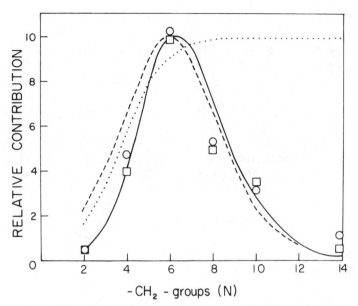

Figure 5. Experimental and theoretical measures of transmittance signal size as a function of the number (N) of —CH2— groups between the thiobarbituric acid nucleus nitrogen and the terminal methyl group of M540 analogues. ○, Transmittance signal obtained with oxidized cholesterol membranes (i.e., this work), dye concentration 5 μg/mL. □, Signal-to-noise ratio obtained with squid axons (Ross et al., 1977). Calculated signal level: dotted line, Eq. (2); dashed line, Eq. (3); solid line, Eq. (1). Curves normalized as described in text.

First consider the saturation binding of dye to the membrane as a function of N described by $B(N)$. The binding of drugs, metabolites, dyes, and other molecules to membranes and other surfaces has been addressed by many authors, and the quantitative descriptions take many forms (Haynes, 1974; Herz and Helling, 1966; Bashford et al., 1979; McLaughlin and Harary, 1976; Diamond and Katz, 1974). An ideal isotherm for quantifying binding of charged dye molecules (like M540) to membranes would take into consideration the large surface potential imparted to the membrane when the amount of dye bound is high. The surface charge would tend to limit the extent of dye binding. However, as a first approximation we have chosen to use the simple Langmuir isotherm to model the partition of merocyanine dye analogues between the membrane and the bathing solution (Davies, 1958). Surface charge effects are ignored in this model:

$$B(N) = \frac{B_s c K_a(N)}{1 + c K_a(N)} \qquad (2)$$

In this equation, B_s is the amount of dye bound to the membrane under saturating conditions. Ordinarily, B_s would be expressed as moles/cm^2. Since, however, we wish only to compare relative contributions to the signal, B_s is assigned to be 1.0 units on an arbitrary scale in Fig. 5. B_s is assumed to be independent of N. The bulk solution molar dye concentration, c, was 5 μM in all the experiments shown in Fig. 5. $K_a(N)$ is the membrane association constant for each of the M540 analogues and can be related to the association constant for M540, $K_a(6)$, by

$$K_a(N) = K_a(6)\exp [(N-6)\Delta G_b/RT] \qquad (3)$$

Again, N is the number of $-CH_2-$ groups between the thiobarbituric acid nitrogen atoms and the terminal methyl groups of the alkyl chains (Fig. 1). Dragsten and Webb (1978) estimate the *dissociation* constant for M540 in the presence of oxidized cholesterol membranes to be 0.5μg/mL. This number converts to a value for $K_a(6)$ of $1.1 \times 10^6 M^{-1}$. ΔG_b is the Gibbs free energy change for transfer of a $-CH_2-$ group from buffer to an "average" binding site on the membrane. Typical values for transfer of $-CH_2-$ groups from water to hydrocarbon solvents and to membranes range from 0.3 to 0.8 kcal/mole (Herz and Helling, 1966; Tanford, 1980).

$B(N)$ is plotted as a dotted line in Fig. 5 using a ΔG_b of 0.6 kcal/mol. Since a high dye concentration was used in the experiment of Fig. 5, $B(N)$ is relatively insensitive to changes in N for the more hydrophobic analogs of M540. $B(N)$ is small, however, for low values of N, which means that analogues with few $-CH_2-$ groups will not have large signals because they have little affinity for the membrane.

Now, let us consider the second $-CH_2-$ dependent term in the equation for ΔA. The term $C(N)$ represents the dependence of signal size on the relative populations between which dye molecules move during potential changes. According to Waggoner and Grinvald (1977),

$$C(N) = \frac{1}{[2 + R(N) + 1/R(N)]} \qquad (4)$$

Where $R(N)$ is the equilibrium ratio of dye molecules in the two membrane-bound populations that are affected by membrane potential changes. At low dye concentrations, populations I and II are mainly involved, and at high dye concentrations, populations I and III are mainly involved according to the model in Fig. 6. We expect that $R(N)$, and hence $C(N)$, will depend upon N, because the longer alkyl groups on the thiobarbituric acid nucleus will pull this end of the molecule toward the center of the membrane, thus favoring population I relative to the other two populations. The extent of this equilibrium shift will be in proportion to the free-energy difference, ΔG_r, between the two environments per $-CH_2-$ group added to the dye structure. Thus, $R(N)$ can be written as

$$R(N) = \frac{[\text{population I}]}{[\text{population II}]} = \exp \frac{(N-6)\Delta G_r}{RT} \qquad (5)$$

As $R(N)$ increases or decreases from 1, the signal size, which is proportional to $C(N)$, should diminish. In calculating the values of $C(N)$ for Fig. 5, we assumed that

$R(N)$ equals 1 for M540, the most sensitive dye of the series (i.e., $N = 6$). A value of 0.4 kcal/mol was used for ΔG_r. The normalized results are plotted as a dashed line in Fig. 5. This function matches the shape of the experimental data better than the dye-binding function, $B(N)$, which is shown by the dotted line.

Since we have proposed that the dependence of signal size on the number of —CH_2— groups should be proportional to the product of $B(N)$ and $C(N)$, we have plotted this product as a solid line in Fig. 5. The product function fits the experimental data better than either $B(N)$ or $C(N)$ alone.

4. DISCUSSION

A model for the mechanism of the potential sensitivity of M540 is shown in Fig. 6. Aspects of this model have been presented in earlier work by Ross et al. (1974, 1977), Tasaki and Warashina (1976), Waggoner and Grinvald (1977), and Dragsten and Webb (1978). The absorption difference spectra presented in this paper lend strong support to the model. The major effect of a membrane potential change is rotational movement of M540 dipoles until a new equilibrium distribution is reached. Rotation occurs about a pivot point that can be a negatively charged sulfonate group, as in M540, or a positively charged tetraalkyl amine group, for as Fig. 3c shows, the latter dye gives essentially the same dichroic difference spectra as M540. Presumably any membrane-impermeant anchoring group, perhaps even a carbohydrate moiety, could serve as a pivot point. It is essential that some group be present to prevent movement of the dye molecule across the membrane, since signals from dye on two sides of the membrane cancel one another (Ross et al., 1977; Dragsten and Webb, 1978).

The model requires that dye molecules involved in the population redistribution have different absorption and fluorescence properties. In fact, dimers are much less fluorescent and absorb at shorter wavelengths than monomers attached to the

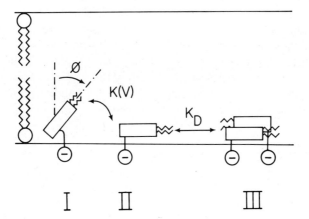

Figure 6. Model for Monomer–Dimer rotation mechanism for M540.

membrane (Ross et al., 1977; Waggoner and Grinvald, 1977). The involvement of dimers in the mechanism suggests that optical changes should be concentration-dependent. Dragsten and Webb (1978) used the concentration dependence of fluorescence signals from oxidized cholesterol bilayers to infer the formation of nonfluorescent dimers and higher dye concentrations. The absorption difference spectra in Fig. 3d–f of this work show the increasing contribution of the dimer population III to the signal at higher dye concentrations.

Ross et al. (1977) were the first to notice a strong dependence of signal size on the nature of hydrocarbon groups attached to M540 analogues. This dependence has been seen with numerous dyes in different chromophore classes (Ross et al., 1977; Gupta et al., 1981; Nyirjesy et al., in preparation). In this work we have shown that a reasonable explanation for a sharp optimum in signal size when additional —CH_2— groups are added to M540 analogues includes two effects. First, additional hydrophobic groups on the chromophore increase the amount of dye bound to the membrane, up to a point of saturation. This suggests that signals from less hydrophobic dyes can be increased by increasing their concentration. The second effect is important, because it explains the lower signals from very hydrophobic dyes that would otherwise be expected to bind very tightly to the membrane and give larger signals. Optimal response to a potential change requires that both populations involved in the potential-dependent redistribution be in a roughly one-to-one equilibrium. We propose that long-alkyl groups attached to the thiobarbituric acid nucleus pull most of the membrane-bound dye into population I, perhaps even saturating binding sites in this region of the membrane. Under this condition, the amount of dye that can move during a potential change is reduced, and the optical signal is small. If this explanation is correct, it may be possible to increase the magnitude of optical signals by adding alkyl groups to locations on the chromophore structure in a way that will enhance binding to the membrane but that will not affect the optimal distribution of dye between the potential-dependent populations that are localized on the membrane. For example, it may be advantageous to leave butyl groups on the thiobarbituric acid but to add extra hydrocarbon at the other end of the molecule, attached, for example, to the benzoxazole ring.

While membrane-permeant cyanine and oxonol dyes (Waggoner et al.,1977) do not share the monomer–dimer rotation mechanism described in the model of Fig. 6, other classes of merocyanine dyes may. For example, WW375 is a useful absorption dye (Ross et al., 1977). It yields a dichroic absorption difference spectrum on squid axons that is somewhat similar to that of M540. Unfortunately, WW375 is too unstable to be easily studied on spherical bilayers.

REFERENCES

Bashford, C. L., B. Chance, J. C. Smith, and T. Yoshida, (1979) *Biophys. J.*, **25**, 63–80.

Davies, J. T. (1958) *Proc. R. Soc. Lond.*, **A245**, 417–443.

Diamond, J. M., and Y. Katz, (1974) *J. Membr. Biol.*, **17**, 121–154.

Dragsten, P. R. and W. W. Webb (1978) *Biochemistry,* **17,** 5228-5240.

Gupta, R. K., B. M. Salzberg, A. Grinvald, L. B. Cohen, K. Kamino, S. Lesher, M. B. Boyle, A. S. Waggoner, and C. H. Wang, (1981) *J. Membr. Biol.,* **58,** 123–137.

Haynes, D. (1974) *J. Membr. Biol.,* **17,** 341–366.

Herz, A. H., and J. O. Helling, (1966) *J. Coll. Interface Sci.,* **22,** 321–403.

Kushner, I. M., and C. P. Smyth, (1949) *J. Am. Chem. Soc.,* **71,** 1401.

Lelkes, P. I., and I. R. Miller, (1980) *J. Membr. Biol.,* **52,** 1–15

McLaughlin, S., and H. Harary, (1976) *Biochemistry,* **15,** 1941–1947.

Nyirjesy, P., E. B. George, M. Basson, and A. S. Waggoner, in preparation.

Ross, W. N., B. M. Salzberg, L. B. Cohen, and H. V. Davila, (1974) *Biophys. J.,* **14,** 983–986.

Ross, W. N., B. M. Salzberg, L. B. Cohen, A. Grinvald, H. V. Davila, A. S. Waggoner, and C. H. Wang, (1977) *J. Membr. Biol.,* **33,** 141.

Tanford, C. (1980) *The Hydrophobic Effect,* 2nd ed., Wiley–Interscience, New York.

Tasaki, I., and A. Warashina, (1976) *Photochem. Photobiol.,* **24,** 191–207.

Waggoner, A. S., and A. Grinvald, (1977) *Ann. N.Y. Acad. Sci.,* **303,** 217–241.

Waggoner, A. S., C-H. Wang, and R. L. Tolles (1977) *J. Membr. Biol.,* **33,** 109-140.

CHAPTER 8

SIMULTANEOUS MONITORING OF ACTIVITY OF MANY NEURONS FROM INVERTEBRATE GANGLIA USING A MULTIELEMENT DETECTING SYSTEM

J. A. LONDON
D. ZEĆEVIČ
L. B. COHEN

Department of Physiology
Yale University School of Medicine
New Haven, Connecticut

1.	INTRODUCTION	116
2.	PHARMACOLOGICAL AND PHOTODYNAMIC EFFECTS OF THE DYE	117
3.	DETECTOR CONFIGURATION	118
4.	SPATIAL RESOLUTION	119
	4.1 SUPRAESOPHAGEAL GANGLION	121
	4.2 BUCCAL GANGLION FROM *NAVANAX*	121
5.	RECORDING FROM A BEHAVING ANIMAL	124
6.	SUMMARY	129
7.	FUTURE DIRECTIONS	130
	REFERENCES	130

1.   INTRODUCTION

The activity of individual neurons has been studied with the long-range goal of describing how this activity underlies behavior, how changes in the activity of these neurons occur as a result of learning and how these changes in activity reflect the encoding of memory. Techniques that use microelectrodes to measure this activity are limited in that they can observe single-cell activity in only as many cells as one can impale. In the first attempt with optical methods (Salzberg et al., 1973), we were fortunate to be able to monitor activity in a single leech neuron. Now, however, optical methods are available in which the activity of hundreds of individual neurons may be recorded.

In this chapter we will describe optical monitoring techniques for simultaneously recording from a large number of neurons in isolated invertebrate ganglia as well as from ganglia in minimally dissected, behaving preparations. We also describe the extent of the pharmacological and photodynamic damage effected by the dyes, and discuss some of the problems associated with light scattering through the preparation. For a detailed description of methods see Cohen and Lesher, Chapter 6 of this volume.

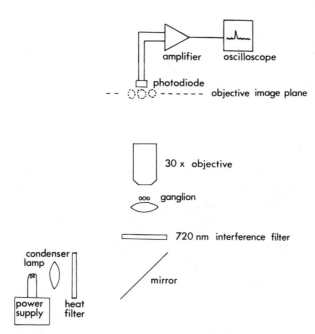

Figure 1.   Schematic diagram of optical monitoring system. Apparatus used for monitoring absorption changes in ganglia. A DC power supply is used for the 12-V, 100-W tungsten–halogen filament lamp. The light is heat-filtered and made quasi-monochromatic with an interference filter. The image of the ganglion appears at the objective image plane, the plane at which the photodiodes are also situated. The signals are fed into current-to-voltage converters and amplified. From Salzberg et al. (1977).

In general, the optical monitoring technique is based on a system for monitoring changes in the absorption (Fig. 1) or fluorescence of a dye that has stained a piece of nervous tissue. In the present experiments a ganglion is placed in a recording chamber that is positioned on the stage of a microscope. The microscope is equipped with a lamp housing (a 12-V, 100-W tungsten–halogen bulb) and a bright-field condenser, and the condenser is adjusted for maximum light intensity. The light is passed through a heat filter and an interference filter. The selection of the peak transmission wavelength of the interference filter is dependent on the particular dye used. A long working distance objective forms a magnified real image of the preparation on a photodiode array (10 × 10 or 12 × 12). Each of the 1.4 × 1.4 mm^2 active elements of the array is separated from its neighbors by 100 μm of insulating regions. The output of each photodiode was amplified; the signals were high-frequency-filtered, and the amplifiers were AC-coupled, so that only changes in intensity were recorded. The amplifier outputs could be passed to an A–D converter and subsequently stored and analyzed in a computer.

Changes in light absorption by a stained neuron from the barnacle supraesophageal ganglion mirrored sub- and suprathreshold events (Fig 2). On the left, a small subthreshold membrane potential change as a result of injection of depolarizing current is recorded electrically (bottom trace). A small change in the light absorption (top traces) of the cell is recorded simultaneously in a single trial. When signal averaging was employed, potential changes as small as 1 mV were detected. When the cell is stimulated (shown on the right) to give an action potential, the light trace also reflects this activity.

2. PHARMACOLOGICAL AND PHOTODYNAMIC EFFECTS OF THE DYE

The pharmacological effects of many dyes and the effects of photodynamic damage and bleaching have been examined in several preparations. Grinvald et al. (1981) studied the polysynaptic reflex response of the isolated ocellus and supraesophageal ganglion of the barnacle. A decrease in the illumination on the ocellus elicits an increase in spike activity monitored by suction electrode recordings from peripheral nerves and connectives. This response is referred to as the off response. By screening different dyes and applying different concentrations, it was possible to find a concentration (0.1–1.0 mg/mL) that did not effect the off response, suggesting that the pharmacologic effects of the dyes could be controlled.

It was noted that the off response of the preparation did decline after a long recording session, indicating that there was some photodynamic damage. In addition, bleaching of the dye would occur after several minutes of constant illumination. However, both of these effects could be minimized by using short recording periods. In later experiments with *Navanax,* and with a different dye, photodynamic damage and bleaching appeared to be insignificant (London et al., 1984a,b; Zečević et al., 1985). Constant illumination of a preparation for 10 min resulted in little if any change in the pattern of neuronal activity or in signal size. The pharmacological

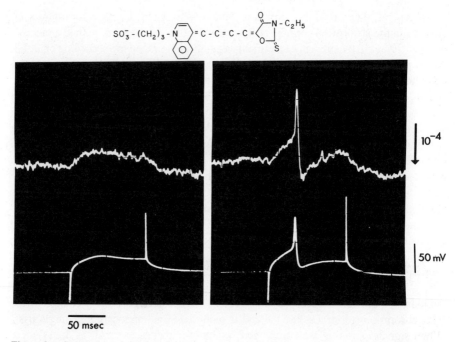

Figure 2.   Comparison of simultaneous changes in light absorption at 720 nm with voltage changes in a stained barnacle neuron. The structure of the merocyanine dye used in this experiment is shown at the top. For both subthreshold and action potential activity the optical signals (top traces) have a shape that is essentially identical to the voltage signals (bottom trace), ignoring the capacity artifacts on the voltage signals. The direction of the arrow adjacent to the optical trace indicates the direction of an increase in absorption; the length of the arrow represents $\Delta A/A$, the change in absorption over the resting absorption. From Salzberg et al. (1977).

effects of dyes were examined in behavioral experiments. Responses to food and subsequent feeding were compared using animals whose buccal ganglia either had or had not been stained. Little, if any, change in behavior was observed to result from staining.

3.   DETECTOR CONFIGURATION

Optical methods to study neuronal ensembles have been progressing steadily. Early multisite experiments employing individual photodiodes glued directly to light guides (Salzberg et al., 1977) recorded simultaneous activity in individual neurons of the barnacle supraesophageal ganglion. Absorption changes exhibited by a merocyanine–oxazolone dye were recorded. Using clad glass rods, light from the images of cells was conducted to individual photodetectors. The maximum number of photodiodes employed using this technique was 14, and the number of cells from which activity was recorded simultaneously was 8. Without signal averaging, a postsynaptic signal as small as 4 mv could be recorded, although for a signal that small the recording conditions had to be "optimal." However, it was evident that the number of monitored neurons could not be increased substantially with the light-guide approach, mainly because positioning the light guides becomes increasingly difficult with an increase in their number. Optical methods were further developed (Grinvald et al., 1981) with the introduction of a 10 × 10 or a 12 × 12 array of photodetectors. This array allowed a large increase in the number of neurons that could be monitored.

4. SPATIAL RESOLUTION

Several experiments were performed to determine the spatial resolution of the detectors in recording neuronal activity, using the barnacle and the leech nervous systems. Light guides were positioned adjacent to one another, and a single neuron was stimulated. A large signal was recorded only in the detector positioned directly over the image of the stimulated cell body, indicating that the optical signal was confined to the expected region of the image plane. In another experiment, the activity of two different but adjacent cells was recorded by one detector (Fig. 3A).

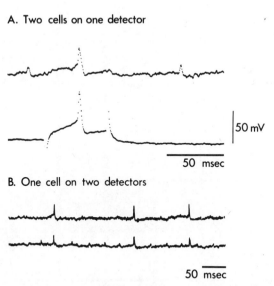

Figure 3.   Resolution capabilities of photodiodes. (*A*) The top trace is taken from a photodiode; the bottom trace is a microelectrode recording. The photodiode exhibits three spikelike events of two different amplitudes; the microelectrode records only one of these events. This indicates that the small optical events result from the activity of a different neuron, adjacent to the cell from which the electrical recording was taken. (*B*) Two traces from adjacent photodiodes. The three action potentials are recorded simultaneously by both detectors, suggesting that the two adjacent photodiodes are receiving light transmitted by the same neuron. From Salzberg et al. (1977).

Again, adjacent detectors that did not receive the projected images of the neurons did not record the cells' activity. If a cell was so large that its image was projected onto two different detectors, the activity of that cell would be recorded simultaneously on both detectors (Fig. 3B). From the relative size of the glass rods and cell images it is expected that a light guide could collect signals from more than one neuron and that two light guides could monitor signals from a single neuron, and this was demonstrated (Salzberg et al., 1977).

A schematic drawing of the projection of the ganglion onto the array (Fig. 4) illustrates how the activity of one cell may be recorded on several different detectors and generate duplicate activity. Each square represents the surface area of the preparation to which a single photodiode is sensitive. The light of several cells can be seen to extend beyond the boundary of a single detector. These cells' activity will be recorded simultaneously on several detectors, and are referred to as duplicate signals. Some cells cover the entire area of the diode, and the signal recorded from these cells on that detector will therefore be relatively large. Several detectors have only a small part of the cells' image projected onto them, and the signal from these cells will be small. In addition, it can be seen that one detector may record the activity of several

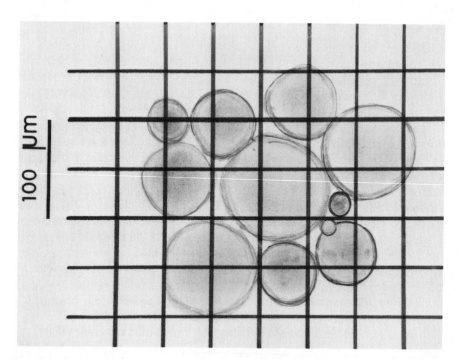

Figure 4.   Drawing of the projected image of an idealized cluster of cells onto the array. Note how various-size areas of one cell may fall on different detectors, giving rise during a recording to duplicate spike activity of different size on neighboring detectors. Note also how areas of more than one cell project their images onto a single detector; thus one detector may record the activity of several cells. We thank Dr. A. Grinvald for allowing us to use this drawing.

cells—for example, when sections of three different cells are projected onto one detector.

4.1   SUPRAESOPHAGEAL GANGLION

The concepts suggested by the schematic drawing of neurons and the array in Fig. 4 were demonstrated with the barnacle supraesophageal ganglion and the 10×10 array (Grinvald et al., 1981). A peripheral nerve was electrically stimulated, and an optical recording from the ganglion made (Fig. 5). The outputs of all functioning elements of the array are shown. On several traces, there are changes in light intensity that correspond to action potentials from different neurons. Some traces exhibit spikes of different sizes. This result confirms what is expected, since many detectors could receive light from different-size portions of two or more cells. The results from Fig. 5 suggest that the apparatus could now monitor action potential activity in several hundred neurons if the cell bodies were large (> 30 μm) and the cell somata were fully invaded by the action potential.

4.2   BUCCAL GANGLION FROM *NAVANAX*

A more detailed demonstration of the detection of duplicate signals on adjacent elements of the array is shown in Fig. 6 using the 12×12 array and a buccal ganglion from *Navanax inermis* (London et al., 1984a,b; Zećević et al., 1985). In the top section the activity of the same cell is recorded on four adjacent detectors. Smaller signals from a second, different cell also appear; these signals are recorded on only two detectors, detectors 18 and 19. The first spike from this second cell occurs between the first and second spike of the large cell. In the bottom section of this figure, a different set of detectors illustrates the same point—two spikes from the same cell may be seen to occur simultaneously on four adjacent detectors. Two of the problems associated with optical recordings are also illustrated by Fig. 6. One problem arises from the signal-to-noise ratio. The high background noise could obscure small action potentials. For example, there may also be spike activity simultaneously on detectors 47 and 48, but the signals are not much larger than the noise. A second problem arises from the slow changes in light intensity, which are probably movement artifacts. These slow changes could also obscure some of the neuronal activity.

Another experiment was conducted to determine how well a single detector could discriminate between cells in the same *xy* position that were not in the same focal plane. The size of the optical signal was compared when the neuron was moved above and below the plane of focus. The signal size was reduced by 50% when the neuron was moved $\pm$ 300 μm away from the plane of focus (Fig. 7). This change in signal size as a function of focus will depend on the numerical aperture of the objective; with a larger numerical aperture, there will be a more rapid decrease in signal size for the same change in focus. Therefore, in a ganglion of 600-μm diameter, one could focus

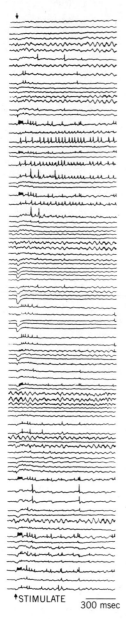

Figure 5. Optical recordings using a 10 × 10 array from a barnacle supraesophageal ganglion. Each trace is the activity from 1 detector. A 5-ms stimulus was delivered (at the arrow) to a peripheral nerve. Many traces exhibit spike activity; Some spikes occur simultaneously but on different detectors, and some traces exhibit spikes of different amplitudes, suggesting that more than 1 neuron projects an image onto that detector. From Grinvald et al. (1981).

↑STIMULATE   300 msec

in the middle of the ganglion and record signals from the activity of neurons at the top and at the bottom of the ganglion with a reduction in signal size of only 50%.

We were concerned about the effects of light scattering on our estimate of the size of the neurons, which we obtain from the number of detectors on which the activity of one cell was seen. Since light is refracted and dispersed as a result of interacting with the ganglion, the light from an individual cell would become more and more scattered

NAVANAX   ANALYSIS

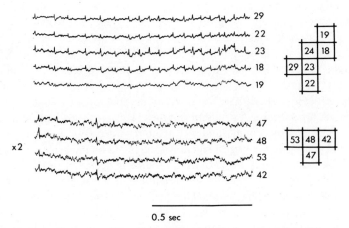

0.5 sec

Figure 6.   Optical recordings using a 12 × 12 array taken from the buccal ganglion of *Navanax inermis*. The number to the right of each trace identifies the detector from which the trace was taken. The drawings to the right represents the relative position of the detectors whose activity is displayed. In the top section, 4 detectors exhibit simultaneous spike activity, probably from the same neuron. The bottom section is an amplified recording × 2 of 4 detectors which indicates some of the difficulties of this technique. Note the slow oscillations occurring at the end of the recordings, which are probably due to movement artifact. In addition, there is significant noise which may mask small action potentials. From London et al. (1984a,b) and Zećević et al. (1985).

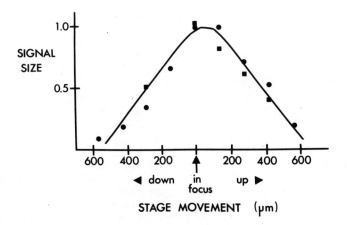

Figure 7.   Size of the optical signal depends on the position of the cell body relative to the plane of focus. A 50% decrease in signal size is seen when the soma is moved 300 μm either up or down from the plane of focus. From Salzberg et al. (1977).

as it passed through the ganglion. It would be possible that the light from a cell at the bottom of the ganglion whose image would normally fall on only one detector would become dispersed enough to result in a blurred image that would fall on several detectors. This would lead to an erroneous estimate of the actual cell size. Substantial scattering would also decrease the signal-to-noise ratio by spreading the light that would otherwise reach a single detector over many detectors. To measure the scattering a small spot of light was focused on the object plane (Fig. 8) (London et al., 1984a,b,; Zećević et al., 1985). The image of the spot fell on a single detector of the array (Fig. 8, top section). A ganglion was then placed in the path of the light, and the number of detectors receiving light was again recorded (Fig. 8, bottom section). It can be seen that while the largest signal was recorded on the detector that originally had the light focused on it, some of the surrounding detectors also recorded smaller levels of light. However, because the amplitude of the scattered light is less than one-third the amplitude of the original light signal, the signals from the scattered light might be lost in the recording noise. Thus, scattering in *Navanax* ganglia is substantially less than that found in vertebrate tissue (Orbach and Cohen, 1983) and apparently does not seriously distort estimates of cell sizes.

5.   RECORDING FROM A BEHAVING ANIMAL

We have recently been able to use the diode array to monitor activity from the nervous system of a minimally dissected preparation in which the animal behaves. Activity of

LIGHT SCATTERING, NAVANAX

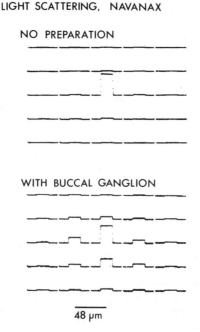

48 µm

Figure 8.   Measurement of light scattering by *Navanax* ganglia. The top section is a measurement of the response of 25 adjacent detectors when a small spot of light was focused in the object plane on a single detector. The bottom section is the response of the same 25 detectors, but now a buccal ganglion has been placed between the object plane and the objective. From Zećevi ĕt al. (1985).

the buccal ganglion of the marine slug, *Navanax inermis,* was monitored while the animal responded to food and engaged in feeding activity (London et al., 1984a,b: Zećević et al., 1985).

Navanax is a carnivore that feeds on several species of opisthobranchs including other *Navanax,* and some species of fish (Paine, 1963). Roughly, the feeding movements consist of the protraction of the mouth, prey contact and grasping with the lips, followed by an explosive expansion of the pharynx. This expansion creates a negative pressure so that the prey is sucked into the pharynx, where it is subsequently digested. Neurons controlling the expansion have been described and reside in the buccal ganglion (Wollacott, 1974; Spira et al., 1980).

To observe the neuronal activity in the buccal ganglion that occurs during feeding, a minimally dissected ''whole'' animal preparation was developed (Fig. 9). An

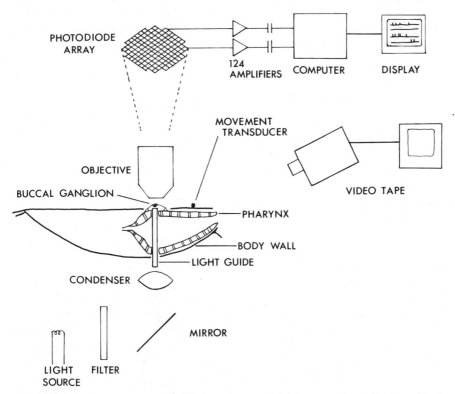

Figure 9. Schematic drawing of whole-animal recording apparatus. The light is made quasi-monochromatic by an interference filter and focused by a condenser onto the light guide. The light guide was constructed from a 1-mm clad quartz cylinder of glass encased in a metal tube; at the top of the light guide was a platform which suspended the ganglion over the glass. The image of the ganglion was projected onto the 12 × 12 photodiode array. The signals from individual diodes were fed to individual amplifiers, A–D converted, stored in a computer, and later displayed. A movement transducer was placed on the pharynx to record feeding movements. A video camera was also used to record the behavior during the optical monitoring. From London et al. (1984a,b) and Zećević et al. (1985).

incision was made in the skin on the ventral surface above the buccal ganglion. The skin was retracted to expose the buccal ganglion. A 1-mm clad glass light guide was used to carry the light from the condenser to the ganglion. The top one-third of the guide is surrounded by a Sylgard platform (Dow). The guide is passed through incisions in the animal's dorsal skin and dorsal and ventral pharynx such that the buccal ganglion is positioned over the light guide and is supported by the platform. After staining, the ganglion was pinned to the platform and covered with a 3% solution of agar dissolved in seawater to decrease movement artifacts. The movements of the anterior portion of the animal were recorded with a camera and video tape. The optical measurements were synchronized with the video recording by flashes from a light emitting diode that signalled the beginning and end of the optical recording. In addition, pharyngeal expansions were monitored with a force transducer, consisting of a glass rod connected to a displacement recorder (Grass Instrument Corp.) that was placed on the pharynx. The signals from the force transducer were fed to the computer and to a chart recorder. The beginning and ending of the optical recording session were also marked on this recording.

The activity data from these experiments were displayed in a raster diagram (see Figs. 10, 11). Each vertical line represents the occurrence of one action potential. The bottom trace is a recording from the movement transducer that had been positioned on the pharynx. An upward deflection indicates expansion. The dotted line represents the position of the transducer, and therefore the state of the pharynx, at rest. The numbers just to the left of the spike traces represent the number of detectors on which this particular neuron's activity was recorded. The numbers to the far left represent the identification number of the detector used to enter this neuron's activity into the computer and to generate the raster trace. The generation of the raster diagram requires several steps. Twenty detectors' traces are displayed; these detectors are nearest neighbors so that duplicates (i.e., the same cell's activity seen on more than one detector) may be identified in the data and the occurrence of the duplicate can be recorded and then removed from the data. The times of occurrence of the identified action potentials are entered into the computer. These times are then used to generate the raster display. The results shown here are preliminary; more detailed identification of the neurons is in progress.

Some 25 dyes were tested on isolated *Navanax* ganglion; the best dye seemed to be a pyrazolone oxonal, RH 155 (synthesized by R. Hildesheim and A. Grinvald). This dye gave the best signals (the largest S/N ratio was approximately 10:1). Animals in this recording situation exhibited spontaneous and food-induced expansions of the pharynx. During spontaneous expansions, activity in 6–19 cells was detected. Fig. 10 shows an example where nineteen cells were active. In this example, there is a large unit that is active during the beginning of the expansion; this cell's activity is recorded on 13 detectors. There is also a large cell whose activity was observed on 8 detectors and which is active after the initial expansion occurs. Several cells are active during the expansion and become quiet during the maintained expansion and contraction phase. Other cells, like those displayed on detectors 57 and 15, become active at the end of the expansion. There are 13 small cells that are active during this spontaneous expansion; their activity is recorded on only 1 detector.

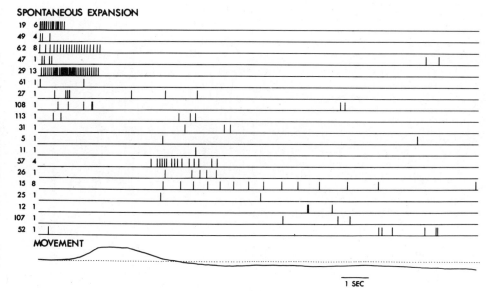

Figure 10. Raster diagram generated from an optical recording from the buccal ganglion during spontaneous (feeding-like) activity. All the neurons whose activity was detected are shown. The number just to the left of a trace is the cell size number, the number of detectors that recorded this activity (i.e., on how many detectors this cell's activity was seen). The number to the left of the cell size number is the cell's identification number. This number is the number of one detector on which the cell's activity is observed; this detector was used to enter the cell's activity into the computer. Usually, a detector is chosen for a particular cell because it recorded the largest signal for that cell. The bottom trace is a record of the movement of the pharynx; an upward deflection indicates expansion. From London et al. (1984a,b) and Zečević et al. (1985).

During feeding (Fig. 11) the pattern of muscular activity is similar in form to that of the spontaneous expansions; however, there are more neurons that are active during feeding than during the spontaneous expansion. During feeding, usually 15–30 cells are active. In the experiment shown, 27 cells are active. As in the spontaneous expansion, during this feeding cycle, there is a burst of activity reflected as increased spike activity prior to the expansion. Several large-size cells, on detectors 67 and 43, are active early during the expansion; one of these, on detector 67, may be the same large cell that is active during the spontaneous expansion. There are more medium-size neurons active in this feeding experiment (cells whose activity is seen on 2–3 detectors) than were active in the spontaneous expansion shown. However, it should be noted that the two figures come from experiments on different preparations.

One question we addressed was whether or not we were recording from all the cells that were active during a behavior, or were we only recording a small fraction of the active cells. To answer this question we compared the number of cell bodies in the ganglion to the number of cells we could identify by their optical activity when many of the cells in the ganglion were made to fire action potentials. To determine the

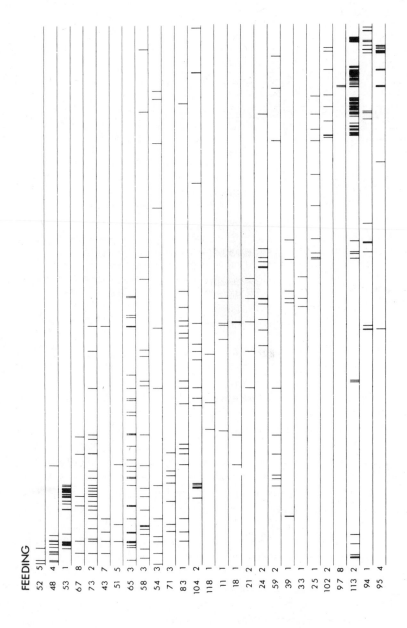

Figure 11. Raster diagram generated from an optical recording of the buccal ganglion during feeding. Same as Fig. 10.

number of cell bodies, ganglia were stained with methylene blue and the number of cell bodies was counted using a computer-interfaced microscope (Macagno, 1978; Boyle et al., 1983). An image of the ganglion is superimposed on a digitizing tablet. A cursor is used to mark the cell bodies in the XY plane, while an encoder in the microscope shaft measures the Z-axis point. A program later removes duplicate entries and lists the corrected number of cells counted. We found that there are 200 ± 10 cells in the ganglion ($N = 16$). To obtain the number of cells whose activity could be detected optically, cells were activated by stimulating the peripheral nerves and connectives with suction electrodes. In the best experiment we were able to distinguish 157 different neurons that were active. The range of the number of cells that we identified from three stimulation experiments was 112–140 neurons. This was between 55 and 79% of the total number of neurons in the three ganglia. This percentage may be an underestimate of the number of cells active for several reasons. One possibility is that we may not correctly separate the activity of small cells. These cells' activities may occur on the same detector and have small optical action potentials that appear too similar to allow us to differentiate between them. Another possibility is that the stimulation may not activate all the neurons, causing an underestimate of the percentage of cells from which we are recording activity.

6.   SUMMARY

The optical monitoring system has reached a level of sophistication where activity from a network of cells can be recorded in a minimally dissected behaving animal. We monitored activity in the buccal ganglion of the mollusk, *Navanax inermis*, during spontaneous expansions and during feeding. Deleterious pharmacological effects and photodynamic damage appear to be negligible in this preparation. The S/N is large enough to detect individual cell's activity without signal averaging, provided the cell soma is of sufficient size ($> 20 \ \mu m$). Indeed, if the S/N is large enough, it is possible to see graded or subthreshold potentials without the use of signal averaging. The activity of adjacent cells in the same focal plane could be detected as different cells by a single detector. In addition, cells that were out of focus but whose projected images fell on a detector also have their activity recorded as a significant signal by that detector. This finding suggested that recording from all the cells in a ganglion was possible. Preliminary evidence suggests that our recordings from *Navanax* buccal ganglia are at least 70% complete. It appears that the assumption that most of the cells' acitivity is being recorded is reasonable. A problem that could prevent correct identification of activity was the effect of light scattering on the signal. A signal dispersed owing to light scattering could give an erroneous estimate as to the cell size. However, the effect of light scattering on *Navanax* signals is small, and we feel that our estimates of cell size are accurate.

7.   FUTURE DIRECTIONS

A global objective is to continue to improve the optical technique. We hope that better dyes will become available, ones that will produce larger signals in response to voltage changes in the membrane. With better dyes and with increased S/N we would be able to record synaptic potentials consistently and therefore map synaptic interactions. Many of the present limitations that occur result from the small signal size.

We also plan to further characterize the cells participating in the behavior. One way in which we hope to do this is by monitoring the direction of spike propagation in the peripheral nerves and matching their timing with soma recordings. Activity of cells with peripheral nerve spikes proceeding outward (centrifugal) are of possible motoneuron origin; activity of axons spikes proceeding inward (centripetal) may be from sensory neurons. Soma activity with no concomitant peripheral nerve activity could be from an interneuron. With improvements in cell characterization and identification, and with the recording of synaptic potentials, we hope to apply this technique to the testing of models of the neural basis of behavior and learning.

ACKNOWLEDGMENTS

We thank R. Hildesheim, Dr. A. Grinvald, Dr. A. Waggoner, and Nippon Kankok-Shikiso-Kenkyusho Co. for supplying us with dyes. We thank Dr. A. Grinvald for permission to use the drawing in Fig. 4. This study was supported in part by a Grass Foundation Fellowship to J.A.L., a Fulbright grant, 83-04765, to D.Z., and Public Health Service grants NS08437 and NS07169.

REFERENCES

Boyle, M. B., L. B. Cohen, E. R. Macagno, and H. Orbach, (1983) Number and size of neurons in the CNS of gastropod molluscs and their suitability for optical recording of activity. *Br. Res.*, **266**, 305–317.

Grinvald, A., L. B. Cohen, S. Lesher, and M. B. Boyle, (1981) Simultaneous optical monitoring of activity of many neurons in invertebrate ganglia using a 124-element photodiode array. *J. Neurophysiol.*, **45**, 829–839.

London, J. A., D. Zećević, L. B. Cohen, (1984a) Optical recording of neuronal activity from buccal ganglia during pharyngeal expansion (feeding) in a minimally dissected *Navanax. Neuroscl. Abstr*, 10, 508.

London, J. A., D. Zećević, and L. B. Cohen (1984b) Optical recording of neuronal activity of the buccal ganglion of *Navanax* during feeding. *Biol. Bull.*, **167**, 529.

Macagno E. (1980) Number and distribution of neurons in leech segmental ganglia. *J. Comp. Neurol.*, **190**, 283–302.

Orbach, H. S., and L. B. Cohen, (1983) Optical monitoring of activity from many areas of the *in vitro* and *in vivo* salamander olfactory bulb. A new method for studying functional organization in the vertebrate central nervous system. *J. Neurosci.*, **3**, 2251–2262.

Paine, R. T. (1961) Food recognition and predation on opisthobranchs by *Navanax inermis*. *Veliger*, **6**, 1–9.

Salzberg, B. M., H. Davila, and L. B. Cohen (1973) Optical recording of impulses in individual neurons of an invertebrate central nervous system. *Nature*, **246**, 508–509.

Salzberg, B. M., A. Grinvald, L. B. Cohen, H. V. Davila, and W. N. Ross, (1977) Optical recording of neuronal activity in an invertebrate central nervous system: Simultaneous monitoring of several neurons. *J. Neurophysiol.*, **40**, 1281–1291.

Spira, M. E., D. C. Spray, and M. V. L. Bennett, (1980) Synaptic organization of expansion motoneurons of *Navanax inermis*. *Br. Res.*, **195**, 241–269.

Woolacott, M. (1974) Patterned neural activity associated with prey capture in *Navanax*. *J. Comp. Physiol.*, **94**, 69–84.

Zećević, D., J. A. London, and L. B. Cohen, (1985) Simultaneous monitoring of activity of many neurons in the buccal ganglia of *Navanax* during feeding: a demonstration. *J. Gen. Physiol.*, in press.

CHAPTER 9

OPTICAL STUDIES OF EXCITATION AND SECRETION AT VERTEBRATE NERVE TERMINALS

BRIAN M. SALZBERG
ANA LIA OBAID
University of Pennsylvania
Philadelphia, Pennsylvania

HAROLD GAINER
National Institutes of Health
Bethesda, Maryland

1. INTRODUCTION 134
2. EQUIVALENCE OF OPTICAL AND ELECTRICAL MEASUREMENTS OF MEMBRANE POTENTIAL 135
3. OPTICAL RECORDING OF ACTION POTENTIALS FROM NERVE TERMINALS OF THE FROG *XENOPUS* 137
4. PROPERTIES OF THE ACTION POTENTIAL IN THE NERVE TERMINALS 140
5. IONIC BASIS OF THE DEPOLARIZING PHASE OF THE ACTION POTENTIAL 143
6. INTRINSIC OPTICAL CHANGES THAT ACCOMPANY SECRETION FROM MAMMALIAN NERVE TERMINALS 148
7. EFFECTS OF EXTRACELLULAR CALCIUM 151

8. DEUTERIUM OXIDE DECREASES THE SIZE
    OF THE LIGHT-SCATTERING RESPONSE
    TO STIMULATION   155
9. THE EFFECT OF INCREASING THE EXTRACELLULAR
    VOLUME FRACTION WITH HYPERTONIC MEDIUM   157
10. CONNECTION OF THE LARGE INTRINSIC OPTICAL
    SIGNAL TO SECRETORY EVENTS AT THE   157
    NERVE TERMINALS
    REFERENCES   160

1. INTRODUCTION

Optical techniques for the detection and analysis of transmembrane electrical events have found a variety of applications over the past decade because they offer certain advantages over more conventional measurements. Because the membranes of interest are not mechanically violated, the methods may be relatively noninvasive. Spatial resolution is limited only by microscope optics and noise considerations; it is possible to measure changes in membrane potential from regions of a cell having linear dimensions on the order of 1 μm. Temporal resolution is limited by the physical response of the probes and by the bandwidth imposed upon the measurement, again by noise considerations, and response times faster than any known membrane time constant may be achieved. Because mechanical access is not required, unusual latitude is possible in the choice of preparation, and voltage changes may be monitored in membranes that are otherwise inaccessible. Finally, since no recording electrodes are employed and the measurement is actually made at a distance from the preparation—in the image plane of an optical apparatus—it is possible to record changes in potential simultaneously from a large number of spatially separated sites. Several recent reviews have surveyed the literature on optical measurement of membrane potential (Cohen and Salzberg, 1978; Waggoner, 1979; Freedman and Laris, 1981; Salzberg, 1983; Grinvald, 1985), and they should be consulted by the reader interested in these techniques per se.

Here we shall begin by recalling the evidence that optical methods are equivalent, at least in a limited sense, to electrode measurements, but we will not consider technical details concerned with signal-to-noise enhancement or with the selection of appropriate molecular indicators. These are treated at length elsewhere (Cohen et al., 1974; Ross et al., 1977; Salzberg et al., 1977, 1983; Gupta et al., 1981; Grinvald et al., 1982) and are considered in the chapters by Cohen and Lesher and by Grinvald et al. in this volume. We will then proceed to illustrate the application of optical techniques to a problem that has proved difficult to approach by alternative means—namely, the direct recording of the action potential from the nerve terminals of a vertebrate, and the study of its ionic basis. This example of optical measurement of membrane potential is interesting because it provides information that is currently unobtainable by other means. In the course of these investigations, we happened upon a large and rapid change in the intrinsic optical properties of some mammalian neurosecretory terminals, and this change in light scattering now appears to be intimately associated with the physiological events that attend the release of the neuropeptides vasopressin and oxytocin. We will conclude by showing some of the evidence that this optical signal is related to secretion.

2.   EQUIVALENCE OF OPTICAL AND ELECTRICAL MEASUREMENTS OF MEMBRANE POTENTIAL

An electrode measurement can provide continuous information about the absolute value and time course of a varying transmembrane voltage difference. Under the circumstances encountered most frequently, noise in the electrical recording may be reduced to negligible proportions, and the bandwidth is generally broad enough to introduce little or no distortion of the time course or frequency spectrum of the biological events of interest. The degree to which an optical measurement of membrane potential approximates this ideal is best examined in a preparation that permits simultaneous recordings by both techniques and that allows for both spatial and temporal control of the membrane potential. This is most readily achieved in the giant axons of the squid, where the large diameter of the fibers permits the insertion of electrodes that reduce the longitudinal resistance of the cell and that, as part of a feedback circuit, uniformly control the transmembrane voltage over the large membrane area that is monitored optically. The apparatus used to study the relation of the optical changes to electrical events occurring in and across the membrane of the squid axon is illustrated schematically in the chapter by Cohen and Lesher (this volume), and a modified version of this arrangement, in which the axon is mounted horizontally in a voltage clamp chamber that is incorporated into the stage of a compound microscope, is used in our laboratory to study optical changes in perfused as well as intact giant axons (Salzberg, 1978).

Extrinsic (dye-dependent) fluorescence, absorption, and birefringence changes measured in voltage-clamped squid giant axons stained with molecular probes (Cohen et al., 1974; Davila et al., 1974; Ross et al., 1977; Gupta et al., 1981) have established unambiguously that these voltage-sensitive dye molecules respond linearly to changes in membrane potential but are insensitive to even the large changes in membrane conductance and to the ionic currents that flow during an action potential. The response times of some of these probes are extremely short; certain merocyanine–rhodanine (dye XVII of Ross et al., 1977), merocyanine–oxazalone (NK 2367, Salzberg et al., 1977), and styryl dyes (di-6-ASPPS of Loew et al., 1985), for example, change their absorption within 2 μs of a step change in membrane potential (B. M. Salzberg, F. Bezanilla, and A. L. Obaid, unpublished observations). The linearity of the voltage dependence and the accuracy of the temporal response imply that molecular indicators of membrane potential such as merocyanine and styryl dyes (Salzberg et al., 1972; Davila et al., 1973; Cohen et al., 1974; Ross et al., 1977; Gupta et al., 1981; Loew and Simpson, 1981; Grinvald et al., 1982) are capable of monitoring synaptic and electrotonic potentials as well as action potentials, and they have now found wide application in a variety of systems (Cohen and Salzberg, 1978; Waggoner, 1979; Freedman and Laris, 1981; Salzberg et al., 1983; Grinvald, 1985). (Indeed, the fidelity of the optical response to the true transmembrane potential recently provided the basis for an optical determination of the series resistance, by Ohm's law, in the squid giant axon [Salzberg and Bezanilla, 1983].)

Some caveats are required, however. In certain respects, optical measurement of membrane potential by means of these molecular transducers is not equivalent to an electrode measurement. Although the optical changes are frequently linear functions of the transmembrane potential, the slope of the response depends on a multitude of additional factors that are difficult to estimate, and the optical signals can therefore provide no measure of the absolute magnitude of the potential change without a separate calibration. This is an important limitation on the equivalence of optical and electrical measurements of membrane potential. For example, the size of a potential-dependent absorption change will be proportional to the product of the extinction change of the dye, $\Delta\epsilon$, and the number of dye molecules that participate in the extinction change, Δn. Fluorescence changes depend, in addition, on changes in the quantum efficiencies of the dye molecules as well as on nonspecific fluorescence from dye molecules, free in solution and bound to sites where the potential does not change (Waggoner and Grinvald, 1977). Thus the size of the optical signal depends on the membrane area from which it is measured, possibly the membrane topology, the amount and state of aggregation of bound probe, the sites of probe binding, and the sensitivity of the probe to other changes that may accompany the change in potential, as well as on the collection efficiency and quantum efficiency of the apparatus (Salzberg et al., 1977; Cohen and Salzberg, 1978; Salzberg, 1983).

Even if an electrode calibration of the absolute magnitude of the potential change associated with a given optical signal can be obtained, it will remain valid only transiently, since photolysis and changes in bound dye concentration may occur rapidly. In a multicellular preparation, this difficulty is compounded when the number of active elements may vary from trial to trial, and such is the case in the neurohypophyseal system considered in this chapter. Finally, optical measurements of membrane potential are frequently, though not always, AC coupled in order to subtract electronically the large background light levels. As a consequence, knowledge of DC potential levels is lost, even if an absolute calibration were available. (Probes that are truly electrochromic (Platt, 1961) might offer the possibility of an absolute voltage calibration if the wavelength shift in a given membrane system could be demonstrated to depend solely on potential. However, no sensitive potentiometric probe has yet been shown unequivocally to respond electrochromically to membrane potential changes in a natural membrane (see Loew et al., 1985)).

In many circumstances, however, a faithful optical transcription of the time course of a transmembrane potential change may provide useful information that is otherwise obtainable only with difficulty, if at all. This seems clearly the case in the experiments on amphibian nerve terminals that are illustrated in the sections that follow, and in some of the applications described by others in this volume.

3. OPTICAL RECORDING OF ACTION POTENTIALS FROM NERVE TERMINALS OF THE FROG *XENOPUS*

Our understanding of the details of synaptic transmission in the vertebrates has been circumscribed by our inability to monitor directly the action potential in the nerve terminals themselves. The giant synapse of the squid has provided a very successful invertebrate model for the study of the release of transmitter substances (Bullock and Hagiwara, 1957; Katz and Miledi, 1967; Llinás et al., 1976), because in this preparation both the postsynaptic and presynaptic membranes are accessible for electrophysiological investigation including voltage clamp. Much of our present knowledge of vertebrate synaptic physiology is derived from the study of the frog neuromuscular junction (Katz, 1969), but in this instance "the small size of the nerve terminal prevents a direct measurement of the presynaptic potential change" (Katz, 1969).

The vertebrate hypothalamo–neurohypophyseal system represents an alternative model for the study of excitation–secretion coupling at nerve terminals. Magnocellular neurons located in the hypothalamus (supraoptic and paraventricular nuclei in mammals; preoptic nucleus of lower vertebrates) project their axons as bundles of fibers through the median eminence and infundibular stalk to terminate in the neurohypophysis, where the neurohypophyseal peptides (oxytocin and vasopressin, or their homologues) and proteins (neurophysins) are secreted into the circulation. Indeed, the neurohypophysis has been a classical in vitro preparation for measuring the calcium-dependent release of peptide hormones under various conditions of stimulation (Douglas, 1963; Douglas and Poisner, 1964), and extracellularly recorded compound action potentials are readily measured in this organ (Dreifuss et al., 1971). The analysis of excitation–secretion coupling in the neurohypophysis has been hindered, however, by the circumstance that these nerve terminals are also too small (0.5–1.0 μm) for intracellular measurement of the electrophysiological events that affect release.

In the remainder of this chapter, we will demonstrate the use of optical probes to monitor rapid changes in membrane potential from a population of nerve terminals in the posterior pituitary (neurohypophysis) of the frog *Xenopus*. Resting and slowly changing transmembrane voltages in synaptosomes prepared from rat brain homogenates have already been measured using cyanine (Blaustein and Goldring, 1975) and merocyanine (Kamino and Inouye, 1978) dyes, and action potentials have been recorded from the growth cones of tumor cells in tissue culture using an oxonol dye (Grinvald and Farber, 1981). We describe here a method for recording action potentials from a population of nerve terminals in the intact neurohypophysis of *Xenopus,* and we demonstrate the manipulation of the shape of the action potential by extracellular calcium and other agents known to alter the release of neurohormones and neurotransmitters. We also show that when voltage-sensitive Na channels are blocked by tetrodotoxin (TTX) and K channels are blocked by tetraethylammonium (TEA), direct electric-field stimulation of the nerve terminals evokes large active responses that probably arise from an inward Ca current associated with hormone release from these nerve terminals.

Figure 1b shows an optical recording that represents the intracellular potential changes during the action potentials in a population of synchronously activated nerve terminals in the neurohypophysis of the frog *Xenopus laevis*. This recording was photographed from an oscilloscope screen without signal averaging. The technique used here, multiple site optical recording of transmembrane voltage (MSORTV), is an extension of the method used to record extrinsic absorption changes in the squid giant axon and in many other preparations (Salzberg et al., 1977; Grinvald et al., 1981; Senseman et al., 1983; Salzberg, 1983) and is described in greater detail in Chapter 6 by Cohen and Lesher in this volume. The MSORTV system is illustrated schematically to the left of the optical spike (Fig. 1a). Light from a tungsten–halogen lamp was collimated, made quasimonochromatic with a heat filter (KG-1, Schott Optical Co., Duryea, PA) and an interference filter (700 nm; 70 nm full width at half maximum), and focused by means of a bright-field condenser onto the pars nervosa of the posterior pituitary. The pars nervosa had been excised together with the hypothalamus and the infundibulum, and the entire preparation was vitally stained by incubating it for 25 min in a 100 μg/mL solution of the merocyanine–rhodanine dye, NK 2761 (Kamino et al., 1981; Gupta et al., 1981) (Nippon Kankoh Shikiso Kenkyusho Inc., Okayama, Japan) in *Xenopus* Ringer's solution (112 mM NaCl; 2 mM KCl; 2 mM CaCl$_2$; 15 mM HEPES; 33 mM glucose; pH adjusted to 7.35). Light transmitted by the stained preparation was collected by a high numerical aperture water immersion objective, which formed a real image on a 12 × 12 element silicon photodiode matrix array (MD 144-0; Integrated Photomatrix Inc., Mountainside, NJ) located in the image plane of a compound microscope (UEM, Carl Zeiss, Inc., Oberkochen, West Germany). The photocurrents generated by the central 124 array elements were separately converted to voltages and amplified as described previously by Salzberg et al. (1977) and Grinvald et al. (1981).

In most of the experiments, all of the amplifier outputs were passed to a data acquisition system based on a PDP 11/34A computer (Digital Equipment Corp., Maynard, MA) capable of acquiring a complete frame every 0.8 ms with an effective resolution of 18 bits. This data acquisition system is similar to that described previously by Grinvald et al. (1981) and employed by Grinvald et al. (1982), Senseman et al. (1983), and Salzberg et al. (1983). In some experiments, the temporal resolution was markedly improved (17 μs per point for a single detector element) when selected outputs of the current-to-voltage converters were also passed in parallel to a 16-channel signal averager (TN 1500; Tracor Northern Inc., Middleton, WI). These digitized output signals could be stored on magnetic tape for later display and analysis.

A map of the preparation was obtained by superimposing a transparent overlay representing the photodiode array elements on a photograph taken through the trinocular tube of the microscope after removing the photodiode array housing. The

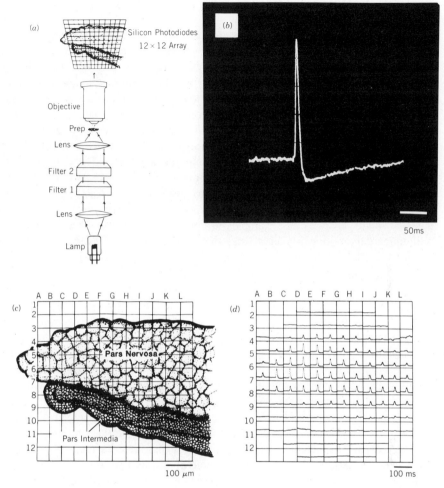

Figure 1. Multiple-site optical recording of transmembrane voltage (MSORTV) from nerve terminals in the neurohypophysis of *Xenopus*. (*a*) Schematic diagram of the optical portion of the MSORTV system. Collimated light from an incandescent source is made quasi-monochromatic with interference and heat filters and focused on the preparation by means of a bright-field condenser with numerical aperture matched to that of the objective. A high numerical aperture water immersion objective projects a real image of a portion of the preparation onto the central 124 elements of a 144-element photodiode matrix array whose photocurrent outputs are converted to voltages, AC coupled and amplified, multiplexed, digitized, and stored in a PDP 11/34A computer under Direct Memory Access (DMA). A full frame is recorded, with an effective resolution of 18 bits, every 800 μs. (*b*) A photograph of an oscilloscope recording of the change in the transmitted intensity monitored by one channel (element E5) of the MSORTV system, following a single brief shock to the hypothalamus. This element monitored intensity changes from a region of the posterior pituitary entirely within the pars nervosa. The fractional change in intensity during the action potential was approximately 0.25%; single sweep, 722 ± 21 nm; rise time of the optical system (10–90%) 1.1 ms; AC coupling time constant 1.0 s. (*c*) Drawing of the region of the posterior pituitary imaged on the photodiode matrix array, showing the positions of the individual detector elements with respect to the tissue. Drawn from a photomicrograph of the preparation and a transparent overlay representing the detector array. Objective 20×, 0.33 NA. (*d*) Extrinsic absorption changes obtained in a single sweep, following stimulation of the hypothalamus, superimposed on corresponding elements of the photodiode matrix array. The five elements in each corner of the array were not connected. The largest signals represent fractional changes in transmitted intensity of approximately 0.3%. After Salzberg et al. (1983).

experiment shown in Fig. 1a employed a 20×, 0.33 numerical aperture water immersion objective (Nikon) to image part of the neurointermediate lobe (pars nervosa plus pars intermedia) onto the photodetector array. Figure 1c is a drawing of the projected region of the gland, prepared from the photomicrograph. Figure 1d shows the MSORTV display of the action potentials recorded optically, in a single sweep, from different areas of the pars nervosa, following a single brief stimulus applied to the hypothalamus. It is observed that the optical signals are confined to elements of the grid that correspond to the neurohypophysis; note that a small portion of the neurohypophysis lies under the pars intermedia (row 9). The optical signal illustrated in Fig. 1b represents the analogue output from channel E5. This record, which is typical, illustrates the large signal-to-noise ratio that may be achieved in these experiments. A variety of control experiments showed that these signals represent voltage changes. For example, no signals could be recorded in the absence of illumination or in white light, and the signals could be inverted at shorter wavelengths (Senseman and Salzberg, 1980).

Evidence that the optical signals shown in Fig. 1 do in fact represent the membrane potential changes associated with the invasion of the nerve terminals of the neurohypophysis by the action potential is provided indirectly by some morphometric observations. In the rat, it is estimated that the 18,000 magnocellular neurons in the hypothalamus give rise to approximately 40 million terminal endings in the pars nervosa of the posterior pituitary, and morphometric data suggest that here, secretory endings and swellings comprise 99% of the excitable membrane area, with axons making up the remaining 1% (Nordmann, 1977). Comparably detailed morphometric studies do not yet exist for the neurohypophysis of *Xenopus*, but the ultrastructural resemblance of the anuran neurohypophysis to that of the rat (Gerschenfeld et al., 1960; Rodriguez and Dellman, 1970; Dellman, 1973;) suggests that, in *Xenopus*, it is also very heavily enriched with nerve terminals. We thus infer that the optical signals shown in Fig. 1, originating in the lateral regions of the pars nervosa, are, at the least, dominated by the membrane potential changes in the terminals of the magnocellular neurons. (For additional arguments, see Salzberg et al., 1983.)

4. PROPERTIES OF THE ACTION POTENTIAL IN THE NERVE TERMINALS

The nerve terminal action potentials shown in Fig. 1 exhibit two rather unexpected features. First, they have relatively long durations (i.e., full width at half height of about 6 ms) compared with those recorded from vertebrate axons (e.g., < 2 ms in the frog (Stämpfli and Hille, 1976)). There must be, of course, some temporal dispersion in the population recorded by a single photodetector element, and 6 ms represents, therefore, an upper bound on the spike width. However, internal evidence suggests that, in fact, very little temporal dispersion is present, since the spike width is constant over a considerable range of objective magnifications. Also, the value for the width of the action potential is in good agreement with the 5- to 6-ms duration reported for the action potentials measured in magnocellular neuron somata (Poulain and Wakerley, 1982) and for the compound action potential in the neurohypophysis (Dreifuss et al., 1971).

The more remarkable characteristic of the nerve terminal action potential illustrated in Fig. 1 is its conspicuous after-hyperpolarization that is reminiscent of the after-hyperpolarization frequently associated with a potassium-mediated potassium conductance (e.g., snail neurons (Meech and Strumwasser, 1970); rat myotube (Barrett et al., 1982); guinea pig Purkinje cells (Llinás and Sugimori, 1980a,b)). We found that it was possible to alter dramatically the after-hyperpolarization recorded optically by changing the ionic composition of the bath. Figure 2 shows the results of a series of experiments designed to examine the calcium dependence of this long-lasting hyperpolarization. The traces shown here represent typical outputs from single channels of the 124-channel MSORTV system; each trace was recorded in a single sweep. In each row, the optical signal on the left (NR) was recorded in normal Ringer's solution which contained 2 mM Ca^{2+}. These control records are normalized to the same peak height.

The middle trace in Fig. 2a (high Ca) shows the effect of a 10-min exposure to elevated (5 mM) extracellular Ca^{2+} concentration; the after-hyperpolarization increased by approximately 90%. This probably represents a lower limit, since it is likely that some dye was bleached during the measurements, and some diminution of the signal is almost inevitable. We always found a large increase in the size of the after-hyperpolarization, although the magnitude of the effect was variable. An increase in the height (14% in Fig. 2a) was often noted; this finding suggests that an inward Ca^{2+} current might contribute to the normal rising phase of the action potential (see below). The record on the right in Fig. 2a (NR) shows the optical signal from the same region of the neurohypophysis 10 min after the return to 2 mM Ca^{2+}. The optical traces in Fig. 2b show the effect of reducing extracellular Ca^{2+} to 0.1 mM by replacing the Ca^{2+} with Mg^{2+}. The middle trace (low Ca) shows that after 10 min, the undershoot is completely absent, replaced by a long-lasting depolarization. The magnitude of this effect also was variable from one experiment to the next; however, a reduction of extracellular calcium to 0.1 mM or less never failed to reduce the after-hyperpolarization. Further lowering of the extracellular Ca^{2+} occasionally eliminated the action potential entirely, but we consider this to be a consequence of steady-state depolarization to which the optical measurements are insensitive.

The right-hand trace in Fig. 2b (NR) again illustrates the recovery (10 min following return to 2 mM Ca^{2+}). In this experiment, reduction of the magnitude of the signal resulting from bleaching was more apparent. While the effects of changes in extracellular Ca^{2+} were always reversible, this was not the case with other ionic substitutions. Cadmium ion, for example, is known to block calcium currents in a variety of preparations (e.g., Standen, 1981), and Fig. 2c shows that after 2 min in 200 μM Cd^{2+} (trace on the right, Cd), the after-hyperpolarization was almost entirely eliminated. This effect was only partially reversible. Longer exposure to cadmium ion eliminated the after-hyperpolarization entirely. The peak height was also considerably reduced, but, in the absence of recovery, it is impossible to know whether this is, at least partially, an effect of bleaching. Figure 2d shows that nickel ion, another inorganic calcium blocker (Kaufmann and Fleckenstein, 1965) at

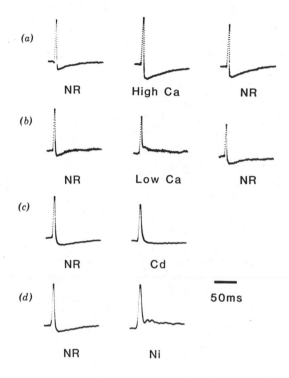

(a)

NR   High Ca   NR

(b)

NR   Low Ca   NR

(c)

NR   Cd

50ms

(d)

NR   Ni

Figure 2. Calcium dependence of the long-lasting after-hyperpolarization of the action potential recorded from nerve terminals of the neurohypophysis. (a) Effect of elevated Ca^{2+} on the nerve terminal action potential. Left-hand trace shows the nerve terminal action potential, recorded in normal Ringer's solution (NR; 2 mM Ca^{2+}). This, and all the records that follow it, were obtained in single sweeps from an element of the photodiode array monitoring potential changes from nerve terminals in the neurohypophysis. The action potential shown in the middle trace (high Ca) was recorded after 10 min in a Ringer's solution containing 5 mM Ca^{2+}. The right-hand trace (NR) shows the nerve terminal action potential 10 min following the return to normal (2 mM Ca^{2+}) Ringer's solution. (b) Effect of lowered extracellular Ca^{2+} on the shape of the nerve terminal action potential. The left-hand trace (NR) shows the control action potential. The action potential in the middle trace (low Ca) was recorded 10 min after replacement of the bathing solution with one containing 0.1 mM Ca^{2+} and 2 mM Mg^{2+}. The right-hand trace (NR) shows the nerve terminal action potential 10 min following return to control (2 mM Ca^{2+}) Ringer's solution. Some bleaching is evident from the reduction in overall signal size. (c) Effect of 0.2 mM Cd^{2+} on the shape of the nerve terminal action potential. Left-hand trace (NR) shows the action potential in normal Ringer's solution. Right-hand trace (Cd) shows the action potential 2 min following the addition of 0.2 mM Cd^{2+} to the normal Ringer's solution. This effect could not be reversed. (d) Effect of 1 mM Ni^{2+} on the shape of the nerve terminal action potential. Left-hand trace (NR) shows the control action potential. Right-hand trace (Ni) shows the action potential 2 min following the addition of 1 mM Ni^{2+} to normal Ringer's solution. This effect was only partially reversible. After Salzberg et al. (1983).

millimolar concentrations, also blocks the spike after-hyperpolarization. In some experiments, Ni^{2+} produced a small but consistent increase in the spike height. The origin of this effect is not clear, and, although a decrease in sodium inactivation is possible, a change in resting potential cannot been ruled out. All of the experiments illustrated in Fig. 2 suggest that calcium plays a significant role in shaping the action potential in the vertebrate nerve terminals studied here. Experiments with other Ca^{2+} antagonists such as Mn^{2+} and Co^{2+} are consistent with this idea, although their potency seems to be less than that of cadmium.

5.   IONIC BASIS OF THE DEPOLARIZING PHASE OF THE ACTION POTENTIAL

Because the entry of calcium plays an important role in excitation–secretion coupling, we attempted to determine whether an inward calcium current contributed measurably to the depolarizing phase of the action potential in the terminals of the neurohypophysis. A priori, calcium is unlikely to be the dominant carrier of the inward current in these nerve terminals, because the very large surface-to-volume ratio in these small secretory endings would then result in the imposition of a large, probably intolerable calcium burden. Figure 3 illustrates an experiment designed to examine this point. Because calcium antagonists leave the depolarizing phase of the action potential largely intact, sodium appears to be carrying the bulk of the inward current. If so, it may be blocked by tetrodotoxin (TTX), and a positive result would provide evidence that the rising phase of the action potential is due primarily to the opening of voltage-dependent sodium channels. To test this possibility it was necessary to change the stimulation conditions, since, in the presence of TTX, block of the sodium action potential in the axons of the infundibulum would prevent the excitation of the nerve terminals in the neurohypophysis by blocking spike conduction.

We were able to circumvent this difficulty by resorting to a form of field stimulation in which the lateral tips of the pars nervosa were held by suction electrodes, and very brief (200–300 μs) currents were passed between the two electrodes. In this fashion, the terminals could be excited directly, without the requirement of axonal transmission. This technique was employed in the experiments shown in Fig. 3. The MSORTV record on the left in Fig. 3a shows the control, in 2 mM Ca^{2+} Ringer's solution. The amplitude of the optical spike increased with field strength, as expected, since a large and potentially variable number of nerve terminals were stimulated and monitored. A maximal response was obtained typically with stimuli of 100–200 V and was largely independent of stimulus polarity. As with the optical signals elicited by stimulation of the hypothalamus, the amplitude of the response was approximately 0.3% of the transmitted light intensity at 700 nm. The

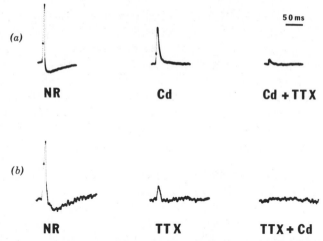

Figure 3.   Effects of sodium and calcium channel blockers on the rising phase of the nerve terminal action potential elicited by direct-field stimulation of the neurohypophysis. (*a*) Left-hand trace (NR) shows the nerve terminal action potential in normal Ringer's solution. The middle trace (Cd) shows the action potential 18 min after the addition of 0.2 mM Cd^{2+} to the normal Ringer's solution. The right-hand trace (Cd + TTX) shows the optical signal recorded from the same population of nerve terminals 6 min after the addition of 5 μM TTX to normal Ringer's solution containing 0.2 mM Cd^{2+}. (*b*) Left-hand trace (NR) shows the nerve terminal action potential in normal Ringer's solution. The middle trace (TTX) shows the residual potential change 6 min following the addition of 2 μM TTX to the Ringer's solution bathing the preparation. The right-hand trace (TTX + Cd) shows the elimination of this TTX-resistant response when calcium channels are blocked by Cd^{2+}. For this record, the preparation had been bathed in a Ringer's solution containing 2 μM TTX and 0.2 mM Cd^{2+} for 5 min. All traces are single sweeps, 700 nm. Rise time of the light-measuring system (10–90%) was 1.1 ms.

middle trace (Cd) was recorded from the same population of nerve terminals following 18 min exposure to 200 μM Cd^{2+} to block any inward calcium current. The reduction in the magnitude of the action potential is obvious, along with a considerable broadening of the spike and the complete loss of the after-hyperpolarization, the latter effects most likely resulting from the elimination of the large calcium-mediated potassium conductance.

The extent of the reduction in the height of the action potential may be exaggerated, however, by factors other than blockage of calcium channels (e.g., dye bleaching). The right-hand trace in Fig. 3*a* (Cd + TTX) shows that 6 min after application of 5 μM TTX, the Cd^{2+}-resistant portion of the terminal action potential was completely eliminated. The residual signal seen in this record is the passive depolarization produced by the field stimulation; it was reversed with reversal of stimulus polarity, and it exhibited the same wavelength dependence as that of the optical recording of the normal action potential. (It should be noted, however, that the size of the passive electrotonus was very variable from element to element.)

The complementary experiment is shown in Fig. 3*b*. Here, the sodium current was first blocked with TTX (middle trace, Fig. 3*b*), and the small residual active response was eliminated by Cd^{2+} (right-hand trace, TTX + Cd) and substantially decreased in size by 1 m*M* Ni^{2+} (not shown). Thus, the experiments illustrated here and others suggest that the rising phase of the nerve terminal action potential, in the neurohypophysis of *Xenopus,* is mediated predominantly by sodium, although these data also suggest at least a small calcium component. The magnitude of this calcium component is difficult to assess quantitatively, particularly because the TTX blockade of the normal regenerative sodium depolarization may limit the opening of voltage-dependent calcium channels. The important role for calcium entry in excitation–secretion coupling is fully consistent with its apparently small contribution to the terminal action potential (Katz and Miledi, 1969). In adrenal chromaffin cells, for example, where calcium influx is the critical event in

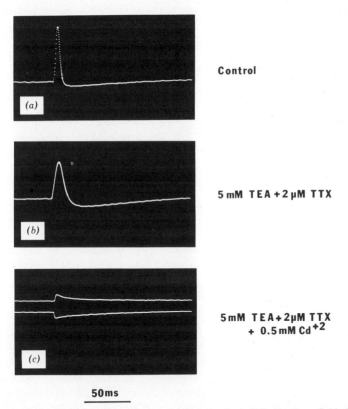

Control

5 mM TEA + 2 μM TTX

5 mM TEA + 2 μM TTX
+ 0.5 mM Cd^{+2}

50 ms

Figure 4.   Cadmium sensitivity of active responses (calcium spikes) elicited by direct-field stimulation of the neurohypophysis in the presence of TTX and TEA. Optical recording of action potentials from nerve terminals of the *Xenopus* neurohypophysis following staining for 25 min in 0.1 mg/mL NK 2761. Outputs of a single channel from the photodiode matrix array in the MSORTV system. (*a*) Action potential recorded in normal Ringer's solution. (*b*) Active response of the nerve terminals following 20-min exposure to 5 m*M* TEA plus 2 μ*M* TTX in normal Ringer's solution. (*c*) Passive response (electrotonus) remaining 15 min after the addition of 0.5 m*M* Cd^{2+} to the TEA–TTX Ringer's solution bathing the preparation, upon stimulation with normal and reversed polarity; 700 nm. Rise time of the light measuring system (10–90%) was 1.1 ms. Temperature 18–22°C. After Obaid et al. (1985).

excitation–secretion coupling (Douglas, 1978), the Ca^{2+} component of the action potential is also small (Brandt et al., 1976).

Nonetheless, under conditions in which the terminal membrane depolarization is prolonged, sufficient inward calcium current may develop to give rise to a relatively large calcium action potential. The records shown in Fig. 4 are from one of 24 posterior pituitaries in which TEA was added, in addition to TTX, in order to block the voltage-dependent potassium conductance and thereby prolong the depolarization of the nerve terminal membrane. Figure 4a shows a control action potential from a population of nerve terminals. The action potential exhibits the characteristic rapid upstroke that we attribute primarily to the effect of a fast inward sodium current, a rapid repolarizing phase resulting from K^+ efflux, and the typical after-hyperpolarization resulting from the Ca^{2+}-mediated increase in K^+ conductance. When 5 mM tetraethylammonium (TEA) was added to the Ringer's solution (not shown), the spike was prolonged, and, in most experiments, a hump appeared on the falling phase, possibly owing to repetitive firing of some of the terminals. The addition of a high concentration of TTX, sufficient to block voltage-sensitive sodium channels, abolished the hump and unmasked a large after-hyperpolarization that presumably results from the enhanced activation of a Ca^{2+}-mediated K^+ conductance. Figure 4b shows the optical response recorded from the same population of terminals represented in Fig. 4a, following 20 min exposure to 5 mM TEA and 2 μM TTX. The regenerative response illustrated is smaller and slower than the normal action potential, and appears to depend entirely upon the influx of calcium and the subsequent rise in a TEA-insensitive K^+ conductance. The further addition of 0.5 mM Cd^{2+} to the bath (Fig. 4c) completely blocked the remaining active responses. (We have already seen, in Figs. 2 and 3, that such low levels of cadmium do not block the action potentials in the nerve terminals of the neurohypophysis in the absence of TTX.) The residual optical signals shown in Fig. 4c represent the purely passive electrotonus, as indicated by their symmetric response to stimulus polarity. The combined effects of TEA and Cd^{2+} could often be reversed, but it was never possible to reverse completely the effect of TTX.

Although the optical signal recorded in Fig. 4b has the appearance of a calcium action potential, it is possible that the major contribution to the rising phase of the response, in TTX, comes from the influx of Na^+ through Ca^{2+} channels, which are not blocked by the toxin. It was possible, in these experiments, to assess qualitatively the relative contributions of Ca^{2+} and Na^+ to the active response by recording optical signals at different $[Ca^{2+}]_o$, at high and low $[Na^+]_o$, and at different $[Na^+]_o$, at high and low $[Ca^{2+}]_o$. These experiments were all carried out in the combined presence of 5 mM TEA and 1 μM TTX. Figure 5 illustrates the principal results. Figure 5A shows that at normal $[Na^+]_o$, the amplitude and rate of rise of the upstroke, as well as the amplitude of the after-hyperpolarization, increased as $[Ca^{2+}]_o$ increased. This is, of course, consistent with the idea that Ca^{2+} carries inward current in the presence of both TTX and TEA. At lowered extracellular concentrations of Na^+ (sucrose replacement, Fig. 5B), a decrease in $[Ca^{2+}]_o$ from normal 2 mM (trace a) to 0.2 mM (trace b) eliminated virtually all of the active response. A comparison of trace b in Fig. 5B (low Ca^{2+}, low Na^+) with trace c in Fig. 5A (low Ca^{2+}, normal Na^+),

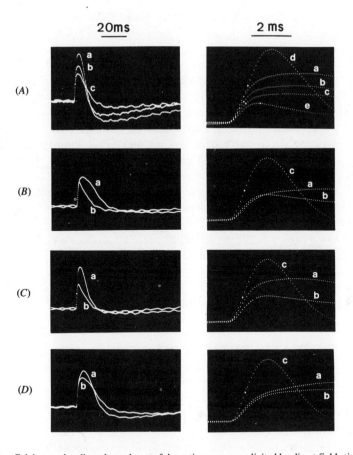

Figure 5. Calcium and sodium dependence of the active response elicited by direct field stimulation of nerve terminals after blocking the voltage-sensitive sodium and potassium channels with TTX and TEA. (A) Left side: optical responses in the presence of 1 μM TTX and 5 mM TEA at *normal* sodium concentration (120 mM) and the following calcium concentrations (for the times indicated): (a) 10 mM (19 min); (b) 2 mM (20 min); (c) 0.1 mM (1.9 mM Mg^{2+} substitution, 20 min). Right side: traces a,b, and c are the same records as those shown on the left side, after expanding the time axis. An initial control response (trace d), obtained in normal Ringer's solution, and the passive response (trace e), obtained 10 min after the addition of 0.2 mM Cd^{2+} to the bathing solution containing 1 μM TTX and 5 mM TEA are shown for comparison. (B) Left side: optical responses in the presence of 1 μM TTX and 5 mM TEA, at *low* sodium concentration (8 mM, sucrose substitution) and the following calcium concentrations (for the times indicated): (a) 2 mM (16 min); (b) 0.2 mM (1.8 mM Mg substitution, 21 min). Right side: traces a and b are the same records as those shown on the left side, after expanding the time axis. An initial control action potential (trace c) obtained in normal Ringer's solution is shown for comparison. (C) Left side: optical responses in the presence of 1 μM TTX and 5 mM TEA, at *low* calcium concentration (0.2 mM, 1.8 mM Mg^{2+}) and the following sodium concentrations (for the times indicated): (a) 120 mM (normal, 21 min); (b) 8 mM (sucrose substitution, 21 min). Right side: traces a and b are the same records as those shown on the left side, after expanding the time axis. An initial control action potential (trace c), obtained in normal Ringer's solution, is shown for comparison. (D) Left side: optical responses in the presence of 1 μM TTX and 5 mM TEA, at *normal* calcium concentration (2 mM) and the following sodium concentrations (for the times indicated): (a) 120 mM (normal, 16 min); (b) 8 mM (sucrose substitution, 21 min). Right side: traces a and b are the same records as those shown on the left side, after expanding the time axis. An initial control action potential (trace c), obtained in normal Ringer's solution, is shown for comparison. The rise time (10–90%) of the light-measuring system was 1.1 ms. Single sweeps. Temperature 18–20°C. After Obaid et al. (1985).

suggests that at low $[Ca^{2+}]_o$, Na^+ might contribute significantly to the inward current conducted by TTX-insensitive channels. On the other hand, at normal $[Ca^{2+}]_o$ (Fig. 5D), varying the $[Na^+]_o$ had little effect on the rate of rise of the optical signal, suggesting that sodium ions are unable to compete effectively with normal concentrations of Ca^{2+}. However, at low $[Ca^{2+}]_o$ (Fig. 5C), the active component of the response was markedly $[Na^+]_o$-sensitive. Thus, it appears that in low $[Ca^{2+}]_o$, sodium ions are able to carry depolarizing membrane current through these cadmium-sensitive (TTX-TEA–insensitive) calcium channels.

The calcium-mediated active responses, recorded optically and described here, appear to share some properties with the local subthreshold action potentials described by Hodgkin (1938) in unmyelinated crab nerve; with TTX blocking any propagated action potential in the axons of the infundibulum, the small calcium currents cannot excite a sufficient length of nerve to produce a propagated action potential. Nonetheless, their dependence on $[Ca^{2+}]_o$ and sensitivity to Cd^{2+} block indicated that they probably result from a voltage-sensitive calcium influx into the terminals (Kostyuk and Krishtal, 1977; Hagiwara and Byerly, 1981; Obaid et al., 1985), which probably mediates hormone release.

6. INTRINSIC OPTICAL CHANGES THAT ACCOMPANY SECRETION FROM MAMMALIAN NERVE TERMINALS

The anatomy of the amphibian posterior pituitary provides a unique opportunity for the study in vitro of excitation–secretion coupling in the vertebrate: There is no postsynaptic excitable membrane to confound the interpretation of the optical signals (Salzberg et al., 1983; Obaid et al., 1985) and no barrier to the collection of its secretory products. However, relatively little is known about the secretion of amphibian neuropeptides and neurohormones, whereas an extensive literature has accumulated on the biochemistry of secretion in the mammalian, and particularly the rodent, posterior pituitary. We felt that it would be important to be able to correlate the electrical events in the terminals with the kinetics of release of the secretory products, and we thought that this might be accomplished by means of optical recording of membrane potential combined with sensitive radioimmunoassays using the CD-1 mouse.

Figure 6a shows an optical recording of a train of seven action potentials, stimulated at 16 Hz, in a population of nerve terminals in the neurohypophysis of the CD-1 (Charles River Breeding Laboratories) mouse. It represents the change in transmitted light intensity at 675 nm, monitored in a single sweep by one channel of the 124-channel MSORTV system. The preparation had previously been incubated in a Ringer's solution (in mM, 154 NaCl, 5.6 KCl, 1 MgCl$_2$, 2.2 CaCl$_2$, 10 glucose, 20 HEPES, adjusted to pH 7.4 with NaOH) containing 0.1 mg/mL of the merocyanine–oxazolone dye NK 2367 (Salzberg et al., 1977). The method used here was virtually identical to that used to record electrical activity from the neurosecretory terminals of the frog neurohypophysis. The trace in Fig. 6a exhibits a

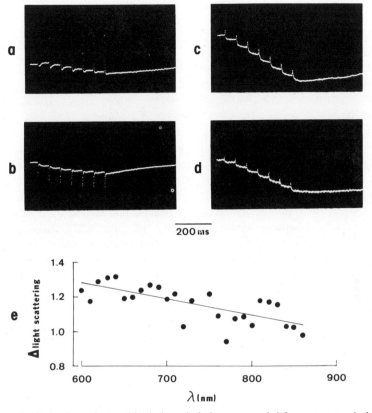

Figure 6. Extrinsic (absorption) and intrinsic optical changes recorded from nerve terminals in the neurohypophysis of the CD-1 mouse. (*a*) Changes in the extrinsic absorption at 675 nm (52 nm full width at half maximum) of nerve terminals following staining for 25 min in 0.1 mg/mL NK 2367. A single train of 500 μs (60 V) stimuli at 16 Hz was delivered to the posterior pituitary through the axons of the infundibulum for 400 ms. The resulting changes in transmitted light intensity are shown, recorded by a single element of the 144-element photodiode matrix array, with an upward deflection of the trace again representing a decrease in transmitted intensity. The AC coupling time constant for the light measurement was 400 ms. The fractional changes in light intensity during the action potential were approximately 0.3%. Single sweep. (*b*) Same experiment as in (*a*), except that the transmitted intensity was monitored at 540 nm (60 nm full width at half maximum). At this wavelength, the voltage-dependent extrinsic absorption change exhibited by the merocyanine-oxazolone NK 2367 is known to be opposite in sign to that observed at 675 nm (Senseman and Salzberg, 1980). (*c*) Intrinsic optical changes recorded at 675 nm (52 nm full width at half maximum) from the nerve terminals of an unstained neurohypophysis. The preparation was stimulated as in (*a*), and the resulting changes in transmitted light intensity recorded by a single element of the photodetector array are shown. Here again, an upward deflection of the optical trace represents a decrease in transmitted light intensity, or an increase in opacity. The AC coupling time constant for the light measurement was 3 s. The intrinsic optical signal, resulting from the train of stimuli shown here, corresponds to a fractional change in transmitted intensity of approximately 0.2%. Single sweep. (*d*) Similar experiment to (*c*), except that the optical signal was DC coupled using a digital sample-and-hold circuit that eliminated the distortion introduced by capacitative coupling. Unstained preparation, 675 nm (52 nm full width at half maximum). Single sweep. (*e*) Wavelength dependence between 600 and 860 nm of the intrinsic optical changes recorded from nerve terminals of the unstained mouse neurohypophysis. The fractional intensity changes have been corrected for the efficiencies of the apparatus and for the gradual deterioration of the preparation. The stimulus trains were delivered at 3-min intervals. All trials were carried out in normal Ringer's solution. The rise time of the light measuring system (10–90%) was 1.1 ms. After Salzberg et al. (1985).

149

gradual decline in baseline, followed by recovery, having the appearance of a transient hyperpolarization.

The constant size of the upstroke, however, suggested to us that this apparent hyperpolarization might not represent a change in membrane potential. Figure 6b shows a single trial recorded by the same detector element, except that the measuring wavelength was 540 nm. At this wavelength, the potential-dependent absorption change exhibited by membrane stained with NK 2367 is known to be opposite in sign to that observed at 675 nm (Senseman and Salzberg, 1980). Notice, however, that although the optical spikes are reversed in sign, the slow component of the signal is essentially unchanged. This weak dependence on wavelength of the slow component of the optical trace suggested that it might reflect a change in light scattering.

Figure 6c illustrates an identical experiment, carried out on another mouse neurohypophysis, before staining. Here, the dye-dependent extrinsic absorption signal that depends on membrane potential is absent, revealing a series of increases in transmitted light intensity (downward deflections of the trace), each preceded by a transient decrease in transparency (upward deflection) occurring at the time of the action potential. This experiment employed a relatively long AC coupling time constant (3 seconds) in order to minimize the distortion of the time course of the optical signal. Nonetheless, the tendency toward recovery following the train of stimuli is largely an artifact of the coupling time constant. A similar experiment is illustrated in Fig. 6d, in which a digital sample-and-hold circuit, consisting of 12-bit analogue-to-digital and digital-to-analogue converters connected in series (Senseman and Salzberg, 1980), is used instead of capacity coupling. This method avoids any AC distortion of the signal, which is observed to remain nearly constant for seconds following the cessation of the stimuli.

Figure 6e illustrates the wavelength dependence of these intrinsic optical signals between 600 and 860 nm, corrected for the spectral efficiency of the apparatus. The wavelength range was selected to exclude major absorption peaks of intrinsic pigments—for example, hemoglobin. The fractional change in transmitted intensity exhibits a gradual decline with increasing wavelength, consistent with a change in light scattering. All components of the optical signals were eliminated when the neurohypophysis was depolarized with KCl or when the Ringer's solution contained micromolar concentrations of TTX.

It must be clear to the reader that optical techniques that employ visible light, especially those that rely upon intrinsic optical changes that require no exogenous probes, offer unique capabilities in terms of resolution in time and space, combined with a gentleness that permits the nondestructive measurement of many cellular and subcellular processes. In particular, light-scattering methods have been applied to biological systems, in vitro and in vivo, since Tyndall (1876), and have played a role in a variety of recent efforts to study changes in neuron structure during action potential propagation and synaptic transmission (Cohen, 1973). Cohen and Landowne (1970) reported small (8×10^{-7}) changes in light scattering in the skate electric organ that appeared to be presynaptic in origin, and Shaw and Newby (1972) detected calcium-dependent movement in a thoracic ganglion of a locust (*Schistocerca gregaria*) by monitoring the power spectrum of "twinkling" (light beating) produced by scattered laser light (see also Piddington and Sattelle, 1975; Englert and Edwards, 1977). These preliminary reports, however, have not been pursued.

In the absence of potentiometric dye staining of the terminals of the neurohypophysis, two intrinsic optical signals, illustrated in Fig. 6c, are evident following each stimulus. The rapid upward deflection of the trace (decrease in transparency), which we have termed the "E-wave" (Salzberg et al., 1985), appears to coincide with the arrival of the action potential at the terminals (cf. Fig. 6a), whereas the subsequent increase in transparency, the "S-wave," appears to correspond to a slower and longer-lasting process occurring in the nerve terminals. Several lines of evidence suggested that this latter intrinsic optical signal (Fig. 6c,d) was closely correlated with the secretory process. If this association is more than accidental, we would anticipate that the later changes in transparency (but not the E-wave) would exhibit certain properties that characterize neurosecretory systems in general, and the release of neuropeptides in particular—namely, dependence on stimulation frequency, with marked facilitation; dependence on extracellular Ca^{2+} concentration; and sensitivity to Ca^{2+} antagonists and various pharmacological procedures known to influence secretion (Salzberg et al., 1985).

The frequency dependence of the transparency changes (intrinsic optical signals) that we have referred to as "light scattering" (Salzberg et al., 1985) is illustrated in Fig. 7. In each panel, the stimuli are delivered to the neurohypophysis for 400 ms, but at different frequencies. Panels a–d show a monotonic increase in the total size of the light-scattering change with number of stimuli. The size of the fractional change in light scattering per stimulus ($\Delta50$–460 ms/N) varies with frequency, reaching a broad maximum at between 8 and 16 Hz. Figure 7d demonstrates the tendency of the total optical change to saturate at very high stimulation rates. In each instance, a marked facilitation of the optical response is evident between the first and second stimuli. Figure 7e was obtained under identical conditions (16 Hz) to those of Fig. 7c, except that 1 mM $CdCl_2$ was added to the bath for 15 min before the single trial recorded here. In Fig. 7e, Cd^{2+} is seen to block more than 90% of the S-wave (downward deflection of the trace; decrease in opacity) of the scattering change without significantly affecting the rapid upstroke (E-wave) that occurs coincident with the action potential. The plot in Fig. 7f summarizes the dependence of the magnitude of the light-scattering change on the rate and number of stimuli and also includes the effect of barium ions (see below).

7. EFFECTS OF EXTRACELLULAR CALCIUM

Calcium ion profoundly influences neurosecretory activity (e.g., Dodge and Rahamimoff, 1967; Katz, 1969; Douglas, 1975), and the extracellular concentration of this ion would be expected to modulate the size of an intrinsic optical signal related to secretion from the terminals of the neurohypophysis. Figure 8 shows the effect of $[Ca^{2+}]_o$ on the light-scattering signal evoked by stimulation of the infundibulum at 10 Hz for 400 ms. Traces a–f, which are each the average of 16 sweeps, were

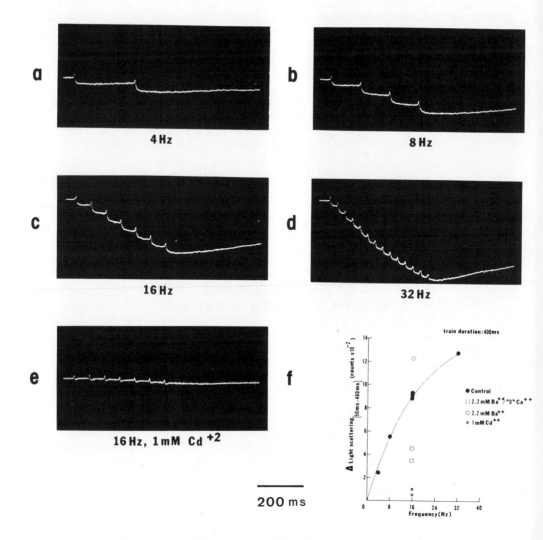

Figure 7. Frequency dependence and Cd^{2+} block of the light-scattering change in nerve terminals of the mouse neurohypophysis. (a), (b), (c), and (d) show the intrinsic optical signals at 675 nm that result from stimulation of the terminals of the mouse pars nervosa at 4, 8, 16, and 32 Hz, respectively, for 400 ms, in normal Ringer's solution. (e) Identical experiment to (c), except that the intrinsic optical signal was recorded 15 min after the addition of 1 mM Cd^{2+} to the Ringer's solution bathing the neurohypophysis. (f) Fractional change in transparency between 50 and 460 ms, as a function of stimulation frequency, following stimulation for 400 ms. The data were not corrected for the absolute number of stimuli. At 16 Hz, data are also plotted for three trials from the same experiment, in which 1 mM Cd^{2+} was added to the bath, 2.2 mM Ba^{2+} was substituted for the Ca^{2+} in the bath, and 2.2 mM Ba^{2+} was added to the normal Ringer's solution. (a)–(e) were all single trials. AC coupling time constant 3 s; rise time of the light measuring system (10–90%) 1.1 ms. Temperature 23°C. After Salzberg et al. (1985).

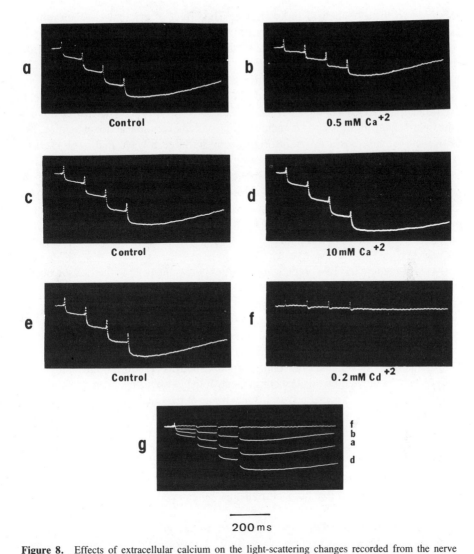

200 ms

Figure 8. Effects of extracellular calcium on the light-scattering changes recorded from the nerve terminals of the mouse neurohypophysis. (*a*), (*c*), and (*e*) Light-scattering changes accompanying stimulation at 10 Hz of an unstained mouse neurohypophysis in normal Ringer's solution (2.2 mM Ca^{2+}). (*b*) Light-scattering changes in the same preparation 20 min following reduction of the extracellular calcium concentration to 0.5 mM by Mg^{2+} substitution. (*d*) Light-scattering changes in the same preparation 40 min following exposure to Ringer's solution containing 10 mM Ca^{2+}. (*f*) Light-scattering changes in the same neurohypophysis 10 min following the addition of 0.2 mM Cd^{2+} to the bathing solution. Records (*a*)–(*f*) are each the average of 16 sweeps and were recorded in the alphabetical order shown. (*g*) Traces (*a*) (control), (*b*) (0.5 mM Ca^{2+}), (*d*) (10 mM Ca^{2+}), and (*f*) (0.2 mM Cd^{2+}) are shown superimposed, after normalization to the height of the first E-wave. AC coupling time constant 3 s; rise time of the light measuring system (10–90%) 1.1 ms. Temperature 24°C. After Salzberg et al. (1985).

obtained in the alphabetical order shown. The effect of calcium concentration is apparent, and the blockade of the optical signal by 0.5 mM Cd^{2+} in Fig. 8f is dramatic. Calcium ion, of course, has other effects, including a direct effect on excitability (Frankenhaeuser and Hodgkin, 1957), and these must be considered. To eliminate the effects of threshold variation, the stimuli were supramaximal in each instance. Examination of the records in Fig. 8 (cf., e.g., 8b and 8d) reveals significant differences in the amplitudes of the E-waves that precede each S-wave in the light-scattering signal and coincide in time with the terminal action potential. The E-wave evidently has a different origin from the large S-wave that is blocked by cadmium. We suggest that this very early intrinsic optical change (E-wave) reflects the arrival of the action potential at the terminals of the neurohypophysis, and we have assumed that as a compound optical signal, its size is roughly proportional to the number of terminals activated at a given time. On this assumption, we have compensated for changes in the invasion of the tissue resulting from changes in extracellular calcium by normalizing the S-wave according to the size of the initial E-wave.

This procedure has two potential difficulties. First, the peak of the E-wave occurs slightly before the peak of the voltage change in the terminals, as determined by the response of the voltage-sensitive dyes. Although the extrinsic absorption signals could be distorted if the dye response exhibited a slow component, the merocyanine–rhodanine and merocyanine–oxazolone dyes, at the concentrations used in these experiments, respond in less than 2 μs to step changes in voltage (B. M. Salzberg, A. L. Obaid, and F. Bezanilla, unpublished observations), and their overall response is dominated by this fast component. If the E-wave depended on the inward current, it might be expected to reach a maximum at the time of the maximum rate of rise of the extrinsic absorption (voltage) signal. In fact, the time-to-peak of this early component of the opacity change lies somewhere between the time-to-peak of the voltage change and the time-to-peak of the time derivative of the absorption (voltage) change. This result suggests that the ultimate interpretation of the E-wave will be complicated and that, although this portion of the intrinsic optical signal is related to excitation, it may exhibit both current and potential dependence (Cohen et al., 1970a,b).

A second possible difficulty involves our assumption that the size of the E-wave, whether related primarily to current or voltage, is proportional to the number of active terminals. This can only be approximately correct, since it will be affected by changes in the magnitude of the upstroke of the terminal action potential, particularly those resulting from changes in the resting membrane potential. It will also be sensitive to changes in the temporal dispersion of excitation of the terminals, and small changes in the size of the factor used for normalizing the optical signals might seriously distort the quantitative analysis of the optical signals. On balance, however, under constant stimulation conditions, the height of the E-wave would seem to provide a rough comparative estimate of the degree of invasion of the tissue and thereby offer a convenient normalization for the larger light-scattering changes that appear to be related to secretion (S-wave). Moreover, none of the qualitative conclusions that we have drawn are affected by this procedure. Accordingly, Fig. 8g shows records obtained in normal Ringer's solution (a), together with records obtained in low

$[Ca^{2+}]_o$ (b), high $[Ca^{2+}]_o$ (d), and normal Ringer's solution to which 0.2 mM Cd^{2+} was added (f), all normalized so as to equate the sizes of the E-waves. In this way, the effect of $[Ca^{2+}]_o$ on the light-scattering signal, per active terminal, is demonstrated most convincingly.

8. DEUTERIUM OXIDE DECREASES THE SIZE OF THE LIGHT-SCATTERING RESPONSE TO STIMULATION

Heavy water (deuterium oxide) depresses excitation–contraction coupling in different muscle types (Kaminer, 1960; Svensmark, 1961; Bezanilla and Horowicz,

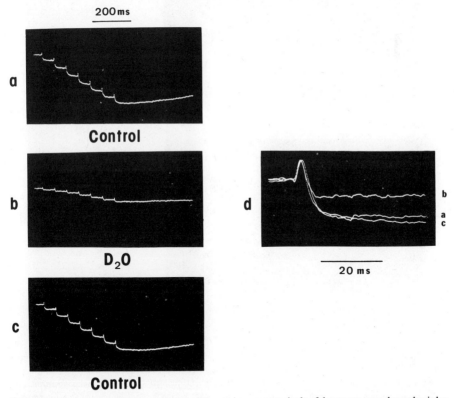

Figure 9.   Depression of the intrinsic optical change in nerve terminals of the mouse neurohypophysis by heavy water. (a) Light-scattering changes at 675 nm following a single train of stimuli delivered to the neurohypophysis at 16 Hz for 400 ms, in control Ringer's solution. Single sweep. (b) Light-scattering changes at 675 nm in the same preparation 31 min following the substitution of more than 98% of the H_2O in the Ringer's solution by D_2O. Single sweep. (c) Same as (a), 41 min following return to normal Ringer's solution. Single sweep. (d) Light-scattering responses to single stimuli, obtained following the corresponding trials. The records are shown superimposed following normalization to the height of the E-wave. Averages of 16 sweeps, stimulated every 5 s. AC coupling time constant 3 s. Rise time of the light measuring system (10–90%) 1.1 ms. Temperature 23°C. After Salzberg et al. (1985).

1975) and excitation–secretion coupling in a variety of systems including pancreatic beta cells (Lacy et al., 1972). In the pars nervosa of the mouse, complete ($> 98\%$) replacement of the water in the Ringer's solution by D_2O resulted in a reduction in the size of the light-scattering signal by about 60%. Figure 9 illustrates the effect of D_2O substitution on the optical response to stimulation of the terminals of the mouse neurohypophysis. Panels *a*, *b*, and *c* show light-scattering changes at 675 nm obtained during single trials in normal Ringer's solution (*a*), 30 min in D_2O-substituted Ringer's solution (*b*), and 40 min following the return to normal

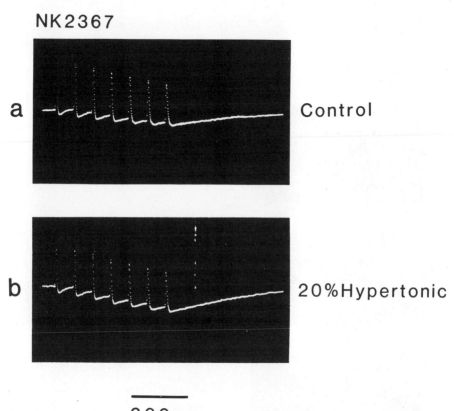

NK2367

a   Control

b   20%Hypertonic

200ms

Figure 10.   Effects on the intrinsic and extrinsic optical signals of increasing the extracellular volume fraction with hypertonic saline. (*a*) Extrinsic and intrinsic optical signals at 675 nm recorded in a single sweep during stimulation at 16 Hz of a mouse neurohypophysis that had been stained with 25 μg/mL NK 2367 for 25 min. Control Ringer's solution. The progressive decline in the height of the action potentials probably reflects the accumulation of potassium in the extracellular space. The downward light-scattering signal masks the Frankenhaeuser–Hodgkin effect on the envelope of the undershoots. (*b*) Same experiment as (*a*), following 10-min exposure to a Ringer's solution made 20% hypertonic by the addition of sucrose. Extracellular accumulation of potassium is less evident here (nearly constant peak heights and reduced Frankenhaeuser–Hodgkin effect), while the light-scattering component of the signal is not decreased. Single sweep. AC coupling time constant 400 ms. Rise time of the light measuring system (10–90%) 1.1 ms. Temperature 19°C. After Salzberg et al. (1985).

Ringer's solution (*c*). The depression produced by D_2O seems to reflect primarily a decrease in the light scattering from each active terminal, consistent with earlier observations of the effects of D_2O (Kaminer, 1960; Svensmark, 1961; Bezanilla and Horowicz, 1975; Lacy et al., 1972), rather than a decrease in the number of active terminals, as judged by the very small variation in the size of the E-wave. This is illustrated in Fig. 9*d*, where the averages of 16 trials are shown, each record obtained immediately following the corresponding single trial. These records have been normalized to the height of the first peak, to compensate for small differences in the number of active terminals. It should be noted that replacement of water with D_2O produces a shift in the true pD, compared to the value measured with a glass pH electrode (Glascoe and Long, 1961), of approximately 0.4 pD units in the "alkaline" direction. To control for the possibility that the depression of the light-scattering change shown in Fig. 9 resulted from a pD shift, we repeated these experiments in normal Ringer's solution with its pH adjusted to 8.0. This procedure had no effect on the normal light-scattering signals.

9. THE EFFECT OF INCREASING THE EXTRACELLULAR VOLUME FRACTION WITH HYPERTONIC MEDIUM

Rendering the Ringer's solution bathing the neurohypophysis hypertonic is expected to increase the volume of the extracellular space by shrinking the terminals. We might be able to detect this effect qualitatively in several different ways. Perhaps simplest would be to note the effect of hypertonicity on the appearance of extracellular potassium accumulation (Frankenhaeuser and Hodgkin, 1956). Figure 10 illustrates this behavior indirectly in a neurohypophysis that had been stained with the merocyanine–oxazolone dye NK 2367 (Salzberg et al., 1977). Panel *a* shows the sum of the extrinsic absorption change, representing membrane potential, and the light-scattering signal during stimulation at 16 Hz for 400 ms in normal Ringer's solution. The decrease in the overall size of the action potential (20% between the first and last spike in the train) suggests that in this preparation, sufficient potassium is accumulating extracellularly to affect the undershoots of the action potentials, but that this effect is masked by the downward deflection produced by the S-wave of the light-scattering signal. When the solution bathing the preparation is made 20% hypertonic by the addition of sucrose, the reduction in spike amplitude is less pronounced (8%), as though the effective extracellular volume was increased by the shrinking of the terminals. The light-scattering signal, however, is not appreciably reduced.

10. CONNECTION OF THE LARGE INTRINSIC OPTICAL SIGNAL TO SECRETORY EVENTS AT THE NERVE TERMINALS

The intrinsic optical signals described above are closely correlated with the secretory activity of the magnocellular neuron terminals located in the neurohypophysis of the mouse. A mechanical artifact associated with contraction of vascular smooth muscle cannot be ruled out unequivocally, and such an effect would indeed be expected to exhibit many of the same properties as an intrinsic signal related to secretion. However, the time course of the S-wave, the component of the light-scattering change that we associate with secretion, would seem to be too fast by at least an order of magnitude to have its origin in the contraction of smooth muscle. The S-wave reaches its half-maximal value within 7–8 ms of the application of the stimulus pulse to the region of the pars nervosa where the infundibular stalk enters (and 3–4 ms following the peak of the extrinsic signal). Even neglecting any conduction latency, this is considerably faster than any known mechanical response of vascular smooth muscle.

Similarly, we are inclined to rule out changes related to the terminal action potential alone, in accounting for the major component of the intrinsic optical signal. For example, extracellular accumulation of potassium ion might give rise to a change in light scattering, mediated by a transport number effect (Girardier et al., 1963; Barry and Hope, 1969; Cohen et al., 1972b) resulting from an increased salt concentration in the restricted space immediately outside the terminals. At least two kinds of evidence argue against such a mechanism. First, the replacement of all of the calcium in the bath by barium ion reduced the light scattering by only about 50% in 10 min (Fig. 7f)—this despite the fact that barium is unlikely to replace calcium in activating a potassium conductance (Standen and Stanfield, 1978; Vergara and Latorre, 1983), and barium is likely, in fact, to reduce the voltage-dependent potassium conductance (Armstrong and Taylor, 1980; Eaton and Brodwick, 1980; Armstrong et al., 1982). Thus, the accumulation of potassium is expected to be severely decreased under these conditions. On the other hand, extracellular barium does replace calcium in some secretory systems (Dicker, 1966; Rubin, 1974), although it is not as effective. Thus, the observation that the addition of 2.2 mM barium to the normal complement of extracellular calcium *increases* the light-scattering signal (Fig. 7f) seems particularly significant, since, again, potassium accumulation might be expected to be reduced. (The increase, however, is smaller than that contributed by the addition of calcium itself, consistent with the lesser efficacy of barium in effecting secretion.) The experiment using sucrose (Fig. 10b) to increase the extracellular volume fraction also suggests that the light-scattering signal does not depend primarily on an alteration of extracellular space, since the effect of repetitive stimulation should be less under conditions in which the terminals are shrunk by the hypertonicity of the bath, but the light-scattering change was not diminished.

On the other hand, the S-wave transparency changes that seem to reflect rapid alterations in light scattering are dramatically affected by many of the same interventions that are known to alter the release of neuropeptides from these same terminals. The dependence on stimulation frequency (Fig. 7) is in good accord with evidence obtained in several laboratories, suggesting that neuropeptide secretion increases with frequency of stimulation (Dreifuss et al., 1971; Nordmann and Dreifuss, 1972; Poulain and Wakerley, 1982; unpublished data on the mouse neurohypophysis from the laboratory of H.G.), and the facilitation of the intrinsic optical response seen in virtually all of the records seems particularly telling. The effect of extracellular calcium concentration on the magnitude of the S-wave of the intrinsic optical signal (Fig. 8) is also in remarkably good agreement with the classical behavior of secretory systems (Douglas, 1978). The effect of replacing all of the water in the Ringer's solution with D_2O, however, had not been reported in the mammalian neurohypophysis before our experiments (Salzberg et al., 1985). Under these conditions, we found (Fig. 9) a depression of the light-scattering signal of approximately 60%. Measurements of secreted vasopressin carried out in one of our laboratories (H.G., manuscript in preparation), using radioimmunoassays, have revealed a nearly identical decrease in vasopressin release from the mouse neural lobes when D_2O is substituted for H_2O under comparable stimulation conditions.

Another series of experiments, carried out in our laboratory together with T. D. Parsons, at the suggestion of David Quastel, has shown that neomycin and other aminoglycoside antibiotics depress the S-wave of the light-scattering signal (Parsons et al., 1985). This finding is significant because, at low concentration, these drugs act presynaptically at the snake neuromuscular junction and depress evoked transmitter release by competing with $[Ca^{2+}]_o$ (Fiekers, 1983). Our observation that, for example, neomycin at 220 μM concentration in normal Ringer's depressed the magnitude of the S-wave by 60% after 11 min provides additional evidence that these agents act presynaptically, since there are no postsynaptic elements in the neurohypophysis, and that they act as competitive antagonists of $[Ca^{2+}]_o$ (Parsons et al., 1985). Further, the compatibility of the aminoglycoside effects on the light scattering from the neurohypophysis, with voltage clamp data reported from the neuromuscular junction (Fiekers, 1983), reinforces the interpretation (Salzberg et al., 1985) that the light-scattering signals reflect processes closely related to neuropeptide release.

We have, as yet, no reason to implicate any particular step among the sequence of events that couples excitation to secretion in the generation of the large intrinsic optical signals described here, and the identity of the physiological event or events remains unclear. The fusion of secretory vesicles during exocytosis should result in the loss of relatively high refractive index particles and should thereby reduce the refractive index gradients in the terminal. However, the very early onset and fast rise of the S-wave suggests that the optical signal might arise as a result of some calcium-dependent process prior to the fusion of secretory vesicles and the release of their contents. For example, it is possible that the light-scattering changes that we have reported reflect a phase transition of the contents of the secretory vesicles, or they may reveal, instead, rapid alterations in the state of intracellular calcium stores following calcium entry but prior to secretion (Neering and McBurney, 1984). Thus, these intrinsic optical changes may be related to the transparency changes that precede tension development which have been observed in cut skeletal muscle fibers (Kovacs and Schneider, 1977; Rios et al., 1983). Identification of the physiological origin of the intrinsic optical changes must await precise measurement of light scattering per se. Laser light-scattering measurements, particularly on simplified model systems (such as secretosomes or vesicle suspensions), may assist in the interpretation of these very interesting signals.

In any case, the weak dependence on wavelength of the intrinsic signal related to secretion, contrasted with the strong wavelength dependence of the extrinsic absorption signals provided by linear potentiometric probes (e.g., merocyanines) (Ross et al., 1974, 1977; Cohen and Salzberg, 1978; Salzberg, 1983), should permit one to monitor simultaneously, in a stained preparation, the voltage changes in the nerve terminals and the time course of events intimately associated with the release of secretory products. The inherently fast responses of the two optical measurements may then improve our ability to resolve early events in the coupling of excitation to secretion.

We have chosen the vertebrate neurohypophysis to illustrate the technique of optical recording of membrane potential from regions of cells that are electrically inaccessible by conventional means. These experiments have led to the use of light-scattering methods for the investigation of excitation–secretion coupling at nerve terminals, and it seems very likely that intracellular indicators for measurement of pH and $[Ca^{2+}]_i$ will also find application in this system. Photoinactivation (e.g., of chelators) for producing rapid changes in free calcium and other components of the neurosecretory apparatus should also play a useful part in increasing our understanding of exocytosis in general, and, of course, image enhancement techniques are certain to improve our ability to visualize the secretory process at the subcellular level. It seems quite clear that optical methods will contribute significantly to the study of cell physiology in the decades to come.

ACKNOWLEDGMENTS

We are grateful to Professors C. M. Armstrong, L. B. Cohen, M. Morad, and R. K. Orkand for valuable discussions, and to Professors Orkand and D. M. Senseman for their participation in some of the experiments discussed. We also thank L. B. Cohen and S. Lesher for providing the software used for data acquisition and display. This research was supported by USPHS grant NS 16824.

REFERENCES

Armstrong, C. M., and S. R. Taylor (1980) Interaction of barium ions with potassium channels in squid giant axons. *Biophys. J.*, **30**, 473–488.

Armstrong, C. M., R. P. Swenson, and S. R. Taylor (1982) Block of squid axon K channels by internally and externally applied barium ions. *J. Gen Physiol.*, **80**, 663–682.

Barrett, J. N., K. L. Magleby, and B. S. Palotta (1982) Properties of single calcium activated potassium channels in cultured rat muscle. *J. Physiol. (Lond.)*, **331**, 211–230.

Barry, P. H., and A. B. Hope (1969) Electroosmosis in membranes: Effects of unstirred layers and transport numbers. *Biophys. J.*, **9**, 700–728.

Bezanilla, F., and P. Horowicz (1975) Fluorescence intensity changes associated with contractile activation in frog muscle stained with Nile Blue A. *J. Physiol. (Lond.)*, **246**, 709–735.

Blaustein, M. P., and J. M. Goldring (1975) Membrane potentials in pinched-off presynaptic nerve terminals monitored with a fluorescent probe: Evidence that synaptosomes have potassium diffusion potentials. *J. Physiol. (Lond.)*, **247**, 589–615.

Brandt, B. L., S. Hagiwara, Y. Kidokoro, and S. Miyazaki (1976) Action potentials in the rat chromaffin cell and effects of acetylcholine. *J. Physiol. (Lond.)*, **263**, 417–439.

Bullock, T. H., and S. Hagiwara (1957) Intracellular recording from the giant synapse of the squid. *J. Gen. Physiol.*, **40**, 565–577.

Cohen, L. B. (1973) Changes in neuron structure during action potential propagation and synaptic transmission. *Physiol. Rev.*, **53**, 373–418.

Cohen, L. B., and D. Landowne (1970) Changes in light scattering during synaptic activity in the electric organ of the skate, *Raia erinacea*. *Biophys. J.*, **10**, 16a.

Cohen, L. B., and B. M. Salzberg. (1978) Optical measurement of membrane potential. *Rev. Physiol. Biochem. Pharmacol.*, **83**, 33–88.

Cohen, L. B., R. D. Keynes, and D. Landowne (1972a) Changes in light scattering that accompany the action potential in squid giant axons: Potential-dependent components. *J. Physiol. (Lond.)*, **224**, 701–725.

Cohen, L. B., R. D. Keynes, and D. Landowne (1972b) Changes in light scattering that accompany the action potential in squid giant axons: Current dependent components. *J. Physiol. (Lond.)*, **224**, 727–752.

Cohen, L. B., B. M. Salzberg, H. V. Davila, W. N. Ross, D. Landowne, A. S. Waggoner, and C.-H. Wang (1974) Changes in axon fluorescence during activity: molecular probes of membrane potential. *J. Membr. Biol.*, **19**, 1–36.

Davila, H. V., B. M. Salzberg, L. B. Cohen, and A. S. Waggoner (1973) A large change in axon fluorescence that provides a promising method for measuring membrane potential. *Nature New Biol.*, **241**, 159–160.

Davila, H. V., L. B. Cohen, B. M. Salzberg, and B. B. Shrivastav (1974) Changes in ANS and TNS fluorescence in giant axons from *Loligo*. *J. Membr. Biol.*, **15**, 29–46.

Dellmann, H. D. (1973) Degeneration and regeneration of neurosecretory systems. *Int'l. Rev. Cytol.*, **36**, 215–315.

Dicker, S. E. (1966) Release of vasopressin and oxytocin from isolated pituitary glands of adult and new-born rats. *J. Physiol. (Lond.)*, **185**, 429–444.

Dodge, F., and R. Rahamimoff (1967) Cooperative action of calcium ions in transmitter release at the neuromuscular junction. *J. Physiol. (Lond.)*, **193**, 419–432.

Douglas, W. W. (1963) A possible mechanism of neurosecretion-release of vasopressin by depolarization and its dependence on calcium. *Nature*, **197**, 81–82.

Douglas, W. W. (1978) Stimulation-secretion coupling: variations on the theme of calcium activated exocytosis involving cellular and extracellular sources of calcium. *Ciba Fdn. Symp.*, **54**, 61–90.

Douglas, W. W., and A. M. Poisner (1964) Stimulus secretion coupling in a neurosecretory organ and the role of calcium in the release of vasopressin from the neurohypophysis. *J. Physiol. (Lond.)*, **172**, 1–18.

Dreifuss, J. J., I. Kalnins, J. S. Kelley, and K. B. Ruf (1971) Action potentials and release of neurohypophysial hormones *in vitro*. *J. Physiol. (Lond.)*, **215**, 805–817.

Eaton, D. C., and M. S. Brodwick (1980) Effects of barium on the potassium conductance of squid axon. *J. Gen. Physiol.*, **75**, 727–750.

Englert, D., and C. Edwards (1977) Effect of increased potassium concentrations on particle motion within a neurosecretory structure. *Proc. Natl. Acad. Sci. USA*, **74**, 5759–5763.

Fiekers, J. F. (1983) Effects of the aminoglycoside antibiotics, streptomycin and neomycin, on neuromuscular transmission. I. Presynaptic considerations. *J. Pharmacol. Exp. Ther.*, **225**, 487–495.

Frankenhaeuser, B., and A. L. Hodgkin (1956) The after-effects of impulses in the giant nerve fibres of *Loligo. J. Physiol. (Lond.)*, **131**, 341–376.

Frankenhaeuser, B., and A. L. Hodgkin (1957) The action of calcium on the electrical properties of squid axons. *J. Physiol. (Lond.)*, **137**, 218–244.

Freedman, J. C., and P. C. Laris (1981) Electrophysiology of cells and organelles: Studies with optical potentiometric indicators. *Int. Rev. Cytol. Suppl.*, **12**, 177–246.

Gerschenfeld, H. M., J. H. Tramezzani, and E. De Robertis (1960) Ultrastructure and function in neurohypophysis of the toad. *Endocrinology*, **66**, 741–762.

Girardier, L., J. P. Reuben, P. W. Brandt, and H. Grundfest (1963) Evidence for anion-permselective membrane in crayfish muscle fibers and its possible role in excitation–contraction coupling. *J. Gen. Physiol.*, **47**, 189–214.

Glascoe, P. K., and F. A. Long (1960) Use of glass electrodes to measure acidities in deuterium oxide. *J. Phys. Chem.*, **64**, 188–190.

Grinvald, A. (1985) Real-time optical mapping of neuronal activity: From single growth cones to the intact mammalian brain. *Ann. Rev. Neurosci.*, **8**, 263–305.

Grinvald, A., and I. C. Farber (1981) Optical recording of Ca^{2+} action potentials from growth cones of cultured neurons using a laser microbeam. *Science*, **212**, 1164–1169.

Grinvald, A., L. B. Cohen, S. Lesher, and M. B. Boyle (1981) Simultaneous optical monitoring of activity of many neurons in invertebrate ganglia, using a 124-element photodiode array. *J. Neurophysiol.*, **45**, 829–840.

Grinvald, A., R. Hildesheim, I. C. Farber, and L. Anglister (1982a) Improved fluorescent probes for the measurement of rapid changes in membrane potential. *Biophys. J.*, **39**, 301–308.

Grinvald, A., A. Manker, and M. Segal (1982b) Visualization of the spread of electrical activity in rat hippocampal slices by voltage-sensitive optical probes. *J. Physiol. (Lond.)*, **333**, 269–291.

Gupta, R. K., B. M. Salzberg, A. Grinvald, L. B. Cohen, K. Kamino, S. Lesher, M. B. Boyle, A. S. Waggoner, and C.-H. Wang (1981) Improvements in optical methods for measuring rapid changes in membrane potential. *J. Membr. Biol.*, **58**, 123–137.

Hagiwara, S., and L. Byerly (1981) Calcium channel. *Ann. Rev. Neurosci.*, **4**, 69–125.

Hodgkin, A. L. (1938) The sub-threshold potentials in a crustacean nerve fibre. *Proc. R. Soc. Lond. B. Biol. Sci.*, **126**, 87–121.

Kaminer, B. (1960) Effect of heavy water on different types of muscle and on glycerol-extracted psoas fibres. *Nature*, **185**, 172–173.

Kamino, K., and A. Inouye (1978) Evidence for membrane potential changes in isolated synaptic membrane ghosts monitored with a merocyanine dye. *Jpn. J. Physiol.*, **28**, 225–237.

Kamino, K., A. Hirota, and S. Fujii (1981) Localization of pacemaking activity in early embryonic heart monitored using a voltage sensitive dye. *Nature*, **290**, 595–597.

Katz, B. (1969) *The Release of Neural Transmitter Substances*. Thomas, Springfield, IL, 193 pp.

Katz, B., and R. Miledi (1967) A study of synaptic transmission in the absence of nerve impulses. *J. Physiol. (Lond.)*, **192**, 407–436.

Katz, B., and R. Miledi (1969) Tetrodotoxin-resistant electrical activity in presynaptic terminals. *J. Physiol. (Lond.)*, **203**, 459–487.

Kaufmann, R., and A. Fleckenstein (1965) Ca^{++}-kompetitive electro-mechanische Entkoppelung durch Ni^{++}- und Co^{++}-Ionen am Warmblütermyokard. *Pflügers Arch. Ges. Physiol.*, **282**, 290–297.

Kostyuk, P. G., and O. A. Krishtal (1977) Separation of sodium and calcium currents in the somatic membrane of mollusc neurons. *J. Physiol. (Lond.)*, **270**, 545–568.

Kovacs, L., and M. F. Schneider (1977) Increased optical transparency associated with excitation–contraction coupling in voltage-clamped cut skeletal muscle fibers. *Nature*, **265**, 556–560.

Lacy, P. E., M. M. Walker, and C. J. Fink (1972) Perfusion of isolated rat islets *in vitro*. Participation of the microtubular system in the biphasic release of insulin. *Diabetes*, **21**, 987–998.

Llinás, R., and M. Sugimori (1980a) Electrophysiological properties of *in vitro* Purkinje cell somata in mammalian cerebellar slices. *J. Physiol. (Lond.)*, **305**, 171–195.

Llinás, R., and M. Sugimori (1980b) Electrophysiological properties of *in vitro* Purkinje cell dendrites in mammalian cerebellar slices. *J. Physiol. (Lond.)*, **305**, 197–213.

Llinás, R., I. Z. Steinberg, and K. Walton (1976) Presynaptic calcium currents and their relation to synaptic transmission: Voltage clamp study in squid giant synapse and theoretical model for the calcium gate. *Proc. Natl. Acad. Sci. USA*, **73**, 2918–2922.

Loew, L. M., and L. Simpson (1981) Charge shift probes of membrane potential. A probable electrochromic mechanism for ASP probes on a hemispherical lipid bilayer. *Biophys. J.*, **34**, 353–365.

Loew, L. M., L. B. Cohen, B. M. Salzberg, A. L. Obaid, and F. Bezanilla (1985) Charge shift probes of membrane potential. Characterization of Aminostyrylpyridinium dyes in the squid giant axon. *Biophys. J.*, **47**, 71–77.

Meech, R. W., and F. Strumwasser (1970) Intracellular calcium injection activates potassium conductance in *Aplysia* nerve cells. *Fed. Proc.*, **29**, 834.

Neering, I. R., and R. N. McBurney (1984) Role of microsomal Ca storage in mammalian neurones? *Nature*, **309**, 158–160.

Nordmann, J. J. (1977) Ultrastructural morphometry of the rat neurohypophysis. *J. Anat.*, **123**, 213–218.

Nordmann, J. J., and J. J. Dreifuss (1972) Hormone release evoked by electrical stimulation of rat neurohypophyses in the absence of action potentials. *Brain Res.*, **45**, 604–607.

Obaid, A. L., R. K. Orkand, H. Gainer, and B. M. Salzberg (1985) Active calcium responses recorded optically from nerve terminals of the frog neurohypophysis. *J. Gen. Physiol.*, **85**, 481–489.

Parsons, T. D., A. L. Obaid, and B. M. Salzberg (1985) Light scattering changes associated with secretion from nerve terminals of the mammalian neurohypophysis are depressed by aminoglycoside antibiotics. *Biophys. J.*, **47**, 447a.

Piddington, R. W., and D. B. Sattelle (1975) Motion in nerve ganglia detected by light-beating spectroscopy. *Proc. R. Soc. Lond. B Biol. Sci.*, **190**, 415–420.

Platt, J. R. (1961) Electrochromism, a possible change in color producible in dyes by an electric field. *J. Chem. Phys.*, **34**, 862–863.

Poulain, D. A., and J. B. Wakerley (1982) Electrophysiology of hypothalamic magnocellular neurons secreting oxytocin and vasopressin. *Neuroscience*, **7**, 773–808.

Rios, E., W. Melzer, and M. F. Schneider (1983) An intrinsic optical signal is related to the calcium transient of frog skeletal muscle. *Biophys. J.*, **41**, 396a.

Rodriguez, E. M., and H. D. Dellman (1970) Hormonal content and ultrastructure of the disconnected neural lobe of the grass frog (*Rana pipiens*). *Gen. Comp. Endocrinol.*, **15**, 272–288.

Ross, W. N., B. M. Salzberg, L. B. Cohen, and H. V. Davila (1974) A large change in dye absorption during the action potential. *Biophys. J.*, **14**, 983–986.

Ross, W. N., B. M. Salzberg, L. B. Cohen, A. Grinvald, H. V. Davila, A. S. Waggoner, and C.-H. Wang (1977) Changes in absorption, fluorescence, dichroism and birefringence in stained giant axons: Optical measurement of membrane potential. *J. Membr. Biol.*, **33**, 141–183.

Rubin, R. P. (1974) *Calcium and the secretory process*. Plenum, New York, pp. 45–46.

Salzberg, B. M. (1978) Optical signals from giant axons following perfusion or superfusion with potentiometric probes. *Biol. Bull.*, **155**, 463–464.

Salzberg, B. M. (1983) "Optical recording of electrical activity in neurons using molecular probes." In J. L. Barker and J. E. McKelvey, Eds., *Current Methods in Cellular Neurobiology*, Vol. 3: *Electrophysiological Techniques*, Wiley, New York.

Salzberg, B. M., and F. Bezanilla (1983) An optical determination of the series resistance in *Loligo*. *J. Gen. Physiol.*, **82**, 807–818.

Salzberg, B. M., H. V. Davila, L. B. Cohen, and A. S. Waggoner (1972) A large change in axon fluorescence, potentially useful in the study of simple nervous systems. *Biol. Bull.*, **143**, 475.

Salzberg, B. M., L. B. Cohen, W. N. Ross, A. S. Waggoner, and C.-H. Wang (1976) New and more sensitive molecular probes of membrane potential: Simultaneous optical recordings from several cells in the central nervous system of the leech. *Biophys. J.*, **16**, 23a.

Salzberg, B. M., A. Grinvald, L. B. Cohen, H. V. Davila, and W. N. Ross (1977) Optical recording of neuronal activity in an invertebrate central nervous system: Simultaneous monitoring of several neurons. *J. Neurophysiol.*, **40**, 1281–1291.

Salzberg, B. M., A. L. Obaid, D. M. Senseman, and H. Gainer (1983) Optical recording of action potentials from vertebrate nerve terminals using potentiometric probes provides evidence for sodium and calcium components. *Nature*, **306**, 36–40.

Salzberg, B. M., A. L. Obaid, and H. Gainer (1985) Large and rapid changes in light scattering accompany secretion by nerve terminals in the mammalian neurohypophysis. *J. Gen. Physiol.*, **86**, 395–411.

Senseman, D. M., and B. M. Salzberg (1980) Electrical activity in an exocrine gland: Optical recording using a potentiometric dye. *Science*, **28**, 1269–1271.

Senseman, D. M., H. Shimizu, I. S. Horwitz, and B. M. Salzberg (1983) Multiple site optical recording of membrane potential from a salivary gland: Interaction of synaptic and electrotonic excitation. *J. Gen. Physiol.*, **81**, 887–908.

Shaw, T. I., and B. J. Newby (1972) Movement in a ganglion. *Biochim. Biophys. Acta*, **255**, 411–412.

Stämpfli, R., and B. Hille (1976) "Electrophysiology of the peripheral myelinated nerve." In R. Llinas and W. Precht, Eds., *Frog Neurobiology*, Springer, Berlin, pp. 3–32.

Standen, N. B. (1981) Ca channel inactivation by intracellular Ca injection into *Helix* neurones. *Nature*, **293**, 158–159.

Standen, N. B., and P. R. Stanfield (1978) A potential- and time-dependent blockade of inward rectification in frog skeletal muscle fibres by barium and strontium ions. *J. Physiol. (Lond.)*, **280**, 169–191.

Svensmark, O. (1961) The effect of deuterium oxide on the mechanical properties of muscle. *Acta Physiol. Scand.*, **53**, 75–84.

Tyndall, J. (1876) The optical deportment of the atmosphere in relation to the phenomena of putrefaction and infection. *Philos. Trans. R. Soc.*, **166**, 27–74.

Vergara, C., and R. Latorre (1983) Kinetics of Ca^{2+}-activated K^+ channels from rabbit muscle incorporated into planar bilayers. Evidence for Ca^{2+} and Ba^{2+} blockade. *J. Gen. Physiol.*, **82**, 543–568.

Waggoner, A. S. (1979) Dye indicators of membrane potential. *Ann. Rev. Biophys. Bioeng.*, **8**, 47–68.

Waggoner, A. S., and A. Grinvald (1977) Mechanism of rapid optical changes of potential sensitive dyes. *Ann. N.Y. Acad. Sci.*, **303**, 217–242.

CHAPTER **10**

REAL-TIME OPTICAL MAPPING OF NEURONAL ACTIVITY IN VERTEBRATE CNS IN VITRO AND IN VIVO

A. GRINVALD
M. SEGAL

Department of Neurobiology
Weizmann Institute of Science
Rehovot, Israel

U. KUHNT

Max-Planck Institut für Biophysikalische Chemie
Goettingen, West Germany

R. HILDESHEIM
A. MANKER
L. ANGLISTER

Department of Neurobiology
Weizmann Institute of Science
Rehovot, Israel

J. A. FREEMAN

Department of Anatomy
Vanderbilt University
Nashville, Tennessee

1.   INTRODUCTION   166
2.   THE PRINCIPLE OF OPTICAL MONITORING OF
     NEURONAL ACTIVITY   167
3.   RECENT APPLICATIONS OF OPTICAL RECORDING   169
     3.1   VISUALIZATION OF SPATIOTEMPORAL
           PATTERNS OF POPULATION ACTIVITY IN
           MAMMALIAN BRAIN SLICES   169
     3.2   OPTICAL RECORDING FROM ISOLATED AND
           INTACT BRAIN STRUCTURES IN VITRO   177
     3.3   REAL-TIME IMAGING OF NATURALLY
           EVOKED RESPONSES IN THE INTACT BRAIN
           IN VIVO   177
4.   TECHNICAL ASPECTS OF OPTICAL MEASUREMENTS   184
     4.1   THE APPARATUS   185
     4.2   LIMITATIONS OF THE PRESENT OPTICAL
           PROBES AND OPTICAL RECORDING   186
5.   DESIGN AND SYNTHESIS OF IMPROVED
     OPTICAL PROBES   188
6.   DIFFICULTIES WITH THE INTERPRETATION
     OF OPTICAL SIGNALS   189
     6.1   AMPLITUDE CALIBRATION   190
     6.2   ANALYSIS OF INTRACELLULAR
           POPULATION ACTIVITY   190
7.   THE SPATIAL RESOLUTION OF OPTICAL RECORDING   190
     7.1   THE MICROSCOPE RESOLUTION   191
     7.2   THE EFFECT OF LIGHT SCATTERING ON
           THE SPATIAL RESOLUTION   191
8.   CONCLUSIONS   192
     REFERENCES   193

1.   INTRODUCTION

Electrophysiological techniques have both the spatial and the temporal resolution required to investigate the real-time function of individual neurons or of neuronal networks. However, they suffer from two major limitations: For the investigation of single neurons, the use of microelectrodes usually cannot provide simultaneous recordings from more than two sites on the neuron's arborization, thus severely hampering understanding of the detailed integration function of even the basic computational elements. Similarly, the study of the cellular basis of the function of any given network requires the simultaneous recording from many individual cells, and this is usually limited to 2–3 cells or a few populations of cells. Other approaches do not share some of the powerful advantages of electrophysiological techniques; for example, the 2-deoxyglucose method (Sokoloff, 1977), which permits high-resolution localization of active neurons (Sejnowski et al., 1980, Lancet et al., 1982), has a time resolution of minutes or hours rather than milliseconds. Positron-emission tomography (Raichle, 1979), which offers three-dimensional localization of active regions in the live brain, lacks both time and spatial resolution. A newly developed magnetic imaging technique (Kaufman and Williamson, 1982) also shares some of these drawbacks.

A powerful alternative approach is offered by monitoring neuronal activity optically utilizing voltage-sensitive dyes. The initial efforts of Tasaki and his collaborators (1968), of Patrick et al. (1971), and primarily of L. B. Cohen and his colleagues have laid the foundation for this promising technique (Davila et al., 1973, 1974; Salzberg et al., 1973, 1977; Cohen et al., 1974; Ross et al., 1974, 1977; Grinvald et al., 1977, 1981a). The availability of suitable voltage-sensitive dyes and arrays of photodetectors has recently facilitated the optical monitoring of electrical activity in the processes of single nerve cells *simultaneously* from more than 100 sites, both in culture (Grinvald et al., 1981b) and in invertebrate ganglia (Ross and Krauthamer, 1984; Grinvald et al., 1986). This method also provides the unique ability of detecting activity in many individual neurons in an entire invertebrate ganglion, while a particular behavioral response is elicited (Salzberg et al., 1977; Grinvald et al., 1981; London et al., this volume). The in vitro activity of individual populations of neuronal elements (cell bodies, axons, dendrites, or nerve terminals) at many neighboring loci in mammalian brain slices (Grinvald et al., 1982c) or isolated but intact brain structures (Orbach and Cohen, 1983; Salzberg et al., 1983) has been investigated. Dynamic patterns of electrical activity evoked in the intact vertebrate (Grinvald et al., 1984) or mammalian brain (Orbach et al., 1983, 1985) by a natural stimulus were also recently monitored optically. By employing computerized optical recording and a display processor, video-displayed images of neuronal elements can be superimposed on the corresponding patterns of the optically detected electrical activity, thus allowing the spatiotemporal patterns of intracellular activity to be visualized in slow motion (Manker, 1982).

Developments and applications of the technique for vertebrate CNS preparations in vitro and in vivo are described below. Much of this topic has been already reviewed (Grinvald and Segal, 1983; Grinvald, 1985). For earlier extensive reviews, see Conti (1975), Waggoner and Grinvald (1977), Cohen et al. (1978), Cohen and Salzberg (1978), Salzberg (1983), and other chapters in this book.

2.   THE PRINCIPLE OF OPTICAL MONITORING OF NEURONAL ACTIVITY

The optical technique is simple in principle. Voltage-sensitive probe molecules are used vitally to stain a preparation. The bath-applied dye molecules bind to the external surface of excitable membrane and act as molecular transducers that transform changes in membrane potential into optical signals. General aspects of the optical recording technique are discussed in more detail by L. B. Cohen and S. Lesher in Chapter 6. The similarity between optical and intracellular recording is illustrated in the inset of Fig. 1, which depicts an optical recording (noisy trace) of an action potential and an electrical recording (smooth trace). Note that the time course of the optical recording is nearly identical to that of the electrode recording. Evidently, by using an array of photodetectors positioned in the microscrope image plane, the activity of many individual targets can be detected simultaneously.

The first computerized optical monitor of neuronal activity from 100 sites was constructed by Cohen and his colleagues (Grinvald et al., 1981). Figure 1 depicts a similar but improved apparatus (Grinvald et al., 1981b, 1982c). The improvements are discussed in detail later.

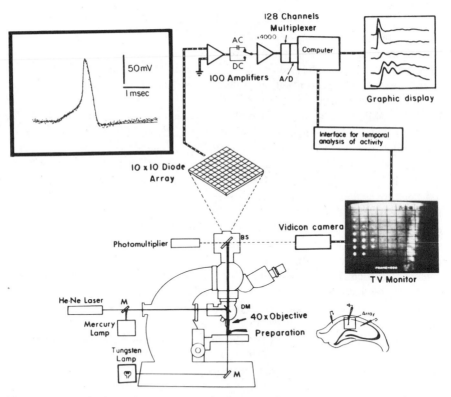

Figure 1. The computer-controlled optical apparatus used to record electrical activity from multiple locations. In transmission measurements, the preparation is illuminated by a tungsten lamp, and optically detected electrical activity is monitored with a 10 × 10 photodiode array. In fluorescence experiments, a He–Ne laser or a 100-W mercury lamp is also employed as a light source. The TV screen shows a slice preparation. The bright hexagon illustrates the spatial pattern of the electrical activity 0.8 ms after the stimulation. *Inset:* Comparison between optical and intracellular recordings from squid giant axon. Modified from Ross et al. (1977) and Grinvald et al. (1982c).

3.   RECENT APPLICATIONS OF OPTICAL RECORDING

3.1   Visualization of Spatiotemporal Patterns of Population Activity in Mammalian Brain Slices

The brain slice preparation offers considerable technical advantages to the physiologist interested in either the study of the electrical properties of single CNS neurons in their native environment or the function of local neuronal circuits in various brain regions. The advantages and limitations of this preparation were recently reviewed (Dingledine, 1983). The slice preparation is especially suitable for optical measurements because (1) it is transparent, thus transmission measurements can be made (in such preparation these are expected to provide better signal-to-noise ratio than fluorescence measurements); (2) thin slices can be prepared thus minimizing light scattering and out-of-focus blurring which hamper the three-dimensional resolution (see Sect. 7); and (3) many physiological, anatomical, and biochemical techniques can be combined simultaneously with the optical measurements from the same slice.

The application of voltage-sensitive dyes to the investigation of population activity in mammalian brain slices was attempted with rat and guinea pig hippocampus slices (Grinvald et al., 1982c; Kuhnt and Grinvald, 1982). The hippocampus was chosen for the first application of the technique in this direction because of the high degree of stratification among its various neuronal elements. Such stratification is readily observed in transverse slices (Yamamoto, 1972). In such a three-dimensional preparation, however, a given photodetector will receive light from several presynaptic and/or postsynaptic neuronal elements. Thus, optical monitoring of slice activity leads to recordings of heterogeneous intracellular population activity. Nevertheless, the high degree of organization in some preparations like the hippocampal transverse slice still permits intracellular population recordings from relatively well-defined populations of neuronal elements (i.e., dendrites, cell bodies, and axons).

Optical Signals From Stained Slices

Figure 2 depicts simultaneous recordings from 10 photodetectors oriented along the long axis of the pyramidal cells in the CA1 region. The stimulation of the stratum (s.) radiatum produced a short latency (2–4 ms), fast (3.5–6 ms) signal in a strip 90–180 μm wide, corresponding to the Schaffer collateral–commissural system (2 top traces of Fig. 2). The signals could be traced in adjacent detectors for a long distance, but only along the collateral system (not shown). These fast signals reflect action potentials in the Schaffer collateral axonal fibers. A second wave of slower excitation was detected in the same region with a latency of 4–15 ms, but this wave "traveled" along the dendritic tree toward the s. pyramidale and s. oriens. These slow signals represent the dendritic excitatory postsynaptic potentials (EPSPs). The dendritic depolarization triggered multiple optical spikes at the pyramidal cell bodies. These optical signals represent the action potential discharges there. The optical signals

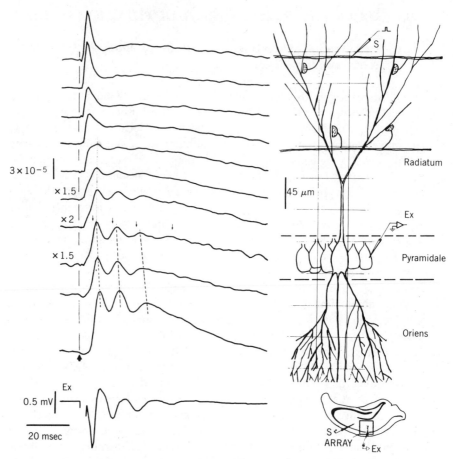

Figure 2. Optical recordings from 10 loci along the long axis of CA1 pyramidal cells. The scheme shows the various areas along the cell axis, the position of the stimulating (S) and recording (Ex) electrodes, and the relative position of the 10 individual photodetectors (light squares). A scheme of the slice in the bottom right corner shows the optically monitored area (square) and the approximate position of stimulating and recording electrodes. Twenty trials were averaged. Modified from Grinvald et al. (1982c).

propagated into the s. oriens with an average conduction velocity of 0.1 m/s. Their passive spread back into the apical dendrites was noticeable only over a short distance. The optical signals detected at the s. oriens probably reflect the activity of the basal dendrites and axons of the pyramidal cells as well as the activity of interneurons there and possibly neuroglial depolarization.

There was good correlation between the electrically recorded field potential and the optical signals from the s. pyramidale. However, it is important to note that although the optical signals are restricted to their site of origin and represent intracellular population activity, the electrically recorded field potential may spread over a large and often unpredictable area, and it represents extracellular current flow.

Light-Scattering Signals

Light scattering signals were first investigated in the squid giant axon and crab nerve. (Cohen et al., 1968, 1972; Tasaki et al., 1968; Cohen and Keynes, 1971) and in cerebral slices (Lipton, 1973). In the hippocampal slice, the light-scattering response to the stimulation had characteristics that were clearly different from the dye responses. The time course of the light-scattering signal was independent of wavelength. It was present also at the wavelength where the dye absorbs and therefore added to the dye response measured at that wavelength. The light-scattering signals produced by the stimulation had a rather slow time course, outlasting any known changes in neuronal membrane potential. The time constant for the decay of the light-scattering signal was estimated to be 250 ms. Evidently, the slow light-scattering signals are not linearly related to membrane potential changes, although their spatial and temporal characteristics were correlated with the electrical activity. The signals were generated first in s. radiatum, had a longer latency in the s. pyramidale, and had the longest in s. oriens. Interestingly, the largest magnitude

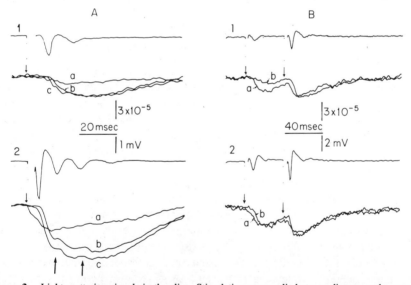

Figure 3.   Light-scattering signals in the slice. Stimulation was applied at s. radiatum, and responses from an unstained slice were recorded. (*A*) The effect of stimulus strength on light-scattering signals. (*A*1) Weak stimulus. Top trace shows the extracellular recordings from the pyramidal layer. Bottom traces are optical recordings from (*a*) the s. radiatum, (*b*) the s. oriens, and (*c*) the s. pyramidale. (*A*2) Stronger stimulus, (*a–c*) same positions as (*A*1). The thick arrows show the onset of second and third peaks of multiple discharges on the slow light-scattering signals. The thin arrows show time of stimulus onset. (*B*) Pulse–pair facilitation of the light-scattering signals. (*B*1) Responses to weak twin pulses; (*a*) is from the s. radiatum 50 μm away from the stimulating electrode; (*b*) is 300 μm away from the stimulating electrode, The electrical field potential recording is from the s. pyramidale. (*B*2) Light-scattering responses to stronger stimuli. Same as (*B*1), except that the facilitation is smaller. Twenty trials were averaged. Modified from Grinvald et al. (1982c).

change was found in the s. pyramidale (Figure 3*A*2). Both the magnitude and latency to onset of the light-scattering response were dependent on the stimulation intensity. With low stimulation intensity, the light-scattering responses were small and had a slow rise time, whereas high stimulation intensity resulted in light-scattering responses with shorter latencies, faster rise times, and larger amplitudes (cf. Fig. 3*A*1 with 3*A*2). The onset of the light-scattering responses in the s. pyramidale had the same latency as that of the population spike (Fig. 3*A*2).

When twin pulse stimulation was applied, the light-scattering response to the second stimulus was often augmented in comparison with that of the first stimulus. This was more noticeable in the s. radiatum far away from the stimulation site than next to the electrode, and only at moderate stimulus strength (Fig. 3*B*). Reversal of the stimulus polarity led to a marked decrease in the size of the light-scattering signals as well as in that of the electrically recorded population spike. Application of tetrodotoxin (TTX) abolished the light-scattering signals.

In summary, the light-scattering signals were triggered by the physiological activity and were not merely an artifact of stimulation. Therefore, their further investigation may provide useful information. A thorough investigation of light scattering changes that accompany secretion by nerve terminals has been reported recently (Salzberg et al., 1985).

Separation of Pre- and Postsynaptic Components

To test if the fast signals represent the presynaptic action potentials and if the long-latency slow signals indeed represent EPSPs generated in the apical dendrites of CA1 pyramidal neurons, a low-Ca^{2+} (0.2 mM), high-Mg^{2+} (4.0 mM) solution was used to block postsynaptic activity. The field potential was monitored continuously and was found to be completely suppressed after 20 min, at which time optical responses were measured (Fig. 4). Whereas the initial fast response was somewhat enhanced by the low-Ca^{2+} medium, the long-latency slow response was completely blocked. The occurrence of fast signals was restricted to the region of the Schaffer collaterals (over an inspected distance larger than 900 μm). The increase in the spike size may reflect recruitment of additional presynaptic elements in the low-Ca^{2+} medium, possibly owing to increased excitability. These effects were reversible, and the original response was restored by perfusion with the normal medium. The application of TTX ($5 \times 10^{-6} M$) abolished the optical responses simultaneously with the disappearance of the electrical responses (Fig. 4). These experiments also indicate that the fast signals are indeed Na^+, and not locally evoked dendritic Ca^{2+}, action potentials.

Properties of the Unmyelinated Axons in the Hippocampus Slice

In the absence of Ca^{2+}, focal stimulation of s. radiatum evoked activity mostly in the unmyelinated axons. Far away from the point of stimulation, activity is detected exclusively from these axons and their terminals. The experiments discussed below demonstrated for the first time that optical recording can readily provide information

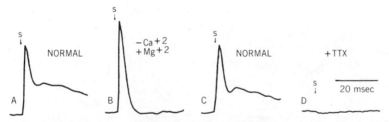

Figure 4.   Optical recording from presynaptic elements. (*A*) Optical and electrical recordings in normal medium from the s. radiatum. (*B*) Optical recordings in a medium containing 0.2 mM Ca^{2+} and 4.0 mM Mg^{2+}. (*C*) Optical recording 20 min after the low-Ca^{2+} medium was replaced by the normal medium. The stimulating electrode was less than 150 μm away from the optical recording site. (*D*) The effects of TTX on the optical responses. Small arrows marked by s: time of stimulation. Modified from Grinvald et al. (1982c).

regarding the intracellular activity even in 0.1 μm axons that cannot be impaled with microelectrodes.

Effects of Tetraethylammonnium.   TEA (2 mM) was applied in the incubation medium, and its effect on the evoked optical responses was assessed. A broadening of the fast responses was observed; there was no change in latency or magnitude of the evoked optical response, but it decayed more slowly than under control conditions. The control response could be restored after 20–30 min of washing with normal medium.

Conduction Velocity.   The conduction velocity along the Schaffer collateral system was estimated by measuring the delay between peaks of the fast responses measured at different locations along the collaterals. The microscope field of view was shifted, and three adjacent fields were sampled, yielding a total length of over 1 mm of the fibers; a conduction velocity of 0.2 m/s was determined.

Refractory Period.   The refractory period of the activated axons was estimated by applying two pulses at various interpulse intervals. There was almost no response to the second stimulus when it was applied less than 4 ms after the first one and only a partial response when it was applied 4 ms after the first one; a full response was obtained with a 7-ms delay between the pulses. It appears that the refractory period is in the range of 3–4 ms.

Visualization of the Spread of Activity in Slices

Figure 5 demonstrates the main advantage of optical recording—that is, the feasibility of simultaneous recording from many (hundreds of) neighboring loci. The figure illustrates the spread of focal excitation along the Schaffer collateral commissural axons. Subsequent postsynaptic spread down the apical dendrites of the CA1 neurons initiated the multiple discharge at the axonal hillock region of the pyramidal cells, which continued toward the s. oriens. Occasionally, the dendritic depolarization was

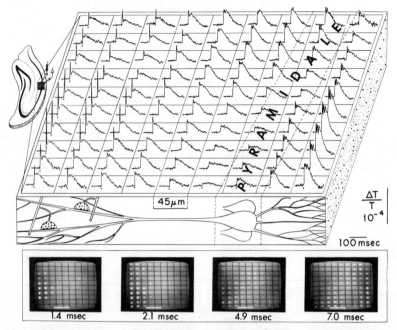

Figure 5. Imaging of electrical activity in hippocampal slice. *Top left:* Scheme of the slice showing the area monitored by the array. *Top:* Optical traces from 9 × 10 detectors are shown at their appropriate locations in the field of view. Activity was evoked by stimulation of the Schaffer collateral–commissural system. Twenty trials were averaged. Each detector samples an area of 45 × 45 μm. *Bottom:* The four video frames show the imaging of neuronal activity performed by the display processor. The appropriate time is marked below each frame. The superimposed video picture of the preparation was omitted for clarity. Modified from Grinvald et al. (1982c).

not the largest in size at the location of the synapses in the s. radiatum (e.g., top of columns 5–7). One possible explanation is that Ca^{2+} action potentials were evoked by the EPSPs. The slow after-hyperpolarization and the shape of the dendritic responses support this interpretation, in line with other recent observations (Schwartzkroin and Slawsky, 1977; Wong et al., 1979; Llinas and Sugimori, 1980). The experiment illustrated in Fig. 5 indicates that optical recording has facilitated for the first time the detection of dynamic patterns of electrical activity of cell populations; moreover, these patterns are heterogeneous even in a relatively small area in the highly ordered CA1 region. Fast analysis of the data presented in Fig. 5 to obtain spatiotemporal information was facilitated by the display controller, which affords an on-line, slow-motion visualization of the electrical activity. The imaging of electrical activity at four different time intervals is shown in the bottom of Fig. 5.

Other findings of these studies (Grinvald et al., 1982c; Kuhnt and Grinvald, 1982) are as follows: (1) 20 μM of 4-aminopyridine (4-AP) induces broadening of the presynaptic action potentials in s. radiatum. This broadening depends on the external Ca^{2+} concentration. In normal solution it led to a dramatic increase in the postsynaptic response and to its facilitation by a second stimulus, (2) 4-AP also broadens the action potentials in the myelinated axons in the alveus and oriens, suggesting that potassium channels are active at the nodal region, (3) Picrotoxin increases postsynaptic response at the s. pyramidale as well as in the s. radiatum, suggesting that tonic inhibitory synapses are located there also, (4) In preliminary experiments it was found that long-term potentiation (LTP) is correlated with both pre- and postsynaptic changes as well as reduction of the inhibition (M. Segal and A. Grinvald, unpublished results).

Local circuits can probably be investigated in slices by optical recording of the responses to intracellular stimulation of single identified cells in a manner similar to the investigations performed on invertebrate ganglia (Grinvald et al., 1981a; Ross and Krauthamer, 1984). Such potentially useful experiments were not yet reported, however.

Iontophoretic Injection of Fluorescent Voltage-Sensitive Probes; Recording From Processes of Single Neurons

To record electrical activity or synaptic responses optically from the site of the synapses of single nerve cells of intact CNS preparations, voltage-sensitive dyes should be intracellularly injected, thereby staining only the cell under investigation. (Signals may be obtained when the membrane-bound probe is inside or outside, but such signals are of opposite sign (Cohen et al., 1974; Gupta et al., 1981.) This approach was pioneered by Salzberg; however, none of the dyes that he tested were adequate for the recording of optical signals from the processes. To this end we have designed and synthesized voltage-sensitive fluorescent probes optimized for iontophoretic injection (Agmon et al., 1982; Grinvald et al., 1982a). The experiments, to be described below, were performed on leech ganglia. In principle, however, this approach is also suitable for the slice preparation. We have tried only fluorescence rather than absorption probes because theoretical calculations (Agmon, 1982) predict that fluorescence measurements would provide better signal-to-noise ratio (SNR).

Leech neurons were injected with such a probe (e.g., RH-461). The dye diffused through the processes without adversely affecting the cell viability, electrical properties, or synaptic input. Changes in membrane potential in a given illuminated segment of the neuronal arborization in the neuropil were monitored optically during cell body stimulation. Furthermore, synaptic responses, such as those at the site of the synapses between the P sensory cell and the injected L or AE motor neurons, could also be recorded (Fig. 6). As expected, the synaptic potential recorded from the neuropil has a faster rise time (A. Grinvald, B. M. Salzberg, R. Hildesheim and V. Lev-Ram, submitted). This technique is more difficult to use than recording from the processes of isolated culture cells, because the background fluorescence from the intracellular binding sites reduces the SNR 10- to 20-fold. In addition, autofluorescence of some preparations may further decrease the relative size of the signal. Nevertheless, intracellular dye injection may be indispensable for the investigation of remote dendrites, which might exhibit semiautonomous electrical behavior, not reflected at all in the distant cell body.

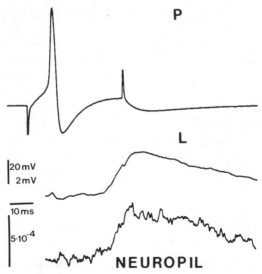

Figure 6.   Fluorescence recording of synaptic potential from the processes of a single cell. The L cell in the leech segmental ganglion was iontophoretically injected with RH-461 for 15 min (0.3 nA). The dye diffused in the processes for 1 h. Then the presynaptic P sensory cell in the posterior ganglion was impaled and stimulated (top trace) while a synaptic response was recorded from the cell body of the injected L cell using a second microelectrode. The noisy trace is the fluorescence recording from the contralateral neuropil, 200 μm away from the L cell body. Modified from Grinvald et al. (1986).

Fluorescence Recording From Hippocampal Slices

Large optical signals were recently recorded from hippocampal slices stained with the fluorescent probe RH-414 or other fluorescent probes. The signal-to-noise ratio obtained in those experiments was about 100-to-150 (RMS noise) without signal averaging. The fractional change in fluorescence was 2-6 $\times$ 10^{-3} (Grinvald et al., 1985). Fluorescence recording from slices may have two advantages: (1) Thick slices can be investigated because transparency of the slices is no longer an essential requirement. However, the optical signals will be recorded only from the surface of the slice facing the objective, up to depth of 200–600 μm. (2) The signal-to-noise ratio can be readily improved by 2- to 10-fold by increasing the illumination intensity (e.g., Grinvald et al., 1983). Thus it is predicted that activity of single cells and their arborization can be monitored in fluorescence experiments with the present dyes. Transmission experiments should, in principle, be even more suitable for such experiments (see p. 298, Grinvald, 1985); however, it seems that the present transmission probes used in slices still do not provide the optimal signal size. Indeed, in recent transmission experiments with RH 155 we achieved somewhat better signal to noise ratio relative to the fluorescence measurements. The fractional change in transmission was 5-6 $\times$ 10^{-4}. The recent improvements in both transmission and fluorescence experiments reported here, is also attributed to improvements in the procedures for cutting slices with the egg slicer developed by Dr. L. Katz, and to their better maintenance. (In the recent studies the field potentials vary from 6 to 20 mV, whereas in the past experiments they were only 0.5 to 2 mV.)

3.2. Optical Recording from Isolated and Intact Brain Structures in Vitro

Experiments similar to those described above were also carried out on isolated, relatively transparent brain structures. In such preparations one would expect signals to be even larger than in slices, because a larger fraction of cells and synaptic connections are intact. Recently, Orbach and Cohen (1983) demonstrated that the isolated salamander olfactory bulb can be investigated using both transmission and fluorescence measurements. Salzberg et al. (1983) investigated the ionic mechanisms underlying the electrical activity of a population of nerve terminals in the isolated neurohypophysis of *Xenopus*. This preparation offers a unique opportunity for the study of both excitation–secretion coupling and electrical properties of vertebrate nerve terminals, because the majority of its excitable membrane area is composed of the terminals themselves. Very large optical signals were observed in these experiments. They are described in detail by Salzberg et al. in Chapter 9.

3.3. Real-Time Imaging of Naturally Evoked Responses in the Intact Brain In Vivo

The detection of patterns of population activity in mammalian brain slices suggested that the optical technique could be applied to the investigation of mammalian brain tissue and that intracellular population recordings still provide meaningful information (Grinvald et al. 1982c). Orbach and Cohen (1983) were the first to demonstrate that large fluorescence signals can be obtained from the salamander olfactory bulb in vivo, after topical application of styryl dyes. For the in vivo experiment, the top of the skull was removed from midway along the telencephalic hemispheres forward, and the salamander was put on the stage of the microscope. A suction electrode was used to stimulate the cut olfactory nerve. Signals from in vivo bulbs were obtained using both transmission (not shown) and epi-illumination, using the fluorescence of dyes RH160 and RH414. The traces in Fig. 7 illustrate the outputs of a portion of the diode array from an experiment using RH414. The signals measured from the in vivo bulb have sizes and time courses similar to those measured in vitro. The in vivo measurements (Fig. 7) did have somewhat more low frequency noise than the in vitro results, but the signal-to-noise ratio remains relatively large in single trials.

The traces shown in Fig. 7 were scaled by dividing the changes in intensity by the resting fluorescence at each detector. Thus the size of the signal in each trace represents the relative size of the fractional change in fluorescence, $\Delta F/F$ ($\sim 1.3 \times 10^{-2}$. These results suggested that the optical method could prove a powerful approach for studying functional organization in the intact mammalian brain as well. Orbach et al. exploited the technique to investigate the rat visual and the

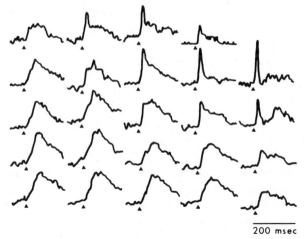

200 msec

Figure 7.   Fluorescence measurements from 25 detectors from an in vivo salamander olfactory bulb. The olfactory nerve was stimulated with a suction electrode using a 0.3-ms pulse at the times indicated by the *triangles*. A $\times 7$, 0.2-NA objective was used. The resting light intensity reaching the photodetectors was about $7 \cdot 10^7$ photons/ms, approximately 100 times less than the resting intensity in absorption measurements. From Orbach and Cohen (1983).

somatosensory cortex (1985). Furthermore, they have shown that meaningful optical signals can also be detected in response to a natural physiological stimulus. Responses to an odorant stimulus were also recently obtained from the salamander olfactory bulb (Kauer et al., 1984).

Mapping of Electrical Activity in the Mammalian Cortex

In vivo measurements on the mammalian cortex are expected to be difficult because of the large noise from movements of the brain due to heartbeat and respiration and because of the relative opacity of the cortex. The dense packing of the neuronal and glial elements may prevent proper staining with sensitive but hydrophobic optical probes. The very large noise from the heart pulsation ($\Delta I/I = 10^{-2}-10^{-3}$) was much greater than the evoked optical signals. However, this noise is synchronized with the heartbeat and the electrocardiogram (ECG). It was therefore relatively easy to reduce it by subtracting the result of a trial with a stimulus from a subsequent trial without a stimulus. (Both trials were triggered by the peak of the electrocardiogram.) This procedure reduced the noise by about a factor of 10. (Fig. 8*b*). Noise from breathing movements was reduced by holding the rat head tightly with a snout clamp and ear bars. In signal-averaging experiments, further improvement was achieved by a computer program that allowed the rejection of exceptionally noisy trials from the accumulated average (Grinvald et al., 1982b; Orbach et al., 1985).

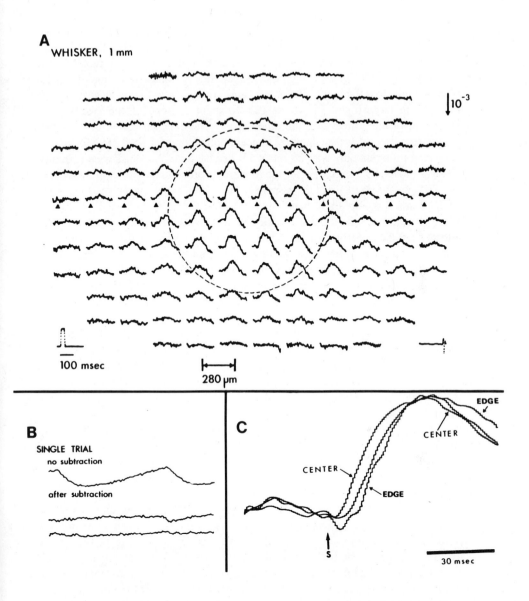

Figure 8. (A) Optical detection of a whisker barrel in the rat somatosensory cortex that results from a 1-mm whisker movement. The inset at lower left shows the timing of the current to the galvanometer (stimulation). The inset at lower right shows the results of the electrocardiogram subtraction. Thirty-two trials were averaged. The interval between stimuli was 9 s. Modified from Orbach et al. (1985). (B) Removal of the heart beat noise by subtraction. (C) Comparison of the time course of the optical signals in the center of the barrel and its periphery.

The whisker barrels (see Woolsey and Van der Loos, 1970) in the rat somatosensory cortex proved to be amenable to optical investigation. Relatively large signals were obtained in response to a small whisker deflection of 0.3 mm. Under the best conditions, such signals could be detected without signal averaging, and about 200 trials could be reliably repeated in the same cortical area. Figure 8 illustrates the results of an experiment in which 124 adjacent regions of the cortex were simultaneously monitored with the diode array. The exposed cortex (4 × 4 mm diameter craniotomy) was stained by topical application of 1 mM RH-414 for 120 min. (The transparent dura was left intact.) When the tip of the whisker B1 (Simons, 1978) was moved by 1 mm, optical signals were recorded in the center of the field of view. The average diameter of the response area in such experiments was about 1300 μm. This area is much larger than that of the whisker barrels in layer IV (300–600 μm). Nevertheless, these are the expected results if the optical signals indeed originate from the underlying barrel. First, most of the optical signals originate from layers I–III, and it is known that the processes from the somata in a single barrel extend to at least two neighboring barrels; second, stimulation of one whisker evoked activity of neurons in neighboring barrels (Simons, 1978); third, light scattering and out-of-focus fluorescence tend to increase the apparent area of the evoked activity (Salzberg et al., 1977; Orbach and Cohen, 1983).

Stimulation of two whiskers that are relatively far apart (e.g., A1 and D4) evoked responses in two circumscribed areas. A large overlap in the response area was obtained when neighboring whiskers were stimulated, but the responses could be resolved if they were activated at different times. Other findings of this study are as follows: (1) In most experiments, signals in the center zone were faster than those in the periphery (see Fig. 8), suggesting that the periphery signals indeed originate from processes extending out of the barrel, and from interneurons or postsynaptic cells in neighboring barrels; (2) larger amplitudes of whisker deflection evoked larger activity, which also spread over larger cortical areas; (3) activity evoked by one whisker often inhibited the activity evoked by a neighboring whisker when the interstimulus interval was about 20–120 ms.

The optical signals from the rat cortex were relatively slow and smooth in contradistinction to the optical signals originating from visually evoked reponses in the frog optic tectum (see below) or the electrically evoked responses in the salamander olfactory bulb (Orbach and Cohen, 1983). Thus, at present, optical signals recorded from mammalian cortex have far less information content than those recorded from lower vertebrates. The slow and smooth nature of the optical signals could be explained by (1) assuming that the average intracellular population activity in the cortex should appear smeared and slow because of the number of relay stations involved or (2) assuming that a large contribution from glia depolarization exists. Distinction between these two alternatives has not been made. Slow signals that are of glial origin rather than axonal origin were recently recorded from rat optic nerve stained with RH-414 (Lev-Ram and Grinvald, 1984).

We also attempted to determine whether optical imaging of neuronal activity in the cortex could be useful for the investigation of spontaneous epileptic activity in the cortex. Such experiments are more demanding because the epileptic activity must be

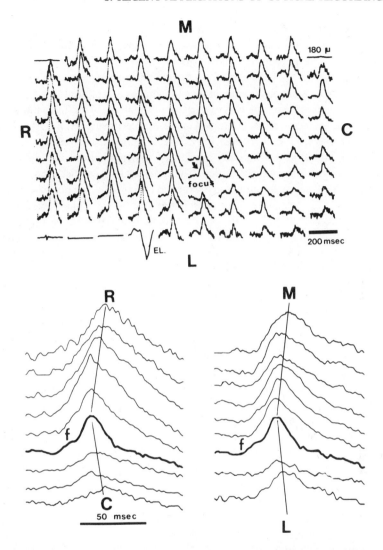

Figure 9.   Detection of an epileptic focus in the rat visual cortex. Epileptic-like seizure was induced by topical application of 0.1 mM bicuculline. The spontaneous epileptic discharge was detected from the visual cortex without signal averaging. The detected epileptic focus is marked with an arrow. *Bottom:* The spread of activity from the presumed focus (f) along the rostral caudal axis (left side) and along the medial lateral axis (right side). Modified from Orbach et al. (1985).

recorded without signal averaging. When spontaneous epileptiform discharges were induced in the rat visual cortex by topical application of bicuculline they could be detected optically (see Fig. 9). Furthermore, because of the time resolution of the technique, the locus of the epileptic focus could be detected as well.

In preliminary experiments on the rat visual cortex (Orbach et al., 1982), fluorescence signals were detected from a large area of the cortex when a diffused flash of white light or a flashing checkerboard was used as a stimulus. The response was relatively uniform in the entire field of view, even when a 10° light flash was used. Because more appropriate visual stimuli were not tried, it is impossible to conclude whether the method can be useful for the investigation of the rat visual cortex. More appropriate physiological stimuli were used in preliminary experiments on the cat visual cortex (moving bars of variable orientation, black and white stripes). Reliable optical signals were not obtained in these experiments (H. S. Orbach, L. B. Cohen, A. Grinvald, C. Gilbert, and T. N. Wiesel, unpublished results). However, when cortical inhibition was partially blocked by topical application of bicuculline, relatively large responses were recorded. These results suggest that even though the sensitivity of the dye may be similar in the rat and cat cortices, the noise level and the sensitivity of the dye tested were not sufficient to resolve the signals from the small fraction of neurons that changed their activity in the cat experiments. Indeed, more recently the optical mapping of visually evoked patterns of activity in the cat and the monkey cortex, was substantially improved. This was achieved by testing about 40 new dyes and implementing several technical improvements for the staining and recording procedures, including software and hardware modifications. The observed patterns were clearly sensitive to various parameters of the visual stimuli used (Grinvald et al., 1985).

Mapping of Visually Evoked Responses in the Frog Optic Tectum

Optical studies of the intact frog optic tectum (Anglister and Grinvald, 1983; Grinvald et al., 1984) suggest that at present such a preparation may be more suitable for optical experiments partly because of its relative transparency. When a flash of light was presented to the frog's eye, large fluorescence signals were detected from the exposed and topically stained contralateral tectum (without signal averaging; see Fig. 10). Display of light patterns to the frog's eye, on an oscilloscope screen (moving or stationary spots, bars, or annuli) evoked complex response patterns in the tectum. The optical signals from a single detector were highly structured; the number of resolved components from a single detector was similar to the number of components in the corresponding electrically recorded evoked potential (Fig. 10B). Moreover, in this preparation, optical recording of the activity at different depths, by simply changing the microscope focus, allowed a partial laminar analysis to be performed (Fig. 10C).

The retinotectal connections are known to possess a large degree of topographical order. A point stimulus in the visual field evokes activity in a restricted area in the tectum. We characterized the present spatial resolution and sensitivity of the technique by presenting discrete and weak stimuli: A spot of light was first positioned on a corner of a monitor screen for 100 ms and then after a delay of 210 ms in the opposite corner (26° displacement) for another 150 ms. Figure 11 illustrates that two excitation foci (circled areas) at the tectum could be resolved. Both had a latency of ~ 100 ms with respect to the onset of each light stimulus. Additional experiments showed that inhibitory interactions shaped the size of the excitation vertical column;

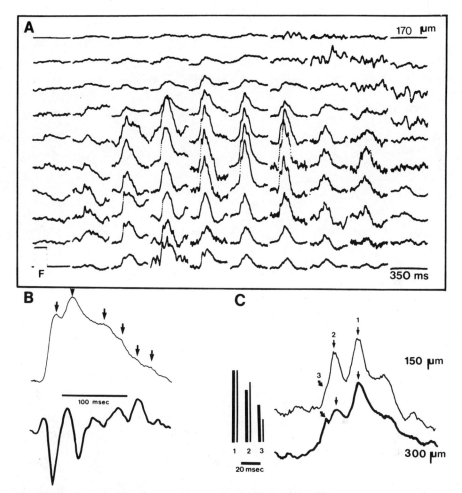

Figure 10. (A) Fluorescence recording of naturally evoked responses in the tectum. The outputs of 96 photodetectors are shown at their relative locations. Signal averaging was not used. A 150-ms light flash stimulus (shown in F) was presented to the contralateral eye. (B) Comparison of optical intracellular population recording with the electrically recorded, surface-evoked potential. (C) Partial laminar analysis. Comparison between the output of a photodetector in two consecutive experiments at two different depths of 150-μm and 300-μm. A 40× water immersion objective was used (Modified from Grinvald et al., 1984).

blocking of inhibitory interactions with drugs led to a considerable increase (10-fold or more) of the excitation area. The conduction velocity of the bicuculline-induced spread of excitation from the excited foci was found to be faster along the rostrocaudal axis than along the mediolateral one. This behavior may reflect asymmetry in the long-range synaptic interactions in the tectum. Other explanations have not yet been ruled out (Grinvald et al., 1984).

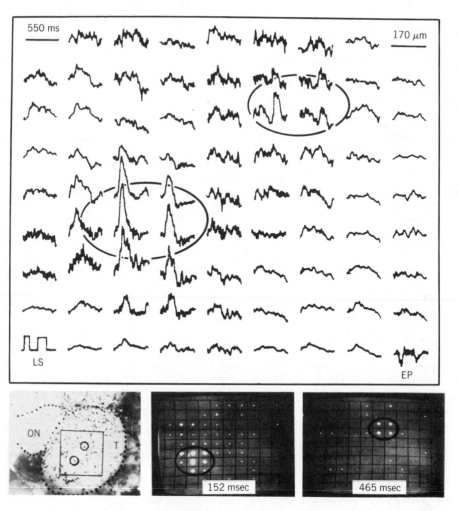

Figure 11. Real-time mapping of visually evoked responses in the intact frog tectum. *Top:* Responses from 9 × 9 photodetectors positioned over the tectum are shown. Two loci of activity could be resolved corresponding to the two different light spots presented one after the other to the contralateral eye. Forty trials were averaged. *Bottom:* Visualization of the spatial patterns of activity at two different times, depicting the two loci of activity. The superimposed picture of the tectum was omitted for clarity and is shown at the left. ON, optic nerve; T, tectum. From Grinvald et al. (1984).

4.   TECHNICAL ASPECTS OF OPTICAL MEASUREMENTS

The development of the optical technique has required multidisciplinary efforts involving organic chemistry, spectroscopy, optics, electronics, computer programming, and neurophysiology. In practice the use of the technique is not always easy. Technical aspects of optical measurements and further improvements are described below.

4.1. The Apparatus

The apparatus has been discussed elsewhere (Grinvald et al., 1981b, 1982c; Salzberg et al., 1983; Orbach and Cohen, 1983). Below we discuss only the recent improvements. In order to be able to measure a DC signal, we had a bank of new amplifiers constructed. The simple analogue circuit used was described elsewhere (Grinvald et al., 1982b). Each individual amplifier contains an analog sample-and-hold circuit to measure the light intensity at the onset of the measurements after the optical shutter is opened. This value is then differentially subtracted from the light signal before it is further amplified ($\times$ 1–4000). In this way a high-gain auto-DC-offset recording can be obtained rather than an AC recording. Furthermore, the shutter can be opened only for the duration of the measurements, thus minimizing bleaching and photodynamic damage (discussed below). The analogue sample-and-hold circuit suffers from droop, which becomes significant when small signals are measured. (The droop can be corrected by a subtraction procedure that must be implemented anyway in DC recordings, in order to correct for the bleaching time course.) Another alternative is to use a digital sample-and-hold circuit utilizing, back-to-back, analog-to-digital (A/D), and digital-to-analog (D/A) conversion (Senseman and Salzberg, 1980). Amplifiers with an auto-zero tracking A/D conversion are now being designed (J. Meyer and W. Singer, personal communication). The outputs of the amplifiers are multiplexed and digitized by two multiplexer 8-bit A/D converter cards and deposited in the PDP 11/34 computer memory. The time resolution thus obtained is 0.6 ms for each trace. Twelve-bit A/D conversion has also been used, but it requires twice the computer memory (for 16-bit computers). Unless a digital multiplexer is used, it is also two times slower. An improved time resolution (0.3 ms) was recently achieved by the construction of a digital multiplexer collecting sequential data from four 32-channel A/D cards (A. Grinvald, unpublished results) or by the use of faster A/D multiplexers (e.g., Data Translation; W. N. Ross, personal communication). In some low-light-level fluorescence experiments, a single photodiode or a photomultiplier was used instead of the array (Salzberg et al., 1973; Grinvald et al., 1982b, 1983). A fivefold reduction in the photodiode-amplifier noise was obtained by the use of a large feedback resistor on the first stage of the current-to-voltage converter (200 MΩ–1 GΩ; see Grinvald et al., 1983; J. Pine and A. Grinvald, unpublished observation).

To combine electrophysiological recordings with the optical experiment, up to four hydraulic Narashige micromanipulators are mounted on a supporting plate attached to the movable stage of the Universal microscope. Recently, we have also used an inverted microscope for transmission (Kuhnt and Grinvald, 1982; U. Kuhnt, in preparation) and fluorescence (Grinvald et al., 1983) experiments. The inverted microscope was mounted on an X–Y positioner (e.g., Calvet and Calvet, 1981) and its stage replaced by a large, flat stage rigidly attached to the vibration isolation table top. This arrangement offers four advantages:

(1.) Heavy manipulators can be mounted on the fixed stage. The insertion of microelectrodes into the preparation is done with the aid of a swing-in stereomicroscope that replaces the swing-out condenser of the inverted microscope. With a 0.63 numerical aperture (NA) condenser, the microelectrode angle can be as large as 45° and the working distance is 10 times larger (instead of the 27° and 1.6 mm working distance permitted by the water immersion objective used in the Universal microscope).

(2.) It is easy to change the objectives under the preparation to attain the optimal magnification without interfering with intracellular recordings.

(3.) High numerical aperture objectives with relatively short working distances can be used, because the electrode impalements are made from the side of the condenser. (Custom-made condensers with a larger working distance can also be obtained.) The use of such objectives provides an improvement both in the SNR and in the spatial resolution (see below).

(4.) Under these conditions, the optical measurements can be made during fast perfusion of the preparation with various physiological solutions. Our experience suggests that the use of the inverted microscope is preferable whenever applicable (e.g., for tissue cultures, brain slices, small and thin invertebrate ganglia). However, for optical recording from the intact brain of an anesthesized animal or from large invertebrate ganglia, an upright microscope should be used.

In the optical experiments, it is important to know which part of the preparation is viewed by each photodetector element. Therefore, the preparation is also viewed with a Vidicon camera; the video image is stored on video tape to further identify the location of the elements of the photodiode array with respect to the preparation image. Inspection of the large amount of optical data is time-consuming and slows the experiment. For fast analysis of the observed patterns of activity, a computer-to-video imaging processor having video frame buffer was designed (Manker, 1982; Grinvald et al., 1982c). It permits visualization of the pattern of electrical activity on a TV screen in slow motion. The display processor projects a calibrated outline of the array on the TV monitor, superimposed on the television picture of the preparation. In addition, a bright symbol is displayed in each of the 100 elements of the picture. The sizes of these symbols are proportional to the amplitude of the electrical activity at a given time. The interface displays the temporal activity from all the detectors "simultaneously" for each time interval, thus providing the imaging of the detected activity. The display processor was interfaced with the PDP-11/34 by a DR11K I/O interface. Detailed information about the hardware and software used for the various optical measurements is available (L. B. Cohen, B. M. Salzberg, W. N. Ross, and A. Grinvald, personal communication).

4.2. Limitations of the Present Optical Probes and Optical Recording

There are some difficulties with the use of the optical technique that warrant detailed discussion. Intimate understanding of them is imperative for successful and optimal use of the technique.

Signal Size

The size of the optical signals is often small; therefore, whenever intracellular recording is easy, it is preferable. For example, in the present brain-slice experiments, spontaneous activity of single cells could not be unequivocally identified. The fractional change in transmission $\Delta T/T$, for the detection of an action potential from a single mammalian neuron maintained in culture, or in invertebrate single cells, was in the range of 2-4 $\times$ 10^{-4} (Grinvald et al., 1981a, b). But $\Delta T/T$ for action potentials from a population of pyramidal cells was only 6 x 10^{-4}. In transmission measurements, one would expect a much larger fractional change when many cells are active (e.g. Salzberg et al., 1983). It has been reported that the sensitivity of an optical probe may depend on the preparation (Ross and Reichardt, 1979; Gupta et al., 1981). Thus, more sensitive optical probes probably already exist for mammalian brain slices and more dyes should therefore be tested.

Light Scattering

If the light absorption or fluorecence signals are small, the activity-dependent (dye-unrelated) light-scattering signals from the preparation (Cohen et al., 1968, 1971, 1972; Lipton, 1973) may distort the voltage-sensitive optical signals. Large and slow light-scattering signals were observed in unstained hippocampal slices (Grinvald et al., 1982c) and in the frog optic tectum. Such signals were observed in at least some loci in the salamander olfatory bulb (Orbach and Cohen, 1983). A solution to this problem was provided by substracting the light-scattering signals from the optical response (Grinvald et al., 1982c). Furthermore, in the guinea pig experiment with the more sensitive dye RH-155, the light-scattering signals were small relative to the optical signals (Kuhnt and Grinvald, 1982; U. Kuhnt, in preparation). In fluorescence experiments the factional change is about 10-fold larger than in transmission experiments, and therefore the contribution of light-scattering is negligible.

Pharmacological Side Effects

Light-independent pharmacological side effects can be expected whenever extrinsic probe molecules are bound to the neuronal membrane, especially if high concentrations of dye are used. The dye binding may change the threshold, the specific ionic conductances, synaptic transmission, membrane resistance, etc. After careful choice of the proper dye, significant pharmacological side effects were not observed in many of the above studies. However, more stringent tests using intracellular recordings are required to assess pharmacological effects, and careful controls are always essential. More than 200 voltage-sensitive probes are now available with very different chemical structures and net charges. It is unlikely that all of them will cause similar pharmacological side effects. An example has indeed been reported (Grinvald et al., 1981a) in which proper choice of the voltage-sensitive probe eliminated pharmacological side effects on a behavioral reflex mediated by a polysynaptic pathway. It was also reported that pharmacological side effects may be minimized if the staining is done with a modified saline solution containing high concentrations of divalent cations (Senseman et al., 1983).

Photodynamic Damage

The dye molecules, in the presence of intense illumination, sensitize the formation of reactive singlet oxygen and other free radicals. These reactive radicals attack membrane components and damage the cells (Pooler, 1972; Cohen et al., 1974; Ross et al., 1977). Such photodynamic damage limits the duration of the experiments. Using the present transmission probes, continuous measurements can be made for 1–5 min without marked damage. If the duration of each trial is 50 ms (DC recording only), then 1200–6000 trials can be carried out before significant damage occurs. This is precisely the reason it is important to construct DC amplifiers as described in the instrumentation section. In fluorescence experiments with brighter light sources (lasers, mercury lamp, etc.), the duration of a reliable experiment is often limited to a total of 5–10 s only. The use of antioxidants (radical scavengers) has been suggested (Oxford et al., 1978), but in one tested case it also reduced the signal size (Salzberg, 1978). Because very large numbers of radical scavengers exist, including enzymes, but only a few were tested, this approach is still promising. Thus, a systematic evaluation of the usefulness of radical scavengers in optical experiments may be rewarding. It is important to note that the extent of photodynamic damage (or bleaching) may depend also on the preparation (cf., e.g., Ross and Reichardt, 1979, or Grinvald et al., 1981b, with Ross and Krauthamer, 1984, or Salzberg et al., 1973, with Salama and Morad, 1976).

Bleaching during the optical measurement also limits the duration of reliable experiments. To minimize bleaching, the exposure time of the preparation to light should be reduced to a minimum. The dye bleaching may also affect the time course of the optical signals, especially in fluorescence experiments when bright light sources are used. A solution to this problem has been described (Grinvald et al., 1982b). Subtraction procedures of the type used to remove the heartbeat noise are effective as well.

5. DESIGN AND SYNTHESIS OF IMPROVED OPTICAL PROBES

Because of the above difficulties, the key to the success of optical monitoring of neuronal activity has been the design of adequate voltage-sensitive probes. Of more than 1800 dyes already tested (Cohen et al., 1974; Ross et al., 1977; Gupta et al., 1981; Loew et al., 1981; Grinvald et al., 1982b, 1983), fewer than 200 have proved to be sensitive indicators of membrane potential while causing minimal pharmacological side effects or light-induced photochemical damage to the neurons. Considerable improvement has recently been achieved in the quality of fluorescent probes, especially, styryl dyes (Cohen et al., 1974; Grinvald et al., 1978, 1982b, 1983; Loew et al., 1978, 1979, 1984). For example, the best fluorescence dye for neuroblastoma cells is designated RH-421; the change in fluorescence intensity with this dye is 25%/100 mV of membrane potential change (Grinvald et al., 1983). Furthermore, the photodynamic damage with this dye was reduced by a factor of 200 relative to that of merocyanine 540 (Salzberg et al., 1973). Another useful family of probes are asymmetrical oxonol dye analogues of WW-781 designed recently (Grinvald et al., 1978; Gupta et al., 1981). In the design of improved transmission voltage-sensitive probes, progress has been slower, presumably because for some preparations the sensitivity of the best probes is already close to the theoretical maximum (Waggoner and Grinvald, 1977). A significant limitation of the optical probes is that a dye's sensitivity can vary from one preparation to the next and even among different species of the same genus (Ross and Krauthamer, 1984). Thus for a new preparation, careful selection of the best probe is an important prerequisite. The availability of several "kits" of close analogues of the best types of dyes has greatly facilitated such a selection. However, additional efforts of synthesis are worthwhile for both absorption and fluorescence.

Because the quality of the voltage-sensitive probes is the limiting factor for the widespread application of this technology, it is unfortunate that only in the laboratory of Dr. A. Waggoner have voltage-sensitive dyes been synthesized in the past (about 500), and at present only our laboratory is synthesizing many such dyes (about 400). Much faster progress might be expected if (1) more laboratories would synthesize probes; (2) synthesis of dyes having different structures were attempted (other than cyanines, merocyanines, and oxonols); (3) dye screening would be done on the relevant preparations rather than mostly on squid giant axons (e.g., the useful dye RH-414 provides only a very small fluorescence signal in the squid experiment); (4) synthesis and testing were done simultaneously for quick feedback; (5) theoretical approaches for the design of probes of the type introduced by Loew et al. (1978) were incorporated (Loew's theoretical approach was developed for electrochromic absorption probes, but so far none of the styryl dyes have proven useful in absorption experiments. Other styryl dyes that give large fluorescence signals are probably not electrochromic probes exclusively (A. Grinvald, L. B. Cohen, unpublished results; Loew et al., 1985); (6) theoretical work on the underlying mechanisms and their relation to signal size were to be performed (e.g., Waggoner and Grinvald 1977); and (7) more investigations of the biophysical mechanisms underlying the probe sensitivity were carried out (e.g., Ross et al., 1977; Waggoner and Grinvald, 1977; Dragsten and Webb, 1978; Loew and Simpson, 1981; Loew et al., 1985).

6. DIFFICULTIES WITH THE INTERPRETATION OF OPTICAL SIGNALS

Optical signals and intracellular electrical recordings follow the same time course (ignoring the series resistance problem in electrical recordings (see Brown et al., 1979; Salzberg and Bezanilla, 1983; Ross and Krauthamer, 1984). However, the resting potential cannot readily be measured or manipulated. The inability to evaluate reversal potentials for various EPSPs and IPSPs limits the interpretation of some observed responses. Furthermore, difficulties exist with the interpretation of intracellular population recordings.

6.1   Amplitude Calibration

The size of optical signals is related to the membrane area, the extent of binding, and the sensitivity of the dye for a given membrane. For example, in hippocampal slices, differences exist in "concentrations" of membrane elements across the slice. In distal dendrites, many processes exist, and therefore there is a larger membrane area. There is much less membrane in the somata layer, and the amplitude of the optical signals from the s. pyramidale is estimated to be roughly 3–4 times smaller than those from equal potential changes within the s. radiatum. Furthermore, a lower density of bound dye molecules in some parts of the tissue will result in a smaller signal size. There is no evidence that dye molecules bound to different cell types, or even different segments of the same cell, will provide equal signals. Thus, direct comparison of the amplitudes of optical signals in different regions may not be straightforward; only comparison of signals recorded from the same area under different experimental conditions is reliable. Otherwise the interpretation must rely principally on the time course of the signals. In the investigation of single cells, however, information about the amplitude can be interpreted (see e.g., Grinvald et al., 1981b; Krauthamer and Ross, 1984).

6.2   Analysis of Intracellular Population Activity

The analysis of the intracellular population recording may be difficult in three-dimensional preparations with heterogeneous neuronal elements viewed by a single detector. For example, although the hippocampus was selected because of its clear stratification, neurons of various types coexist within strata, and signals from these neurons reaching the same detector can undoubtedly obscure signals from any particular population of neuronal elements (i.e., axons, dendrites, pyramidal somata, etc.). Thus, the activity detected at the s. oriens reflects the activity of the CA1 pyramidal axons and the basal dendrites as well as that of the interneurons there. However, proper manipulation of stimulus size, location, frequency, and pharmacological treatment (or varying the ionic composition, etc.,) may permit a separation of the various components. Furthermore, in such preparations, changes in membrane potential in glial cells may contribute significantly to the optical recording. Orbach and Cohen (1983) concluded that in their salamander experiment, the contribution from glia was small. However, data to substantiate that claim for their slow signals were not presented. In other cases possible contribution of glial signal was considered (Grinvald et al., 1982c; Salzberg et al., 1983). Lev-Ram and Grinvald (1984) have recently recorded very large slow signals (longer than 1 s) from the rat optic nerve stained with RH-414. These slow signals were sensitive to the extracellular potassium concentration and represent glial depolarization (Lev-Ram, 1985). Thus caution should be exercised in the interpretation of slow signals. Development of optical probes that are specific to a given cell type, or the possible iontophoretic injection of suitable fluorescent dye into single cells, or the investigation of evoked activity in single cells impaled with a microelectrode should resolve some of these difficulties.

7. THE SPATIAL RESOLUTION OF OPTICAL RECORDING

7.1   The Microscope Resolution

The spatial resolution of optical recording from thin two-dimensional preparations is excellent ($\sim$1 μm). However, that for three-dimensional preparations is relatively poor (Salzberg et al., 1977; Orbach and Cohen, 1983; Grinvald and Segal, 1983). For example, in the experiments on the frog optic tectum (Grinvald et al., 1984), the spatial resolution is about 200 μm with a 10 $\times$ objective and about 80 μm with a 40 $\times$ objective. These results are not surprising, because conventional microscopes do not resolve images well in a thick preparation, and therefore the three-dimensional resolution of the optical signals is also hampered. It has been suggested that (Grinvald et al., 1984) the spatial resolution may be substantially improved at least in three ways: (1) by the design of custom-made, long-working-distance objectives with high numerical aperture; (2) by mathematical deconvolution of results obtained from measurements at different focal planes and by the use of the mathematical equation for the point spread function (i.e., the defocus blurring function) of a given objective (e.g., Agard and Sedat, 1983); (3) by the use of a confocal detection system (e.g., Egger and Petran, 1967) and optical recording with focal laser microbeam (Grinvald and Farber, 1981) and scanning (Dillon and Morad, 1981) in three dimensions instead of continuous illumination of the whole field under investigation. No one has yet implemented these existing optical approaches and mathematical analysis to improve the three-dimensional resolution, but at least two laboratories are attempting to perform this important task (P. Saggau and G. ten Bruggencate, personal communication; G. Gerstein, personal communication.)

7.2.   The Effect of Light Scattering on the Spatial Resolution

Light scattering from cellular elements leads to the deterioration of images resolved by conventional microscope optics. The light scattering both blurs the images of individual targets and causes an expansion of the apparent area of detected activity. These effects of light scattering on optical recordings were recently investigated (Orbach and Cohen, 1983).

A solution was described for this problem that is quite significant for the optical investigation of cortexlike structures. Egger and Petran (1967) constructed a modified microscope for confocal imaging; if only a small spot in the preparation is illuminated at a given time and coincident detection is employed at the image plane, only from a small spot where the unscattered image should appear, then the effect of light scattering is considerably diminished. Using their modified microscope, Egger and Petran (1967) were able to visually resolve single neurons residing over the roof of the ventricle in the optic tectum, 500 μm below the surface. This approach can be implemented for optical recording as well by using laser scanning and coincident random access detection with a photodiode array.

Another, simpler solution to the problem of three-dimensional resolution is to stain only a very small area in the preparation by iontophoretic application or by pressure injection of the dye. Specific staining restricted to the deep layer below the surface would also increase depth of the loci susceptible to the optical measurements. Alternatively, if the mathematical equation for the light-scattering point-spread function were determined, iterative deconvolution procedures could be used to refine the data.

8. CONCLUSIONS

Optical methods have several inherent advantages in comparison with electrical recording. First, optical recording can be obtained from very small neuronal elements, provided a suitable voltage-sensitive dye can bind to its membrane. (However, a large number of such elements may have to be synchronously active, or, alternatively, a large number of responses may have to be averaged, in order to detect the responses.) Second, this method is a noninvasive one: There is no need to impale the membrane and risk injuring it in order to record. For this reason, a recording can be obtained from the same elements for a considerable length of time (e.g., 6 h). Finally, the major advantage is that a very large number of detectors can be placed side by side to monitor simultaneously the electrical patterns of activity from hundreds of loci, thus affording real-time imaging of neuronal activity and localization of functional units. Considerable progress has also been made in designing arrays of extracellular electrodes that provide complementary information (Thomas et al., 1972; Pickard and Welberry, 1976; Gross, 1979; Pine, 1980; Freeman, 1977; Kruger and Bach, 1981; Bowler and Llinas, 1982). The combination of optical and electrical techniques would most probably provide another dimension to the investigation of brain function.

Many difficulties have been associated with optical measurements. The four groups that have been active in this field have focused mostly on improvements of the technology and its exploitation in new directions. The technique is now far more developed for widespread usage. The cost of the specific equipment is decreasing despite its improved performance, and the number of optical monitors in use has tripled during the past year.

Some difficulties still remain, and possible solutions have been outlined; new methods to handle and analyze the large amount of data should be developed as well as novel conceptual frameworks for the interpretation of such data. The design of improved voltage-sensitive probes should not be neglected. These difficulties call for additonal multidisciplinary efforts. If the above improvements are successful, then the optical recording technique will provide a tomographylike tool for imaging neuronal activity in three dimensions (up to a depth of 400–1000 μm). The submillisecond time resolution and 30- to 60 μm spatial resolution that can be provided by such methods could be especially helpful in promoting understanding of the development and organization of the CNS and the information processing performed by it as well as for investigation of single cells and local circuits in simpler preparations.

ACKNOWLEDGMENTS

We thank Drs. H.S. Orbach and L.B. Cohen who carried out the rat cortex experiments, at the Weizmann Institute and Woods Hole, together with A.G. We also thank Drs. I. Z. Steinberg, V. I. Teichberg, B. M. Salzberg, and W. N. Ross for their critical and constructive comments, and A. S. Waggoner for a generous gift of large kits of voltage-sensitive dyes. Our work was supported by grants from the NIH (NS 14716), the U.S.–Israel Binational Science Foundation, the Muscular Dystrophy Association, the Israel Academy of Sciences, the Psychobiology Foundation, and the March of Dimes.

REFERENCES

Agard, D. A., and J. W. Sedat (1983) Three-dimensional architecture of a polytene nucleus. *Nature* **302**, 676–681.

Agmon, A. (1982) Optical recording from iontophoretically injected neurons with voltage sensitive dyes. M. S. Thesis, Feinberg Graduate School, Weizmann Institute of Science, Rehovot, Israel.

Agmon, A., R. Hildesheim, L. Anglister, and A. Grinvald (1982) Optical recordings from processes of individual leech CNS neurons iontophoretically injected with new fluorescent voltage-sensitive dye. *Neurosci. Lett.*, **10**, S35.

Anglister, L., and A. Grinvald (1983) Real time visualization of the spatiotemporal spread of electrical responses in the optic tectum of vertebrates. Presented at the Annual Meeting Israeli Physiological Society; described in Grinvald and Segal (1983).

Bowler, J., and R. Llinas (1982) Simultaneous sampling and analysis of the activity of multiple, closely adjacent purkinje cells. *Neurosci Abstr.*, **8**, 830.

Brown, J. E., H. H. Harary, and A. S. Waggoner (1979) Isopotentiality and optical determination of series resistance in *Limulus* ventral photoreceptor. *J. Physiol.*, **296**, 357–372.

Calvet, J., and M. C. Calvet (1981) A simple device for making a standard inverted phase-contrast microscope moveable. *J. Neurosci. Methods* **4**, 105–108.

Cohen, L. B. and R. D. Keynes (1971) Changes in light-scattering associated with the action potential in crab nerve. *J. Physiol.*, **212**, 259–275.

Cohen, L. B. and B. M. Salzberg, (1978) Optical measurement of membrane potential. *Rev. Physiol. Biochem. Pharmacol.*, **83**, 35–88.

Cohen, L. B., R. D. Keynes, and B. Hille (1968) Light scattering and birefringence changes during nerve activity. *Nature*, **218**, 438–441.

Cohen, L. B., R. D. Keynes and D. Landowne (1972) Changes in axon light-scattering that accompany the action potential: Current dependent components. *J. Physiol.*, **224**, 727–752.

Cohen, L. B., B. M. Salzberg, H. V. Davila, W. N. Ross, D. Landowne, A. S. Waggoner, and C.-H. Wang (1974) Changes in axon fluorescence during activity; molecular probes of membrane potential. *J. Membr. Biol.*, **19**, 1–36.

Cohen, L. B., B. M. Salzberg, and A. Grinvald (1978) Optical methods for monitoring neuron activity. *Annu. Rev. Neurosci.*, **1**, 171–182.

Conti, F. (1975) Fluorescent probes in nerve membranes. *Annu. Rev. Biophys. Bioeng.*, **4**, 287–310.

Davila, H. V., B. M. Salzberg, and L. B. Cohen (1973) A large change in axon fluorescence that provides a promising method for measuring membrane potential. *Nature New Biol.*, **24**, 159–160.

Davila, H. V., L. B. Cohen, B. M. Salzberg, and B. B. Shrivastav (1974) Changes in ANS and TNS fluorescence in giant axons from *Loligo. J. Membr. Biol.*, **15**, 29–46.

Dillon, S., and M. Morad (1981) Scanning of the electrical activity of the heart using a laser beam with acousto-optics modulators. *Science*, **214**, 453–456.

Dingledine, R. (Ed.) (1983) *Brain Slices*, Plenum, New York.

Dragsten, P. R., and W. W. Webb (1978) Mechanism of membrane potential sensitivity of fluorescent membrane probe merocyanine 540. *Biochemistry*, **17**, 5228–5240.

Egger, M. D. and M. Petran (1967) New reflected light microscope for viewing unstained brain and ganglion cells. *Science*, **157**, 305–307.

Freeman, J. A. (1977) Possible regulatory function of acetylcholine receptor in maintenance of retinotectal synapses. *Nature*, **269**, 218–222.

Grinvald, A. (1984) Real time optical imaging of neuronal activity. *Trends Neurosci.*, **7**, 143–150.

Grinvald, A. (1985) Real time optical mapping of neuronal activity; From single growth cones to the intact mammalian brain. *Annu. Rev. Neurosci.*, **8**, 263–305.

Grinvald, A., and I. Farber (1981) Optical recording of Ca^{2+} action potentials from growth cones of cultured neurons using a laser microbeam. *Science*, **212**, 1164–1169.

Grinvald, A., and M. Segal (1983) "Optical monitoring of electrical activity; Detection of spatiotemporal patterns of activity in hippocampal slices by voltage-sensitive probes," in R. Dingledine, Ed., *Brain Slices*, Plenum, New York, pp. 227–261.

Grinvald, A., B. M. Salzberg, and L. B. Cohen (1977) Simultaneous recording from several neurons in an invertebrate central nervous system. *Nature*, **268**, 140–142.

Grinvald, A., K. Kamino, S. Lesher, and L. B. Cohen (1978) Larger fluorescence and birefringence signals for optical monitoring of membrane potential. *Biophys. J.*, **21**, 82a.

Grinvald, A., W. N. Ross, I. Farber, D. Saya, A. Zutra, et al. (1980) "Optical methods to elucidate electrophysiological parameters," in U. Z., Littauer, Ed., *Neurotransmitters and Their Receptors*, Wiley, New York, pp. 531–546.

Grinvald, A., L. B. Cohen, S. Lesher, and M. B. Boyle (1981a) Simultaneous optical monitoring of activity of many neurons in invertebrate ganglia, using a 124 element 'photodiode' array. *J. Neurophysiol.*, **45**, 829–840.

Grinvald, A., W. N. Ross, and I. Farber (1981b) Simultaneous optical measurements of electrical activity from multiple sites on processes of cultured neurons. *Proc. Natl. Acad. Sci. USA*, **78**, 3245–3249.

Grinvald, A., R. Hildeshiem, A. Agmon, and A. Fine (1982a) Optical recording from neuronal processes and their visualization by iontophoretic injection of new fluorescence voltage-sensitive dyes. *Neuroscience.* **8**, 491.

Grinvald, A., R. Hildesheim, I. C. Farber, and L. Anglister (1982b) Improved fluorescent probes for the measurement of rapid changes in membrane potential. *Biophys. J.*, **39**, 301–308.

Grinvald, A., A. Manker, and M. Segal (1982c) Visualization of the spread of electrical activity in rat hippocampal slices by voltage sensitive optical probes *J. Physiol.*, **333**, 269–291.

Grinvald, A., A. Fine, I. C. Farber, and R. Hildesheim (1983) Fluorescence monitoring of electrical responses from small neurons and their processes. *Biophys. J.* **42**, 195–198.

Grinvald, A., L. Anglister, J. A. Freeman, R. Hildesheim, and A. Manker (1984) Real time optical imaging of naturally evoked electrical activity in the intact frog brain. *Nature*, **308**, 848–850.

Grinvald, A., B. M., Salzberg, R. Hildesheim, and V. Lev-Ram (1986). Optical recording from neuropil processes iontophoretically injected with a new voltage-sensitive probe. (Submitted)

Grinvald, A., C.D. Gilbert, R. Hildesheim, E. Lieke, and T.N. Wiesel (1985) Real time optical mapping of neuronal activity in the mammalian cortex in vivo and in vitro. Neurosci Abstr. 11, 18.

Gross, G. W. (1979) Simultaneous single unit recording in vitro with a photoetched laster deinsulated gold multimicroelectrode surface. *IEE Trans. Biomed. Eng. BEE*, **26**, 273–279.

Gupta, R., B. M. Salzberg, L. B. Cohen, A. Grinvald, K. Kamino, et al. (1981) Improvements in optical methods for measuring rapid changes in membrane potential. *J. Membr. Biol.*, **58**, 123–138.

Kauer, J. S., D. M. Senseman, and L. B. Cohen (1984) Voltage-sensitive dye recording from the olfactory system of the tiger salamander. *Neurosci. Abstr.*, **10**, 846.

Kaufman, L., and S. J. Williamson (1982) Magnetic location of cortical activity. *Ann. N.Y. Acad. Sci.*, **388**, 197–213.

Krauthamer, V., and W. N. Ross (1984) Regional variations in excitability of barnacle neurons. *J. Neurosci.*, **4**, 673–682.

Kruger, J. and M. Bach (1981) Simultaneous recording with 30 microelectrodes in monkey visual cortex. *Exp. Brain Res.*, **41**, 191–194.

Kuhnt, U., and A. Grinvald (1982) 4-AP induced presynaptic changes in the hippocampal slices as measured by optical recording and voltage-sensitive probes. *Pflugers Arch.* (Suppl.), **394**, R45.

Lancet, D., C. A. Greer, J. S. Kauer, and G. M. Shepherd (1982) Mapping of odor related neuronal activity in the olfactory bulb by high resolution 2-deoxyglucose autoadiography. *Proc. Natl. Acad. Sci. USA*, **79**, 670–674.

Lev-Ram, V. and A. Grinvald (1984) Is there a potassium dependent depolarization of the paranodal region of myelin sheath? Optical studies of rat optic nerve. Neurosci. Abstr. 10, 948.

Lev-Ram, V. (1985) A potassium depolarization of the oligodendrocyte paranodal region in rat optic nerve. Neurosci. Abstr. 11, 87.

Lipton, P. (1973) Effects of membrane depolarization on light scattering by cerebral slices. *J. Physiol. (Lond.)*, **231**, 365–383.

Llinas, R., and M. Sugimori (1980) Electrophysiological properties of in vitro Purkinje cell dendrites in mammalian cerebellar slices. *J. Physiol. (Lond.)*, **305**, 197–213.

Loew, L. M., and L. L. Simpson (1981) Charge shift probes of membrane potential. *Biophys. J.*, **34**, 353–363.

Loew, L. M., G. W. Bonneville, and J. Surow (1978) Charge shift probes of membrane potential:theory. *Biochemistry*, **17**, 4065–4071.

Loew, L. M., S. Scully, L. Simpson, and A. S. Waggoner (1979) Evidence for a charge shift electrochromic mechanism in a probe of membrane potential. *Nature*, **281**, 497–499.

Loew, L. M., L. B. Cohen, B. M. Salzberg, A.L. Obaid, and F. Bezanilla (1985) Charge shift probes of membrane potential. Characterization of Aminostyrylpyridinium dyes in the squid giant axon. *Biophys. J.*, **47**, 71-77.

Manker, A. (1982) Design of a display processor to visualize neuronal activity. M.Sc. thesis, Feinberg Graduate School, Weizmann Institute of Science, Rehovot, Israel.

Orbach, S. H., and L. B. Cohen (1983) Simultaneous optical monitoring of activity from many areas of the salamander olfactory bulb. A new method for studying functional organization in the vertebrate CNS. *J. Neurosci.*, **3**, 2251–2262.

Orbach, H. S., L. B. Cohen, and A. Grinvald (1982) Optical recording of evoked activity in the visual cortex of the rat. *Biol. Bull.*, **163**, 389.

Orbach, H. S., L. B. Cohen, A. Grinvald, and R. Hildesheim (1983) Optical monitoring of neuron activity in rat somatosensory and visual cortex. *Neurosci. Abstr.*, **9**, 39.

Orbach, H. S., L. B. Cohen, and A. Grinvald (1985) Optical monitoring of neuronal activity in the mammalian sensory cortex. *J. Neurosci.* **5**, 1886-1895.

Oxford, G. S., J. P. Pooler, and T. Narahashi (1977) Internal and external application of photodynamic sensitizers on squid giant axons. *J. Membr. Biol.*, **36**, 159–173.

Patrick, J., B. Valeur, L. Monnerie, and J. P. Changeux (1971) Changes in extrinsic fluorescence intensity of the electroplax membrane during electrical excitation. *J. Membr. Biol.*, **5**, 102–120.

Pickard, R. S. and T. R. Welberry (1976) Printed circuit microelectrodes and their application to honeybee brain. *J. Exp. Biol.*, **64**, 39–44.

Pine, J. (1980) Recording action potentials from cultured neurons with extracellular microcircuit electrodes. *J. Neurosci. Methods*, **2**, 19–31.

Pooler, J. P. (1972) Photodynamic alteration of sodium currents in lobser axon. *J. Gen. Physiol.*, **60**, 367–387.

Raichle, M. E. (1979) Quantitative in vivo autoradiography with positron emission tomography. *Brain Res. Rev.*, **1**, 47–68.

Ross, W. N., and V. Krauthamer (1984) Optical measurements of potential changes in axons and processes of neurons of a barnacle ganglion.

Ross, W. N., and L. F. Reichardt (1979) Species-specific effects on the optical signals of voltage sensitive dyes. *J. Membr. Biol.*, **48**, 343–356.

Ross, W. N., B. M. Salzberg, L. B. Cohen, and H. V. Davila (1974) A large change in dye absorption during the action potential. *Biophys. J.*, **14**, 983–986.

Ross, W. N., B. M. Salzberg, L. B. Cohen, A. Grinvald, H. V. Davila, et al. (1977) Changes in absorption, fluorescence, dichroism and birefringence in stained axons: Optical measurement of membrane potential. *J. Membr. Biol.*, **33**, 141–183.

Salama, G., and M. Morad (1976) Merocyanine 540 as an optical probe of transmembrane electrical activity in the heart. *Science*, **191**, 485–487.

Salzberg, B. M. (1978) Optical signals from squid giant axons following perfusion or superfusion with potentiometric probes. *Biol. Bull.* **155**, 463–464.

Salzberg, B. M. (1983) "Optical recording of electrical activity in neurons using molecular probes," in *Current Methods in Cellular Neurobiology*, Vol. 3, J. L. Barker and J. F. McKelvy, Ed., Wiley, New York. pp. 139–187.

Salzberg, B. M., and F. Bezanilla (1983) An optical determination of the series resistance in *Loligo*. *J. Gen. Physiol.* **82**, 807–817.

Salzberg, B. M., H. V. Davila, L. B. Cohen (1973) Optical recording of impulses in individual neurons of an invertebrate central nervous system. *Nature*, **246**, 508–509.

Salzberg, B. M., A. Grinvald, L. B. Cohen, H. V. Davila, and W. N. Ross (1977) Optical recording of neuronal activity in an invertebrate central nervous system; Simultaneous monitoring of several neurons. *J. Neurophysiol.*, **40**, 1281–1291.

Salzberg, B. M., A. L. Obaid, D. M. Senseman, and H. Gainer (1983) Optical recording of action potentials from vertebrate nerve terminals using potentiometric probes provides evidence for sodium and calcium components. *Nature*, **306**, 36–39.

Salzberg, B.M., A.L. Obaid. and H. Gainer (1985) Large and rapid changes in light scattering accompany secretion by nerve terminals in the mammalian neurohypophysis. J. Gen. Physiol., **86**, 395–411.

Schwartzkroin, P. A., and M. Slawsky (1977) Probable calcium spikes in hippocampal neurons. *Brain Res.*, **135**, 157–161.

Sejnowski, T. J., S. C. Reingold, D. Kelley, and A. Gelperin (1980) Localization of [H^3]-2-deoxyglucose in single molluscan neurons. *Nature*, **287**, 449–451.

Senseman, D. M., and B. M. Salzberg (1980) Electrical activity in an exocrine gland; Optical recording with a potentiometric dye. *Science*, **208**, 1269–1271.

Senseman, D. M., H. Shimizu, I. S. Horwitz, and B. M. Salzberg (1983) Multiple-site optical recording of membrane potential from a salivary gland. *J. Gen. Physiol.*, **81**, 887–908.

Simons, D. J. (1978) Response properties of vibrissa units in rat SI somatosensory cortex. *J. Neurophysiol.*, **41**, 798–820.

Sokoloff, L. (1977) Relation between physiological function and energy metabolism in the central nervous system. *J. Neurochem.*, **19**, 13–26.

Spencer, W. A., and E. R. Kandel (1961) Electrophysiology of hippocampal neurons. IV. Fast prepotentials. *J. Neurophysiol.*, **24**, *272–285*.

Stuart, A. E., and D. Ortel (1978) *Neuronal properties underlying processing of visual information in the barnacle. Nature,* **275**, 187–190.

Tasaki, I., and A. Warashina (1976) Dye membrane interaction and its changes during nerve excitation. *Photochem. Photobiol.*, **24**, 191–207.

Tasaki, I., A. Watanabe, R. Sandlin, and L. Carnay (1968) Changes in fluorescence, turbidity, and birefringence associated with nerve excitation. *Proc. Natl. Acad. Sci. USA,* **61**, 883–888.

Thomas, C. A., Jr., P. A. Springer, C. E. Loeb, Y. Berwald-Netter, and L. M. Okun (1972) A miniature microelectrode array to monitor the bioelectric activity of cultured cells. *Exp. Cell Res.,* **74**, 61–66.

Vergara, J., and F. Bezanilla (1976) Fluorescence changes during electrical activity in frog muscle stained with merocyanine. *Nature,* **259**, 684–686.

Waggoner, A. S. (1976) Optical probes of membrane potential. *J. Membr. Biol.,* **27**, 317–334.

Waggoner, A. S. (1979) Dye indicators of membrane potential. *Annu. Rev. Biophys. Bioeng.,* **8**, 47–63.

Waggoner, A. S. and A. Grinvald (1977) Mechanisms of rapid optical changes of potential sensitive dyes. *Ann. N.Y. Acad. Sci.,* **303**, 217–242.

Wong, R. K. S., D. A. Prince, A. I. Basbaum (1979) Intradendritic recordings from hippocampal neurons. *Proc. Natl. Acad. Sci. USA,* **76**, 986–990.

Woolsey, T. A., and H. Van der Loos (1970) The structural organization of layer IV in the somatosensory region (SI) of mouse cerebral cortex. *Brain Res.,* **17**, 205–242.

Yamamoto, C. (1972) Intracellular study of seizure-like after-discharges elicited in thin hippocampal sections in vitro. *Exp. Neurol.,* **35**, 154–164.

CHAPTER 11

OPTICAL DETERMINATION OF ELECTRICAL PROPERTIES OF RED BLOOD CELL AND EHRLICH ASCITES TUMOR CELL MEMBRANES WITH FLUORESCENT DYES

PHILIP C. LARIS
Department of Biological Sciences
University of California
Santa Barbara, California

JOSEPH F. HOFFMAN
Department of Physiology
Yale University School of Medicine
New Haven, Connecticut

1.	INTRODUCTION	200
2.	THEORY	200
3.	FLUORESCENCE/MEMBRANE POTENTIAL CALIBRATION	201
4.	MEMBRANE RESISTANCE AND THE ELECTROGENIC SODIUM PUMP	204
5.	ELECTROGENIC Na/AMINO ACID COTRANSPORT	205
6.	OTHER TRANSPORT PROCESSES	206
7.	CONCLUSION	207
	REFERENCES	207

1. INTRODUCTION

Measurements of the electrical properties of the plasma membranes of red blood cells with optical methods were initiated shortly after the discovery that a fluorescent merocyanine dye would give a relatively large signal during the passage of an action potential in the giant axons of the squid (Davila et al., 1973). In the initial studies with red cells the fluorescence of a carbocyanine dye was shown to be altered when a change was induced in the membrane potential of these cells (Hoffman and Laris, 1974). Since this finding, studies of fluorescence have been an important means of monitoring membrane potentials of cells, organelles, and vesicles (Cohen and Salzberg, 1978; Freedman and Laris, 1981; Cohen and Hoffman, 1982).

In recent years a number of techniques have been employed to estimate membrane potentials in red cells and Ehrlich ascites tumor cells. While the direct methods of electrophysiology may be the obvious choice, no one has yet successfully recorded potentials from red cells except for the giant *Amphiuma* red cell (Hoffman and Lassen, 1971; Lassen, 1972), and it is uncertain in Ehrlich cells whether the peak of a transient change seen immediately following impalement with a microelectrode (Lassen et al., 1971) or a stable potential (Smith and Levinson, 1975; Smith and Vernon, 1979) is the true measure of the potential for this cell. In addition, rapid and transient changes in the membrane potential of Ehrlich cells such as those expected after the addition of ionophores, Na-dependent amino acids, or ouabain have not yet been measured in a cell impaled with a microelectrode.

2. THEORY

In theory, the membrane potential can be determined indirectly from measurements of the ratio of the cytoplasmic/medium activities of any permeant ion distributed across the membrane at equilibrium, for then E_m, the membrane potential, equals the equilibrium potential for-the ion as given by the Nernst equation,

$$E_m = \frac{RT}{ZF} \ln \frac{a_o}{a_i}$$

where R, T, Z, and F have the usual meanings and a_o and a_i are the activities of the ion in the medium and cells, respectively. This approach to measuring membrane potentials has been applied to red cells using the naturally occurring ions, protons and chloride and to the Ehrlich cell with chloride and synthetic ions such as tetraphenyl phosphonium. Proton distribution can be determined with NMR studies using ^{31}P (Moon and Richards, 1973) or ^{19}F (Taylor et al., 1981), and changes in this distribution can be followed with NMR or by measurements of the external pH of suspensions of red cells in unbuffered media in the presence of a protonophore (Macey et al., 1978). It has been pointed out, however, that the protons may not always be at equilibrium in red cells because of Cl–hydroxyl exchange movements (Wieth et al., 1980). Similarly, Cl in the Ehrlich cell and other cells may not always be at equilibrium since Cl participates in cotransport systems of the type where K + Cl or Na + K + Cl are moved together, and hence the distribution of Cl may be influenced by other ions that are transported by different mechanisms. Furthermore, the conductive fluxes of Cl in red cells and Ehrlich cells are slow, and hence net movements for redistribution following a change in the potential are slow.

Some of the problems encountered with naturally occurring ions can be avoided by the use of lipid-soluble synthetic ions. Since they are not transported, they would be expected to reach equilibrium, and because they are lipid-soluble, their net movements across membranes can be relatively rapid. The calculation of the membrane potential, however, requires values of cytoplasmic activities, and it may thus be necessary to correct measurements of the cell concentrations of the ions for binding and for the uptake by organelles also capable of generating electrical potential differences across their membranes. These corrections involve additional assumptions that complicate determinations of membrane potential by this approach.

The cell/medium distribution ratio of charged lipophilic fluorescent dyes such as carbocyanines is also influenced by the membrane potential. For example, the cell/medium ratio of the positively charged dye, 3,3′dipropylthiadicarbocyanine [diS-C_3-(5)], has been shown in red cell suspensions to increase when the membrane is hyperpolarized and decrease when it is depolarized (Sims et al., 1974). For this reason these dyes have been called redistribution or slow dyes. Because the relative fluorescence intensity of this dye is different (quenched) in cells when compared to that of the bulk phase, a change in membrane potential can change the partitioning of the dye, causing a change in the overall fluorescence intensity. Hence, the fluorescence intensity of the dye in cell suspensions has been used as a means of monitoring membrane potentials in cells and organelles.

The basic technique employed to initiate studies with these dyes has not changed; it remains empirical. The cell/dye ratio is selected by finding the ratio that will yield maximum change in fluorescence for a given change in E_m. The alteration of the membrane potential by the addition of an ionophore such as valinomycin or by a change in the ionic composition of the medium requires some knowledge of the permeability properties of the cell type being studied. Knowledge of ion fluxes in the presence and absence of the dye will also be useful in determining whether or not interpretations of the fluorescence signals are reasonable. The cell/dye ratio not only influences the magnitude of the change in fluorescence intensity with a change in potential but also can determine the direction of the change. Generally, systems have been used where a decrease in the fluorescence intensity of a cyanine-type dye is associated with a hyperpolarization and an increase is associated with a depolarization of the membrane. Usually the changes are largest with this arrangement, but this kind of response may not obtain in all cases.

3. FLUORESCENCE/MEMBRANE POTENTIAL CALIBRATION

A number of different approaches have been used in attempts to calibrate fluorescence intensity in terms of mV of membrane potential. Single values of fluorescence intensity have frequently been calibrated using the "null-point" method (Hoffman and Laris, 1974). In this method a concentration of external K is found for which there is no change in fluorescence upon the addition of valinomycin. At this concentration of external K, $E_m = E_K$, provided that P_K in the presence of valinomycin is a dominant factor in determining E_m. For *Amphiuma* red cells, this approach yielded values of E_m essentially the same as those determined by direct impalement with microelectrodes (Hoffman and Lassen, 1971; Lassen, 1972). For the human red cell the null-point method yielded values of E_m commensurate with expectations based on the Donnan potential where $E_m = E_{Cl}$ (Hoffman and Laris, 1974, 1984). With the Ehrlich cell the membrane potential estimated by the null-point method after a preliminary incubation in Krebs–Ringer solution is generally -25 to -30 mV (Laris et al., 1976) in the same range as the potentials reported by Lassen et al. (1971) and somewhat more negative than the stable potentials of Smith and Vernon (1979) and the estimates from the Cl equilibrium potential (Laris et al., 1976). Smith and Robinson (1981) have pointed out a problem with the use of the null-point method in cells with internal compartments. Because valinomycin can influence the membrane potentials of both mitochondria and the plasma membrane, the null point for a cell like the Ehrlich cell might be that external K concentration for which the sum of the changes in fluorescence at the inner membrane of the mitochondrion and the plasma membrane is zero. Actually, the membrane potential of the mitochondria may not always be a problem, because cyanine dyes block oxidation at complex I (Waggoner, 1976). The mitochondrial potential probably drops rapidly after the addition of the dye to the cells, for in the absence of glucose, ATP levels fall significantly in the time period during which the dye is equilibrating with the cells (Laris et al., 1978). In fact, the null point was not shifted when the cells were first treated with antimycin A or other mitochondrial inhibitors (Laris, unpublished observations). Nevertheless, Philo and Eddy (1978) have routinely added cyanide and oligomycin to Ehrlich cells to avoid problems concerned with mitochondrial potentials.

The question of the origin of the changes in fluorescence intensity observed with fluorescent dyes in eukaryotic cells is a critical problem. This question has been addressed in the Ehrlich cell in studies with cytoplasts derived from Ehrlich cells and devoid of internal compartments (Henius et al., 1979). Qualitatively, all the fluorescence or membrane potential changes, including those with Na-dependent amino acids, the Na/K pump, ionophores, and propranolol and K movements (Henius and Laris, 1979; Thornhill 1983) that have been observed in intact cells are seen with cytoplasts. And quantitatively, the null-point determinations of membrane potential are similar in the cytoplasts and intact cells. This kind of problem should be considered for each cell type and perhaps even the line of Ehrlich cells being studied. The latter point is made because optical techniques used to measure membrane potentials have demonstrated marked differences in Ehrlich cells obtained from different sources. While hyperpolarizations with glucose that were attributed to the action of the Na/K pump were observed by Laris et al. (1978), Heinz et al. (1981) reported marked hyperpolarizations with glucose that were not ouabain-sensitive and appeared to be the result of a proton pump. When cells from the two sources were studied at the same time in the same laboratory (Laris, unpublished observations), the original observations of both groups were repeated. Hence, it is apparent that there are marked and interesting differences between the two lines of cells.

Methods for converting fluorescence intensity or changes in fluorescence intensity to mV for red cells have been made in a number of ways. One method relates the fluorescence intensity to the membrane potential that was calculated from the constant field equation (Goldman, 1943; Hodgkin and Katz, 1949),

$$E_m = \frac{RT}{F} \frac{\alpha[K]_o + [Cl]_i}{\alpha[K]_i + [Cl]_o}$$

where $\alpha = (P_K)/(P_{Cl})$. In this instance α was found to equal 20 in the presence of valinomycin (Hoffman and Laris, 1974; Freedman and Hoffman, 1979). Another method utilized the Donnan equilibrium potential,

$$E_m = \frac{RT}{F} \ln \frac{[Cl]_i}{[Cl]_o}$$

established when nonpermeant anions replaced Cl (Freedman and Hoffman, 1979). Still other investigators (Macey et al., 1978; Freedman and Novak, 1983) have compared the difference in fluorescence intensity to the difference in E_m in two sets of conditions and have determined ΔE_m by measuring the change in external pH in the presence of an uncoupler and an inhibitor of anion exchange, DIDS (4,4'-diisothiocyano-2,2'stilbene disulfonate). If protons are assumed to be at equilibrium, then

$$E_m = \frac{RT}{F} \ln \frac{[H^+]_o}{[H^+]_i}$$

and if the cellular pH does not change, then

$$\Delta E_m = - \frac{RT}{F} \ln \Delta pH_o$$

Finally, since redistribution dyes are permeant, they would be expected to reach an equilibrium distribution commensurate with the membrane potential such that

$$E_m = \frac{RT}{F} \ln \frac{a_o}{a_i}$$

where a_o and a_i respectively, are the external and cellular activities of the dye. Hladky and Rink (1976) calibrated their system by determining the external dye concentration required to maintain a_i constant at two different membrane potentials. Then

$$\Delta E_m = \frac{RT}{F} \ln \frac{a'_o}{a''_o}$$

where a_o' and a''_o are the external activities of the dye for which there is the same amount of dye associated with the cells and thus the same cellular activity. The relationship between fluorescence intensity and membrane potential calculated in these ways can be shown to vary from -75 to $+50$ mV and is linear over a limited range (Freedman and Hoffman, 1979; Hladky and Rink, 1976). In *Amphiuma* red cells this relationship is curvilinear in the range -50 to -10 mV where the change in fluorescence intensity was compared to membrane potentials measured directly with microelectrodes (Pape, 1982). In the human neutrophil fluorescence intensity was directly proportional to the membrane constant field equation (Simchowitz et al., 1982). Thus, empirically, this relationship is found to be linear over a limited but useful range in a number of cases, even though the basis for this result is not at present understood.

4. MEMBRANE RESISTANCE AND THE ELECTROGENIC SODIUM PUMP

Estimates of the membrane resistance, R_m, of red cells and the Ehrlich cell can also be made from studies based on estimates of membrane potentials as measured with fluorescent dyes and the electrogenicity of the Na/K pump (Hoffman et al., 1979). If the ouabain-sensitive portion of the membrane potential, E_m^{ouab}, and the ouabain-sensitive Na efflux are known, R_m is given by Ohm's law,

$$R_m = \frac{E_m^{ouab}}{I_{pump}}$$

where I_{pump} represents the pump current, taken to be one-third of the ouabain-sensitive Na efflux. The value of R_m calculated in this manner is $1\text{–}5 \times 10^5$ ohm/cm^2 for normal human red cells and $1\text{–}5 \times 10^4$ ohm/cm^2 for Ehrlich ascites tumor cells. These values generally agree with other estimates made by considering ion fluxes. R_m for red cells will essentially be equal to the reciprocal of the chloride conductance, and the estimates presented above fall in the same range as those based on indirect determinations of chloride conductance (Knauf et al., 1977). For the Ehrlich cell the membrane resistance calculated from estimates of Na, K, and Cl conductance (Hoffman et al., 1979) is 2.4×10^4 ohm/cm^2, a value within the range calculated by means of optical techniques.

If R_m is known, it is possible to make predictions about the influence of ion movements on the membrane potential and test models for these ion movements. For example, in the normal mode the stoichiometry of the human red cell Na/K pump is 3Na:2K:1ATP, and the pump is electrogenic (Hoffman et al., 1979, 1980a). A question was raised about a possible net movement of charge through the pump when it operates in the so-called "uncoupled" mode. This uncoupled Na efflux is an ouabain-sensitive flux that the pump mediates in the complete absence of external Na and K (Garrahan and Glynn, 1967). Taking values of R_m for cell suspensions containing SO$_4$ rather than Cl (a circumstance that increases R_m) and measurements of the ouabain-sensitive Na efflux, a change of 3–4 mV depolarization would be expected upon the addition of ouabain if the pump was electrogenic in the uncoupled mode. No potential change, however, was observed with the addition of ouabain in this situation, although a change of 3–4 mV would have been readily detected by fluorescence measurements (Dissing and Hoffman, 1983). Subsequent investigations

(Dissing and Hoffman, 1983, 1984) revealed that in a sulfate medium there is an ouabain-sensitive SO_4 efflux that makes the ouabain-sensitive Na efflux electrically neutral. Recently the uncoupled Na efflux has been studied in a resealed red cell ghost system in which the impermeant ions tartrate and TRIS are the major ions present (Dissing and Hoffman, 1984). With these ghosts a small but detectable decrease in fluorescence intensity is observed upon the addition of ouabain, indicating that in the absence of permeant ions the uncoupled Na efflux is electrogenic. The small depolarization discernible with ouabain in the tartrate ghost system requires internal Na, Mg, and ATP and is observed only under conditions where an ouabain-sensitive Na efflux is observed in parallel measurements. The uncoupled Na efflux is inhibited in this preparation, presumably because there are no permeable anions present to balance the outwardly directed charge during the operation of the pump. The important point for the present purposes is that the methods using optical determinations in parallel with flux measurements have demonstrated that mobile anions can move in conjunction with uncoupled Na pump activity, even though it is not yet clear how this movement is accomplished.

5. ELECTROGENIC Na/AMINO ACID COTRANSPORT

Marked changes are observed in the fluorescence intensity of Ehrlich cells as a result of changes in the concentrations of external or internal amino acids. These changes are dependent upon the presence of Na and are therefore thought to be due to changes in membrane potential resulting from the activity of the Na/amino acid cotransport system. Given the estimation of R_m for Ehrlich cells as presented above and measurements of amino acid fluxes in these cells, would one expect the membrane potential of these cells to be influenced by the activity of the Na/amino acid cotransport system? According to an electrical model of proton-dependent cotransport of glucose in Neurospora (Slayman and Slayman, 1974), there are two relatively simple limiting cases for the behavior of the cotransport system modeled as a Na battery and resistance in series. In the first case the cotransport system would behave like a current source depolarizing the membrane with a negligible decrease in resistance. In the second, the activated cotransport system would behave like a shunt resistance. Though critical measurements to distinguish between these two possibilities are not available, the feasibility of the two models can be explored. With maximal conditions of external Na and amino acid, the change in the membrane potential associated with the activation of cotransport is estimated to be about 25 mV. The current required for the observed change of approximately 25 mV would be equivalent to a flux of 120–600 mmol/L cells × h, assuming a stoichiometry of 1 Na:1 amino acid and a resistance of $1–5 \times 10^4$ ohm/cm^2. V_{max} estimations for AIB (aminoisobutyric acid) uptake are in this range, 400 mmol/L cells × h (Johnstone and Laris, 1980). Taking the limiting case of a shunt resistance, the total membrane resistance during cotransport would be proportional to the fraction of the potential remaining. According to this scheme, R_m would have to fall to 20% or less. Estimates of conductance (Hoffmann et al., 1979) and Na/amino acid cotransport suggest that the second limiting case is unlikely if R_m remains constant with changes in E_m.

When Ehrlich cells are first removed from the abdominal cavities of mice and incubated in physiological salt solutions lacking amino acids, fluorescence studies indicate that the cells are hyperpolarized and gradually depolarize over a period of 30–60 min at 37°C (Laris et al., 1976). Two events that could influence the membrane potential are occurring during this period: One is that cell Na is reduced by the activity of the Na/K pump, and the other is that there is an efflux of amino acids from the cells. Since the initial hyperpolarization is not blocked by ouabain and is seen with low ATP levels, this hyperpolarization cannot be attributed to the activity of the Na/K pump. In the second possibility the coefflux of Na with amino acids into a medium lacking amino acids could hyperpolarize the cells. According to this idea, when the amino acid pool and Na concentrations are high initially, the flux is high, and as a result E_m is more negative. The potential would then gradually become less negative as the Na and amino acids concentrations in the cells diminished because of efflux. In fact, Ehrlich cells that were incubated with a Na-dependent amino acid, AIB, so as to maintain a high cellular amino acid pool, hyperpolarized when subsequently placed in an amino acid free medium (Laris et al., 1978). Furthermore, the level of fluorescence intensity was inversely proportional to the cellular amino acid concentration, indicating that the membrane potential is more negative with higher cellular AIB concentrations. Thus, when the amino acids are not in a steady-state distribution, the membrane potential of these cells, as monitored by fluorescence, is sensitive to both internal and external amino acid concentrations. Flux measurements are available to test the idea that Na amino acid coefflux could influence E_m. For example, the values of V_{max} for glycine (high-cell Na) and AIB (normal-cell Na) are 1100 mmol/L cells × h (Johnstone, 1978) and 540 mmol/L cells × h, respectively (Johnstone, unpublished observations). Using these values of the fluxes, together with a stoichiometry of 1 Na:1 amino acid and the average value of R_m as presented above, E_m for the case of a current source model, would be expected to be 50 mV more negative at V_{max} conditions with glycine and 25 mV more negative with AIB, consistent with the results obtained with fluorescent dyes.

The current source models for the Na/K pump and the Na amino acid cotransport in Ehrlich ascites tumor cells pose another kind of problem for estimation of the membrane potential in those cells using the null-point method. As has been pointed out by Simchowitz et al. (1982), R_m will be reduced with the addition of valinomycin, and this change will lessen the influence of the pump and cotransport currents on E_m. Thus, in these circumstances E_m will be changed (reduced) by the agent used in the determination of E_m.

6.   OTHER TRANSPORT PROCESSES

The cyanine dyes have been useful in determining the electrical characteristics of a number of other ion transport processes and the extent to which they are conductive. An early example of this kind of use concerned the Ca-dependent K flux (Gardos effect) in the human red cell. The conductive nature of this ion flux was first demonstrated with a cyanine dye (see Hoffman et al., 1980b) even though the cyanine dyes are known to inhibit this transport system. This measurement was possible because the dye is slower in inhibiting the transport system than it is in sensing the membrane potential. Studies with dog red cells have demonstrated that an addition of ATP externally resulted in changes in Na and K permeability that were large enough to influence the membrane potential as measured with diS-C$_3$(5) (Parker et al., 1977). In these red cells and also in cat red cells, shrinkage caused an increase in P_{Na} that resulted in a change in membrane potential again determined with a fluorescent-dye technique (Castranova et al., 1979). Thus, in this instance it was found that the change in anion transport that occurred with cell shrinkage was the result of the alteration in membrane potential rather than in cell volume per se.

Fluorescent dyes have also been employed to determine whether the ion fluxes associated with volume regulatory mechanisms are conductive or electrically silent. For example, after a rapid (osmotic) swelling in hypotonic media, many cells shrink toward their initial volume by turning on a mechanism that releases KCl and water. These KCl losses have been shown to be electrically neutral in duck red cells (Kregenow, 1977; McManus et al., 1985) and in the Ehrlich cell (Thornhill and Laris, 1984) and conductive in the human lymphocyte (Grinstein et al., 1982). Duck red cells also possess a cotransport system that, after rapid (osmotic) shrinkage in hypertonic media, is activated to regulate a volume increase by mediating the net movement of the ion complex Na + K + 2Cl. This cotransport system has also been shown to be electrically silent in these cells by the use of fluorescent dye techniques (Kregenow, 1977; McManus et al., 1985).

7. CONCLUSION

By way of summary, it should be apparent that studies of optical measurements to monitor membrane potentials in cell suspensions have proved to be valuable since their initial employment a little over 10 years ago. Analysis of electrical parameters of membranes with fluorescent dyes is best made when the cell's permeability characteristics to ions are also known. This type of correlative information provides a way of evaluating whether or not the conclusions drawn from the optical studies are reasonable. Although many problems may be encountered with the application of optical methods, they are relatively easy to sort out, given close attention to appropriate controls, and the results obtained with these methods are comparable to those obtained with other methods.

ACKNOWLEDGMENTS

This work was supported by NIH grants HL 09906 and AM 17433.

REFERENCES

Castranova, V., M. J. Weise, and J. F. Hoffman (1979) Anion transport in dog, cat and human red cells. *J. Gen. Physiol.*, **74**, 319–334.

Cohen, L. B., and J. F. Hoffman (1982) "Optical measurements of membrane potential," in *Techniques in Cellular Physiology, P118*, Elsevier/North-Holland, New York, pp. 1–13.

Cohen, L. B., and B. M. Salzberg (1978) "Optical Measurement of membrane potential" *Rev. Physiol. Biochem. Pharmacol.*, **83**, 35–88.

Davila, H. V., B. M. Salzberg, L. B. Cohen, and A. S. Waggoner (1973) A large change in axon fluorescence that provides a promising method for measuring membrane potential. *Nature New Biol.*, **241**, 159–160.

Dissing, S., and J. F. Hoffman (1983) Anion-coupled Na efflux mediated by the Na/K pump in human red blood cells. *Curr. Top. Membr. Transport*, **19**, 693–695.

Dissing, S., and J. F. Hoffman (1984) *Fourth International Conference on Na,K-ATPase.* Cambridge University. Abstract #89.

Freedman, J. C., and J. F. Hoffman (1979) The relation between dicarbocyanine dye fluorescence and the membrane potential of human red blood cells set at varying Donnan equilibria. *J. Gen. Physiol.*, **74**, 187–212.

Freedman, J. C., and P. C. Laris (1981) Electrophysiology of cells and organelles: Studies with optical potentiometric indicators. *Int. Rev. Cytol.* (Suppl.), **12**, 177–246.

Freedman, J. C., and T. S. Novak (1983) Membrane potentials associated with Ca-induced K conductance in human red blood cells: Studies with a fluorescent oxonol dye, WW781. *J. Membr. Biol.*, **72**, 59–74.

Garrahan, P. J., and I. M. Glynn (1967) The sensitivity of the sodium pump to external Na. *J. Physiol. (Lond.)*, **192**, 175–188.

Goldman, D. E. (1943) Potential, impedance and rectification in membranes. *J. Gen. Physiol.*, **27**, 37–60.

Grinstein, S., C. A. Clarke, S. Dupre, and A. Rothstein (1982) Volume-induced increase of anion permeability in human lymphocytes. *J. Gen. Physiol.*, **80**, 801–823.

Heinz, A., G. Sachs, and J. A. Schafer (1981) Evidence for activation of an active electrogenic proton pump in Ehrlich ascites tumor cells during glycolysis. *J. Membr. Biol.*, **61**, 143–153.

Heinz, E., P. Geck, and C. Pietrzyk (1975) Driving forces of amino acid transport in animal cells. *Ann. N.Y. Acad. Sci.*, **264**, 428–441.

Henius, G. V., and P. C. Laris (1979). The Na^+ gradient hypothesis in cytoplasts derived from Ehrlich ascites tumor cells. *Biochem. Biophys. Res. Commun.* **91**, 1430–1436.

Henius, G. V., P. C. Laris, and J. D. Woodburn (1979) The preparation and properties of cytoplasts from Ehrlich ascites tumor cells. *Exp. Cell Res.*, **121**, 337–346.

Hladky, S. B., and T. J. Rink (1976) Potential difference and the distribution of ions across the human red blood cell membrane: A study of the mechanism by which the fluorescent cation, diS-C_3(5) reports the membrane potential. *J. Physiol. (Lond.)*, **263**, 287–319.

Hodgkin, A. L., and B. Katz (1949) The effect of sodium ions on the electrical activity of the giant axon of the squid. *J. Physiol. (Lond.)*, **108**, 37–77.

Hoffman, J. F., and P. C. Laris (1974) Determination of membrane potentials in human and *Amphiuma* red blood cells by means of a fluorescent probe. *J. Physiol. (Lond.)*, **239**, 519–552.

Hoffman, J. F., and P. C. Laris (1984) "Membrane electrical parameters of normal human red blood cells," in M. P. Blaustein and M. Lieberman, Eds., *Electrogenic Transport: Fundamental Principles and Physiological Implications*, Raven, New York, pp. 287–293.

Hoffman, J. F., and U. V. Lassen (1971) Plasma membrane potentials in *Amphiuma* red cells. Abstract, XXV Int. Congr. Physiol. Sci., Munich, **9**, 253.

Hoffman, J. F., J. H. Kaplan, and T. J. Callahan (1979) The Na:K pump in red cells is electrogenic. *Fed. Proc.*, **38**, 2440–2441.

Hoffman, J. F., J. H. Kaplan, T. J. Callahan, and J. C. Freedman (1980a) Electrical resistance of the red cell membrane and the relation between net anion transport and the anion exchange mechanism. *Ann. N.Y. Acad. Sci.*, **341**, 357–360.

Hoffman, J. F., D. R. Yingst, J. M. Goldinger, R. M. Blum, and P. A. Knauf (1980b) "On the mechanism of Ca-dependent K transport in human red blood cells," in U. V. Lassen, H. H. Ussing, and J. O. Wieth, Eds., *Membrane Transport in Erythrocytes*, Alfred Benzon Symp. 14, pp. 178–192.

Hoffmann, E. K., L. O. Simonsen, and A. Sjøholm (1979) Membrane potential, chloride exchange, and chloride conductance in Ehrlich mouse ascites tumor cells. *J. Physiol. (Lond.)*, **296**, 61–84.

Johnstone, R. M. (1978) The basic asymmetry of Na^+-dependent glycine transport in Ehrlich cells. *Biochim. Biophys. Acta*, **512**, 199–213.

Johnstone, R. M., and P. C. Laris (1980) Bimodal effects of cellular amino acids on Na-dependent amino acid transport in Ehrlich cells. *Biochim. Biophys. Acta*, **599**, 715–730.

Knauf, P. A., G. F. Fuhrmann, S. Rothstein, and A. Rothstein (1977) The relationship between anion exchange and net anion flow across the human red blood cell membrane. *J. Gen. Physiol.*, **69**, 363–386.

Kregenow, F. M. (1977) "Transport in avian red cells," in J. C. Ellory, and U. L. Lew, Eds., *Membrane Transport in Red Cells*, Academic, New York, p. 383–426. (See Fig. 12, p. 411.)

Laris, P. C., and G. V. Henius (1982) Influence of glucose on Ehrlich cell volume, ion transport, and membrane potential. *Am. J. Physiol.*, **242**, C326–C332.

Laris, P. C., H. A. Pershadsingh, and R. M. Johnstone (1976) Monitoring membrane potentials in Ehrlich ascites tumor cells by means of a fluorescent dye. *Biochim. Biophys. Acta*, **436**, 475–488.

Laris, P. C., M. Bootman, H. A. Pershadsingh, and R. M. Johnstone (1978) The influence of cellular amino acids and the Na^+:K^+ pump on the membrane potential of the Ehrlich ascites tumor cell. *Biochim. Biophys. Acta*, **512**, 397–414.

Lassen, U. V. (1972) "Membrane potential and membrane resistance of red cells," in M. Rorth and P. Astrup, Eds., *Oxygen Affinity of Hemoglobin and Red Cell Acid Base Status*, Academic, New York, pp. 291–304.

Lassen, U. V., A. M. Nielsen, L. Pape, and L. O. Simonsen (1971) The membrane potential of Ehrlich ascites tumor cells. *J. Membr. Biol.*, **6**, 269–288.

Macey, R. I., J. S. Adorante, and F. W. Orme (1978) Erythrocyte membrane potentials determined by hydrogen ion distribution. *Biochim. Biophys. Acta*, **512**, 284–295.

McManus, T. J., M. Haas, L. C. Starke, and C. Y. Lytle (1985) The duck red cell model of volume-sensitive chloride-dependent cation transport. *Ann. N.Y. Acad. Sci.*, in press.

Moon, R. B., and J. H. Richards (1973) Determination of intracellular pH by ^{31}P magnetic resonance. *J. Biol. Chem.*, **248**, 7276–7278.

Pape, L. (1982) Effect of extracellular Ca^{2+}, K^+ and OH^- on erythrocyte membrane potentials as monitored by the fluorescent probe 3,3'-dipropylthiodicarbocyanine. *Biochim. Biophys. Acta*, **686**, 225–232.

Parker, J. C., V. Castranova, and J. M. Goldinger (1977) Dog red blood cells: Na and K diffusion potentials with extracellular ATP. *J. Gen. Physiol.*, **69**, 417–430.

Philo, R. D., and A. A. Eddy (1978) The membrane potential of mouse ascites tumour cells studied with the fluorescent probe 3,3'-dipropyloxadicarbocyanine. *Biochem. J.*, **174**, 801–810.

Simchowitz, L., I. Spilberg, and P. DeWeer (1982) Sodium and potassium fluxes and membrane potential of human neutrophils. *J. Gen. Physiol.*, **79**, 453–479.

Sims, P. J., A. S. Waggoner, C.-H. Wang, and J. F. Hoffman (1974) Studies on the mechanism by which cyanine dyes measure membrane potential in red blood cells and phosphatidylcholine vesicles. *Biochemistry*, **13**, 3315–3330.

Slayman, C. L., and C. W. Slayman (1974) Depolarization of the plasma membrane of neurospora during active transport of glucose. Evidence for a proton dependent cotransport system. *Proc. Natl. Acad. Sci. USA*, **71**, 1935–1939.

Smith, T. C., and C. Levinson (1975) Direct measurements of the membrane potential of Ehrlich ascites tumor cells: Lack of effect of valinomycin and ouabain. *J. Membr. Biol.*, **23**, 349–365.

Smith, T. C., and S. C. Robinson (1981) The membrane potential of Ehrlich ascites tumor cells: An evaluation of the null point method. *J. Cell Physiol.*, **106**, 399–406.

Smith, T. C., and K. D. Vernon (1979) Correlation of the effect of Ca^{++} on Na^+ and K^+ permeability and membrane potential of Ehrlich ascites tumor cells. *J. Cell. Physiol.*, **98**, 359–369.

Taylor, J. S., C. Deutsch, G. G. McDonald, and D. F. Wilson (1981) Measurements of transmembrane pH gradients in human erythrocytes using ^{19}F NMR. *Anal. Biochem.*, **114**, 415–418.

Thornhill, W. B. (1983) KCl transport and cell volume changes in the Ehrlich ascites tumor cell. Ph.D. thesis, University of California, Santa Barbara.

Thornhill, W. B., and P. C. Laris (1984) KCl loss and cell shrinkage in the Ehrlich ascites tumor cell induced by hypotonic media, 2-deoxyglucose and propranolol. *Biochim. Biophys. Acta.*, **773**, 207–218.

Waggoner, A. (1976) Optical probes of membrane potential. *J. Membr. Biol.*, **27**, 1–18.

Wieth, J. O., J. Brahm, and J. Funder (1980) Transport and interaction of anions and protons in the red blood cell membrane. *Ann. N.Y. Acad. Sci.*, **341**, 394–418.

CHAPTER 12

AN ACOUSTO-OPTICALLY STEERED LASER SCANNING SYSTEM FOR MEASUREMENT OF ACTION POTENTIAL SPREAD IN INTACT HEART

MARTIN MORAD
Department of Physiology
University of Pennsylvania
Philadelphia, Pennsylvania

STEPHEN DILLON
Department of Pharmacology
College of Physicians and Surgeons
Columbia University
New York, New York

JAMES WEISS
Division of Cardiology
University of California—Los Angeles
Health Center
Los Angeles, California

1. INTRODUCTION   212
2. RESULTS   212
   2.1 DESIGN OF THE LASER SCANNER   215
   2.2 DEFLECTION OF THE LASER BEAM   215
   2.3 MODE OF OPERATION OF THE LASER SCANNER   215
   2.4 COMPARISON OF ACTIVATION MAPS OBTAINED WITH THE LASER SCANNER AND MULTISITE ELECTROGRAMS   216

    2.5   ELECTRICAL ACTIVATION OF INTACT FROG
        VENTRICLE   219
    2.6   LASER SCANNING OF THE MAMMALIAN HEART   220
  3.   SUMMARY AND CONCLUSION   222
     REFERENCES   225

1.   INTRODUCTION

Optical signals representing changes of membrane potential have been measured in a variety of tissues (Senseman and Salzberg, 1980; Davila et al., 1973; Hoffman & Laris, 1974; Vergara and Bezanilla, 1976; Nakajima and Gilai, 1980; Salzberg et al., 1973; Morad and Salama, 1979). In the past few years we have measured the membrane action potential from the heart of a variety of species using not only voltage-sensitive dyes (Salama and Morad, 1976; Morad and Salama, 1979) but also the intrinsic birefringence of the membrane (Weiss and Morad, 1981, 1983).

The magnitude and the spectral response of the optically measured action potential appear to depend not only on the staining of the tissue but on its intrinsic absorption and also on the chemical structure of the dye used (Ross and Reichardt, 1979). In heart muscle, the merocyanine family of dyes have the highest fractional absorption change in measuring the action potential (Morad and Salama, 1979; Fujii et al., 1981). The high signal-to-noise ratio of these dyes makes it possible to measure the changes of membrane potential with good discrimination without signal averaging. A different dye, WW781 (Gupta et al., 1981), in our experiments shows the highest signal-to-noise ratio ($\Delta F/F = 6$–10%) in the fluorescence mode in heart muscle. This dye did not bleach noticeably and had no significant phototoxic or pharmacologic effect on the heart muscle.

2.   RESULTS

Figure 1 represents the time course of contraction and scattered and fluorescent light in a frog ventricular preparation prior to staining, immediately after staining with merocyanine 540, and after contraction is suppressed by omitting Ca^{2+} from the perfusing solution. After staining with merocyanine 540, the motion-induced fluorescence signal (measured at longer wavelengths) not only increased in magnitude but also developed a discontinuity in the signal coincident with the upstroke of the action potential. Suppression of tension by omission of Ca^{2+} from perfusate uncovered a fluorescence signal with a shape characteristic of the cardiac action potential. Although the fluorescence signal appeared quite similar in general shape and time course to that recorded with a microelectrode, in some preparations

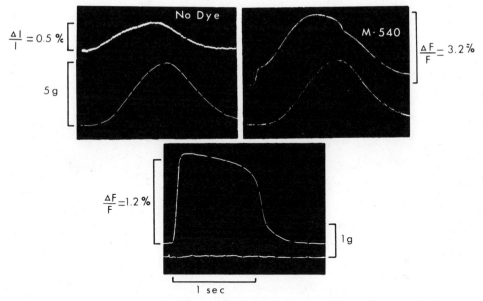

Figure 1. The 585-nm scattered light of an unstained ventricle (upper trace) induced by contraction has a time course similar to that of contraction (lower trace). In the upper right panel, muscle is stained with merocyanine 540 in 1 mM Ca^{2+} containing Ringer. The optical signal (upper trace) increases rapidly by 1.2% before the onset of tension. The fluorescence signal is then distorted as tension develops. In the bottom panel, contraction is completely suppressed by prolonged perfusion in Ca^{2+}-free solution (lower trace). The fluorescence increase of 1.2% has now the characteristic time course of the ventricular action potential. From Salama and Morad (1976).

there were significant differences, especially in the rate of rise of the action potential. These differences appeared to result from the averaging of the optical signal over about 0.5 mm light spot. Decreasing the size of the illuminated spot or the size of the preparation increased the rate of rise of the optically recorded action potential. It could be shown that the optically measured action potential in heart muscle was not contaminated by electrical signals arising from the voltage drop across the series resistance of the extracellular space (Morad and Salama, 1979). Thus voltage-sensitive fluorescence or absorption dyes may be used with great fidelity to follow the time course of membrane depolarization in heart muscle.

Development of new dyes with high signal-to-noise ratios made it possible to map the spread of electrical activity in intact and contracting heart. In our attempt to measure the spread of electrical activation wave in heart, we chose to sweep a laser beam rapidly and repeatedly over the points of interest and measure the fluorescence signal resulting from these points with a single photodiode. This approach, when applied to a large, three-dimensional contracting object, was felt to be ultimately more flexible than the multiple photodetector method used by others (Sawanobori et al., 1981; Fujii et al., 1981).

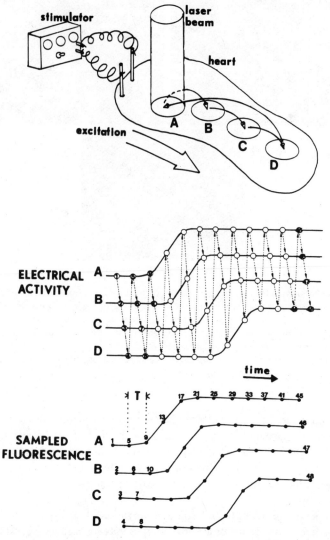

Figure 2. A technique for sampling the action potential "simultaneously" at different sites. The laser beam is moved sequentially between four theoretical sites—A, B, C, and D—on a heart stained with WW781 dye. As the action potential propagates from left to right, it would sequentially activate tissue at the indicated sites. The continuous traces represent action potential upstrokes recorded with multiple microelectrodes. These traces could also represent the fluorescence signals from A to B to C and to D within a short time interval *T*. During this sequence of spot movements the stimulated fluorescence was sampled and stored. After site D is sampled, the laser spot returns to site A to begin another cycle. This sampling sequence is indicated by dotted arrows which connect circled numbers on the continuous traces. The sample fluorescence traces for each site are reassembled as shown at the bottom of the figure. Sample numbers 1,5,9,13,...,41,45 comprise trace A, 2,6,10,...,46 comprise trace B, and so on. The results described in the figures to follow were obtained from 128 sites where the cycle time *T* was 4 ms.

2.1   Design of the Laser Scanner

The principle underlying the design of laser scanning system was to move the laser beam rapidly and repeatedly between discrete sites on the heart and to sample the fluorescence change occurring at each site as caused by membrane depolarization. This idea is illustrated schematically in Fig. 2 with four hypothetical recording sites—A, B, C, and D. As the action potential is conducted from left to right, across each of the four sites, the upstroke of the action potential, representing the moment of activation of that spot, occurs with increasing delay. To sample the fluorescence activity at each of these sites, the laser spot was moved stepwise to spots A, B, C, and D, halting at each spot long enough to permit a fluorescence measurement. After site D was sampled, the laser beam returned to site A, repeating the sampling sequence for a duration determined by the experimenter. If the sampling process is carried out rapidly, then the fluorescence traces would have the same activation sequence as those recorded electrically. As can be seen in Fig. 2, the fluorescence upstroke at spot A would consist of measurement 1, 5, 9, 13, 17, ... With this arrangement, a single photodiode combined with a movable laser spot could be used to create multiple-site optical recordings.

2.2   Deflection of the Laser Beam

The laser beam was moved rapidly in two dimensions with two perpendicularly aligned acousto-optical deflectors. The acousto-optical deflector consists of a voltage-controlled oscillator (50–90 MHz) and series of spherical and cylindrical lenses. The deflector is composed of a glass substrate with ultrasonic transducers bonded to one end. The oscillator signal is transduced into planar acoustic wave fronts that traverse the length of the glass. The compressions and expansions of the glass by the peaks and troughs of the sound wave create a periodic variation in the refractive index across the material. In effect, a thick grating with capability of diffracting visible light is created. This process is analogous to "Bragg" reflection of X-ray light in crystals.

The laser beam, introduced at a small angle into the aperture of the device, is efficiently diffracted out (79–90%) at a new angle. The beam steering is accomplished by changing the frequency input oscillator, which creates a new grating spacing and different exit angle for the beam (Adler, 1967). By using a pair of such devices, the laser spot can be positioned in two dimensions. It takes the deflection system 5 μs to respond to a new set of control signals. When the control signals are held steady, the spot remains stationary.

2.3   Mode of Operation of the Laser Scanner

A minicomputer (PDP 11/23) was used to control the beam position and store the fluorescence measurements. The measurement cycle during scanning starts with presentation of the first set of site coordinates, via a digital-to-analog converter to the acousto-optic devices. After a 5-μs delay the spot appears at the desired site in the heart. During the next 25 μs the photodiode settles to the new fluorescence level being emitted at this spot on the heart. A sample-and-hold amplifier then records this signal as the computer sends the coordinates for the second spot to be sampled. As the new spot is illuminated, the previous fluorescence signal is digitized and stored by the computer. This deflection and measurement cycle is repeated every 31.25 μs, so that 128 sites can be measured every 4 ms. This process may be repeated until the computer memory is filled. Such a system can be operated four times faster, but problems arising from bandwidth-related background noise and a requirement for large computer memory weighed against the advantages of additional speed. The 4-ms sampling time was sufficiently fast to permit construction of activation maps that can depict the spread of the electrical impulse in the heart. The spatial resolution was fixed at 16,384 sites set on a square grid of 128 spots on a side regardless of the spot size. Though the acousto-optical deflector is capable of continuous beam movement, it is intrinsically limited to resolving 150 distinct spots in each dimension, so the spot coordinates are quantized to a conservative 128 values per axis. Sites can be addressed in a random access manner; the time delay between any two movements is fixed at 5 μs.

To create a list of scan sites, a pair of positioning knobs was used to maneuver the laser beam. If a spot was to be included in the scan pattern, a button was depressed, and the spot remained illuminated as the beam was moved elsewhere. Thus a list of spots with independent coordinates was recorded. Since each point was sequentially illuminated at high speed, a continuous projection of the scan pattern could be seen. Figure 3 shows the spots to be scanned by the laser beam on a rabbit heart that is perfused by two of its coronary arteries and stained with the dye WW781. The grid of the laser spots was produced by the rapid movement of single laser beam, giving the impression that the spots are frozen in place. The detector is positioned such that it picks up the signal uniformly from over the entire preparation.

2.4   Comparison of Activation Maps Obtained with the Laser Scanner and Multisite Electrograms

To confirm the reliability of the laser scanning device in measuring the spread of electrical activity, we used the laser scanner simultaneously with 16 extracellular unipolar electrodes. We chose the thin sheet of isolated bullfrog atrium so that both sides of the atrium could be well illuminated by the laser beam for fluorescence measurements. Figure 4 shows a simultaneous laser scan and multisite electrogram recorded from a thin atrial sheet. Electrode activation times (indicated by stippled circles) correspond fairly closely to the isochrones obtained by detailed (128-site) laser scanner. The activation waves of the atrium appear to spread from the site of stimulation (asterisk, Fig. 4), dividing into two wave fronts around the venous opening at the center of the atrial sheet, merging finally with each other at the lower corner of the preparation (see inset, Fig. 4). Altering the position of the stimulating electrode changed the pattern of activation, but the correlation between the activation

A.

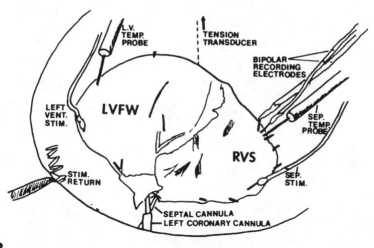

B.

Figure 3. (*A*) An isolated, stained rabbit ventricle being scanned at 128 sites on the epicardial surface. Although the grid of spots appears to be formed by a stationary projection, it is actually created by a rapidly moving laser spot, which has refreshed the pattern many times during the photographic exposure. The number and positions of these spots are under computer control and are determined by the experimenter. Software allows the pattern to be created or changed very easily, thus permitting great flexibility in the choice of scan sites. (*B*) Drawing illustrating the layout of the preparation and the chamber. The left ventricle is isolated from a rabbit heart and cut open along the posterior septum. The left ventricular free wall (LVFW) and the ventricular septum (RVS) are independent regions which are perfused through two separate branches of the left coronary artery. This allows the creation of regional ischemia, ionic manipulation, or drug exposure. Under normal conditions oxygenated Tyrode solution is pumped through each (septal and left) cannula at 2 mL/min. A warm, moist nitrogen atmosphere was maintained over the surface of the heart in order to ensure anoxic surface conditions during ischemia. Ischemia was produced by cutting off flow through the appropriate cannula. Several electrodes were placed at the tissue margins to stimulate and record from the preparation.

times measured with the laser scanner and extracellular electrodes remained true (bottom 2 activation maps, Fig. 4).

To test whether the activation pattern of atrium was markedly distorted by motion-induced light scattering, the contractility of the atrial sheet was changed by elevation of Ca^{2+} or addition of epinephrine to the perfusate. The activation pattern of the atrium was found not to change significantly either in the presence of these

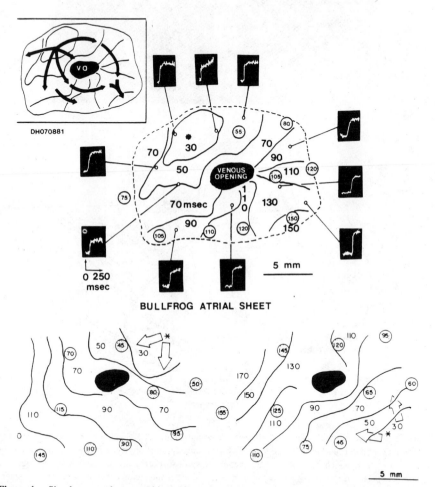

Figure 4. Simultaneous electrographic and laser-scanned recording of atrial activation. A thin sheet of a bullfrog atrium was stretched over a 16-unipolar electrode array. A multiplexing circuit was used to record simultaneously from 16 electrodes. The central map of the figure shows the conduction pattern of an impulse stimulated at a point indicated by asterisk as determined by a 128-site laser scan. The stippled circles indicate the positions and timing from which electrographic activation times were determined. The electrographic times correspond well with the optically derived isochrones. Insets show the optically recorded action potential upstrokes taken from various sites on the sheet. Two other maps show the conduction patterns that resulted from a change in stimulation site. The good agreement between activation times determined optically and electrically verifies the accuracy of the laser scanning technique.

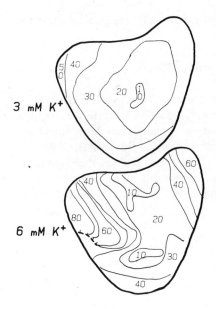

3 mM K⁺

6 mM K⁺

Figure 5. Activation maps of self-pacing frog ventricle. The two maps show activation of the epicardial surface in two different $[K^+]_o$'s. A 128-site scan pattern on the ventricular surface recorded the sequence of activation as the impulse was generated in the atrium and was conducted to the ventricle. The upper map shows that the impulse initially appeared in the center and spread outward to activate the apex and the base. Elevation of external potassium not only caused a reduction in conduction velocity and impulse block but did not alter the general epicardial breakthrough pattern. Earliest activation is delayed by 10 ms.

agents or when the tissue contraction was suppressed in the presence of 50 μM Ca^{2+} and diltiazem (a Ca^{2+} antagonist). These results confirm the reliability of activation maps constructed using the laser scanning technique.

2.5   Electrical Activation of Intact Frog Ventricle

Figure 5 shows the activation sequence in an intact bullfrog ventricle, driven by its own pacemaker. The activation times illustrated were calculated relative to the time of appearance of the first action potential on the surface of the heart. The 20-ms isochrones thus constructed show that epicardial surface excitation in a normally driven ventricle first appeared at the center and then spread outward to activate the apex and the base of the heart. The spread of activation proceeds at about 0.1 m/s, consistent with earlier measurement of spread of excitation in this tissue (Hoffman and Cranefield, 1960). Since the excitation wave was initiated by the normal pacemaker, conducted through the atrium and atrioventricular ring to the ventricle, the appearance of the first action potential at the center of the ventricle may suggest preferential conduction pathways on the endocardium of the frog ventricle (Lewis, 1915). Measurements of epicardial and endocardial activation in a ventricular preparation that was split open to allow laser scanning of both sides of the same preparation confirmed the more rapid spread of activation front on the endocardial side of the heart (Dillon and Morad, 1981). The endocardial surface appeared to activate at a rate twice as fast as the epicardial surface. We could not determine whether this difference in conduction time resulted from electrophysiologic or structural specialization of the endocardium. Elevation of $[K^+]_o$ from 3 to 6 mM

slowed the spread of epicardial activation but did not alter the general pattern of spread of the electrical wave front.

2.6   Laser Scanning of the Mammalian Heart

In order to map normal as well as arrhythmic patterns of activation in mammalian hearts, we made use of an isolated, arterially perfused rabbit heart (Fig. 3). In these preparations the left ventricle was isolated, cut open, and pinned down in a heated chamber. Each half of the left ventricle, the left ventricular free wall (LVFW), and the ventricular septum (RVS) were independently perfused via the appropriate branches of the left coronary artery. Figure 3 shows this arrangement in a heart stained with WW781, with a 128-site scan pattern projected on it by the laser beam. Such preparations were then scanned under control conditions, after addition of

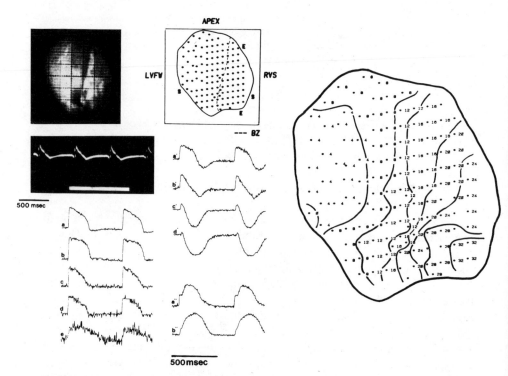

Figure 6.   Laser scan of 128 sites was used to construct a map of activation for a control driven beat. A fluorescence image, upper left corner, was used to select the sites to be scanned. An outline of the preparation, upper right corner, indicates the positions of the chosen scan sites. On this outline the positions of recording electrodes and stimulating electrodes are designated by E and S, respectively. The border between the two perfusion zones is indicated by a dashed line. The white bar on the electrogram recording indicates the duration of laser scanning. Traces a–e are optical records taken from various sites during the scan. Trace a shows the full time course of the cardiac action potential with virtually no distortion, whereas trace e is markedly altered by the background noise. Traces a' through d' and a'' and b'' show the distorting effects of motion artifact. A 4-ms isochrone shows the progression of the activation wave from left to right.

drugs, and during regional ischemia. Single or multiple extrastimuli were administered to induce arrhythmias.

Figure 6 shows 4-ms isochrones constructed after scanning such a heart under control conditions. Note that sample optical action potentials, taken from various regions, all exhibit a distinct upstroke despite the presence of noise or motion artifact. The timing of these upstrokes makes it possible to construct activation maps. The shape of the action potential, however, is often less reliable, since contraction-induced light scattering may add to or subtract from the slower phases of the action potential signal.

Figure 7 illustrates the changes brought about in the conduction pattern when an extrastimulus was introduced into a basic rhythm. The basic driven beats occurring

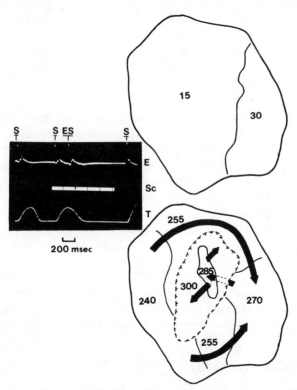

Figure 7.   Optical scan of a control and extrasystolic beat. The extra beat is generated by a stimulus (ES) delivered 225 ms after the second of the three basic driven beats (s) (deflections on the recorded bipolar electrogram, E). The tension trace (T) shows prolongation of contraction accompanying the ES beat. The action potential propagation of the S and ES beat was scanned during the 1-s interval indicated by the white bar (SC). The resultant maps are shown at the top and bottom of the figure. Activation by the basic driven beat occurs rapidly (30 ms, top map). The extra beat entering tissue that is relatively refractory is conducted more slowly than the basic driven beat owing to diminished excitatory current (lower map). In the central area (zone surrounded by T-shaped black symbols), the advancing wave front was unable to activate the tissue immediately. Activity broke through in the middle at 285 ms (45 ms post-extrastimulus), and activation was complete within 300 ms.

every 600 ms, marked (S), were interrupted by an extrastimulus (ES). Although no independent contraction accompanied the ES beat (inset, Fig. 7), the contraction of the extra beat was prolonged and slightly enhanced, suggesting additional activation of some segment of the heart. A slight potentiation of the beat following extrasystole can also be seen. Conduction maps of the normal driven (S) beat depict rapid left-to-right conduction of the impulse. The ES beat, elicited 220 ms after the S beat, however, appeared to encounter a segment of heart where the impulse is delayed, partially blocked, and shunted around. The impulse then appeared to break through the center via a presumed subsurface pathway indicated by dashed arrows. The premature impulse required about 50% more time to traverse the heart than the control beat. These results suggest that significant activation delays may exist in anatomically adjacent locations when a premature beat arises in the heart. In constructing activation maps using optical techniques, the possibility of "hidden conduction" should always be considered, since the laser scanning beam does not penetrate more than 0.5 mm below the epicardial surface.

It could be shown that in preparations that were made ischemic by interrupting coronary flow, premature stimuli led to sustained arrhythmia if the conduction of the extrastimulus was sufficiently slow and the path length sufficiently long as to permit reentry. The reentrant impulse appeared to continue propagating through the heart via other paths depending on which part of the tissue was still refractory. Figure 8 shows the activation maps constructed for the control autopaced beat, first extra beat (delivered at RVS, conducted retrograde), and the second extra beat, which encounters an area of block near the base, as it is conducted in a circular manner. This propagation wave is then broken up and appears to reenter the previously depolarized segments (R_1, R_2, R_3). Reentry is made possible because of an area of unidirectional block and slow conduction of the impulse over the surface of the heart.

The laser scanner system described above can also distinguish between various types of arrhythmia known to occur in the heart. Figure 9 shows the scan of arrhythmia induced by cardiac glycoside acetylstrophanthidin (ACS). Note that unlike the reentrant-type arrhythmia, the activation map of ACS-induced tachycardia has always the same regular pattern, although occurring very rapidly. This pattern is in sharp contrast to that found in the reentrant arrhythmia, where the propagation wave appears to fractionate and multiple foci of shifting block occur.

3. SUMMARY AND CONCLUSION

An optical scanning device that combines a voltage-sensitive dye and an acousto-optically steered He–Ne laser beam is described. This device is capable of scanning 128 sites every 4 ms and recording and storing the fluorescence signals for a duration of up to 1 s (several beats). Comparison of an activation map constructed from laser scanning to those obtained from multiple extracellular electrodes suggests

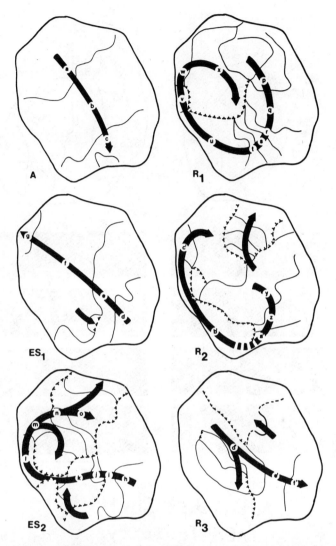

Figure 8. The conduction patterns of induced tachycardia caused by impulse reentry. Map A shows the activation of a control driven beat from a site in the apical portion of the heart. Two extrastimuli (ES1 and ES2) are delivered at the right basal portion of the heart. The first extra beat is conducted with a diminished velocity and small, temporary zone of block. The second, premature beat is conducted at a considerably slower rate, showing many zones of block which markedly delayed total tissue activation (zones *h* through *o*). The impulse, however, after reaching the 12 o'clock position does not die out, because the cells ahead of it were no longer refractory. The impulse thus makes another sweep through the heart (beat R_1). The impulse continued to reenter as shown in maps R_2 and R_3 for the duration of the 1-s scan. Multiple spontaneous impulses, generated by this reentrant mechanism, can be distinguished from those due to automaticity on the basis of their activation patterns.

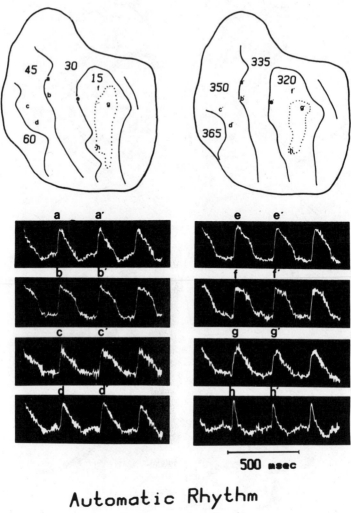

Automatic Rhythm
(ACS Toxicity)

Figure 9. Laser scan of tachycardia induced by acetylstrophanthidin. Analysis of the propagation maps revealed a repetitive pattern of impulse spread from a fixed focus on the preparation. The top of the figure shows two consecutive beats where the zones of earliest activity have been indicated by dotted isochrones. This pattern is unlike that shown in Fig. 8 in that the next spontaneous beat was not dependent on the previous impulse. A gap of some 260 ms separates the beats. Automaticity appears stable and regular. Eight traces of optical action potentials are shown as they were recorded from the lettered sites on the maps. Traces *g* and *h* were examined for signs of diastolic depolarization, which would mark them as pacemaking sites, since they resided in the earliest activity zone. There was no indication of such "phase 4" deflection.

that this technique is highly reliable. Although motion-induced light scattering appears to alter the shape of the action potential, the upstroke can be distinguished quite reliably even in a vigorously contracting muscle. This technique provides high resolution (up to 50 μm) and high flexibility (i.e., the scanned sites can be concentrated over a small or very large area) in measuring the spread of activation in heart muscle. By having only one excitation and one measurement element, the approach offers simplicity and high flexibility to the user.

We have shown that this system can be readily applied to the task for which it was intended—probing the mechanisms of arrhythmias in the mammalian myocardium. It has been demonstrated, for example, that arrhythmias due to automaticity can be readily distinguished from those due to reentry through the mapping capability of the laser scanner. In addition, the ability of laser scanner to measure membrane depolarization directly during arrhythmias may make this technique superior to conventional electrocardiographic mapping techniques.

REFERENCES

Adler, R. (1967) Interaction between light and sound. *IEEE Spectrum*, **4**, 42–54.

Davila, H. V., B. M. Salzberg, and L. B. Cohen (1973) A large change in axon fluorescence that provides a promising method for measuring membrane potential. *Nature New Biol.*, **241**, 159–160.

Dillon, S., and M. Morad (1981) A new laser scanning system for measuring action potential propagation in the heart. *Science*, **214**, 453–456.

Fujii, S., A. Hirota, and R. Kamino (1981) Action potential synchrony in embryonic precontractile chick heart: Optical monitoring with potentiometric dyes. *J. Physiol. (Lond.)*, **319**, 529–541.

Gupta, R. K., B. M. Salzberg, A. Grinvald, L. B. Cohen, K. Kamino, S. Lesher, M. B. Boyle, A. S. Waggoner, and C. H. Wang (1981) Improvements in optical methods for measuring rapid changes in membrane potential. *J. Membr. Biol.*, **58**, 123–137.

Hoffman, J. F., and P. C. Laris (1974) Determination of membrane potentials in human and *Amphiuma* red blood cells by means of a fluorescent probe. *J. Physiol. (Lond.)*, **239**, 519–552.

Hoffman, B., and P. F. Cranefield (1960) *Electrophysiology of the Heart*. McGraw-Hill, New York, p. 79.

Lewis, T. (1915) The spread of the excitatory process in the toad's ventricle. *J. Physiol. (Lond.)*, **49**, p36–p37.

Morad, M., and G. Salama (1979) Optical probes of membrane potential in the heart. *J. Physiol. (Lond.)*, **292**, 267–295.

Nakajima, S., and A. Gilai (1980) Radial propagation of the muscle action potential along the tubular system examined by potential sensitive dyes. *J. Gen. Physiol.*, **76**, 751–762.

Ross, W. N., and L. F. Reichardt (1979) Species-specific effects on the optical signals of voltage-sensitive dyes. *J. Membr. Biol.*, **48**, 343–356.

Salama, G., and M. Morad (1976) Merocyanine 540 as an optical probe of transmembrane electrical activity in the heart. *Science*, **191**, 485–487.

Salzberg, B. M., H. V. Davila, and L. B. Cohen (1973) Optical recording of impulses in individual neurons of an invertebrate central nervous system. *Nature*, **246**, 508–509.

Sawanobori, T., A. Hirota, S. Fujii, and R. Kamino (1981) Optical recording of conducted action potential in heart muscle using a voltage sensitive dye. *Jpn. J. Physiol.*, **31**, 369–380.

Senseman, D. M., and B. M. Salzberg (1980) Electrical activity in an exocrine gland: Optical recording with a potentiometric dye. *Science*, **208**, 1269.

Vergara, J., and F. Bezanilla (1976) Fluorescence change during electrical activity in frog muscle stained with merocyanine. *Nature*, **259**, 684–686.

Weiss, R., and M. Morad (1981) Intrinsic birefringence signal preceding the onset of contraction in heart muscle. *Science*, **213**, 663–666.

Weiss, R., and M. Morad (1983) Birefringence signals in mammalian and frog myocardium: E–C coupling implications. *J. Gen. Physiol.*, **82**, 79–117.

PART 3

INTRACELLULAR INDICATORS

CHAPTER 13

PRACTICAL ASPECTS OF THE USE OF PHOTOPROTEINS AS BIOLOGICAL CALCIUM INDICATORS

JOHN R. BLINKS
EDWIN D. W. MOORE
Department of Pharmacology
Mayo Graduate School of Medicine
Rochester, Minnesota

1. INTRODUCTION   230
2. FREEDOM FROM MOTION ARTIFACTS: A SALIENT ADVANTAGE OF PHOTOPROTEINS FOR STUDIES ON MUSCLE   230
3. RELATION BETWEEN $[Ca^{2+}]$ AND LIGHT INTENSITY: AEQUORIN, OBELIN, AND ACETYLATED AEQUORIN   232
4. SPEED OF RESPONSE: THE EFFECT OF PREEQUILIBRATION WITH Mg^{2+} 233
5. INTRODUCTION INTO CELLS: ALTERNATIVES TO MICROINJECTION   235
6. CONCLUSIONS   236
    REFERENCES   237

Original work described in this paper was supported by USPHS grant HL 12186.

1. INTRODUCTION

The substances under discussion in this presentation are naturally occurring intracellular calcium indicators that are responsible for the bioluminescence of a variety of marine organisms, mostly coelenterates. In many of these animals the photoproteins are stored in specialized cells known as photocytes from which light emission takes place. Within the photocytes the intensity of luminescence is regulated by changes in cytoplasmic calcium ion concentration, the photoproteins serving as calcium-regulated effector proteins in much the same sense as troponin C or calmodulin. The photocytes appear to regulate their intracellular calcium concentration over the same range, and apparently by similar mechanisms, as many other kinds of cells (4). A major difference between the photoproteins and other calcium-regulated proteins is that the photoproteins can exert their physiological action without interacting with any substance other than Ca^{2+}. All of the components required for luminescence (the apoprotein, a low-molecular-weight chromophore, and oxygen) are bound together and behave as a single macromolecule of about 20,000 daltons. This makes it relatively easy to isolate and purify the luminescent system, and to transfer it to the cytoplasm of cells that were not provided with it as original equipment. However, an undesirable corollary of this convenient packaging feature is that the photoprotein reacts only once, since the energy required for photon emission comes from the degradation of the built-in chromophore.

The use of calcium-regulated photoproteins as intracellular $[Ca^{2+}]$ indicators is now well established: in the last decade aequorin (from the jellyfish *Aequorea*) has been used in more than 50 different types of cells, and obelin (from the hydroid *Obelia*) in a number of others. Several comprehensive reviews dealing with the properties and applications of the photoproteins are available (2–5); it is not our purpose here to repeat a large amount of information that is available in those places. It seems preferable to confine our attention to a few recent developments that bear on the suitability of photoproteins for use as cytoplasmic $[Ca^{2+}]$ indicators in living cells. This will be done in the context of a discussion of some of the salient advantages and disadvantages of the photoproteins in relation to other intracellular $[Ca^{2+}]$ indicators.

2. FREEDOM FROM MOTION ARTIFACTS: A SALIENT ADVANTAGE OF PHOTOPROTEINS FOR STUDIES ON MUSCLE

Probably the outstanding advantage of the photoproteins over other available $[Ca^{2+}]$ indicators is their relative immunity from motion artifacts. This, of course, makes them particularly useful in muscle cells and follows from the nature of the signal that they provide—light emission against a background of total darkness. It seems evident that indicators that simply emit light will be less susceptible to motion artifacts than those that create a signal by altering the character of an incident light beam. However, the extent to which this will be true must depend on the experimental conditions, and

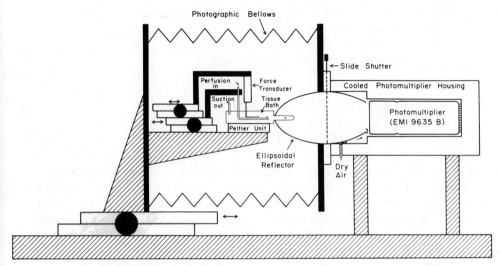

Figure 1. Apparatus for recording very low light intensities from tissues injected with photoproteins. The heart of the apparatus is the ellipsoidal reflector, which has one point of focus in a glass tube projecting from the tissue bath, and the other at the photocathode of the photomultiplier. The tissue is suspended between two hooks, which can be advanced together into the glass tube by means of a micromanipulator movement. A photographic bellows can be closed to make the apparatus completely light-tight. A slow flow of dry air keeps moisture from accumulating on the photomultiplier or dynode chain under humid conditions or when the photomultiplier is cooled. (From ref. 2, with permission of the publisher.)

no indicator is likely to provide signals that are absolutely immune from mechanical artifacts. In the case of the metallochromic dyes, artifacts due to motion of the cell of interest have proved to be so severe that most workers have found it necessary to prevent contraction by one means or another during studies of the calcium transients of muscle cells. Needless to say, this can be a major handicap if one's goal is to relate calcium transients to the contractile process of muscle.

Motion artifacts have not been noted in studies with Ca^{2+}-regulated photoproteins, but this of course is no proof that they do not occur. The best means we have been able to think of to assess the likely magnitude of the problem in studies with photoprotein-injected muscle preparations is to "tattoo" the surface of similar preparations with luminescent particles scraped from luminous watch dials, so as to produce a distribution of luminescence similar to that in the preparations that contain the photoprotein. Light signals are then recorded from the tattooed muscles in exactly the same way as in the photoprotein-injected muscles to see whether the kind of motion encountered in a particular type of experiment produced any change in the light signal recorded from the tattooed preparation. So far we have not found that it did, even in muscles undergoing lightly loaded isotonic contractions, but one should be alert to the possibility that the result might not be the same under all circumstances.

In the pertinent experiments we used an ellipsoidal mirror system (see Fig. 1) to reflect the light from the preparation to the photocathode of the photomultiplier. One of the advantages of this system is that it is symmetrical around the axis of the muscle, and thus rotation of the muscle during contraction should not have a major effect on the efficiency of light gathering. Systems in which the photomultiplier gathers light from the side of the preparation might be more susceptible to motion artifacts, particularly if the aperture of the system is small.

3. RELATION BETWEEN [Ca^{2+}] AND LIGHT INTENSITY: AEQUORIN, OBELIN, AND ACETYLATED AEQUORIN

One feature of the photoproteins that is frequently cited as a disadvantage from the standpoint of quantitative interpretation is the nonlinear nature of the relation between [Ca^{2+}] and light emission. As is illustrated in Fig. 2, the Ca^{2+} concentration-effect curves for aequorin and obelin have very similar shapes, though obelin is appreciably less sensitive to Ca^{2+}. (The difference between the two curves would be slightly greater if they had been determined in solutions of the same salt concentration.) The curves are exceedingly steep: on log–log plots such as those shown here, the midportions of both curves have slopes of approximately 2.5, and the slopes are appreciably above 1.0 throughout most of the useful range of the indicators. The fact that signal strength changes more steeply than does [Ca^{2+}] considerably complicates the quantitative interpretation of light signals recorded from cells containing photoproteins, particularly under circumstances in which gradients of [Ca^{2+}] may exist within the space containing the indicator. Under such conditions, the signal will

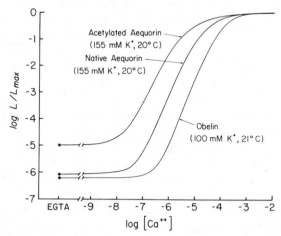

Figure 2. Ca^{2+} concentration-effect curves for native aequorin, acetylated aequorin, and obelin. The first two curves are original and were determined in 155 mM KCl, 5 mM PIPES, pH 7.0, to which CaCl$_2$ (above 10^{-6} M Ca^{2+}) or CaEGTA buffers (below 10^{-6} M Ca^{2+}) were added (total [EGTA] 1 mM). The curve for obelin is redrawn from ref. 24, and was determined in solutions containing 20 mM EGTA or nitrilotriacetate, 20 mM TES, pH 7.0, and 100 mM K^{+}. Double logarithmic plot. L refers to the light intensity measured at a particular [Ca^{2+}]; L$_{max}$ is the peak light intensity for an identical aliquot of photoprotein injected into saturating [Ca^{2+}]. Each curve is normalized to its own L$_{max}$.

be dominated by contributions from those regions where the $[Ca^{2+}]$ is highest, and any calculations of absolute $[Ca^{2+}]$ made from the overall signal can be regarded only as upper limits on the average $[Ca^{2+}]$ within the preparation. Of course, it should be noted that this property of the photoproteins can sometimes work to the experimenter's advantage, as when microscopic image intensification is used to discern spatial differences in $[Ca^{2+}]$ within the cytoplasm.

Shimomura and Shimomura (20) recently reported that the properties of aequorin could be altered in significant ways by chemical modifications of the protein. In particular, they found that the Ca^{2+} concentration-effect curve for acetylated aequorin appeared to be considerably less steep than that for the native protein. This development attracted notice particularly because it was hoped that it might herald the availability of a "linear" aequorin, the signals from which might be more readily interpreted than those of native photoproteins. We took a careful look at this question (16), and found that while acetylation did indeed reduce the slope of the Ca^{2+} concentration-effect curve, it did not reduce it to 1.0. In our hands, the midportion of the curve for acetylated aequorin has a slope between 1.8 and 1.9 (see Fig. 2).

The most striking effect of acetylation on the Ca^{2+} concentration-effect curve for aequorin is an order-of-magnitude increase in the level of the Ca^{2+}-independent luminescence. Although Fig. 2 shows a modest leftward shift of the "foot" of the concentration-effect curve as a result of acetylation, this change practically disappears in the presence of physiological concentrations of Mg^{2+} because acetylated aequorin is somewhat more sensitive to the inhibitory effect of Mg^{2+} than is the native protein. Thus, acetylation does not appreciably alter the lowest $[Ca^{2+}]$ that can be detected with aequorin inside a living cell. What it does do is to reduce the amount of photoprotein that must be introduced into the cell in order to achieve detectable light levels in the presence of very low Ca^{2+} concentrations. This advantage comes with a price tag, as we shall see a little later.

4.   SPEED OF RESPONSE: THE EFFECT OF PREEQUILIBRATION WITH Mg^{2+}

A disadvantage of many of the available methods for measuring intracellular $[Ca^{2+}]$ is that the indicator responds too slowly to follow the $[Ca^{2+}]$ transients of certain excitable cells without distortion. Some methods, such as those involving nuclear magnetic resonance of ^{19}F-labeled chelators (21) and Ca^{2+}-selective microelectrodes, are intrinsically so slow that in their present state of development one would not even consider trying to use them to measure rapid $[Ca^{2+}]$ transients such as those occurring during the twitches of skeletal muscle. The fluorescent indicator Quin2 (27, 28) is inherently moderately fast (19), but cannot be used to measure rapid Ca^{2+} transients for an entirely different reason: It usually must be used in such high concentrations that it buffers the intracellular $[Ca^{2+}]$ and greatly attenuates any rapid transients that the cell might try to produce (28). Of the indicators currently available, only the metallochromic dye antipyrylazo III is capable of responding quickly enough to track the fastest biological $[Ca^{2+}]$ transients without distortion. Baylor et al. (1) have shown that antipyrylazo III is distinctly preferable to arsenazo III in this respect, and both of these dyes are probably capable of responding more quickly than aequorin under physiological conditions. Although we once believed that aequorin would respond about as fast as arsenazo III under the conditions likely to prevail inside most kinds of cells (see ref. 5 for discussion), we have recently found that we (and everyone else) had overlooked an important property of the photoprotein that forces us to alter that conclusion.

The original kinetic studies with aequorin were carried out in experiments in which magnesium-free aequorin was rapidly mixed with solutions containing various concentrations of Ca^{2+} or mixtures of Ca^{2+} and Mg^{2+} (10). Under those conditions the presence of Mg^{2+} was found not to influence the kinetics of the response to a sudden change in $[Ca^{2+}]$, though of course it has long been appreciated that the Ca^{2+} sensitivity of the photoprotein is reduced in the presence of Mg^{2+}. However, we have recently found that if aequorin is *preequilibrated* with millimolar concentrations of Mg^{2+}, its response to a step change in $[Ca^{2+}]$ is significantly slowed (15). Specifically, after preequilibration with 3 mM Mg^{2+} (one of the higher estimates of cytoplasmic $[Mg^{2+}]$ (11)), the initial rate of rise of the aequorin signal on rapid mixing with Ca^{2+} is reduced by more than half (see Fig. 3). In the light of current knowledge about the rate constants governing the binding of magnesium to other

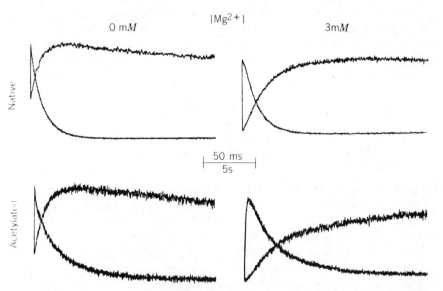

Figure 3. The influence of preequilibration with Mg^{2+} on the kinetics of native (upper panels) and acetylated (lower panels) aequorin. Rapid mixing stopped-flow records of the rise of light intensity after the photoprotein was mixed with a saturating concentration of Ca^{2+}. Each panel contains superimposed traces recorded at two sweep speeds. All solutions contained 150 mM KCl, 5 mM PIPES, pH 7.0. Equal volumes were mixed from syringes containing (a) the photoprotein and (b) 10 mM CaCl$_2$. For the records in the right-hand panels, both syringes contained 3 mM MgCl$_2$ as well. The records were obtained with a Gibson-type stopped-flow apparatus; the dead time was approximately 4 ms. Temperature 20°C.

calcium-binding proteins, this finding seems hardly surprising. However, the possibility that the rate at which aequorin binds Ca^{2+} might be limited by the rate of dissociation of Mg^{2+} seems not to have been considered previously.

Acetylated aequorin responds nearly as fast as native aequorin in the absence of Mg^{2+}, but after preequilibration with 3 mM Mg^{2+}, the initial rate of rise of light intensity after rapid mixing with Ca^{2+} is only half as great for acetylated aequorin as for the native photoprotein (15). For any particular experimental application this potential disadvantage must be weighed against any possible advantage that might be derived from the brighter signals provided by the acetylated photoprotein. Obelin responds nearly three times as fast as aequorin (24) and for this reason may be preferable to aequorin in circumstances in which its lower sensitivity to Ca^{2+} (see Fig. 2) is adequate and a rapid response is required. To our knowledge, the effect of preequilibrating the photoprotein with Mg^{2+} on obelin kinetics has not yet been studied.

5. INTRODUCTION INTO CELLS: ALTERNATIVES TO MICROINJECTION

Another property of the photoproteins that makes them hard to work with stems from their large molecular size. Photoproteins are difficult to get into most kinds of cells, although they do have a gratifying tendency to stay in the compartment where you put them once you have managed to get them there. Unlike the metallochromic dyes, the photoproteins do not lend themselves to iontophoresis, and they have an annoying tendency to plug the very small-tipped pipettes that must be used for microinjection in cells of ordinary size. (Perhaps one cannot fully appreciate the significance of Roger Tsien's developing the labile ester technique (26) for loading tetracarboxylate calcium indicators into cells until one has worked for a while at microinjecting photoproteins!) Microinjection is possible, even in relatively small cells, but it is difficult and time-consuming, and it is only natural that continuing efforts are being made to develop easier techniques for getting photoproteins into cells, particularly populations of small cells. We are aware of three different methods that have been clearly shown to work, and to leave photoprotein-loaded cells that seem to function reasonably normally. All depend on producing reversible damage to the surface membrane while the cells are exposed to a solution containing the photoprotein.

Obviously, the cells must be incubated in a solution that is essentially Ca^{2+} free while they are being exposed to the photoprotein, or the photoprotein would be discharged too rapidly to be of much use. This in itself may be responsible for some of the membrane damage that is produced by all of the methods, but one of the new techniques depends entirely on that mechanism. That is the EGTA loading technique first reported by Sutherland et al. (25), subsequently developed by Morgan et al. (17, 18) for use in mammalian vascular and cardiac muscle, and then applied to suspensions of cultured cells by Snider et al. (22). In this technique the cells are first made hyperpermeable by incubating them in a solution of very low divalent cation concentration. They are then exposed to the photoprotein and finally resealed by first increasing the $[Mg^{2+}]$ (to about 10 mM), and then gradually increasing the $[Ca^{2+}]$ in the bathing medium to the normal extracellular level.

A second technique, in which the cells are made temporarily leaky by exposing them briefly to hypotonic solutions, was first used by Campbell et al. (8, 9) in studies on erythrocyte ghosts, and later refined by Snowdowne and Borle (6, 23) for use in populations of apparently intact renal and hepatic cells.

A third technique, recently described by McNeill et al. (13, 14), depends on the observation that macromolecules gain access to the cytoplasm of cultured cells during almost any maneuver that causes the cells to detach from a substrate to which they have been adhering (13). Useful amounts of aequorin can be introduced into populations of cultured cells simply by scraping them off the bottom of a culture dish with a rubber policeman while they are bathed in an aequorin-containing solution (14).

Although we are not yet aware of any examples of its having been used successfully for the purpose, it seems reasonable to suppose that the temporary membrane disruption produced by the brief application of high-voltage electrical fields might be developed into a technique for introducing photoproteins into populations of dissociated cells (see 12). Liposomes containing photoproteins offer another seemingly attractive approach; while some initial results with adipocytes were interepreted as giving encouraging results, the investigators involved later concluded that although some photoprotein might have entered the cells, it probably did not have access to the cytoplasmic space (9). We are aware of several other unpublished attempts to introduce photoproteins into cells by the liposome technique, all of which have been unsuccessful.

6.   CONCLUSIONS

The recent developments we have discussed in this chapter are a mixed bag from the point of view of the relative desirability of the photoproteins as intracellular Ca^{2+} indicators. Previous assumptions about the relative immunity of photoprotein signals to motion artifacts seem to have been valid, at least under the conditions of our experiments. Acetylated aequorin seems not to be clearly preferable to the native protein as a calcium indicator; it has some advantages and some disadvantages, but it is certainly not a "linear" indicator. We now realize that under the conditions likely to prevail inside most living cells, aequorin responds somewhat less rapidly than had previously been believed. With the proliferation of new methods for introducing photoproteins into cells, the former obstacles to using these substances in populations of small cells seem to be disappearing. A closing note of caution seems appropriate, however: Cells whose membranes have been torn up and then resealed may not be as normal as they appear. Microinjection probably has a continuing role to play, even if only as a means of checking on the results obtained with easier techniques.

REFERENCES

1. Baylor, S. M., M. E. Quinta-Ferreira, and C. S. Hui (1983) Comparison of isotropic calcium signals from intact frog muscle fibers injected with arsenazo III or antipyrylazo III. *Biophys. J.* **44**, 107–112.

2. Blinks, J. R. (1982) "The use of photoproteins as calcium indicators in cellular physiology," in P. F. Baker, Ed., *Techniques in Cellular Physiology*, Vol. P1, Part II of *Techniques in the Life Sciences*, Elsevier–North Holland, Shannon.

3. Blinks, J. R., P. H. Mattingly, B. R. Jewell, M. van Leeuwen, G. C. Harrer, and D. G. Allen (1978) Practical aspects of the use of aequorin as a calcium indicator: Assay, preparation, microinjection, and interpretation of signals, *Methods Enzymol.*, **57**, 292–328.

4. Blinks, J. R., F. G. Prendergast, and D. G. Allen (1976) Photoproteins as biological calcium indicators. *Pharmacol. Rev.*, **28**, 1–93.

5. Blinks, J. R., W. G. Wier, P. Hess, and F. G. Prendergast (1982) Measurement of Ca^{2+} concentrations in living cells. *Prog. Biophys. Mol. Biol.*, **40**, 1–114.

6. Borle, A., and K. W. Snowdowne (1982) Measurement of intracellular free calcium in monkey kidney cells with aequorin. *Science* **217**, 252–254.

7. Campbell, A. K., and R. L. Dormer (1975) The permeability to calcium of pigeon erythrocyte "ghosts" studied by using the calcium-activated luminescent protein, obelin. *Biochem. J.*, **152**, 255–265.

8. Campbell, A. K., and R. L. Dormer (1978) Inhibition by calcium ions of adenosine cyclic monophosphate formation in sealed pigeon erythrocyte "ghosts." A study using the photoprotein obelin. *Biochemistry*, **176**, 53–66.

9. Campbell, A. K., M. B. Hallett, R. A. Daw, M. E. T. Ryall, R. C. Hart, and P. J. Herring (1981) "Application of the photoprotein obelin to the measurement of free Ca^{2+} in cells," in M. A. DeLuca and W. D. McElroy, Eds,. *Bioluminescence and Chemiluminescence. Basic Chemistry and Analytical Applications*, Academic, New York, pp. 601–607.

10. Hastings, J. W., G. Mitchell, P. H. Mattingly, J. R. Blinks, and M. van Leeuwen (1969) Response of aequorin bioluminescence to rapid changes in calcium concentration. *Nature (Lond.)*, **222**, 1047–1050.

11. Hess, P., P. Metzger, and R. Weingart (1982) Free magnesium in sheep, ferret, and frog striated muscle at rest measured with ion-selective micro-electrodes. *J. Physiol. (Lond.)*, **333**, 173–188.

12. Knight, D. E. (1981) "Rendering cells permeable by exposure to electric fields," in P. F. Baker, Ed., *Techniques in Cellular Physiology*, Vol. P1, part I of *Techniques in the Life Sciences*, Elsevier–North Holland, Shannon.

13. McNeil, P. L., R. F. Murphy, F. Lanni, and D. L. Taylor (1984) A method for incorporating macromolecules into adherent cells. *J. Cell Biol.*, **98**, 1556–1564, 1984.

14. McNeil, P. L., and D. L. Taylor (1985) Aequorin entrapment in mammalian cells. *Cell Calcium*, **6**, 83–93.

15. Moore, E. D. W. (1984) Influence of pre-equilibration with Mg^{++} on the kinetics of the reaction of aequorin with Ca^{++}. *J. Gen. Physiol.*, **84**, 11a.

16. Moore, E. D. W., G. C. Harrer, and J. R. Blinks (1984) Properties of acetylated aequorin revelant to its use as an intracellular Ca^{++} indicator. *J. Gen. Physiol.*, **84**, 11a.

17. Morgan, J. P., T. T. DeFeo, and K. G. Morgan (1984) A chemical procedure for loading the calcium indicator aequorin into mammalian working myocardium. *Pflügers Arch.*, **400**, 338–340.

18. Morgan, J. P., and K. G. Morgan (1982) Vascular smooth muscle: The first recorded Ca^{2+} transients. *Pflĝers Arch.*, **395**, 75–77.

19. Quast, U., A. M. Labhardt, and V. M. Doyle (1984) Stopped-flow kinetics of the interaction of the fluorescent calcium indicator quin2 with calcium ions. (*Biochem. Biophys. Res. Commun.*, **123**, 604–611.

20. Shimomura, O., and A. Shimomura (1982) EDTA-binding and acylation of the Ca^{2+}-sensitive photoprotein aequorin. *FEBS Lett.*, **138**, 201–204.

21. Smith, G. A., R. T. Hesketh, J. C. Metcalfe, J. Feeney, and P. G. Morris (1983) Intracellular calcium measurements by ^{19}F NMR of fluorine-labeled chelators. *Proc. Natl. Acad. Sci. USA*, **80**, 7178–7182.

22. Snider, R. M., M. McKinney, C. Forray, and E. Richelson (1984) Neurotransmitter receptors mediate cyclic GMP formation by involvement of arachidonic acid and lipoxygenase. *Proc. Natl. Acad. Sci. USA*, **81**, 3905–3909.

23. Snowdowne, K. W., and A. B. Borle (1984) Measurement of cytosolic free calcium in mammalian cells with aequorin. *Am. J. Physiol.*, 247, C396–C408.

24. Stephenson, D. G., and P. J. Sutherland (1981) Studies on the luminescent response of the Ca^{2+}-activated photoprotein obelin. *Biochim. Biophys. Acta*, **678**, 65–75.

25. Sutherland, P. J., D. G. Stephenson, and I. R. Wendt (1980) A novel method for introducing Ca^{++}-sensitive photoproteins into cardiac cells. *Proc. Aust. Physiol. Pharmacol. Soc.*, **11**, 160P.

26. Tsien, R. Y. (1981) A non-disruptive technique for loading calcium buffers and indicators into cells. *Nature (Lond.)*, **290**, 527–528.

27. Tsien, R. Y. (1983) Intracellular measurements of ion activities. *Annu. Rev. Biophys. Bioeng.*, **12**, 91–116.

28. Tsien, R. Y., T. Pozzan, and T. J. Rink (1982) Calcium homeostasis in intact lymphocytes: Cytoplasmic free calcium monitored with a new, intracellularly trapped fluorescent indicator. *J. Cell Biol.*, **94**, 325–334.

CHAPTER 14

STRATEGIES FOR THE SELECTIVE MEASUREMENT OF CALCIUM IN VARIOUS REGIONS OF AN AXON

L. J. MULLINS

Department of Biophysics
University of Maryland School of Medicine
Baltimore, Maryland

1.	INTRODUCTION	239
2.	TECHNIQUES USING AEQUORIN	241
	2.1 AEQUORIN CONFINED TO THE CENTER OF THE CELL	241
	2.2 INJECTED AEQUORIN	245
	2.3 AEQUORIN AND PHENOL RED INJECTED AXONS	247
3.	COMPARISON OF AEQUORIN WITH ARSENAZO	249
4.	COMPARISON OF AEQUORIN WITH Ca-ION-SENSITIVE ELECTRODES	251
	REFERENCES	254

1. INTRODUCTION

Unlike sodium and potassium ions, whose distribution in cells approximates that expected for free diffusion, calcium ions have a propensity for binding to various intracellular substances that is virtually unique. They also are accumulated by a variety of membrane-enclosed structures in the cell so that the possibility of internal release as contrasted with plasma membrane-based Ca transport has to be kept in mind.

A corollary of this state of affairs is that a measurement of intracellular Ca concentration may only apply to a very selective region of a cell where the sensor for Ca is located, or, if the sensor is located everywhere throughout the axoplasm, it may give a misleading impression of the actual situation with respect to ions such as Ca. Why bother with the measurement of intracellular Ca? There is much evidence to suggest that $[Ca]_i$ is a controlling factor in many cell processes and that this control is exerted in a highly local site. This being so, it is quite important that measurement of Ca be localized as well.

My task in this chapter is to put before you the various pieces of evidence that indicate how intracellular Ca buffering is brought about, what relationship there may be between stored Ca and ionized Ca, and how one may define a steady state with respect to ionized Ca. These considerations will be followed by a description of the techniques used to measure Ca in axoplasm and how such measurements may be localized to specific regions inside the axon.

We have used two optical methods for the estimation of $[Ca]_i$, the first involving the measurement of light emitted by aequorin when it comes in contact with Ca and the second involving the spectrophotometric measurement of the dye arsenazo III at several wavelengths. Both of these optical techniques have been compared with what Ca-sensitive, ion-specific electrodes indicate about intracellular Ca.

Aequorin is by all odds the most sensitive indicator of the very low concentrations of [Ca] that are found in squid axoplasm, and were it not for its highly nonlinear relationship between [Ca] and light emission, it would surely be an ideal substance for all sorts of studies.

Arsenazo III is entirely linear over the range of [Ca] in axoplasm of hundreds of nM, and it is very useful when one seeks to measure substantial increases in net Ca flux. It is insensitive to some of the very small changes in internal [Ca] that are to be dealt with, and it is not really possible to get a satisfactory value for $[Ca]_i$ in axoplasm using this dye. Another disadvantage of arsenazo is that changes in pH and changes in [Ca] cannot really be separated and, since Ca entry involves change in pH_i, some care has to be taken that experimental records are corrected for possible pH changes.

Ca-sensitive electrodes are the least sensitive of the techniques for the measurements of $[Ca]_i$, although it is not quite clear why this should be so. We have used such electrodes in axons that have also been injected with either arsenazo III or with aequorin, and we find in all cases that the optical indicator indicates substantial changes in [Ca] while the electrode does not. Sensitivity of the electrode for Ca in axoplasm starts somewhere around 150 nM and continues to high levels (tens of μM) where optical indicators either become nonlinear or simply burn up.

2.   TECHNIQUES USING AEQUORIN

2.1   Aequorin Confined to the Center of the Cell

This volume has already had a splendid presentation of the properties and use of aequorin as an indicator of intracellular Ca, and I shall not expand further on its properties except as they relate to the particular strategies we have used to get information about the [Ca] in specific regions of the axon.

We have used aequorin in three quite distinct ways: (1) confined to a dialysis capillary located at the center of an axon, (2) injected into an axon, and (3) injected into an axon together with enough phenol red dye to mask light emission from all but

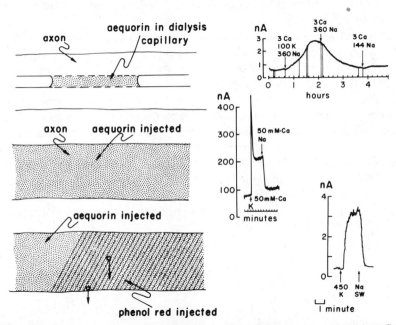

Figure 1.   Three separate methods of using aequorin are compared. At the top, the aequorin is confined to a porous plastic dialysis tube located in the center of a squid axon. On the right is shown the light response to a depolarization of the axon with 100 mM K seawater. Note that the time scale is in hours and the photocurrent in units of 1–2 nA. The center shows an aequorin-injected squid axon. Now aequorin molecules are uniformly distributed, and, as expected, the response to K depolarization (shown at the right of the center panel) is both fast (time scale in minutes), and large (photocurrent in hundreds of nA). At the bottom of the figure is an axon that has been injected with both aequorin and phenol red. The response is to 3 Ca, 450 mM K seawater (hence, the absence of a spike). The time scale is again in minutes, but the photocurrent is in units of 1–3 nA.

the most peripheral parts of the axoplasm. These various arrangements are shown in Fig. 1 together with a record of the aequorin light response to a depolarization of the axon with an elevated KCl concentration in seawater. Note that in the aequorin confined to a dialysis capillary case, it takes about 90 min for the light to reach a maximum whereas in the case of injected aequorin the peak of light emission is reached in about 0.5 min. The response in a phenol red axon (aequorin injected) is about the same in terms of time course, suggesting that the measurements from the injected axon pertain principally to the periphery in both cases. The magnitude of the signal in the aequorin–phenol-red axon is much smaller because the region in which Ca is being measured is also much smaller. The very long time course for the rise of [Ca]$_i$ in the dialysis capillary is a reflection of the strong buffering and accumulation of Ca by intracellular entities. When a nonbinding dye is injected into the center of an axon, one observes that mixing is complete in 15 sec; the movement of Ca is more than 300 times slower.

Let us start with axons with a dialysis tube containing aequorin inside the axon. This was the first technique we used (DiPolo et al., 1976) to get at the question of what the [Ca] in axoplasm might be. The idea was to get as far away from the surface membrane where much local Ca traffic takes place and into a region that might be representative of cytoplasm in general. At the time this work was done, the comprehensive studies of Allen et al. (1977) on the relationship between [Ca] and light emission of aequorin had not been done. We were aware of claims that light emission went as the square of the [Ca], and there also seemed to be some data in favor of a linear relationship. To get around this uncertainty about the actual relationship, we decided to try to match the light emission of the aequorin in the dialysis tube with that of a CaEGTA/EGTA buffer ratio. To make this measurement apply to the whole of axoplasm we needed to find a [Ca] in seawater where the axon

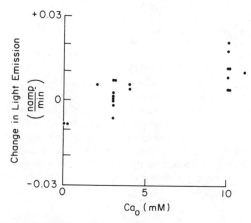

Figure 2. Change in light emission for axons immersed in artificial seawater of various [Ca]. Slope was measured from the linearly changing portion of the light emission curve approximately 15 min after a solution change. Rate of increase of light emission in 10 mM Ca seawater corresponds to about 0.4 nM/min increase in free Ca (measured on axis of fiber). From DiPolo et al. (1976), with permission.

was neither gaining nor losing Ca. A study of this sort is shown in Fig. 2, where we measured the rate of change in aequorin glow in axons with various concentrations of Ca in seawater. A value of 3 mM was selected as one that led to a steady state with respect to [Ca]$_i$ as indicated by aequorin glow. Later on, a similar conclusion was reached when we made analytical measurements of total Ca in axons.

Having an appropriate [Ca]$_o$, we then measured the aequorin glow of an axon as shown in Fig. 3, removed the aequorin from the dialysis capillary, and replaced it with a CaEGTA/EGTA buffer mixture and allowed this to equilibrate with the axoplasm for 30 min. The buffer mixture was then removed and replaced by new aequorin. In the top of Fig. 3 the glow of a particular mixture matches very closely the initial aequorin glow, and the buffer mixture had a [Ca] of 25 nM. A second experiment of a measurement of resting glow and a match with another buffer mixture

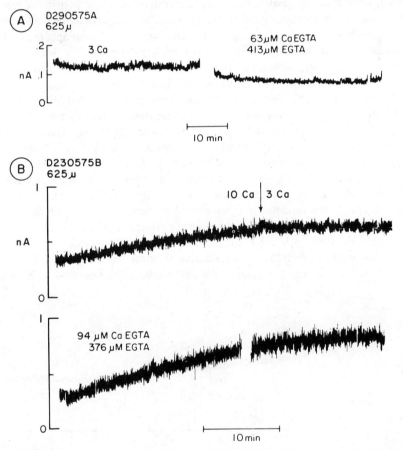

Figure 3.   Composite figure to show the effect on resting glow of equilibrating axons with EGTA buffers containing about 500 μM total chelator. The traces are not continuous in the two figures but have been interrupted for about 30 min to allow buffer equilibration in the axoplasm. Nominal ionized [Ca] in the axoplasm after equilibration, 23 nM (A) and 38 nM (B). From DiPolo et al. (1976), with permission.

is shown in the lower part of the figure. From a number of such measurements we concluded that resting [Ca] in axoplasm in the absence of Ca gradients was 30 nM.

Another area where aequorin in a dialysis capillary was useful was in separating mitochondrial buffering of Ca from the buffering of other entities. Analytical measurements of [Ca] in axoplasm from freshly isolated axons shows this to be 60 μM. If all this were contained in the mitochondria (and mitochondrial volume is 1% of total axoplasmic volume), then the release of this Ca would lead to a [Ca] of 0.6 μM, a value some 20 times larger than normal. An experiment to test whether the mitochondria do contain the bulk of stored Ca is shown in Fig. 4. The axon with a resting glow of 0.6 nA was treated with the uncoupler FCCP, which is known to rapidly collapse the proton gradient in mitochondria and leads to a release of Ca. The light emission rose about three-fold and then declined with time as the released Ca was pumped out by the Ca extrusion mechanisms of the surface membrane. The point of the experiment is to emphasize that mitochondria have only a very small fraction of the known Ca content of axons and most Ca is stored in quite another buffer system.

A final area where aequorin in a dialysis tube is useful is in examining the change of axoplasmic buffering with time. It is a common experience in working with aequorin that in a freshly isolated axon, the response to a substantial Ca load is a very small increase in aequorin light emission. The most reasonable explanation for this finding is that buffering is very strong. As time progresses, the axon becomes more sensitive to Ca loads, and eventually even slight imbalances in the Ca fluxes produce such large increase in aequorin light emission that an appreciable fraction of the aequorin is now being consumed and the experiments necessarily must be terminated. An experiment illustrating this change in buffering is shown in Fig. 5, where at 2 h the axon with a dialysis capillary with aequorin was treated with Na-free seawater. This is known to produce a large, sustained increase in Ca influx, and the result is an increase in aequorin glow with time that is terminated by a return to Na seawater. Seventeen hours later the axon was subjected to an identical treatment, and one can note the great difference in the slope by which [Ca] rises with time. It is of interest to note that the resting glow of the axon did not change during this time (the axon was in 3 mM Ca seawater) and, hence, was presumably in a steady state with respect to Ca. What did

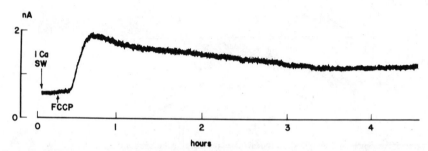

Figure 4. An axon is treated with FCCP in 1 Ca (Na) seawater. The response is a rapid release of Ca from internal stores and the extrusion of some of the Ca over 3 h. Final [Ca]$_i$ is about twice the initial concentration. From Requena et al. (1977).

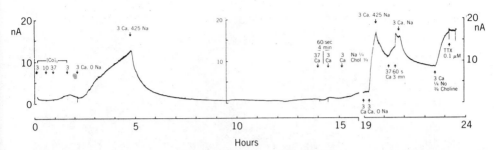

Figure 5.   An axon with aequorin confined to a dialysis capillary is treated (at 2 h) with 3 Ca, Na-free seawater and again at 19 h. Note the difference in slope, reflecting a decline in Ca buffering. Ordinate is photo output from aequorin in nanoamperes. From Requena et al. (1977).

change was the way that Ca was buffered by axoplasm. If we recall that the dialysis capillary containing aequorin is some 175 μm from the surface membrane, then the slow rise of aequorin glow is a reflection of the uptake of most of the Ca as it diffuses from the surface to the center of the axon. Eliminate the buffering, and one has a very rapid rise in glow.

2.2   Injected Aequorin

When aequorin is injected into a squid axon, it rapidly diffuses throughout the axoplasm and yields an axon that can rapidly respond to Ca flux imbalances. As an example, Baker et al. (1971) were able to detect the response to six action potentials when delivered at 250 s^{-1}.

We have used injected axons mainly to study the response of the axon to steady depolarization produced by elevated K solutions. An example is shown in the middle record of Fig. 1 where there are a spike and a plateau. We have been able to show (Mullins and Requena, 1981) that virtually all of this response can be abolished by tetanizing the axon for 15 min in Li seawater. This treatment has been shown to reduce $[\text{Na}]_i$ by inserting a Na-sensitive electrode into the axon before stimulation. It therefore seems reasonable to ascribe the response of the axon to steady depolarization as a manifestation of Na–Ca exchange. The theory of Na–Ca exchange predicts that Ca entry will be given by the product $[\text{Ca}]_o[\text{Na}]_i^n$, where n is the number of Na coupled to the movement of Ca. If n is a large number, Ca entry will be very sensitive to $[\text{Na}]_i$, as indeed it appears to be.

Another use for injected aequorin has been in exploring the effect of pH_i on the entry of Ca by Na–Ca exchange. It is known that the exchange process is highly pH-sensitive and in the direction of making it run faster as pH rises. With a pH-sensitive glass electrode in the axon in addition to aequorin, it has been possible to show that a change in axoplasmic pH from 7.1 to 7.6 increases the amplitude of the aequorin response 14-fold. An experiment is shown in Fig. 6.

Finally, aequorin has proved most useful in examining the response to repetitive voltage clamp pulses of a duration similar to that of an action potential. The reason for

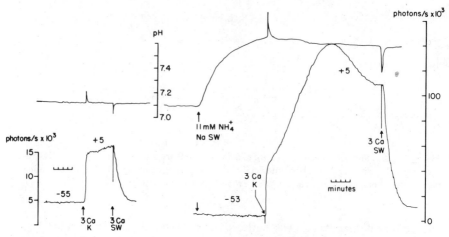

Figure 6. An axon was injected with aequorin and kept in Ca-free (Na) seawater (50 mM Mg) for 3 h. The axon also had electrodes for the measurement of pH and membrane potential inside; the pH (upper trace) after the Ca-free treatment was 7.12. Membrane potential was not recorded on the chart but was displayed digitally, and values were transcribed to the record as shown. A depolarization in 3 mM Ca (50 mM Mg) 450 mM K seawater changed the potential from -55 to $+5$ mV and led to the aequorin record shown on the left-hand lower trace. After recovery in 3 mM Ca (Na) seawater, the axon was treated with 11 mM $NH_4{}^+$, 3 mM Ca (Na) seawater, and the value of pH_i changed to 7.62. A depolarization now led to the aequorin record shown on the right (lower trace). Note that the light-emission scales are quite different in the two recordings, so that the aequorin responses to depolarization differ by a factor of 14 (axon 060281A). From Mullins et al. (1983).

undertaking such studies is that it is known that the Na channel allows the passage of Ca with a permeability of 1/100 that for Na (Baker et al., 1971), so the effect of repetitive stimulation is that some Ca will enter via these channels. If we poison with TTX and use a voltage clamp to produce pulses, then there is another TTX-insensitive entry that has been claimed to be a Ca entry via a Ca channel. Now it has proved somewhat difficult to separate these two sorts of claims, because while it is clear that Ca entry with steady depolarization requires $[Na]_i$, one could imagine that this is because of a channel inactivation early in the depolarization. We have therefore carried out repetitive voltage clamp pulses (Mullins and Requena, 1985) on aequorin-injected axons when the Ca in seawater is 3 mM and the axon is treated with TTX externally and tetraethylammonium ion internally to minimize Na and K currents. Under these conditions there is no entry of Ca under repetitive voltage clamp pulses when the external solution is Na seawater, but there is an easily detectable entry when the external solution is Li seawater. A further finding is that there is no detectable entry either in Li or Na seawater if the internal Na of the axon is low. A record is shown in Fig. 7. It is also of interest that Ca entry continues to increase as one goes from pulse amplitudes of 60 to 180 mV (i.e., from zero membrane potential to $+120$). Since one is now close to the Ca equilibrium potential, this result is explainable only on the basis that Ca entry is by Na–Ca exchange.

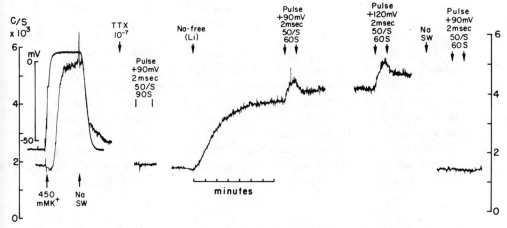

Figure 7. The ordinate is counts per second for photons detected by the photomultiplier, and time in minutes is also shown. The upper trace is membrane potential, and the scale immediately to the left of this tracing indicates the value of membrane potential. The lower trace is light output. An aequorin + TEA injected axon was stimulated at 60/s for 900 s in a Ca-free (Na) seawater to load the axons with Na. The left panel shows that a depolarization with high K solution containing TTX was applied, and repetitive voltage clamp pulses were applied as indicated (holding potential − 58 mV). A change to 3 mM Ca (Li) seawater containing TTX was then made, and the usual increase in resting glow occurred. Pulses were applied as before as indicated on the record, and the final change was to 3 Ca (Na) seawater containing TTX, where the result was the same as at the start of the experiment. From Mullins and Requena (1985).

2.3   Aequorin and Phenol Red Injected Axons

In an axon with little Ca load, the resting glow is so low that even if the total glow is included in the measuring system, Ca entries are detectable as increases in glow of the fiber that are transient. One is, in fact, measuring a rather large increase in light over a background that is small. As Ca loading of the fiber proceeds, core glow comes to represent a larger and larger fraction of the total light, and it becomes difficult to measure changes in Ca entry.

One way to get around this difficulty is to screen the core of the axoplasm by the use of a dye that selectively absorbs aequorin photons. It is possible to show (Mullins and Requena, 1979) that half the photons emitted by aequorin come from a rim of axoplasm 25 μm in thickness. The bottom of the illustration in Fig. 1 shows that photons produced deep in the core of the axoplasm are likely to be absorbed whereas those produced near the axon periphery have a far greater probability of escaping from the axon and being measured.

What new sort of information can be obtained from phenol red axons? First, we have stimulated axons in 50 mM Ca seawater both with and without phenol red. Under these circumstances, if the seawater contains CN, both sorts of axons give about the same response. If CN is omitted from seawater, then there is little change in the aequorin-injected axon, but in the phenol red + aequorin-treated axon the response to stimulation falls to 5% of control levels. Clearly, to detect reasonable Ca entry, CN must be present in phenol red axons. How can one account for this difference in behavior?

One way is, suppose that some relatively small fraction of the Ca entering with stimulation is taken up by the most peripheral mitochondria. Since in the phenol red axon, this is the only region of measurement, the response to stimulation in such axons is totally abolished by mitochondrial buffering. In a non–phenol-red axon, by contrast, the diffusion of Ca from the periphery to the core is capable of producing a light response that just compensates, in CN-treated axons, for the loss of Ca.

Further support for this view is to be found in an experiment shown in Fig. 8. Here an axon was depolarized in high-K solution and gave no response in terms of increased light emission in a phenol red axon. In an axon without phenol red there

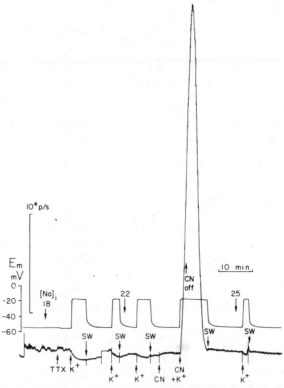

Figure 8.   A phenol red and aequorin injected squid axon was treated with TTX and depolarized with 450 mM K solutions containing 3 mM Ca. The apparent fall in light signal is, in fact, a result of the refractive index difference between Na and K solutions and not a real change in light emission. The axon had a Na-sensing electrode inside and initial [Na]$_i$ was 18 mM. Subsequent depolarizations were made when [Na]$_i$ was 22 mM, and then 2 mM CN was added to seawater. This produced no change in resting glow but an application of high K solution. There was a large increase in aequorin glow, one that was completely reversed by removing CN. A subsequent test depolarization in the absence of CN is also shown. From Requena and Mullins (unpublished data).

would have been a measurable response. If CN were included in the seawater, then the response is very large, a finding that suggests that peripheral mitochondria can actively take up entering Ca.

These findings suggest (1) that most of the Ca entry with stimulation can produce a very high concentration of Ca in the most peripheral part of the axoplasm if CN is present to prevent the uptake of much of this Ca, and (2) that in the normal axon one measures a fraction of the entering Ca that escaped capture by the peripheral mitochondrial buffers, and this fraction is not visible in the phenol red axon. This accounts qualitatively for the findings, but not quantitatively. The matter is complex because there are, in the course of Ca buffering, pH changes in axoplasm, and these changes affect further Ca entry. Another factor is that buffering is undoubtedly [Ca]-dependent, and we know little about the kinetic parameters of buffering.

Support for the idea that peripheral mitochondria do take up Ca during physiological activity is obtained by measuring the effect of FCCP on Ca in phenol red axons. From axons with a dialysis capillary, it was shown that there was little Ca to release from mitochondria (Fig. 4), while the reverse is true for a phenol red axon, since when $[Ca]_o$ is made zero and FCCP is applied, there is a 200-fold increase in light emission. This change is transient, as the released Ca either is pumped from the fiber or diffuses into that part of the axoplasm shielded from measurement.

3. COMPARISON OF AEQUORIN WITH ARSENAZO

Arsenazo is useful as a check on aequorin results and for more quantitative estimates of actual Ca entry. One of the first uses we have made of the dye is to examine the response to depolarization resulting from the exposure of an axon to seawater with elevated [K]. This comparison is shown in Fig. 9 where two axons, one injected with aequorin and the other with arsenazo, were treated with 50 mM Ca,450 mM KCl solutions, and the resulting change in optical signal followed. The aequorin signal is one that is typical in 50 mM Ca solutions, but the arsenazo signal is clearly different.

If the seawater [Ca] is reduced to 3 mM, then the response of an aequorin axon and that of an arsenazo axon are the same. The conclusion one draws from such observations is that there is an additional process of Ca entry with depolarization that is activated by high $[Ca]_o$, and this sums with the Ca entry that is dependent on Na–Ca exchange. Why the spike on the aequorin record? It would appear that the extra Ca entry from 50 mM Ca seawater increases the local Ca concentration just inside the membrane so that one goes out of the linear range of light emission versus [Ca]; hence one gets more light initially than during the plateau. Such a behavior cannot be expected for the arsenazo record.

Another use of arsenazo has been in measuring the Ca entry with depolarization in a quantitative manner. As noted in the paragraph above, aequorin glow changes are difficult to relate to Ca entry because unless these changes are very small, one may be in nonlinear regions of the light–[Ca] curve. With arsenazo there is every reason to believe that the initial slope with which arsenazo absorbance rises with time is a measure of the net flux of Ca into the cell. Thus, to relate Ca entry, for example, to

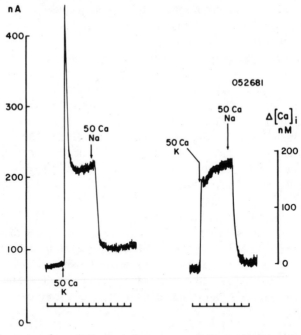

Figure 9. A comparison is made between the response of an aequorin-injected axon (ordinate, photocurrent in nA) and one injected with arsenazo III (ordinate change in [Ca]$_i$) to depolarization (a transfer from 50 mM Ca (Na) to 50 mM Ca (K) seawater) and to subsequent repolarization. Time marks on the abscissa are minutes. Both axons had been isolated for 4–5 h and hence could be expected to have gained substantial Na. Since, from previous measurements with aequorin, the resting [Ca]$_i$ was about 60 nM, we can note that the initial resting glow (75 nA) was increased with depolarization and the plateau was 210 nA, a threefold change corresponding to an increase in ionized Ca by 150 nM. This agrees reasonably well with the observation with arsenazo III. From Mullins et al. (1983), with permission.

[Na]$_i$, arsenazo is the method of choice for such measurements. Figure 10 shows the results obtained with two axons that had been injected with arsenazo and had been impaled with Na-sensitive electrodes. It was thus possible to relate [Na]$_i$ to rate of Ca entry, as the curve shows. Note that the relationship is very steep, and, indeed, the Hill coefficient fitting the curve is 7. This sort of relationship is important physiologically because it allows [Na]$_i$ to have a very sensitive control of Ca entry with depolarization. It is also true that Ca entry with depolarization can be effected via Ca channels—these apparently have other modalities for their control.

An additional use for arsenazo in measuring the Ca entry with depolarization is to look at how Ca entry with steady depolarization and steady [Na]$_i$ depends on membrane potential. Our findings (Mullins et al., 1983) are that the rate of Ca entry increases exponentially with depolarization and that the curve approximates an *e*-fold increase in Ca influx for a 25-mV depolarization. This finding is important because for both theoretical and empirical reasons the effect of membrane potential on Ca influxes can be described as

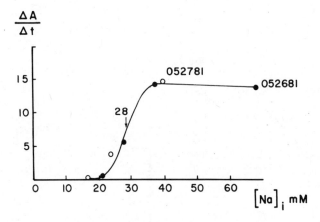

Figure 10. The initial rate of increase in [Ca]$_i$ upon the application of 3 mM Ca, 450 mM K is plotted (ordinate) as a function of [Na]$_i$ for two axons injected with arsenazo III. From Mullins et al. (1983), with permission.

$$\text{influx} = k \exp (EF/RT), \quad \text{efflux} = k' \exp (-EF/RT)$$

where E is membrane potential.

4.   COMPARISON OF AEQUORIN WITH CA-ION-SENSITIVE ELECTRODES

Since comparisons of results obtained with aequorin and results obtained with other techniques are always desirable, we decided to introduce both aequorin and a Ca-sensitive electrode into a squid giant axon. The electrode was a capillary about 125 μm in diameter, and it, together with a reference, was introduced into an aequorin-injected axon.

The results obtained are shown in Fig. 11 where, on the left, the lower trace at the start of the recording is aequorin light measured as photomultiplier counts per second and the upper trace is the difference between the Ca electrode potential and that of the membrane potential. Notice that a change in aequorin glow brought about by depolarization of the fiber with a high K solution was not reflected in any corresponding change in the Ca electrode.

The axon was repolarized and placed in seawater containing 50 mM Ca. In this solution it continued to gain Ca as indicated by increasing aequorin counts, which went from 1,000/s to 100,000/s without any corresponding change in the [Ca] as indicated by the Ca electrode. When the count had reached 10^5/s, 2 mM CN was added to the seawater, and shortly thereafter the Ca stored in mitochondria was released, and the Ca as indicated by the electrode and the aequorin count rose together. Some substantial experience with Ca electrodes has led us to the conclusion that they appear to lose sensitivity to Ca in axoplasm when the concentration falls

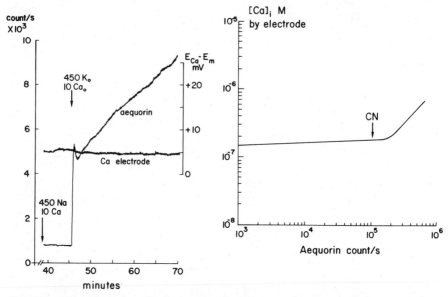

Figure 11. The initial lower trace is aequorin glow measured as count/s; the upper trace is the difference between an electrode measuring E_m and a Ca electrode (E_{Ca}). This axon was stimulated by 75,000 impulses to raise $[Na]_i$ and was then depolarized as indicated with 450 mM K. The aequorin light emission rose from 800 s^{-1} to 9700 s^{-1} without any change in the output of the Ca electrode. The electrode had a slope of 22 mV between aequorin light emission and electrode indication over the entire range of light emission obtained in the experiment. The application of CN to the axon led, with a delay, to a count that increased to the level of 500,000 s^{-1}; the Ca electrode responded when the count reached 200,000 count/s. Data of Requena et al. (1984).

below about 150 nM even though in calibrations with CaEGTA/EGTA buffers there is still useful sensitivity at a [Ca] of 10 nM. Part of the reason for this loss of sensitivity in axoplasm may be that axoplasm does not have diffusible buffers to carry Ca to the electrode membrane, and diffusion at [Ca] of 10 nM may be too slow to produce equilibration in a reasonable time. Another difficulty may be that Ca may leak out of the electrode and contaminate the measuring membrane when [Ca] is low and no buffer is present.

The Ca electrode is a useful device for measuring in the range of 150–1000 nM, as the results in Fig. 12 show. Here an axon was kept depolarized with a high K solution that also contained CN. The electrode showed the [Ca] to be stable at a value of 1 μM, and when CN was removed the [Ca] fell rapidly to a value of about 100 nM. The record appears to be truncated, as though Ca were still falling but the electrode was not following the reaction. The axon was then returned to high K solution, and upon recovering the initial baseline, the solution was changed to one that was both CN-free and Na-containing. In this Na-containing solution, Ca; starts its decline somewhat sooner, presumably reflecting the ability of Na$_o$ to extrude Ca at once rather than having the start of extrusion delayed by the requirement that CN diffuse from the

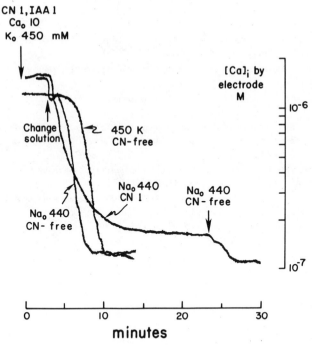

Figure 12. An axon previously loaded with Ca and kept in 450 mM K, 10 mM Ca solution containing 1 mM each of CN and iodoacetamide was transferred at the point indicated by "change solution" to several different solutions, and after [Ca]$_i$, as indicated by the electrode, had reached a steady level, the axon was returned to the original solution and the change to another solution carried out. The results show that the initial rate of decline of [Ca] is about the same whether the axon is in a CN-containing or in a CN-free solution that contains Na. Repolarization is not necessary to bring Ca down to low levels. All recovery solutions contained Ca = 10 mM. Data of Requena et al. (1984).

fiber. A final test on this axon was to recover the original baseline and then change to Na-containing and CN-containing seawater. Under these conditions, mitochondria cannot accumulate Ca, and the fall of [Ca]$_i$ is the result of the pumping out of Ca across the surface membrane. Note that now the electrode appears to follow Ca quite reasonably, and it indicates a value of 150 nM for the axon in CN-containing Na seawater. Removal of CN leads the electrode trace to reach the same value as it did in the other two cases, and this again appears truncated.

A summary of our experience with Ca electrodes is that they indicate axoplasmic [Ca] between 100 and 200 nM in freshly isolated axons, but the aequorin count in the axon has to change at least 50-fold before the Ca electrode begins to show a change.

REFERENCES

Allen, D. G., J. R. Blinks, and F. G. Pendergast (1977) Aequorin luminescence: Relation of light emission to calcium concentration—A calcium-independent component. *Science*, **195**, 996–998.

Baker, P. F., A. L. Hodgkin, and E. B. Ridgway (1971) Depolarization and calcium entry in squid axons. *J. Physiol. (Lond.)*, **218**, 709–755.

DiPolo, R., J. Requena, F. J. Brinley, Jr., L. J. Mullins, A. Scarpa, and T. Tiffert (1976) Ionized calcium concentration in squid axons. *J. Gen. Physiol.*, **67**, 433–467.

Mullins, L. J., and J. Requena (1979) Calcium measurement in the periphery of an axon. *J. Gen. Physiol.*, **74**, 393–413.

Mullins, L. J., and J. Requena (1981) The "late" Ca channel in squid axons. *J. Gen. Physiol.*, **78**, 683–700.

Mullins, L. J., T. Tiffert, G. Vassort, and J. Whittembury (1983) Effects of internal sodium and hydrogen ions and of external calcium ions and membrane potential on calcium entry in squid axons. *J. Physiol. (Lond.)*, **338**, 295–319.

Mullins, L. J. and J. Requena (1985) Ca entry in squid axons during voltage clamp pulses is mainly Na/Ca exchange. Proc. Natl. Acad. Sci. USA. **82**, 1847–1851.

Requena, J., R. DiPolo, F. J. Brinley, Jr., and L. J. Mullins (1977) The control of ionized calcium in squid axons. *J. Gen. Physiol.*, **70**, 329–353.

Requena, J., J. Whittembury, T. Tiffert, D. A. Eisner, and L. J. Mullins (1984) A comparison of measurements of intracellular Ca by Ca electrode and optical indicators. Biochim. Biophys. Acta, **805**, 393–404.

CHAPTER **15**

POLYCHROMATOR FOR RECORDING OPTICAL ABSORBANCE CHANGES FROM SINGLE CELLS

STEPHEN J. SMITH

Section of Molecular Neurobiology
Yale University School of Medicine
New Haven, Connecticut

1.	INTRODUCTION	255
2.	OPTICAL ARRANGEMENT	256
3.	ELECTRONIC SIGNAL AMPLIFICATION	258
4.	INSTRUMENTAL RESOLUTION LIMITS	259
	REFERENCES	260

1. INTRODUCTION

An instrument for recording optical absorbance of single living cells was demonstrated. The apparatus is based on a grating polychromator and measures absorbance simultaneously at several different wavelengths. It was designed for high-resolution recording while leaving cells accessible to multiple microelectrode impalement. The instrument has been used on nerve cells containing the indicator dye arsenazo III to detect cytosolic calcium concentration transients. The apparatus will be described here, and some of the considerations influencing the design will be discussed.

An instrument with multiple spectral channels, such as that described here, may be preferred to a simpler, fixed-wavelength or manually tuned photometer for several reasons. Spectral data may be inherently necessary for a particular analysis, such as discrimination between effects of calcium and protons on the absorbance spectrum of arsenazo III (Brown, et al., 1977; Thomas, 1982). Spectral information may also help to recognize and ameliorate sources of interference to the desired optical measurement. For example, it is often difficult to completely prevent small movements in living cell preparations, and residual movements may cause changes in light transmittance of a magnitude comparable to indicator dye signals. Differential recording at a properly chosen pair of wavelengths can significantly reduce artifacts due to object movement. The improved speed, accuracy, and convenience of spectral measurement with a multichannel instrument, compared to a manually tuned one, can be decisive in work with living specimens of limited experimental lifetime.

2. OPTICAL ARRANGEMENT

Optical components of the recording system are shown in schematic form in Fig. 1. The arrangement consists of an illuminator to project an intense beam of broad-band-filtered light onto the specimen under study (here labeled ganglion), a single optical fiber to collect light transmitted through some part of the specimen, and a grating-based polychromator with silicon photodiodes for light measurement. All components are mounted on a tabletop vibration isolating bench within a closable acoustic and electrical shielding enclosure.

Specimens are Kohler-illuminated with a maximum numerical aperture of 0.7. The field and aperture stops are omitted for clarity in Fig. 1 but are useful to reduce stray light and facilitate adjustment. Direct current is delivered to a 12-V, 100-W

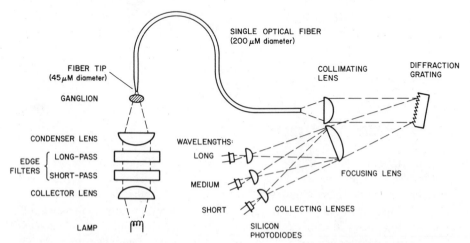

Figure 1. Schematic diagram of polychromator-based apparatus for spectral measurement of optical absorbance in single living cells.

tungsten–halogen lamp by an electronic power supply. The lamp housing is water cooled with minimum air volume between the quartz bulb envelope and an aluminum cooling jacket. This arrangement minimizes fluctuations in filament temperature and light output due to air convection currents. Water cooling has additional advantages in minimizing heating of mechanical components and reducing air drafts within the experimental enclosure, both of which contribute to stable recordings and shorten warm-up stabilization periods.

The edge filters shown in Fig. 1 are used to restrict illuminating light to the broad wavelength band of interest. Possible unwanted effects of illumination at other wavelengths, such as the heat input from longer wavelengths or photochemical actions from shorter wavelengths, are thus minimized. For measurement of calcium transients with arsenazo III, a colored glass long-pass filter with a 530-nm cutoff is used; a combination of heat-absorbing glass and interference short-pass filters are used to eliminate wavelengths longer than 725 nm.

Light transmitted through a restricted specimen area is collected by positioning the tip of a single optical fiber over the desired area using a micromanipulator. The tip is lowered as close as practical to the surface of the specimen. The fiber used is of the graded index type produced for communication purposes and has an outer diameter of 200 μm. In the short lengths employed here (1–2 m), such a fiber has an effective light-accepting area equal to this outer fiber diameter. Smaller recording apertures were produced by drawing fiber tips into a taper with heat. Silver was deposited onto the sides of the tapered tip to eliminate stray light coupling. The silver was overcoated with varnish to prevent toxicity and deterioration. A fiber tip used in many studies had a tip 45 μm in diameter and accepted light in a cone of 20° semiangle, corresponding to a numerical aperture of 0.35.

The polychromator consists of the arrangement of components shown at the right of Fig. 1. The collimating lens is a camera lens of 50 mm focal length and $f = 1.8$. The end of the optical fiber is positioned at the back focal plane of this lens so that light exiting the fiber is collimated for projection onto the diffraction grating. The grating is a replica ruled 1200 lines/mm and blazed at 500 nm. The grating is oriented such that the desired wavelengths (530–725 nm for arsenazo III studies) of the first-order diffracted light on the high-efficiency side of the blazed grating pattern fall onto the focusing lens. The focusing lens has a 300-mm focal length and a 30-mm clear aperture. A spectrally dispersed image of the exit tip of the fiber, enlarged 6×, is thus formed along a curved surface at a radius of 300 mm centered on the focusing lens. An array of up to five detector modules (of which only 3 are shown in Fig. 1, for clarity), are positioned along this surface.

Each detector module consists of a narrow aluminum housing holding an aspheric collecting lens (8-mm focal length, 12-mm diameter) and a silicon photodiode. These elements are positioned such that a reduced image of the 30-mm focusing lens falls entirely on the 1-mm-diameter active area of the photodiode. The use of a small-area photodetector minimizes detector noise power, which is proportional to detector active area. Use of a collecting lens of diameter larger than the mounting package in which the photodiode chip is supplied reduces the spectral "dead" band between adjacent detectors at minimum spacing. With the optics specified, each detector module samples a spectral band approximately 20 nm in width, while adjacent channel-band center spacing can be as little as 30 nm. These parameters can be adjusted by changing the focal length of the focusing lens or by masking the detector module aperture.

Detector channels are tuned to the desired spectral region by adjusting the lateral position of individual detector modules along the polychromator output focusing surface. A narrow band interference filter of known peak wavelength is inserted in the illuminator path to facilitate placement of detectors. Spectral tuning is then calibrated by temporarily replacing the normal collecting fiber with a similar fiber coupled to a calibrated monochromator light source. Tuning curves are generated for each channel by reading photodiode output as a function of a monochromator wavelength setting.

3. ELECTRONIC SIGNAL AMPLIFICATION

Amplification and processing of the photocurrent analog signal from silicon photodiodes are carried out in several stages. Low-level preamplification takes place at a transimpedance amplifier (current-to-voltage converter) located very near the photodiode to reduce electromagnetic interference noise pickup and stray capacitance. Signal switching, scaling, and differential amplification take place at a rack-mounted control amplifier located outside the isolating enclosure.

The transimpedance amplifier is based on an FET operational amplifier with a bias current of less than 1 pA and low input voltage noise. The feedback network consists of a parallel RC combination. A 500-MΩ feedback resistor (matching the photodiode shunt resistance) is used unless a smaller value is required to avoid exceeding the amplifier voltage output limit when photocurrent is large. Parallel capacitance is adjusted to obtain a critically damped transient response (using an LED to produce a step change of light input).

The control amplifier allows gain adjustment of each photocurrent signal. To produce an optical absorbance change signal of constant calibration, the gain of each photocurrent channel is adjusted to make the "resting" or reference photocurrent signal equal to a fixed reference level of 10.00 V. The optical absorbance signal of interest normally takes the form of some small deviation from this reference level. For instance, Ca–arsenazo III signals in an excitable cell may produce peak changes in light transmittance on the order of one part in 1000. Thus, if photocurrent gain is adjusted to provide a 10.00-V signal prior to a stimulus, the signal might change to 9.99 V as the result of that stimulus. To permit efficient recording of such small changes, a differential amplifier is arranged to subtract the 10.00-V reference voltage from the photocurrent signal and amplify the result by another factor of 10. The output is then zero for the resting photocurrent level and 1 V per 1% change in reference to that level. For small changes in light transmittance, such signals are readily converted to absorbance change units by a fixed multiplicative constant (e.g., see Smith and Zucker, 1980).

Finally, the control amplifier contains a unity gain differential amplifier to provide an output proportional to the difference between any two switch-selected spectral channels. For small changes in light signals, the output of this amplifier is proportional to changes in the ratio of transmittance at the two selected wavelengths, or differential absorbance. Such a signal is useful for emphasizing absorbance changes of known spectral character, such as a change in indicator dye absorbance, while tending to reject many interfering signals such as changes in tissue light scattering or some instrumental instabilities.

4. INSTRUMENTAL RESOLUTION LIMITS

The usefulness of optical absorbance measurement on single living cells is often constrained by resolution or signal-to-noise ratio limitations. The minimum resolvable signal may be determined by instrumental resolution limits or by factors inherent to a specimen, such as the tendency of living tissues toward stimulus-evoked or spontaneous movement or light-scattering change. While limitations of the latter class often prove the most serious, they are difficult to discuss in any general way. Differential absorbance recording is usually helpful, but solutions to individual problems are otherwise likely to be specific to the specimens and investigators involved. Instrumental resolution limits, on the other hand, are governed by well-known physical laws. Some of these limits will be discussed here. More complete treatments of some the topics raised can be found in general instrumentation texts, such as that of Strobel (1973), and in the excellent monograph on calcium measurement by Thomas (1982).

Instrumental noise sources can be considered in two categories: dark noise and light noise. These are, quite simply, the apparent output noise when the illuminating source is off and the extra noise when the source is on, respectively. Dark noise is characteristic only of the photodetector device and its associated amplifier, whereas light noise is associated with the illuminating light itself and possibly other factors. Light noise may have two components. One component, which can (at least in theory) always be reduced or compensated for, is due to such factors as fluctuations in lamp output, mechanical instabilities of the optical arrangement, or fluctuations in amplifier gain. The other component of light noise is due to the graininess or quantum nature of light. Arrival of photons at a detector and, more importantly, the individual photoelectric events they cause are random events with the statistics of a Poisson process: In any given period of time the number of such events will have a variance equal to the mean number of such events.

The instrument described here was optimized for operation in a region where the statistical photon noise is the primary resolution-limiting factor. With the 45-μm fiber-tip aperture and 20-nm bandwidth, photodiode currents are on the order of 10 nA, or approximately 10^8 photoelectrons/ms. At the millisecond time resolution level, the corresponding ratio of mean effective photon count to r.m.s. effective photon noise, or shot noise, is therefore 10^4. All sources of dark noise are approximately a factor of 10 below this level. Optimization was therefore based on attempting the largest possible photocurrent, since the ratio of photocurrent to shot noise is proportional to the square root of the photocurrent. Illumination brightness is limited by the surface brightness of the incandescent filament. Other sources such as arcs or lasers could illuminate the target with much greater brightness but would entail other complexities and greater cost. Large photocurrents were therefore sought by maximizing detector efficiency.

The dispersive spectral detector described here achieves large average photocurrents from a multiplexing advantage: Signals at all wavelength bands are integrated continuously. In comparison, the chopped-beam type of spectral detector (Chance et al., 1975), integrates each wavelength for only a small fraction of each measurement interval and is therefore less suitable for the smaller optical apertures where photon flux is limited. The polychromator arrangement shown in Fig. 1 provides an overall efficiency for in-band light of about 30%. In comparison, a chopped-beam detector, where a given interference filter of 50% efficiency is in the beam about 10% of the time, would yield an overall efficiency closer to 5%.

Within limits, it is possible to trade off between temporal and spatial resolution limits to absorbance measurement. Longer integration times will thus allow reduction in size of the optical aperture and improved spatial resolution, and vice versa. This trade-off may fail with integration times of many seconds or longer, however, owing to $1/f$ noise in preamplifiers. Under these circumstances, beam chopper stabilization or photomultiplier tube (PMT) detectors might be desirable in spite of the consequent losses in photoelectron count (PMTs have only about 10–20% the quantum efficiency of silicon photodiodes, even though they offer vastly quieter preamplification).

If excellent accessibility for micromanipulation is not required, temporal and spatial resolution might be improved by increasing the numerical apertures of illumination and light collection. This could be accomplished, for instance, by placing the specimen between oil immersion condenser and objective, and coupling transmitted light collected at the objective image plane, either directly or by an optical fiber, to a polychromator like that described here.

REFERENCES

Brown, J. E., P. K. Brown, and L. H. Pinto, (1977) Detection of light-induced changes of intracellular ionized calcium concentration in *Limulus* ventral photoreceptors using arsenazo III. *J. Physiol. (Lond.)*, **267**, 299–320.

Chance, B., V. Legallais, J. Sorge, and N. Graham, (1975) A versatile time-sharing multichannel spectrophotometer, reflectometer and fluorometer. *Anal. Biochem.*, **66**, 498–514.

Smith, S. J. and R. S. Zucker, (1980) Aequorin response facilitation and intracellular calcium accumulation in molluscan neurones. *J. Physiol. (Lond.)*, **300**, 167–196.

Strobel, H. A. (1973) *Chemical Instrumentation: A Systematic Approach*. Addison-Wesley, Reading, MA.

Thomas, M. V. (1982) *Techniques in Calcium Research*. Academic, London.

CHAPTER 16

CALCIUM TRANSIENTS IN FROG SKELETAL MUSCLE FIBERS INJECTED WITH Azo1, A TETRACARBOXYLATE Ca^{2+} INDICATOR

S. HOLLINGWORTH
S. M. BAYLOR

Department of Physiology
University of Pennsylvania
Philadelphia, Pennsylvania 19104

1.	INTRODUCTION	262
2.	METHODS	263
	2.1 Azo1 CALIBRATIONS	263
	2.2 EXPERIMENTAL ARRANGEMENT	268
	2.3 KINETIC COMPUTATIONS	270
3.	RESULTS	270
	3.1 MEASUREMENTS FROM SPATIAL REGION 1	271
	3.2 MEASUREMENTS FROM SPATIAL REGION 3	276
4.	DISCUSSION	279
	REFERENCES	282

1. INTRODUCTION

The intracellular use of metallochromic indicator dyes such as arsenazo III and antipyrylazo III represents a promising technique for investigating the changes in cytoplasmic free calcium concentration, $[Ca^{2+}]$, that control a variety of cell functions. For example, in vertebrate skeletal muscle fibers these dyes can be introduced into the myoplasmic space, and absorbance signals can be measured during activity which reflect the very brief Ca^{2+} transients that activate the contractile response (Miledi et al., 1977, 1982; Kovacs et al., 1979, 1983; Palade and Vergara, 1982; Baylor et al., 1982b,c, 1983a,b, 1985). However, significant difficulties have been noted in attempts to interpret the signals from these dyes in muscle (Palade and Vergara, 1981, 1982; Baylor et al., 1982b,c, 1983b, 1985). Some of these difficulties appear to be directly attributable to the complexity of the dye chemistry as observed in in vitro studies, for example, complexities related to Ca^{2+}–dye kinetics (Ogawa et al., 1980; Palade and Vergara, 1981; Dorogi et al., 1983) and steady-state stoichiometries (Thomas, 1979; Rios and Schneider, 1981; Palade and Vergara, 1983). In addition, these dyes are subject to the widely recognized problem of reacting with ions other than Ca^{2+}, of which the most important under intracellular conditions are probably H^+ and Mg^{2+}.

Recently a new class of Ca^{2+} indicator dyes with one-to-one stoichiometry for Ca^{2+} and a very high degree of selectivity over H^+ and Mg^{2+} has been described by Tsien (1980, 1983). These "tetracarboxylate" indicators are chemical analogues of EGTA (ethylene glycol bis(β-aminoethyl ether)-N, N'-tetraacetic acid) to which chromophore groups have been added, resulting in compounds having a range of optical properties and chemical affinities. Azo1 (see Tsien, 1983, and Fig. 1 for chemical structure) is a relatively high affinity member of this class of compounds, having a dissociation constant for Ca^{2+} of about 3 μM, and undergoes an absorbance change in the ultraviolet and visible regions of the spectrum upon binding Ca^{2+}. Because of the simplicity of the chemistry of Azo1 in comparison with arsenazo III and antipyrylazo III, it was of interest to test its use as a Ca^{2+} indicator in frog muscle.

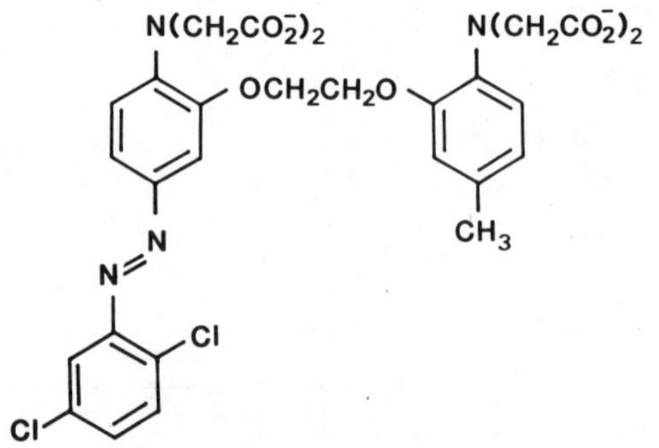

Figure 1.   Structure of Azo1. The molecule has a valence of -4 and a molecular weight of 659.5. After Tsien (1983).

This chapter describes the first intracellular use of Azo1 to measure Ca^{2+}. Several features of the signals from this dye are indicative of a pattern that is uncomplicated and quantitatively interpretable. For example, no evidence has been detected for the existence of a dye-related "dichroic" signal, as was previously found in fibers injected with arsenazo III, antipyrylazo III, and dichlorophosphonazo III (Baylor et al., 1982b,c, 1983b, 1985). Moreover, the absolute absorbance spectrum of Azo1 in an unstimulated fiber is similar to the in vitro spectrum of the dye in its Ca^{2+}-free form. Importantly, the absorbance change of the dye during fiber activity appears to reflect a single temporal component whose amplitude has the spectral sensitivity expected if a fraction of the dye changes to its Ca^{2+}-bound form. It seems likely, however, that the time course of the Ca^{2+}–Azo1 signal is somewhat slower than that of the free $[Ca^{2+}]$ transient itself. This delay may have a straightforward explanation—namely, the kinetic limitation expected if the "off" rate for the Ca^{2+}–dye reaction is in the range 200–300 s^{-1}, a value consistent with the relatively high affinity of Azo1 for Ca^{2+}.

A somewhat unexpected finding, however, is that the peak amplitude of the free $[Ca^{2+}]$ transient as calibrated from the Azo1 signal was only 0.5–0.6 μM. This value, which applies to a single twitch or a brief, high-frequency tetanus, is significantly smaller than previous reports based on measurements from intact fibers injected with aequorin (Blinks et al., 1978; Allen and Blinks, 1979), arsenazo III (Miledi et al., 1982), and antipyrylazo III (Baylor et al., 1982b, 1983b). It is not clear at this time whether the unexpectedly small size of the calibrated Azo1 signal in muscle might be indicative of a problem in the intracellular use of Azo1 or in the calibrations of the other indicators.

Some of these results were presented in September 1984 in Woods Hole, Massachusetts, at the symposium "Optical Methods in Cell Physiology" sponsored by the Society of General Physiologists.

2. METHODS

2.1 Azo1 Calibrations

According to Tsien (1983), the Ca^{2+}–Azo1 reaction produces a change in dye absorbance (but not fluorescence) having the following characteristics:

1. an absorbance maximum for the Ca^{2+}-free form of the dye at 473 nm, with an extinction coefficient $\epsilon(473$ nm$) = 2.6 \times 10^4$ M^{-1} cm^{-1};

2. an absorbance maximum for the Ca^{2+}–dye complex at 370 nm, with an extinction coefficient $\epsilon(370$ nm$) = 2.0 \times 10^4$ M^{-1} cm^{-1};

3. a 1:1 stoichiometry for the Ca^{2+}–dye reaction, with a dissociation constant, K_D, equal to 2.8 μM (at pH $= 7.40$ and 22°C in a 0.1 M KCl solution [R. Y. Tsien, personal communication]).

The in vitro calibrations shown in Figs. 2 and 3 were carried out to check these findings. The principal solution for the calibrations consisted of 140 mM KCl, 10 mM

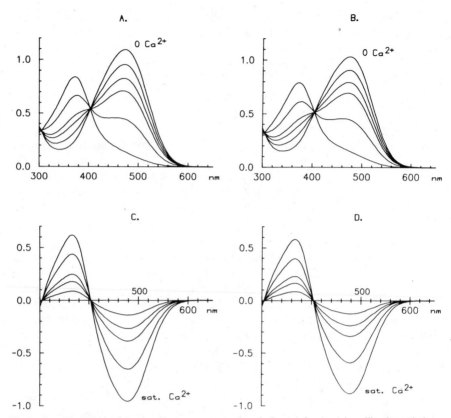

Figure 2. (A), (B) Absolute absorbance versus wavelength (in nm) for Azo1 in calibrating solutions, measured at 21°C on a spectrophotometer. The cuvettes contained either 42 μM dye (1-cm path length) or 398 μM dye (1-mm path length). The curves progress according to increasing amounts of added $[Ca^{2+}]$ from the one with maximum absorbance at 473 nm (no added Ca^{2+}) to the one with maximum absorbance at 370 nm (12 mM added Ca^{2+}). See Section 2 for contents of all cuvettes. (C), (D) "Difference" spectra obtained from the sets of absolute spectra in (A) and (B) by subtracting the 0 Ca^{2+} curves from each of the other curves and plotting the result without further change of units.

PIPES (piperazine-N,N'-bis(2-ethane sulfonic acid)) titrated to pH 6.90 with KOH, 11mM of the Ca^{2+} buffer HEDTA (N-hydroxyethylethylenediaminetriacetic acid, Sigma Chemical Co., St. Louis, MO), as well as various amounts of total Ca^{2+}. Following the additions of Ca^{2+}, all solutions were retitrated to pH = 6.90 (±0.01), the presumed resting pH of frog myoplasm under the conditions of the experiments (Baylor et al., 1982a). Six basic spectra, corresponding to six different total $[Ca^{2+}]$ concentrations (0, 1, 2, 3, 6, and 12 mM) were collected at each of two dye concentrations—42 μM (Fig. 2A) and 398 μM (Fig. 2B). The dye absorbance spectra were measured at 20–22°C on a UV–visible single-beam spectrophotometer (Ultrospec 4050, LKB Instruments Inc.) interfaced to a PDP 11 computer. The raw absorbance data from the instrument were transmitted in digitized form to the

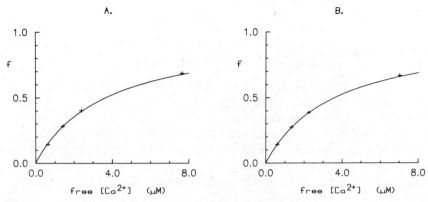

Figure 3.   Determination of the K_D of Azo1 for Ca^{2+} according to the procedure described in Section 2. For each set of data the symbols (+) plot as a function of free $[Ca^{2+}]$ the fractional amplitude f of the difference spectra shown in Fig. 2 (C), (D) relative to the amplitude of the difference spectrum at the highest level of added Ca^{2+}. The continuous curves plot the predicted amplitude as a function of free $[Ca^{2+}]$ assuming the relationships given in Eqs. (1)–(4) in text (which apply if the underlying stoichiometry is 1 Ca^{2+}:1 dye). The curves shown are least-squares fits (carried out separately for each data set) and correspond to $K_D = 3.7\ \mu M$ (A) and $K_D = 3.6\ \mu M$ (B).

computer for storage and subsequent analysis. The measurements were collected between 300 and 700 nm in 2-nm increments and were accurate to within about 0.002 absorbance units. The sample of Azo1, as the tetrapotassium salt, was generously donated by Dr. R. Y. Tsien.

It is clear from Fig. 2A and 2B that, following addition of Ca^{2+}, Azo1 undergoes an absorbance decrease in the visible region and an absorbance increase in the ultraviolet region of the spectrum, with isosbestic points near 404 and 304 nm. Figure 2 C and D plot the "difference" spectra from A and B, respectively. These curves correspond to the absorbance changes produced as increasing fractions of the dye are converted from the Ca^{2+}-free to the Ca^{2+}-bound form. Apart from scaling factors, the spectral shapes in parts C and D are all essentially identical, indicating that the dye forms a single stoichiometric complex with Ca^{2+}. A plot of the scaling factors versus the free $[Ca^{2+}]$ level (computed from equations (1)–(4) as described below) is shown as the symbols (+) in Fig. 3A (for the low dye measurements) and in Fig. 3B (for the high dye measurements). In each case the scaling factors have been normalized relative to the value for the solution in which total Ca^{2+} (12mM) exceeded total HEDTA (11 mM), in which essentially all the dye ($>99\%$) has been converted to the Ca^{2+}-bound form. Thus, for each of the other solutions, the normalized scaling factor reflects the fraction of the dye in the Ca^{2+}-bound form. These factors depend on free $[Ca^{2+}]$ in the manner expected if the underlying stoichiometry is 1 Ca^{2+}:1 dye, as shown by the agreement between the data and the curves in Fig. 3.

The curves in Fig. 3 were calculated under the assumption that three steady-state equations apply to the underlying reactions:

$$[Ca^{2+}][B] = [CaB]K_B \qquad (1)$$

$$[Ca^{2+}][D] = [CaD]K_D \qquad (2)$$

$$[Ca]_T = [Ca^{2+}] + [CaB] + [CaD] \qquad (3)$$

In these equations, $[Ca^{2+}]$, $[B]$, and $[D]$ denote the concentrations of free Ca^{2+}, buffer (HEDTA) not bound with Ca^{2+}, and dye not bound with Ca^{2+}, respectively; $[CaB]$ and $[CaD]$ denote the concentrations of Ca^{2+} bound to HEDTA and dye, respectively, and $[Ca]_T$ denotes total added Ca^{2+}. In addition, $[B]_T$ and $[D]_T$ are used to denote the concentrations of total HEDTA and total dye, respectively. These relationships may then be rearranged to yield an equation in $f = [CaD]/[D]_T$, the fraction of dye in the Ca^{2+}-bound form as a function of $[Ca]_T$, $[B]_T$, $[D]_T$, K_B and K_D:

$$A_3f^3 + A_2f^2 + A_1f + A_0 = 0 \qquad (4)$$

where

$A_0 = -K_B[Ca]_T$

$A_1 = K_D(K_B + [B]_T - [Ca]_T) + K_B([D]_T + 2[Ca]_T)$

$A_2 = K_D(K_D + [D]_T + [Ca]_T - K_B - [B]_T) - K_B(2[D]_T + [Ca]_T)$

$A_3 = [D]_T(K_B - K_D)$

Since K_B, the apparent dissociation constant of Ca^{2+}–HEDTA, is approximately 6.46 μM at pH = 6.90 and 20°C (see, e.g., Rios and Schneider, 1981), each data set in Fig. 3 may be fit using a least-squares procedure to obtain a "best fit" for the value of K_D. The values obtained for K_D, 3.7 μM (Fig. 3A) and 3.6 μM (Fig. 3B), are in good agreement with each other and in reasonable agreement with the value 2.8 μM reported by Tsien at pH = 7.40. The fact that the estimate of K_D from Fig. 3 does not depend on dye concentration is confirmation that the Ca^{2+}–dye stoichiometry is 1:1. For the purposes of calibrating the muscle signals in Section 3, we have used the average value of our two estimates of K_D determined at 21°C and then adjusted this value to that expected at 16°C (the temperature of the muscle experiments) assuming K_D has a Q_{10} of 1.3 (as estimated by Tsien, personal communication). Thus, the final value of K_D used for the muscle calibrations was 3.2 μM.

As can also be determined from Figs. 2 and 3, the maximum absorbance change at 480 nm (i.e., when all the dye has switched from the Ca^{2+}–free to the Ca^{2+}-bound form) is approximately -0.874 times the absolute value of $A(480$ nm$)$ for the Ca^{2+}-free form of the dye. The latter value is slightly smaller (by a factor of 0.99) than $A(473$ nm$)$. Hence, using the value of $\epsilon(473) = 2.6 \times 10^4 M^{-1}cm^{-1}$ quoted above from Tsien (1983), one has that for the Ca^{2+}–dye complex, the change in extinction coefficient $\Delta\epsilon(480) = -0.874 \times 0.99 \times 2.6 \times 10^4 M^{-1}cm^{-1} = -2.25 \times 10^4 M^{-1}cm^{-1}$, the value used for calibrating the data of Figs. 7–10.

Additional spectra (not shown) were measured in Ca^{2+}-free solutions containing 0.25 mM EGTA (instead of HEDTA) and 20 μM dye, in order to check the Mg^{2+} sensitivity of the dye. A change in Mg^{2+} from 0 to 2 mM produced a barely detectable absorbance change. This change had an amplitude and spectral dependence similar to that expected for a change in free $[Ca^{2+}]$ from 0 to 0.04–0.05 μM. Thus, under the assumption that the change in extinction coefficient of the dye is the same for Mg^{2+} binding as for Ca^{2+} binding, the measurements indicate that the Azo1 dissociation constants for Mg^{2+} and Ca^{2+} differ by a factor of 4–5 $\times$ 10^4. Hence, K_D^{Mg} at 16°C is roughly estimated as 120–150 mM. As expected for the tetracarboxylate family of indicators and buffers, Azo1 is highly selective for Ca^{2+} over Mg^{2+} (Tsien, 1980, 1983).

The possible interfering effects of Mg^{2+} on the Ca^{2+}–Azo1 signal from frog muscle may be estimated as follows. If myoplasmic free $[Mg^{2+}]$ at rest is 1 mM, then no more than $1/(1 + 120)$, or less than 1%, of the total dye will be bound with Mg^{2+} at rest, and the effective K_D of the dye for Ca^{2+} will be increased by less than 1%. Moreover, if the change in free $[Mg^{2+}]$ during fiber activity is no more than 2% of the resting level (Baylor et al., 1982a), then the change in Mg^{2+}–Azo1 complex during activity should be no more than 0.02% of $[D_T]$. The latter value is about 1/500 of the change in Ca^{2+}–Azo1 complex seen at the peak of the active signal (see, e.g., Fig. 8B, to be described in Sect. 3) and therefore may be ignored entirely. By the end of the sweep in Fig. 8B, an increase in Mg^{2+}–Azo1 complex of 0.02% of $[D]_T$ could account for as much as 1/10 to 1/5 of the observed slight elevation of the signal above the baseline level. Thus, even at late times, Mg^{2+} interference in the Ca^{2+} signal is probably quite small, although its influence may not be entirely negligible.

The possible effects of proton binding to Azo1 were evaluated in two ways. Absorbance spectra in Ca^{2+}-free solutions containing 0.25 mM EGTA were measured at pH 6.70 and 7.10. No absorbance change was detected, indicating that pH changes in the physiological range have little or no direct effect on the absorbance of the metal-free form of the dye. However, the value of pH in this range might still affect the apparent K_D of Azo1 for Ca^{2+}, without producing its own spectral change. An effect of this kind has, in fact, been reported for Quin2 (Tsien et al. 1982), another tetracarboxylate indicator, and may be explained by a pK ($-\log_{10}$(affinity constant)) for protonation somewhat less than 7 but near the physiological range. It may be noted that the existence of a similar effect with Azo1 is supported by the finding that K_D is 3.65 μM at pH 6.90 (Figs. 2, 3) whereas K_D is reported by Tsien to be 2.8 μM at pH 7.40.

To confirm the presence of this effect, we repeated the measurement of K_D at pH 7.40 using the method of Figs. 2 and 3 and a dye concentration of 30 μM. The value found, K_D = 3.0 μM, is in close agreement with Tsien's measurement and supports the interpretation that Azo1 has one or more pK's for protonation near the physiological range. If one assumes that a single effective pK for protonation of the tetravalent complex explains the measured shift in K_D with pH, then

$$K_D^{app} = K_D^{abs} (1 + 10^{(pK - pH)}) \tag{5}$$

Using our results obtained for K_D^{app} at pH 6.90 and 7.40, this equation may be solved to yield the estimates K_D^{abs} = 2.7 μM and pK = 6.45.

By differentiating Eq. (5) it also follows that:

$$(d/d\text{pH})K_D^{app} = -K_D^{abs}(\log_e 10)\, 10^{(pK - pH)} \tag{6}$$

Substituting the values given above for pK and K_D^{abs}, one obtains the estimate that $\Delta K_B^{app} = -\Delta pH \times 2.21 \ \mu M$ if myoplasmic resting pH is 6.90. Moreover, since the change in pH during a twitch appears to be no more than $+0.004$ (Baylor et al., 1982a), it follows that the effect of the myoplasmic pH change on ΔK_B^{app} should be no more than $-0.009 \ \mu M$. This represents a fractional change in K_D of -0.25%. Furthermore, as given in Baylor et al. (1982a), if f denotes the steady-state fraction of Azo1 bound with Ca^{2+}, the expected change in f, Δf, due to a change in K_D alone (without a change in $[Ca^{2+}]$) satisfies:

$$(\Delta f/f) = -(1-f)(\Delta K_D/K_D) \tag{7}$$

The effect of a 0.25% change in K_D on the muscle Ca^{2+}–Azo1 signal may therefore be estimated as follows. At the peak of the transient in Fig. 8B, when f is large with respect to its value before stimulation but in total less than 0.15, $\Delta f/f$ is less than 0.25%, a change that can be safely ignored. By the end of the sweep in Fig. 8B, however, f may be only slightly elevated above its resting value, possibly as small as a few percent above resting (see Sect. 3). At this time a $\Delta K_D/K_D$ of 0.25% might account for a minor fraction (0.25% of a few percent) of the observed slight elevation of the Ca^{2+}–dye signal above baseline. Thus, as was concluded above for the possible interfering effect of Mg^{2+}, even as late as 160 ms after stimulation the small elevation of the Ca^{2+}–dye signal above baseline should primarily reflect a continued elevation in myoplasmic $[Ca^{2+}]$, with little interference from the small alkalization of myoplasm.

2.2 Experimental Arrangement

The experimental protocol and methods for data acquisition were similar to those previously reported for the experiments with arsenazo III and antipyrylazo III (Baylor et al., 1982a, 1983a). Briefly, an intact single twitch fiber from frog (*Rana temporaria*) was isolated from either the semitendinosus or iliofibularis muscle and mounted on an optical bench apparatus for measuring transmitted light intensities. The bathing solution was normal Ringer's (120 mM NaCl, 2.5 mM KCl, 1.8 mM CaCl₂, and 5 mM Na₂PIPES titrated to pH 7.10) at a temperature of 16°C. In order to eliminate movement artifacts in the optical records, the fiber was stretched to a long sarcomere length (3.6–3.8 μm) and lowered onto a three-pedestal support. Dye was pressure-injected into myoplasm following penetration by a micropipette filled with distilled water plus 14 mM Azo1.

The muscle absorbance data were collected sequentially in time using one of a series of wide-band (30 nm half-band) interference filters (Omega Optical Co.) having peak transmittances every 30 nm between 420 and 630 nm. The region of fiber illumination consisted of either a small spot or slit of light confined within the fiber width. At each wavelength measurements of transmitted light intensity were made simultaneously using two forms of polarized light, denoted 0° (electric vector parallel to the fiber axis) and 90° (electric vector perpendicular to the fiber axis). The raw intensity values have been calibrated in terms of resting absorbance A and changes in

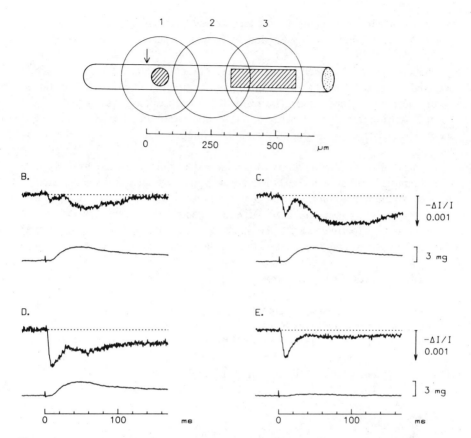

Figure 4. (A) A schematic of the spatial regions of optical recording during the experiment. The downward arrow points to the site of dye injection (relative position = 0 μm). The large-diameter circles with numerical labels indicate the fields of view "seen" by the collecting objective at three different times during the experiment. For the birefringence records shown in (B), (C), or (D) (single sweeps, taken 24–26 min after injection), the entire optical field (i.e., no spot or slit) in regions 1, 2, or 3, respectively, was illuminated with 630-nm light. Before the birefringence record shown in (E) was taken (a single sweep taken 55 min after injection, also whole-field illumination in region 3), the fiber was subjected to a small stretch, which increased the sarcomere spacing from 3.6 to 3.8 μm. The hatched areas of the fiber indicate locations from which absorbance measurements were made, using either a 65-μm diameter spot (region 1) or a 70 $\times$ 300 μm slit (region 3) of light. See text for additional details. Fiber diameter, 87 μm; temperature, 16°C.

absorbance ΔA, using the quantitative relationships previously described (Baylor et al., 1982a). In some cases Beer's law has also been used to convert from absorbance units to dye concentration units, by dividing the former by the product of extinction coefficient and path length. For this purpose the effective path length in myoplasm has been assumed to be 0.7 times fiber diameter (Baylor et al., 1983a). For the measurements of the birefringence signal (e.g., Figs. 4, 9B), the fractional change

($\Delta I/I$) in 630-nm light intensity was recorded with the fiber positioned between crossed polarizers oriented at $+45°$ and $-45°$ with respect to the fiber axis (Baylor and Oetliker, 1975).

Fiber activity was initiated by means of a single action potential, or train of action potentials, elicited by brief shocks from a pair of extracellular electrodes positioned locally near the injection site. The residual tension response was recorded by a sensitive tension transducer attached to one tendon end of the fiber. Both the optical and mechanical traces were digitally sampled, stored, and analyzed on a PDP 11/34 computer.

Because of the limited quantity of available dye, only a few experiments have been attempted so far with Azo1. Moreover, some difficulty was encountered in injecting significant quantities of the dye into muscle fibers by either iontophoresis (as is routinely used in arsenazo III experiments) or pressure (as is used in antipyrylazo III experiments). All the results described in this paper were obtained from a single experiment, on the only fiber to date successfully injected with Azo1.

2.3 Kinetic Computations

The calculation in Fig. 10 of the free [Ca²⁺] transient from the Ca²⁺–dye transient was carried out numerically on the PDP 11 computer by solving Eq. (8) in Section 4 for $[Ca^{2+}](t)$ and $[D](t)$. The derivative of the [CaD] waveform at the ith sampling point, t_i, was computed from the formula $(d/dt)[CaD](t_i) = ([CaD](t_{i+1}) - [CaD](t_{i-1})/(2\,\Delta t)$, where Δt is the time between data samplings, 0.4 ms.

3. RESULTS

Following completion of an apparently successful injection of dye, the "intrinsic birefringence" signal (second component of Baylor and Oetliker, 1975) can be measured to assess possible damage to the fiber as a result of the injection process (Baylor et al., 1982a). For example, at 16°C the amplitude of the birefringence signal is normally $1 - 2 \times 10^{-3}$, and its time course usually reaches a peak 9–12 ms after stimulation. Deviation from these values is usually an indication of injection damage, particularly if the deviation is most marked near the injection site.

The measurements in Fig. 4 (B–E) indicate that some degree of damage was incurred by this fiber owing to the injection of Azo1, even though all observed responses continued to be related to shock strength in an "all-or-nothing" fashion. Figure 4A shows a schematic of a portion of the fiber and the regions of optical recording relative to each other and to the injection site. Two principal sites of recording were used for the absorbance measurements. During the earlier part of the experiment, before the final stretch of the fiber, absorbance measurements were made from spatial region 1 in Fig. 4A, in which the incident illumination was a small spot of light (hatched circle) at one of eight different wavelengths (420–630 nm in steps of 30 nm). This region was only a few tens of microns from the injection site and is presumed to have been damaged somewhat by the injection process (see next paragraph). Moreover, the active records from this region (Fig. 7) were contaminated by small movement artifacts during their falling phases. However, the measurements of resting dye absorbance were well above the values of intrinsic fiber absorbance, and therefore resting dye concentrations could be accurately determined in this region. During the later part of the experiment, both before and after a small additional fiber stretch, absorbances were measured from the optical field denoted 3 in Fig. 4A, in which the incident illumination consisted of a slit of light (hatched rectangle) of either 480 or 630 nm. This region was relatively far from the injection site and is presumed to have been in essentially normal condition (see next paragraph). In addition, the records from this region after final stretch of the fiber were apparently free from movement artifacts. The dye concentration, however, was relatively small (approximately 0.1 mM when averaged over the length of the slit) and therefore not known with great certainty.

Figures 4B–E show examples of the intrinsic birefringence signal and twitch tension in response to a single stimulated action potential. For the birefringence records, light was collected from one of the three fully illuminated fields—region 1 (Fig. 4B), region 2 (Fig. 4C), and region 3. The records from region 3 were taken both before (Fig. 4D) and after (Fig. 4E) the final stretch. In all cases (except perhaps in Fig. 4B), the time-to-peak of the early decrease in birefringence was approximately 9–10 ms, indicative of a normal time course for the second component of the birefringence signal and therefore probably a normal time course for the underlying Ca^{2+} transient (Baylor et al., 1982b). However, the gradient in amplitude of the birefringence signal from its essentially normal value in region 3 to one-fifth that value in region 1 is strongly suggestive of a qualitatively similar gradient in amplitude of the Ca^{2+} transient. In Figs. 4B–D a slower component (third component) of the birefringence signal was also detected. This component was variable in appearance and was undoubtedly related to fiber movement as judged from its variability and the fact that its amplitude was greatly reduced following the stretch of the fiber (Fig. 4E versus 4D). The amplitude of the second component may also have been reduced slightly as a result of the stretch. The conclusion from Fig. 4 is that the absorbance measurements taken from spatial region 1 (to be described below in connection with Figs. 5–7) likely reflect a Ca^{2+} transient of smaller than normal amplitude, while those from spatial region 3 (Figs. 8–10) a Ca^{2+} transient of normal or nearly normal amplitude. Information from both fiber regions contributes to an understanding of the intracellular behavior of Azo1 in this fiber.

3.1   Measurements from Spatial Region 1

The measurements shown in Fig. 5A were made to assess the state of Azo1 in the fiber at rest. The open circles plot the resting spectrum of the dye, $A(\lambda)$, measured at eight wavelengths between 420 and 630 nm and normalized by the time-interpolated value at 480 nm, which decreased steadily during the run as dye diffused away from the injection site (Fig. 6A). The curves in Fig. 5A correspond to the spectral shapes determined from in vitro calibrations when Azo1 was entirely in either the Ca^{2+}-free form (0 Ca^{2+} curve) or the Ca^{2+}-bound form (sat. Ca^{2+} curve). The shape of the

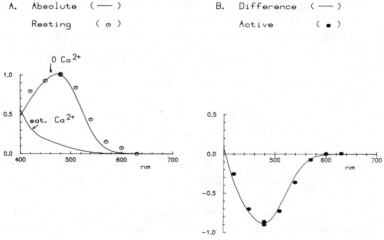

Figure 5. Comparison of normalized absorbance measurements as a function of wavelength for Azo1 signals in vitro and in vivo. (*A*) Continuous curves: normalized cuvette spectra obtained for a solution containing 23 μ*M* Azo1, 150 m*M* KCl, 1 m*M* MgCl, 0.25 m*M* EGTA, 5 m*M* PIPES, pH = 6.90. For the curve labeled 0 Ca²⁺, there was no added Ca²⁺; for the curve labeled sat. Ca²⁺, 1.0 m*M* CaCl₂ was added. The 0 Ca²⁺ curve has been scaled to have a value of 1.0 at 480 nm; the sat. Ca²⁺ curve was scaled by the same absolute factor. The open circles are measurements from a resting muscle fiber (region 1 of Fig. 4) of dye-related absorbance, $A(\lambda)$ (where λ is wavelength in nm), normalized by the simultaneously determined value at 480 nm, as interpolated from the curve in Fig. 6*A*. The dye-related value of $A(\lambda)$ was estimated from the total fiber absorbance measured at wavelength λ, $A_T(\lambda)$, from which the intrinsic absorbance, $A_i(\lambda)$, was subtracted. The relationship used for this latter estimate was $A_i(\lambda) = A_T(630)(\lambda/630)^x$, where $A_T(630)$ is presumed to reflect the intrinsic absorbance alone. In the formula, a value of $x = 1.3$ was used for the 90° absorbance data and a value of $x = 1.1$ for the 0° absorbance data (Baylor et al., 1983b). There were no significant differences between the dye-related 0° and 90° absorbance data. The average value of $A_T(630)$ was 0.047 and $A_T(480)$ was 0.133. (*B*) Continuous curve, Ca²⁺ "difference" spectrum obtained by subtracting the two curves in (*A*), without further scaling. The solid circles are muscle measurements of peak dye-related absorbance changes from records of the type in Fig. 7*B*, normalized by the simultaneously occurring value at 480 nm, as interpolated from the curve in Fig. 6*B*. These data have been scaled to give a "best" fit (as judged visually) to the curve.

muscle resting spectrum is consistent with the interpretation that nearly all of the Azo1 in myoplasm was in the Ca²⁺-free form. This result is expected if resting myoplasmic free [Ca²⁺] is 0.1 μ*M* (Blinks et al., 1982; Lopez et al., 1983) or less (Coray et al., 1980) and the dissociation constant of the Ca²⁺–dye reaction is 3.2 μ*M* (Sect. 2). In this case 3% (= 0.1/(0.1 + 3.2)) or less of the dye should be complexed with Ca²⁺. However, two uncertainties in the data of Fig. 5*A* limit an exact quantitative interpretation. First, in order to calculate the dye-related absorbance in Fig. 5*A* (open circles), it was necessary to correct the total measured absorbance for the fiber's intrinsic absorbance (see legend to Fig. 5). The correction forced $A(630$ nm)/$A(480$ nm) = 0 in Fig. 5*A*, but at 600 nm, a closely adjacent wavelength, the corrected measurement is elevated above the calibration curve. This indicates that there is likely a small error introduced by the correction for intrinsic absorbance in this fiber. Second, the Ca²⁺–dye reaction does not have an isosbestic point for any λ

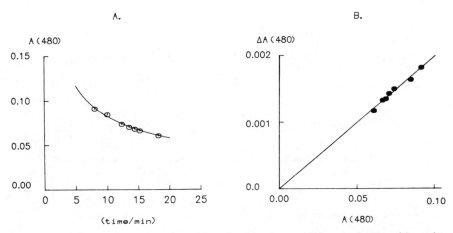

Figure 6. (A) Open circles: resting values of dye-related absorbance at 480 nm as a function of time t after the injection. The continuous curve is the function $A \times (B/t)^{1/2}$, where t is time in minutes and A and B were chosen to be 0.071 and 13.5, respectively. A dependence on $t^{-1/2}$ is expected to describe dye concentration versus time for one-dimensional diffusion of an impulse of dye away from a point source. According to Beer's law (see Sect. 2), a value of $A(480) = 0.071$ corresponds to a total dye concentration of 450 μM. (B) Solid circles: values of the peak change in dye-related absorbance at 480 nm (from records of the type in Fig. 7B) as a function of resting dye absorbance at 480 nm (as determined from the procedure described in the legend to Fig. 5, the continuous curve is a straight line through the origin that provides a "best" fit (judged visually) to the solid circles. The average value of $\Delta A(480)/A(480)$ was 0.020, which, as given in Section 2, is $0.020/0.874 = 0.023$ of the maximum change possible, were all the dye to change from the Ca^{2+}-free to the Ca^{2+}-bound form.

> 404 nm (i.e., within the practical range for making the muscle measurements using the existing apparatus). Therefore, the muscle data cannot be normalized with respect to absorbance measured at an isosbestic wavelength, the preferred procedure for making comparisons with spectral curves measured in cuvette. Thus the data in Fig. 5A do not rule out the possibility that a small fraction of the dye could in fact have been complexed with Ca^{2+} at rest.

In addition to these resting measurements, absorbance changes in response to an action potential were also followed using the same series of eight wavelengths and, at each wavelength, using the 0° and 90° forms of polarized light. A sample of the original records obtained in these measurements is shown in Fig. 7A. The wavelength (in nm) is indicated to the left of each pair of records and, to the right, the value of $A(480$ nm) at the time of the measurement (as interpolated from the curve in Fig. 6A). The signals at 630 nm, a wavelength at which Azo1 does not absorb, were taken to assess the contribution of non-dye-related changes. As expected, these signals are small, although at late times they are in the opposite direction from that usually seen in a completely immobilized fiber without dye (Baylor et al., 1982a, 1983b). They likely reflect the true intrinsic change plus a small movement artifact. The signals at the other wavelengths reflect these components plus a dye-related component that was of equal amplitude in the 0° and 90° records, at least during the rising phase of the signal and through its time-to-peak. The falling phase of the signals at 570, 540, and

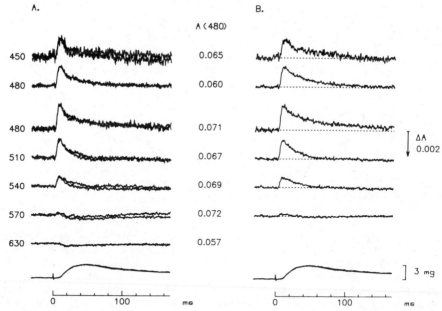

Figure 7.   (A) Original records of absorbance changes in response to a single stimulated action potential measured from region 1 (Fig. 4A) at the wavelengths indicated in nanometers to the left. The two traces superimposed at each wavelength are the 0° and 90° absorbance changes, which were measured simultaneously. The numbers to the right of the traces indicate the value of $A(480)$ at the time of each measurement. The value of $A(480)$ was interpolated from the curve in Fig. 6A. The uppermost 480-nm traces and the 630-nm traces were signal-averaged twice; all other traces are single sweeps. Two tension traces are superimposed, obtained at the beginning and end of the run. (B) The dye-related component of the absorbance changes, obtained by averaging the 0° and 90° traces in part A at each wavelength and subtracting the intrinsic component as estimated by the method described in the text.

510 nm suggests the possibility of a dichroic signal, which by definition is present whenever the 0° and 90° absorbance traces fail to superimpose. However, the amplitude of this component was small and its presence variable from record to record. It is therefore most likely related to a small and variable movement artifact rather than being a true dye-related dichroic signal of the type seen with arsenazo III, antipyrylazo III, and dichlorophosphonazo III (Baylor et al., 1982c, 1983b). This interpretation is supported by the absence of any detectable dichroic signal following reduction of the movement artifact by stretch of the fiber (Fig. 8A).

   The best estimates of the dye-related absorbance changes in the records of Fig. 7A are shown in Fig. 7B, where the 0° and 90° signals have been averaged and a correction applied to remove the non-dye-related component. For this correction, the non-dye-related component was assumed to have the same wave form as recorded in the 630-nm records but to vary in amplitude as $(630/\lambda)^{1.6}$. It is not expected that this correction procedure can successfully remove a variable movement artifact, and the fact that the later time course of the signals in Fig. 7B was variable (e.g., see the two 480-nm records) indicates that this is the case. Within the uncertainty of the

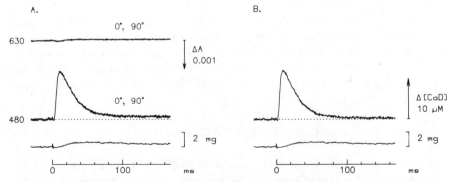

Figure 8. (A) Original records of polarized absorbance changes in response to a single action potential, measured from region 3 (Fig. 4A) at the wavelengths indicated in nanometers to the left. The records were taken after the final stretch of the fiber. $A_T(480)$ was 0.050, and dye-related $A(480)$ was 0.017, corresponding to $[D_T] = 107$ μM. Both the 480-nm and 630-nm traces were signal-averaged three times. (B) The estimated dye-related component of the 480-nm absorbance change in part A, obtained by the same procedure used in Fig. 7B. The records were taken 43–46 min after dye injection.

correction, however, the dye-related changes in Fig. 7B appear to have identical shapes and thus to reflect a single underlying temporal process. This process started shortly after stimulation, reached a peak in 8–10 ms, and returned nearly to baseline by the end of the record.

The nature of the underlying process is clarified by examining its spectral dependence. For this purpose, the peak amplitudes of the absorbance changes in Fig. 7B (and other records not shown) were normalized by their time-interpolated amplitude at 480 nm, analogous to the procedure used in obtaining the plot in Fig. 5A. In addition, the data set so obtained (solid circles, Fig. 5B) was scaled so as to allow comparison with the "difference" spectrum measured in vitro for the absorbance change arising when dye switches from the Ca^{2+}-free to the Ca^{2+}-bound form (continuous curve in Fig. 5B). The close agreement between the muscle data points and the cuvette difference spectrum indicates that the change underlying the signals in Fig. 7B was the transient formation of Ca^{2+}–dye complex. Thus, Azo1, like arsenazo III and antipyrylazo III, appears to be a useful dye for studying the rise and fall of myoplasmic Ca^{2+}.

The data in Fig. 6B supply additional information about the way in which Azo1 responds to the muscle Ca^{2+} transient. The solid circles plot the peak amplitude of the 480-nm absorbance change measured during activity as a function of the resting dye absorbance measured at 480 nm during the run. The theoretical line is the relationship expected if the underlying Ca^{2+} transients did not vary as a function of time or dye concentration and the stoichiometric relationship between dye and Ca^{2+} is 1:1. The data are consistent with this interpretation and indicate that at the peak of the Ca^{2+}–dye signal on average 2.3% of the dye had changed from the Ca^{2+}-free to the Ca^{2+}-bound form. Under the assumption that there is no kinetic delay between changes in free $[Ca^{2+}]$ and changes in Ca^{2+}–dye complex (but see Sect. 4 below), this corresponds to a peak change in free $[Ca^{2+}]$ of 0.08 μM. A Ca^{2+} transient of this amplitude is between 1 and 2 orders of magnitude smaller than values previously calibrated for the measurements made under similar experimental conditions on intact fibers injected with arsenazo III (Miledi et al., 1977) or antipyrylazo III (Baylor et al., 1982b, 1983b). However, as mentioned above in connection with Fig. 4, a smaller than normal Ca^{2+} transient is expected from this functionally modified fiber region near the injection site.

3.2   Measurements from Spatial Region 3

Following the measurements described above, the field of optical recording was moved approximately 450 μm from the injection site and additional absorbance measurements were made both before (Fig. 9A) and after (Fig. 8, 9B, 10) stretch of the fiber to further reduce the movement artifact. The Ca^{2+} transient in this region is presumed to have been nearly normal, as judged from the presence of a normal birefringence signal (Fig. 4D, E). However, the approximately fivefold smaller dye concentration undoubtedly served to limit the amplitude of the dye-related absorbance change measured from this region to a peak value similar to that measured from spatial region 1.

In Fig. 8A examples of 0° and 90° polarized absorbance records are shown at two wavelengths—480 nm, where the absorbance change in response to the formation of Ca^{2+}–dye complex is largest (Fig. 5B), and 630 nm, where only non-dye-related

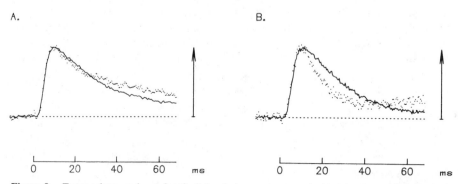

Figure 9.   Temporal comparison of optical signals in response to a single action potential, following scaling of signals to have the same peak value. (A) Comparison of the dye-related absorbance change at 480 nm from region 1 before stretch (dotted trace, average of 8 sweeps) and from region 3 before stretch (continuous trace, average of 6 sweeps). For the plot the continuous trace has been time-shifted by the equivalent of −0.2 ms to compensate for the delay in propagation of the action potential over the extra 400-μm length of fiber. The calibration arrow gives the amplitude of Δ[CaD], which (somewhat fortuitously) was 10.2 μM for both traces. The average total dye concentration during the measurements was 464 μM (dotted trace) and 95 μM (continuous trace). (B) Comparison of the birefringence signal (dotted trace, average of 2 sweeps, 630 nm light) and the dye-related absorbance signal (continuous trace, average of 2 sweeps, 480 nm light) measured simultaneously from region 3 after fiber stretch. The calibration arrow corresponds to a ΔI/I of −0.00082 for the birefringence trace and a Δ[CaD] = 10.1 μM for the absorbance trace.

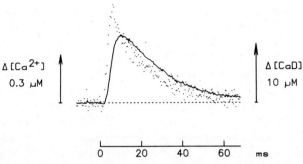

$\Delta [Ca^{2+}]$

0.3 μM

$\Delta [CaD]$

10 μM

0   20   40   60   ms

Figure 10. Temporal comparison of the measured dye-related absorbance change (continuous trace, average of 6 sweeps from region 3, calibrated at right) and the free Ca^{2+} transient (dotted trace, calibrated at left) computed by means of Eq. (8) in text. For the computation, total dye was considered a constant (100.2 μM) and equal to the sum of [CaD](t) and [D](t). The initial condition was set assuming a resting free $[Ca^{2+}] = 0.02$ μM. This corresponds to 0.02/(0.02 + 3.2), or 0.6%, of the dye being in the Ca^{2+}-bound form prior to onset of the transient.

changes are expected. The 630-nm records in Fig. 8A are consistent with the presence of a small intrinsic signal only, free from contamination by movement artifacts (Baylor et al., 1982a). The 480-nm signals, on the other hand, include both the intrinsic signal and the Ca^{2+}–dye signal. The exact superposition of the 0° and 90° records at both 630 and 480 nm implies the absence in this fiber of both an intrinsic dichroic signal and a dye-related dichroic signal. Thus, no evidence was detected from region 3 for any dichroism attributable to Azo1.

Figure 8B shows the dye-related component of the total absorbance change measured at 480 nm in part A under the assumption that the intrinsic component can be removed by subtracting the 630-nm wave form scaled by the factor (630/480)$^{1.6}$. Because the shape of the 630-nm record is closely similar to the standard intrinsic wave form (Baylor et al., 1982a), it is likely that this correction procedure worked significantly better than the same procedure applied to the records in Fig. 7. The dye-related signal in Fig. 8B has been calibrated in terms of the change in concentration of Ca^{2+}–dye complex, based on the value of extinction coefficient, $\Delta\epsilon$(480 nm), given in Section 2. The record indicates a peak increase of about 12 μM Ca^{2+}–dye complex 10 ms after stimulation, with an interval of about 24 ms between the times to half-peak amplitude in the rising and decaying phases. By the end of the sweep the value of the signal had declined to approximately + 1% to + 2% of its peak value. Under the assumption that there is no kinetic delay between the change in $[Ca^{2+}]$ and the change in Ca^{2+}–dye complex (but see Sect. 4), the peak of the signal in Fig. 8B corresponds to a peak increase in free $[Ca^{2+}]$ of about 0.4 μM above resting, with a return to about 4–8 nM above resting 160 ms after stimulation. The peak change in $[Ca^{2+}]$ from region 3 is therefore 5 times larger than that measured from region 1 but, surprisingly, still about an order of magnitude smaller than expected from the arsenazo III and antipyrylazo III results.

It should be pointed out, however, that while the overall shape of the Ca^{2+}–dye wave form in Fig. 8B is presumed to be known with considerably greater accuracy than the analogous waveform in Fig. 7B, the calibration of the amplitude of the underlying $\Delta[\text{Ca}^{2+}]$ is known with less accuracy. This situation arises because of the uncertainty in estimating resting dye concentration in region 3, where dye-related $A(480 \text{ nm})$ was only half of the intrinsic level estimated from $A(630 \text{ nm})$ (see legend, Fig. 8). It might therefore be suggested that the calibration of the peak change in $[\text{Ca}^{2+}]$ is unusually small because of a large overestimate of dye-related $A(480 \text{ nm})$ and therefore total dye concentration, $[\text{D}]_T$. However, two lines of evidence suggest that there has not been a serious overestimate of $[\text{D}]_T$. First, as the estimate of $[\text{D}]_T$ is lowered, there is required an inversely proportional increase in the estimate for the percentage saturation of the dye with Ca^{2+} during the transient. The degree of saturation, in turn, is expected to influence the shape of the [CaD] wave form. For example, at high levels of saturation, the appearance of the top of the [CaD] wave form should be less of a peak and more of a plateau, and its falling phase should be correspondingly prolonged. However, the comparison in Fig. 9A indicates that the [CaD] wave form recorded from region 3 was only marginally less "peaky" than the [CaD] wave form recorded from region 1, where the fractional amount of dye that changed from the Ca^{2+}-free to the Ca^{2+}-bound form was certainly no greater than about 3%. Secondly, at the end of the experiment (80 min after dye injection), a check on resting $A(480 \text{ nm})$ was made within region 3 using the small spot of light to measure absorbance (as for region 1). A gradient of dye absorbance was detected in the direction expected—namely, total absorbance, $A_T(480 \text{ nm})$, at the left extreme of region 3 was 0.032 units higher than $A_T(480 \text{ nm})$ at the right extreme. Thus $[\text{D}]_T$ was apparently 202 μM higher at the left compared with the right extreme, suggesting that when averaged over the whole field dye concentration was at least 101 μM. This number is in fact very close to the average value of $[\text{D}]_T$, 107 μM, estimated from $A(480 \text{ nm})$ measured with the slit of light (legend, Fig. 8). The overall conclusion is that there is unlikely to have been any serious error in calibrating the amplitude of $\Delta[\text{Ca}^{2+}]$ in connection with Fig. 8B due to an overestimate of $[\text{D}]_T$.

Another question that arises from the measurements in region 3 concerns the temporal relationship between the Ca^{2+}–Azo1 signal and the intrinsic birefringence signal. Figure 9B shows a comparison of these two wave forms measured simultaneously. The early rising phases of the two wave forms are closely similar, as has previously been found for the birefringence signal and all other Ca^{2+}–dye signals (Suarez-Kurtz and Parker, 1977; Baylor et al., 1982b). However, in Fig. 9B the birefringence signal reached a slightly earlier peak (9 ms versus 10 ms) and then returned toward baseline distinctly faster than the Azo1 signal. Based on this temporal comparison with birefringence, it is therefore likely that the Ca^{2+}–Azo1 signal is similar to or somewhat faster than the Ca^{2+}–arsenazo III signal measured under identical conditions but somewhat slower than the Ca^{2+}–antipyrylazo III signal (Baylor, Hollingworth, Hui, and Quinta-Ferreira, unpublished). The possible significance of these temporal relations will be taken up in Section 4.

4. DISCUSSION

The experimental measurements reported in this paper indicate that the tetracarboxylate indicator Azo1 may be used to investigate the Ca^{2+} transients that control contractile activity in skeletal muscle fibers. Like arsenazo III and antipyrylazo III, Azo1 can be pressure-injected into myoplasm, and dye-related absorbance changes can be measured during activity that undoubtedly reflect the rise and fall of myoplasmic free $[Ca^{2+}]$. A major advantage of the signal from Azo1 over that from arsenazo III or antipyrylazo III is the apparent absence of components not directly related to the formation of Ca^{2+}–dye complex. For example, based on the insensitivity of Azo1 to H^+ and Mg^{2+} in in vitro calibrations, significant interference from these ions in vivo, at least during a single twitch, may be ruled out with reasonable certainty. Moreover, no evidence has been found for the existence of an Azo1 dichroic signal. Thus, the Azo1 signal in muscle is evidently free of the two non-Ca^{2+} components that have been detected in the arsenazo III and antipyrylazo III signals (Baylor et al., 1982b, 1983b, 1985). In addition, the Ca^{2+} signal from Azo1 and antipyrylazo III share a common advantage when compared with the signal from arsenazo III—namely, the signal from these dyes appears to reflect the formation of one principal stoichiometric complex with Ca^{2+}. The Ca^{2+}–antipyrylazo III signal is primarily a 1 Ca^{2+}:2 dye complex in vitro (Rios and Schneider, 1981), and, as judged from in vivo measurement of its spectral sensitivity and dependence on total dye concentration, the same stoichiometry also holds for the muscle signal (Kovacs et al., 1983; Baylor et al., 1982b, 1983b). The Ca^{2+}–Azo1 signal, on the other hand, is presumed to reflect a pure 1 Ca^{2+}:1 dye complex (Tsien, 1980, 1983; this paper).

A possible disadvantage of the muscle Ca^{2+}–Azo1 signal in comparison with the Ca^{2+}–antipyrylazo III signal is related to the speed with which changes in the Ca^{2+}–dye signal respond to changes in $[Ca^{2+}]$. The temporal comparison in Fig. 9B indicates that the overall shape of the Ca^{2+}–Azo1 signal is somewhat slower than that of the intrinsic birefringence signal, whereas that of the Ca^{2+}–antipyrylazo III signal measured under the same experimental conditions is slightly faster (Baylor et al., unpublished). Under the assumption that there is a unique temporal relationship between the free $[Ca^{2+}]$ transient and the intrinsic birefringence signal, it follows that the Ca^{2+}–Azo1 signal follows a change in $[Ca^{2+}]$ less rapidly than does the Ca^{2+}–antipyrylazo III. For example, the peak of the Azo1 signal would probably be 3–4 ms later than the peak of the antipyrylazo III signal if the signals could be measured simultaneously in the same fiber. It should be noted, however, that some delay is to be expected on theoretical grounds for the way in which a high-affinity Ca^{2+} dye such as Azo1 can respond to changes in free $[Ca^{2+}]$. For a 1 Ca^{2+}:1 dye reaction the general kinetic scheme may be written:

$$Ca^{2+} + D \underset{K_{-1}}{\overset{K_{+1}}{\rightleftharpoons}} CaD$$

where K_{+1} (in units of $M^{-1}s^{-1}$) is the forward rate constant for the Ca^{2+}–dye reaction and K_{-1} (in units of s^{-1}) is the backward rate constant. The kinetic equation implied by this scheme is then:

$$(d/dt)[CaD] = K_{+1}[Ca^{2+}][D] - K_{-1}[CaD] \qquad (8)$$

At steady state this equation reduces to Eq. (2) in Section 2, where K_D, the dissociation constant of the dye for Ca^{2+}, is equal to K_{-1}/K_{+1}. As determined in Section 2, K_D is approximately 3.2 μM for Azo1 at 16°C. It therefore follows that K_{-1} = 3.2 $\mu M \times K_{+1}$. If one assumes the often-quoted value of $10^8 M^{-1}s^{-1}$ for K_{+1} when Ca^{2+} reacts with a site at 25°C (e.g., Johnson et al., 1981) and adjusts this rate to 16°C using a Q_{10} of 1.3, the estimates $K_{+1} = 8 \times 10^7 M^{-1}s^{-1}$ and $K_{-1} = 250s^{-1}$ are obtained. It follows that in response to a step change in free [Ca^{2+}] that is small with respect to K_D, the [CaD] wave form would respond along an exponential time course with a characteristic time $= 1/(250 \ s^{-1}) = 4$ ms. In fact, from the temporal comparisons of the birefringence and Ca^{2+}–dye signals mentioned above, a delay of a few milliseconds in the speed with which Azo1 tracks a muscle free [Ca^{2+}] transient seems probable.

Figure 10 gives a quantitative illustration of the kinetic delay that is imposed between the muscle free [Ca^{2+}] transient and the Ca^{2+}–dye transient if the off rate for the Ca^{2+}–dye reaction is $250s^{-1}$. The continuous trace is the measured Ca^{2+}–Azo1 signal (similar to that shown in Fig. 8B). The dotted trace is the free [Ca^{2+}] transient calculated from Eq. (8) above under the assumptions that $K_{+1} = 8 \times 10^7 M^{-1}s^{-1}$ and $K_{-1} = 250s^{-1}$.

Several conclusions may be drawn from the comparison in Fig. 10. As expected, the free [Ca^{2+}] transient has a somewhat earlier peak (6 ms versus 10 ms) and briefer overall time course than does the Ca^{2+}–dye transient. Moreover, the peak amplitude of the [Ca^{2+}] transient (0.5–0.6 μM) is about one-third larger than its value (0.4 μM) calibrated from the peak of the Ca^{2+}–dye transient using Eq. (2). In addition, it should be noted that the signal-to-noise ratio in the [Ca^{2+}] transient is considerably smaller (by a factor of 3–4) than in the Ca^{2+}–dye transient. The extra noise arises because of the necessity to differentiate the [CaD](t) signal in order to solve for [Ca^{2+}](t) using Eq. (8). Thus, in terms of signal-to-noise ratio, a practical limitation is imposed in going from the Ca^{2+}–dye signal to the free [Ca^{2+}] signal by means of the kinetic equation. Whatever source of noise exists in [CaD](t) will be exaggerated in [Ca^{2+}](t), and the smaller the value of K_{-1} the greater will be the exaggeration. Since the principal source of noise in [CaD](t) in Fig. 10 is the unavoidable shot noise due to random arrival of photons, the only way to avoid the decrease in the signal-to-noise ratio associated with solving Eq. (8) for [Ca^{2+}](t) is to use a dye with a larger value of K_{-1}. Presumably other members of the tetracarboxylate class of dyes (e.g., Azo2, Oxaz1, Oxaz2—see Tsien, 1983) would have a larger value of K_{-1} and would therefore exhibit less delay between [Ca^{2+}](t) and [CaD](t). For a given value of K_{+1}, however, a larger K_{-1} implies a lower affinity of the dye for Ca^{2+}, and therefore a larger quantity of injected dye would be required to achieve the same peak absorbance change, other things being equal.

It should also be noted that once the rapid changes in free $[Ca^{2+}]$ are complete (e.g., by the end of the sweep in Fig. 10), the Ca^{2+}–Azo1 time course is superimposable (apart from a scaling factor) with that of free $[Ca^{2+}]$. At this time in the transient, the use of a high-affinity dye like Azo1 should result in a free $[Ca^{2+}]$ signal with a larger signal-to-noise ratio than would a low-affinity dye (other things being equal), for at this time the steady-state relationship, Eq. (2), rather than Eq. (8), may presumably be applied without loss of information. It thus seems likely that in future experiments with Azo1 it should be possible to make particularly accurate measurements of the time course of the final return of free $[Ca^{2+}]$ to baseline. This in turn should give information at low levels of free $[Ca^{2+}]$ about the turnover rate of the sarcoplasmic reticulum (SR) Ca^{2+} pump and the unbinding of Ca^{2+} from parvalbumin—the processes that presumably determine the time course of restoration of the resting state (see, e.g., Gillis et al., 1982; Baylor et al., 1983a).

Some final comment concerning the unexpectedly small amplitude of the calibrated free $[Ca^{2+}]$ transient observed with Azo1 appears in order. The peak change in free $[Ca^{2+}]$ in Fig. 10, 0.5–0.6 μM, is more than an order of magnitude smaller than the 8-μm peak change during a twitch calibrated from the arsenazo III signal (Miledi et al., 1982; but note the problems in calibrating the arsenazo III signal described in Baylor et al., 1982b) and 4–6 times smaller than that calibrated from the antipyrylazo III signal (Baylor et al., 1982b, 1983b). Moreover, it is also more than an order of magnitude smaller than the steady-state level of $[Ca^{2+}]$ calibrated from the aequorin signal in a stretched fiber during a high-frequency tetanus (Blinks et al., 1978; Allen and Blinks, 1979; and see discussion in Baylor et al., 1982b). The reasons for these differences in the calibrations are not entirely clear at this time, but several possibilities may be proposed. First, the Azo1 calibration is based on a single experiment for which there was evidence of some injection damage. While it appears unlikely that the effect of the damage on the signal from region 3 could have been to reduce its amplitude by more than a factor of 2, determination of an average value must await the completion of more Azo1 experiments. Secondly, the calibration of the arsenazo III and antipyrylazo III signals depends on the value assumed for resting free $[Mg^{2+}]$ in myoplasm (and also resting pH). For example, if free $[Mg^{2+}]$ is closer to 0 than 2 mM, the value of $\Delta[Ca^{2+}]$ calibrated from the antipyrylazo III signal would be reduced by a factor approaching $(1 + 2/6.7)^2 = 1.7$ (see Baylor et al., 1982b). The calibration of the arsenazo III signal would also be reduced, but the factor would depend on the kinetics and relative proportions assumed for the underlying stoichiometric complexes for Ca^{2+} (1:1, 1:2, and 2:2—Palade and Vergara, 1983). Similarly, the calibration of the aequorin signal would also be reduced, and by a factor even larger than the one for the antipyrylazo III signal. Such considerations would narrow but probably not eliminate the discrepancies in the calibrations of the various Ca^{2+}-related signals. Finally, the possibility must be kept in mind that the myoplasmic environment may change the effective K_D's of the dyes (and aequorin) for Ca^{2+}. For example, if the different dyes bind to soluble proteins (as reported for arsenazo III by Beeler et al., 1980) or oriented structures such as the SR or myofilaments (as suggested by the existence of dichroic signals for arsenazo III and antipyrylazo III), the K_D's might change in variable ways. An explanation along these lines has already been suggested to account for the approximately 20-fold range encountered in attempting to calibrate the resting free [Mg^{2+}] level in myoplasm, based on results obtained with three different indicator dyes (arsenazo I, arsenazo III, and dichlorophosphonazo III—Baylor et al., 1982a).

ACKNOWLEDGMENTS

We are grateful to Dr. R. Y. Tsien for donating a sample of Azo1 and providing us with unpublished calibration information. Helpful comments on the manuscript were made by Professor W. K. Chandler and Drs. Y. E. Goldman, C. S. Hui, and M. E. Quinta-Ferreira. Financial support was provided by the U.S. National Institutes of Health (NS 17620 to S.M.B.) and the Muscular Dystrophy Association (postdoctoral fellowship to S.H.).

REFERENCES

Allen, D. G., and J. R. Blinks (1979) "The interpretation of light signals from aequorin-injected skeletal and cardiac muscle cells: A new method of calibration," In C. C. Ashley and A. K. Campbell, Eds., *Detection and Measurement of Free Calcium in Cells*, Elsevier/North Holland, Amsterdam, pp. 159–174.

Baylor, S. M., and H. Oetliker (1975) Birefringence experiments on isolated skeletal muscle fibres suggest a possible signal from the sarcoplasmic reticulum. *Nature (Lond.)*, **253**, 97–101.

Baylor, S. M., W. K. Chandler, and M. W. Marshall (1982a) Optical measurement of intracellular pH and magnesium in frog skeletal muscle fibres. *J. Physiol. (Lond.)*, **331**, 105–137.

Baylor, S. M., W. K. Chandler, and M. W. Marshall (1982b) Use of metallochromic dyes to measure changes in myoplasmic calcium during activity in frog skeletal muscle fibres. *J. Physiol. (Lond.)*, **331**, 139–177.

Baylor, S. M., W. K. Chandler, and M. W. Marshall (1982c) Dichroic components of arsenazo III and dichlorophosphonazo III signals in skeletal muscle fibres. *J. Physiol. (Lond.)*, **331**, 179–210.

Baylor, S. M., W. K. Chandler, and M. W. Marshall (1983a) Sarcoplasmic reticulum calcium release in frog skeletal muscle fibres estimated from arsenazo III calcium transients. *J. Physiol. (Lond.)*, **344**, 625–666.

Baylor, S. M., M. E. Quinta-Ferreira, and C. S. Hui (1983b) Comparison of isotropic calcium signals from intact frog muscle fibers injected with arsenazo III or antipyrylazo III. *Biophys. J.*, **44**, 107–112.

Baylor, S. M., M. E. Quinta-Ferreira, and C. S. Hui (1985) "Isotropic components of antipyrylazo III signals from frog skeletal muscle fibers," In R. P. Rubin, G. Weiss, and J. W. Putney, Jr., Eds., *Calcium in Biological Systems*, Plenum, New York, pp. 339–349.

Beeler, T. J., A. Schibeci, and A. Martonosi (1980) The binding of arsenazo III to cell components. *Biochim. Biophys. Acta*, **629**, 317–327.

Blinks, J. R., R. Rudel, and S. R. Taylor (1978) Calcium transients in amphibian muscle fibres: Detection with aequorin. *J. Physiol.*, **277**, 291–323.

Blinks, J. R., W. G. Wier, P. Hess, and F. G. Prendergast (1982) Measurement of Ca^{2+} concentrations in living cells. *Prog. Biophys. Mol. Biol.*, **40**, 1–114.

Coray, A., C. H. Fry, P. Hess, J. A. S. McGuigan, and R. Weingart (1980) Resting calcium in sheep cardiac tissue and in frog skeletal muscle measured with ion-selective microelectrodes. *J. Physiol. (Lond.)*, **305**, 60–61P.

Dorogi, P. L., C. -R. Rabl, and E. Neumann (1983) Kinetic scheme for Ca^{2+}-arsenazo III interaction. *Biochem. Biophys. Res. Commun.*, **111**, 1027–1033.

Gillis, J. M., D. Thomason, J. Lefevre, and R. H. Kretsinger (1982) Parvalbumins and muscle relaxation: A computer simulation study. *J. Muscle Res. Cell Motil.*, **3**, 377–398.

Johnson, J. D., D. E. Robinson, S. P. Robertson, A. Schwartz, and J. D. Potter (1981) "Ca^{2+} exchange with troponin and the regulation of muscle contraction," In A. D. Grinnell and M. A. B. Brazier, Eds., *Regulation of Muscle Contraction: Excitation-Contraction Coupling,* Academic, New York, pp. 241–259.

Kovacs, L., E. Rios, and M. F. Schneider (1979) Calcium transients and intramembrane charge movement in skeletal muscle fibres. *Nature,* **279**, 391–396.

Kovacs, L., E. Rios, and M. F. Schneider (1983) Measurement and modification of free calcium transients in frog skeletal muscle fibres by a metallochromic indicator dye. *J. Physiol. (Lond.),* **343**, 161–196.

Lopez, J. R., L. Alamo, C. Caputo, R. DiPolo, and J. Vergara (1983) Determination of ionic calcium in frog skeletal muscle fibers. *Biophys. J.,* **43**, 1–4.

Miledi, R., J. Parker, and G. Schalow (1977) Measurement of calcium transients in frog muscle by the use of arsenazo III. *Proc. R. Soc. Lond. B,* **198**, 201–210.

Miledi, R., I. Parker, and P. H. Zhu (1982) Calcium transients evoked by action potentials in frog twitch muscle fibres. *J. Physiol. (Lond.),* **333**, 655–679.

Ogawa, Y., H. Harafuji, and N. Kurebayashi (1980) Comparison of the characteristics of four metallochromic dyes as potential calcium indicators for biological experiments. *J. Biochem.,* **87**, 1293–1303.

Palade, P., and J. Vergara (1981). Detection of Ca^{++} with optical methods. In A. D. Grinnell and M. A. B. Brazier, Eds., *Regulation of Muscle Contraction: Excitation–Contraction Coupling,* Academic, New York.

Palade, P., and J. Vergara (1982) Arsenazo III and antipyrylazo III calcium transients in single skeletal muscle fibers. *J. Gen. Physiol.,* **79**, 679–707.

Palade, P., and J. Vergara (1983) Stoichiometries of arsenazo III–Ca complexes. *Biophys. J.,* **43**, 355–369.

Quinta-Ferreira, M. E., S. M. Baylor, and C. S. Hui (1984) Antipyrylazo III (Ap III) and arsenazo III (Az III) calcium transients from frog skeletal muscle fibers simultaneously injected with both dyes. *Biophys. J.,* **45**, 47a.

Rios, E., and M. F. Schneider (1981) Stoichiometry of the reactions of calcium with the metallochromic indicator dyes antipyrylazo III and arsenazo III. *Biophys. J.,* **36**, 607–621.

Snowdowne, K. W. (1979) Aequorin luminescence in frog skeletal muscle fibers at rest. *Fed. Proc.,* **38**, 1443.

Suarez-Kurtz, G., and I. Parker (1977) Birefringence signals and calcium transients in skeletal muscle. *Nature (Lond.),* **270**, 746–748.

Thomas, M. V. (1979) Arsenazo III forms 2:1 complexes with Ca and 1:1 complexes with Mg under physiological conditions. *Biophys. J.,* **25**, 541–548.

Tsien, R. Y. (1980) New calcium indicators and buffers with high selectivity against magnesium and protons: Design, synthesis, and properties of prototype structures. *Biochemistry,* **19**, 2396–2404.

Tsien, R. Y. (1983) Intracellular measurements of ion activities. *Annu. Rev. Biophys. Bioeng.,* **12**, 91–116.

Tsien, R. Y., T. Pozzan, and T. J. Rink (1982) Calcium homeostasis in intact lymphocytes: Cytoplasmic free calcium monitored with a new, intracellularly trapped fluorescent indicator. *J. Cell. Biol.,* **94**, 325–324.

CHAPTER 17

INTRACELLULAR pH OF *LIMULUS* VENTRAL PHOTORECEPTOR CELLS: MEASUREMENT WITH PHENOL RED

S. R. BOLSOVER
J. E. BROWN
Department of Ophthalmology
Washington University/School of Medicine
St. Louis, Missouri

T. H. GOLDSMITH
Department of Biology
Yale University
New Haven, Connecticut

1. INTRODUCTION   286
2. METHODS   287
   2.1 ABSORBANCE SPECTRA IN VITRO   288
   2.2 OPTICAL RECORDING FROM PHOTORECEPTORS   288
   2.3 ABSORBANCE SPECTRA OF PHENOL RED MEASURED
       IN THE PRESENCE OF ISOLATED RHABDOMS   290
3. RESULTS   291
   3.1 LIGHT-INDUCED ABSORBANCE CHANGES IN THE
       ABSENCE OF DYE   291
   3.2 ABSORBANCE SPECTRUM OF INTRACELLULAR
       PHENOL RED   292

3.3 EFFECT OF PHENOL RED ON THE ELECTRO-
PHYSIOLOGICAL RESPONSES   293

3.4 INJECTION OF pH BUFFERS   295

3.5 ABSORBANCE SPECTRA OF PHENOL RED MEASURED
IN THE PRESENCE OF ISOLATED RHABDOMS   297

3.6 LIGHT-INDUCED CHANGE IN ABSORBANCE
OF INTRACELLULAR PHENOL RED   298

3.7 ABSORBANCE CHANGES OF INTRACELLULAR PHENOL
RED DURING DARK ADAPTATION   300

3.8 SPATIAL SCANNING EXPERIMENTS   302

4. DISCUSSION   304

4.1 ABSORBANCE CHANGES IN THE ABSENCE OF DYE   304

4.2 PHENOL RED AS AN INDICATOR OF
INTRACELLULAR pH   304

4.3 SPATIAL GRADIENT OF pH$_i$ IN DARK-ADAPTED
CELLS   305

4.4 INTRACELLULAR pH BUFFERING CAPACITY   306

4.5 LIGHT-INDUCED CHANGE IN INTRACELLULAR
pH INDICATED BY PHENOL RED   306

4.6 ORIGIN OF THE LIGHT-INDUCED CHANGE IN pH$_i$ 308

4.7 DOES A LIGHT-INDUCED FALL IN INTRACELLULAR pH
PLAY A ROLE IN LIGHT ADAPTATION?   308

REFERENCES   309

1. INTRODUCTION

Indicator dyes have been used to determine changes in intracellular concentration of metal ions in several single-cell preparations. Some of these indicators (e.g., arsenazo III) have marked dependencies on the concentration of hydrogen ions. Therefore, an independent determination of changes of hydrogen ion concentration during physiological events is useful for the interpretation of measurements made with metallochromic indicators. Moreover, changes in intracellular concentration of hydrogen ions have been proposed to signal or modulate physiological events directly. For example, intracellular pH (pH$_i$) has been proposed to modulate physiological responses in invertebrate photoreceptors. Meech and Brown (1976) found that intracellular pH in *Balanus* photoreceptors falls by up to 0.2 pH units after the onset of illumination, and they suggested that this decrease in pH$_i$ may reduce the sensitivity of the cell to light. On the other hand, experiments with both *Limulus* and *Balanus* photoreceptors (Coles and Brown, 1976; Bolsover and Brown, 1982) suggest that the effect of pH_i on the sensitivity of invertebrate photoreceptors is minor. A more exact knowledge of the amplitude, time course, and spatial distribution of the light-induced change in pH_i might further clarify its putative role as a modulator of photoreceptor function.

This chapter describes experiments in which the pH indicator dye phenol red was injected into *Limulus* ventral photoreceptors. The optical absorbance of intracellular dye was measured as a function of wavelength, time, and distance along a linear dimension of the cell. Some of these results have been published in preliminary form (Bolsover et al., 1983).

2. METHODS

The ventral rudimentary eye of *Limulus polyphemus* was desheathed, pinned into a silicon rubber (Sylgard 184, Dow–Corning, Midland, MI) dish, and bathed with 20 mg/mL Pronase (grade B, CalBiochem, La Jolla, CA) in artificial seawater for 1 min. Thereafter the nerve was bathed in artificial seawater (ASW) composed of 422 mM NaCl, 10 mM KCl, 22 mM $MgCl_2$, 26 mM $MgSO_4$, 10 mM $CaCl_2$ and 10 mM TRIS Cl, pH 7.8. To make acetate–ASW, 10 mM sodium acetate was added to ASW; glacial acetic acid was then added until the pH of the ASW was 6.1. To make ammonium–ASW, ammonium hydroxide was added to ASW until the pH of the ASW was 9.0.

Solutions of phenol red (J. T. Baker Chemical Co., Phillipsburg, NJ), bromoxylenol blue (Aldrich Chemical Co., Milwaukee, WI), or bromphenol blue (National Aniline Division of Allied Chemical Corp., New York) were passed through a column of Chelex 100 (Biorad Labs, Richmond, CA) to remove divalent cations; KCl was added to 200 mM, and HCl was added to a final pH of 8.0. The bromphenol blue solution was buffered at pH 8.0 with 5 mM HEPES (N-2-hydroxyethylpiperazine-N'-2-ethanesulfonic acid). A dye solution was injected into single photoreceptor cells from an intracellular micropipette containing 50 mM dye solution by applying pressure (3–20 $\times$ 10^4 Pascal) to the back of the micropipette. Cells were impaled with a second micropipette that contained 2 M KCl for passing voltage-clamp current.

For experiments in which a pH buffer was pressure-injected, cells were impaled with a second micropipette containing a solution made by adding KOH to a 1 M solution of buffer. Buffer solutions used were MOPS at pH 7.2, HEPES at pH 7.7, and HEPES at pH 7.4. The pH 7.4 solution also contained $^{35}SO_4^{2-}$ (New England Nuclear, Boston) at high specific activity. At the end of the experiment, the apparent dimensions of the injected cell were measured under a compound microscope. The preparation was then digested extensively with Pronase (20 mg/ml); Aquasol (NEN, Boston) was added, and the sample was counted in a liquid scintillation counter. The volume of solution injected was calculated from the specific activity of ^{35}S in the sample. The volume of the cell was calculated from the dimensions of the profile of the cell, assuming the cell to be an oblate spheroid. Thus the concentration of buffering substance injected into the cell could be estimated.

To estimate the effect of dilution on the pH of the injected buffer solutions, 1 M MOPS at pH 7.21 was diluted with 625 mM KCl. A tenfold dilution of buffer reduced the pH of the solution to 7.10; a further tenfold dilution reduced the pH to 7.09. A tenfold dilution of 1 M HEPES at pH 7.70 with 625 mM KCl reduced the pH to 7.59; a further tenfold dilution reduced the pH to 7.57.

2.1. Absorbance Spectra *in Vitro*

The absorbance spectrum of phenol red was measured at various values of pH. Each solution contained 27.2 μM phenol red, 400 mM KCl, and 20 mM pH buffer (Good, et al., 1966). Buffers used were MES (pH 4.4–6.1), MOPS (pH 6.1–7.3), TAPS (pH 7.7–8.8), and CAPS (pH 9.6–10.6). Absorbance spectra were measured on a Gilford 250 (Oberlin, OH) spectrophotometer. Ten absorbance spectra covering the pH range from 4.46 to 10.6 were measured. Absorbance maxima were at 430 nm for the acidic form of the dye and 560 nm for the basic form of the dye; the isosbestic wavelength was 480 nm. Calibration curves relating the ratio of absorbances at a pair of wavelengths to the pH were constructed from these absorbance spectra. The absorbance at 560 nm is a simple sigmoid function of pH, because the acidic form of the dye has negligible absorbance at 560 nm. The pK was found to be 7.693. The pH of a sample solution could be related to the absorbances at 560 nm and 480 nm as follows. The absorbance at 560 nm is 6.41 times the absorbance at 480 nm (the isosbestic wavelength) when all dye is in the basic form. The pH of a sample solution was calculated using the equation

$$pH = pK + \log(B/[1-B])$$

where B = fraction of dye in basic form = A_{560}/A^*_{560}; = $A_{560}/(6.41 \times A_{480})$; A_{560} = absorbance at 560 nm; A^*_{560} = maximum absorbance at 560 nm, all dye in the basic form; and A_{480} = absorbance at 480 nm.

To examine the relationship between the absorbance of phenol red at its isosbestic wavelength and dye concentration, four solutions containing 1.1, 5.4, 27.2, and 136 μM dye were prepared in 20 mM MOPS buffer at pH 7.13. Over this 100-fold range the absorbance at 480 nm was linearly related to dye concentration; the absorptivity was 1.17×10^4 L/mol cm. This value was used to estimate the intracellular concentration of phenol red by assuming an average path length through a cell of 90 μm.

2.2. Optical Recording from Photorecptors

The nerve was maintained in darkness between absorbance measurements. The recording setup was enclosed in a five-sided metal box, with the front covered by opaque cloth curtains to reduce stray light. A single beam of light was used both to stimulate the cells and to measure the optical absorbance. The beam was produced by a 100-W tungsten-iodide lamp powered by a current-regulated power supply (model 6267B, Hewlett Packard, Palo Alto, CA). The beam passed through a KG1 infrared blocking filter (Schott, Duryea, PA) and an electromechanical shutter (Vincent

Associates, Rochester, NY); the beam was focused onto the preparation by a microscope condenser. A cover glass was fixed to the chamber immediately above the preparation to eliminate small variations of path length that would otherwise be caused by waves at an air–water interface. Light transmitted through the preparation was collected by a long-working-distance microscope objective and was focused onto the detection system. The light reaching the detection system was limited to that transmitted through a single photoreceptor or other portion of tissue by two pairs of blades that defined a rectangular measuring aperture in the back focal plane of the objective. The entire apparatus was mounted on a vibration isolation table (Newport Research Labs, Fountain Valley, CA). The apparatus was used in three different modes which will be referred to as (1) the single wavelength photometer, (2) the scanning spectrophotometer, and (3) the spatial scanning photometer.

Single Wavelength Photometer. Light illuminating the preparation was limited to a narrow wavelength band by placing an interference filter with center wavelength 610, 560, 498, or 480 nm (10 nm FWHM) in the beam. With these filters in place the light intensity at the preparation was respectively 2.3, 3.5, 0.8, and 1.3 mW/cm^2. The microscope objective (Leitz Wetzlar L20, NA 0.4) focused the light passing through the preparation onto a silicon photodiode (PV100A, E.G. and G., Salem, MA) operated in the unbiased mode. During the 40 ms immediately after the shutter opened, the amplified output of the photodiode was reset to approximately 0 V (using a model 755 digital sample and hold, Hybrid Systems, Burlington, MA). Any subsequent change in optical transmission of the preparation could then be recorded at high gain. The time constant of the photodiode amplifier was 240 μs. The bandwidth of the recording system was limited by that of the tape recorder (0–625 Hz). We call the optical transmission measured at the onset of light (40 ms after the shutter opened) the transmission of a dark-adapted cell, and values calculated from that transmission are referred to as the absorbance of a dark-adapted cell, the pH_i of a dark-adapted cell, and so forth.

Scanning Spectrophotometer. A scanning microspectrophotometer developed by Harary and Brown (1984) was used for these experiments. The preparation was illuminated with white light of intensity 2.3 mW/cm^2. The microscope objective (Leitz Wetzlar UM32, N.A. 0.30) focused the light passing through the preparation onto the input slit of a grating spectrometer (model UFS 200, JY Optical Systems, Metuchen, NJ). The dispersed light was focused in the plane of a linear array of 1024 matched photodiodes (RL1024S, Reticon, Sunnyvale, CA) so that the shortest wavelength light fell on diode number 1 and the longest wavelength light fell on diode number 1024. A subset of the array, consisting of those 512 contiguous diodes corresponding to the wavelength range 384–691 nm, was continuously scanned by an array controller (model 1218, Princeton Applied Research, Princeton, NJ). Each scan of the 512 diodes took 4.4 ms. The signals from the first four diodes were added together to form the output of one pixel. Similarly, each successive set of four diodes formed a pixel. Each scan of the 512 contiguous diodes (= 128 pixels) took a minimum of 4.4 ms. To improve the signal-to-noise ratio at the expense of temporal resolution, the output scan from the array could be delayed to prolong the exposure to light. Usually we chose to scan the array once every 17.6 ms; thus, each pixel integrated the light falling upon it for 17.6 ms. The data ensemble (referred to below as an intensity spectrum) comprised 128 numerical values; each value indicated the intensity of light in a 2.4-nm wavelength band over a 17.6-ms period. Because the array was scanned sequentially from the blue end (pixel 1) to the red end (pixel 128), a single intensity spectrum was not recorded at a single unique time. The values recorded for pixel 1 and for pixel 128 each represent the light intensity during a 17.6 ms period, but the period began 4.4 ms earlier at pixel 1 than at pixel 128. The measurement period at pixels 2 through 127 began at corresponding intermediate times.

During a 12-s flash of light, 80 intensity spectra were recorded. The first 40 were recorded at the maximum rate—that is, one every 17.6 ms. The shutter was closed through most of the first intensity spectrum and was fully open by the end of the second; therefore the third intensity spectrum was the first usable one and represented a period that began more than 17.6 ms but less than about 35 ms after the onset of light. We take this third intensity spectrum to be the spectrum of the dark-adapted cell. The second set of 40 spectra was recorded at 280-ms intervals so that spectrum 80 represents a period that began 11.9 s after the onset of light.

Absorbance difference spectra were calculated from the measured intensity spectra. Phenol red has negligible absorbance in the far red. Therefore, the absorbance difference at pixel 128 ($\lambda = 691$ nm) was constrained to be zero in all calculations by adding a correction value; this same correction value for absorbance was added to all other pixels. This procedure corrected for wavelength-independent fluctuations in light intensity.

Spatial Scanning Photometer. The spatial scanning photometer differed from the scanning spectrophotometer in two respects. First, the measuring beam passed through a narrow bandpass (10 nm FWHM) interference filter before being focused on the preparation. Second, the grating spectrometer was replaced by a mirror and an additional lens, so that an image of the cell was focused onto the central 512 diodes in the array. Each successive set of four contiguous diodes was grouped to form one pixel; the integration time for each pixel was chosen to be 17.6 ms, and the protocol for recording scans was as described for the scanning spectrophotometer. The absorbance difference at pixel 128, which corresponded to a position outside the image of the cell, was constrained to be zero.

All calculated values of absorbance difference and pH_i are given as the mean $\pm$ the standard error of the mean ($s_n - 1/\sqrt{n}$). Experiments were performed at room temperature ($22 \pm 1°C$).

2.3. Absorbance Spectra of Phenol Red Measured in the Presence of Isolated Rhabdoms

Rhabdoms of crayfish (*Procambarus,* Carolina Biological Supply, Burlington, NC) were isolated from dark-adapted animals by macerating the eye in 90 mM potassium phosphate buffer, brought to 330 mOsm with KCl. A drop of this suspension was placed on a dry 22 × 22 mm coverslip that had previously been coated with a film of 0.1% polylysine (Sigma, St. Louis, approximate molecular weight 260,000 daltons). A thin chamber was constructed by placing an 18 × 22 mm coverslip over the first. The chamber was sealed with silicone grease along two edges but remained open at the ends where the lower coverslip projected beyond the upper. With strips of thin Teflon tape as spacers in the silicon grease seals, the internal thickness of the chamber was about 200 μm; without the spacers the fluid was reduced to a layer about 30 μm in depth. The chamber was flushed while mounted on the microscope by placing a drop of solution on the lower coverslip at one end and drawing solution through the chamber with a corner of lens tissue applied at the opposite end. Rhabdoms usually adhered to the polylysine through repeated flushes.

Absorbance spectra of phenol red were measured by microspectrophotometry (Goldsmith, 1978). The measuring beam was 2 x 7 μm in the object plane, much smaller than the 25-μm width of a single rhabdom. Spectra of rhabdom membranes and of clear areas of the chamber were measured in the presence of phenol red (approximately 2 mM) and after flushing out the dye with solution at the same pH. The absorbance spectrum of phenol red in the presence of photoreceptor membranes was then determined by calculating difference spectra, as described in Section 3.

3.   RESULTS

3.1.   Light-Induced Absorbance Changes in the Absence of Dye

When a dark-adapted *Limulus* ventral dye was illuminated for 12 s with bright white light, the intensity of transmitted light reaching the detector of the scanning

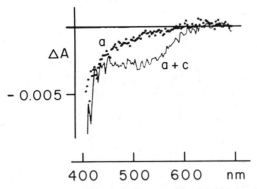

Figure 1.   Light-induced absorbance changes in the absence of dye. Light-induced changes in absorbance were measured using the scanning spectrophotometer. After 10 min in darkness the tissue was illuminated with white light for 12 s. The difference between the absorbance measured at flash onset and 12 s after the onset is plotted as a function of wavelength. (*a*) The measured light passed through axons and no cell bodies. (*a* + *c*) The light passed through one photoreceptor cell body and the underlying axons. The measurement was made before impalement of the cell by a pipette.

spectrophotometer increased during the period of illumination. This change in the optical properties of the tissue can be treated formally as a decrease of optical absorbance. The absorbance fell over the whole wavelength range studied. Trace (*a*) in Fig. 1 (dots) is a light-induced difference spectrum recorded from axons during 12 s of illumination with bright white light. The amplitude of the absorbance change decreased monotonically with wavelength. Trace (*a* + *c*) of Fig. 1 (continuous line) is the difference spectrum recorded when the measuring beam passed through a single photoreceptor cell body that lay on the top surface of the nerve so that the light passed through a layer of axons as well as through a cell body. The difference spectrum (*a* + *c*) can be interpreted as the sum of a monotonically decreasing curve of the same form as that recorded from axons alone (*a*) plus a second absorbance decrease that has a maximum at 530 nm. The absorbance change at 430 nm measured from axons alone was -0.0021 ± 0.0005 absorbance units ($N = 8$); the absorbance change at 430 nm measured from cells lying on axons was -0.0035 ± 0.0003 ($N = 16$). The additional absorbance change at 530 nm measured from the cell bodies was estimated as -0.0012 ± 0.0002 units ($N = 16$). The light-induced absorbance change was smaller during the second of two 12-s periods of illumination 1 min apart but recovered fully after 10 min of darkness (2 cells). Light-induced difference spectra in the absence of dye represent a baseline change which we have subtracted from light-induced difference spectra in the presence of dye.

3.2. Absorbance Spectrum of Intracellular Phenol Red

When phenol red was injected into a photoreceptor cell, the absorbance increased over the wavelength range 384–610 nm. The difference spectrum due to intracellular dye had peaks at 430 and 560 nm that are characteristic of the dye *in vitro* (Fig. 2).

In many preparations, the intracellular pH can be lowered by bathing the preparation in a solution of a weak acid and raised by bathing in a weak base (Roos and Boron, 1981). Addition of acetate ions (acetate–ASW) to the extracellular bath for 5 min produced a change in absorbance of intracellular dye very similar to the change observed on acidification of dye solutions *in vitro* (Fig. 2). Similarly, addition of ammonium ions (ammonium–ASW) to the extracellular bath produced changes of the spectrum of intracellular dye like those observed after making dye solutions alkaline *in vitro*. The apparent isosbestic wavelength determined from the absorbance change produced by acetate–ASW was 490 ± 2 nm ($N = 10$) and by ammonium–ASW was 484 ± 1 nm ($N = 9$). The isosbestic wavelength measured *in vitro* was 483 nm.

The pH_i of dark-adapted cells bathed in ASW was calculated both from the absorbance at 560 nm compared with the absorbance at 480 nm and from the absorbance at 430 nm compared with the absorbance at 480 nm. Although the calculated values for each cell commonly disagreed by ~0.15 pH unit, there was no systematic difference in indicated pH_i. In 10 cells the absorbance at 560 and 480 nm indicated a pH_i of 6.89 ± 0.03, whereas the absorbances at 430 and 480 nm indicated a pH_i of 6.89 ± 0.07.

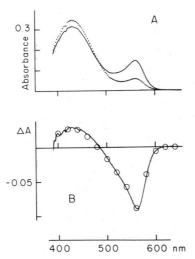

Figure 2.   Absorbance spectrum of intracellular phenol red. (*A*) Absorbance spectra of phenol red inside a dark-adapted cell. Each spectrum is the difference between the absorbance of the cell determined before injection of phenol red and the absorbance after injection. The absorbances were determined at flash onset. The absorbance of intracellular dye increased during the experiment, probably due to dye leaking out of the intracellular micropipette. To correct for this steady increase in dye concentration, we measured the dye absorbance when the preparation was first bathed in ASW, then bathed 5 min in acetate–ASW, and finally 5 min after changing back into ASW. The continuous line is the average of the two measurements made with the preparation bathed in ASW. The dotted line is the measurement made 5 min after substituting acetate–ASW for ASW. Same cell as Fig. 1. (*B*) Continuous line: the difference spectrum of intracellular phenol red due to substituting acetate–ASW for ASW in the bath. The difference between the spectra (*A*) is plotted as a function of wavelength. Circles: the difference spectrum measured in a cuvette due to a large pH change (10.6 to 4.46) plotted at an arbitary absorbance scale for comparison.

The pH_i for dark-adapted cells was also determined using the single wavelength photometer. The absorbance at each of a pair of wavelengths was measured at the onset of separate flashes of monochromatic light. The pH_i was calculated from these initial absorbances to be 6.94 ± 0.02 ($N = 15$) for the 560/480 pair and 6.84 ± 0.02 ($N = 30$) for the 560/498 pair.

3.3.   Effect of Phenol Red on the Electrophysiological Responses

Bright illumination after 5 min or more darkness elicited receptor potentials that sometimes became distorted after injection of phenol red (Fig. 3*B*). Although both the rising phase and peak of the receptor potential were almost unchanged after injection, the decay of the membrane voltage from its peak was markedly slowed. The frequency with which this phenomenon appeared increased with the amount of phenol red injected. The frequency was greater than 50% in cells that had been injected to an absorbance change of at least 0.076 at 480 nm, which corresponds to an intracellular concentration of at least 0.7 m*M* phenol red. Cells that displayed this

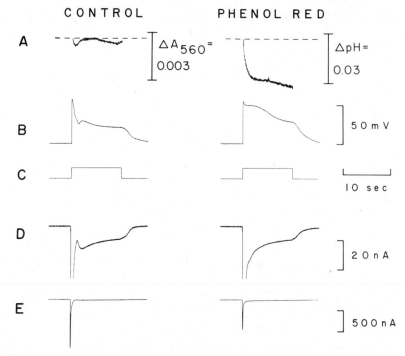

Figure 3.   Effect of phenol red on electrophysiological responses. *Left:* Before injection of phenol red. *Right:* After injection of phenol red (dye absorbance 0.051 at 480 nm). Each stimulus was 3.5 mW/cm^2 at 560 nm for 10 s given after 5 min in darkness. Traces (*A*) and (*B*) were recorded simultaneously; membrane potential was not clamped. (*A*) Absorbance; a horizontal dashed line is drawn through the value of the absorbance measured at the onset of illumination. (*B*) Membrane voltage. (*C*) Light monitor. The cell was voltage-clamped, and light-induced current was recorded at high gain (*D*) and low gain (*E*). Light monitor as (*C*).

distortion of the receptor potential had otherwise normal electrophysiological behavior. The resting voltages in the dark were normal. The receptor potentials elicited by low-intensity illumination were normal, and very dim illumination elicited discrete events (Millecchia and Mauro, 1969). The second and subsequent bright flashes also elicited normal responses. Therefore, we believe that the effect of phenol red injection is not simply due to nonspecific damage.

To characterize better the physiological change responsible for the distorted receptor potentials, we voltage-clamped photoreceptor cells and measured both the light-induced current and the membrane resistance in the dark. Phenol red injection had no significant effect on the slope conductance of cells in darkness. However, the light-induced current decayed more slowly from its peak value after phenol red injection (Fig. 3D). The light-induced current 1 s after the onset of a bright flash was 2.65 ± 0.35 (*N* = 8) times the value before injection.

Photoreceptor cells were injected with two structural analogues of phenol red: bromphenol blue (pK = 4.0) and bromoxylenol blue (pK = 7.0). Bromphenol blue injection did not distort the light response (7 cells). Bromoxylenol blue injection caused a distortion of the light response very similar to that caused by phenol red (4 cells).

3.4.   Injection of pH Buffers

To estimate a possible effect of intracellular components on the pK of phenol red ("protein error"—Clark, 1928), we measured the effect of an injection of 1 M MOPS at pH 7.2 on the pH indicated by intracellular phenol red. The pH was 7.1 after the MOPS solution was diluted 10-fold or 100-fold with 0.625 mM KCl; therefore, the pH of intracellularly injected buffer was probably close to this value. Before injection of MOPS, the absorbance of intracellular phenol red was measured at 560 nm and at a second wavelength, either 480 or 498 nm, using the single-wavelength photometer. The photoreceptor was then repetitively illuminated with 560 nm light, and was injected. Finally, a second absorbance measurement at 480 or 498 nm was made. The change in absorbance at 480 nm was small (ΔA = +0.001 ± 0.005, n = 6). During the injection of MOPS, the absorbance at 560 nm fell and approached an asymptotic value smaller than that before the injection (Fig. 4A). In 6 cells, the pH$_i$ indicated by phenol red was 6.88 ± 0.01 before injection of MOPS; the pH indicated by phenol red immediately after the end of the injection was 6.73 ± 0.03. That is, injection of a buffer at an estimated intracellular pH of 7.1 appeared to cause acidification of the intracellular dye.

HEPES (1 M) at pH 7.7 was also injected into photoreceptors. As the pH of the HEPES solution was 7.6 after a 10- or 100-fold dilution with 0.625 mM KCl, the pH of intracellularly injected buffer was probably close to this value. The injection of this

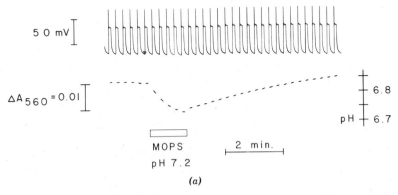

(a)

Figure 4A.   Effect of injecting MOPS buffer. A single photoreceptor cell body was impaled with two pipettes. The first pipette contained phenol red solution and was used to inject dye (dye absorbance 0.064 at 560 nm) and to measure membrane voltage (upper trace). The second pipette contained 1 M MOPS buffer at pH 7.2. Flashes of 560-nm light (5 s long every 15 s) were used to stimulate the cell and to measure the absorbance (lower trace). During the period indicated by the hollow bar, pressure was applied to the back of the pipette containing MOPS. During this period the absorbance at 560 nm fell, indicating a fall in pH$_i$.

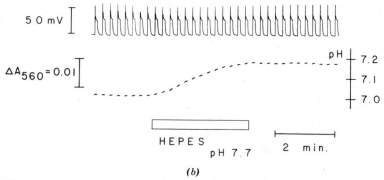

(b)

Figure 4B.   Effect of injecting HEPES buffer. Experimental protocol similar to cell of Fig. 4A. Dye absorbance was 0.033 at 560 nm. The second pipette contained 1 M HEPES at pH 7.7. During injection of buffer the absorbance at 560 nm rose, indicating an increase in pH_i.

buffer increased the absorbance of photoreceptros previously injected with phenol red (Fig. 4B); the pH_i indicated by the dye rose from 7.06 ± 0.04 to 7.16 ± 0.08 ($n = 6$). Thus, injection of a buffer at an estimated intracellular pH of 7.6 appeared to cause alkalinization of the intracellular dye.

The difference between the estimated pH of the buffer solution after injection and the pH indicated by intracellular phenol red immediately after the end of the injection was 0.39 ± 0.04 for all 12 MOPS and HEPES injection experiments. The receptor potentials elicited by repetitive flashes did not change during and after these injections of either MOPS or HEPES buffers.

When photoreceptor cells previously injected with phenol red were bathed in ASW equilibrated with 5% CO_2 plus 95% O_2, the absorbance at 560 nm fell, initially

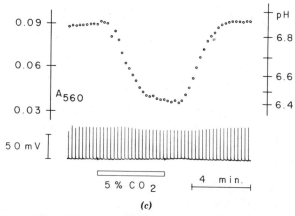

(c)

Figure 4C.   Effect of CO_2–ASW. Dye absorbance was 0.088 at 560 nm, and light flashes were 200 ms in duration. The lower trace shows the membrane potential. During the period indicated by the hollow bar, the solution bathing the preparation was changed from ASW to ASW equilibrated with 95% O_2, 5% CO_2 (pH 6.30); the absorbance at 560 nm fell, indicating a fall in pH_i.

sharply and later more slowly (Fig. 4C). The amplitudes of receptor potentials were slightly reduced. The rapid component of the absorbance change indicated a fall of intracellular pH of 0.27 ± 0.03 units ($n = 8$) (from 6.84 ± 0.01 to 6.57 ± 0.03 in ~3 min). pH_i returned to the normal value after removal of CO_2 from the ASW.

3.5. Absorbance Spectra of Phenol Red Measured in the Presence of Isolated Rhabdoms

To examine whether rhodopsin or other components of rhabdomere membrane could change the absorbance spectrum or pK of phenol red, we made microspectrophotometric measurements of phenol red in the presence of isolated rhabdoms of crayfish. The rhabdoms are spindle-shaped masses of photoreceptor microvilli, about 25 μm in diameter in the middle and about 120 μm long. They are readily detached from the photoreceptor cells (Goldsmith, 1978); most of the cytoplasmic ends of microvilli remain open (T. Sherk and T. Goldsmith, unpublished observations). In an isolated rhabdom there are therefore probably few barriers preventing small solute molecules from coming in contact with both the internal and external surfaces of the microvillar membranes.

To estimate the absorption of phenol red associated with the rhabdom, a difference spectrum was constructed by subtracting the absorbance of the rhabdom alone from the absorbance measured in the presence of phenol red (Fig. 5). Measurements of

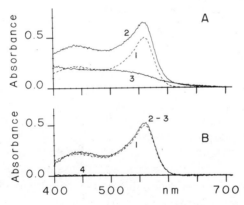

Figure 5. Absorbance spectra of phenol red in the presence of isolated photoreceptor membranes. The absorbance spectrum of a thin chamber containing isolated crayfish rhabdoms was measured using a microspectrophotometer. (A) Curve 1, absorbance of phenol red in free solution. The solution contained 2 mM phenol red, 100 mM KCl, and 90 mM phosphate buffer at pH 7.65. Curve 2, absorbance of a rhabdom in the presence of the same solution of phenol red. Curve 3, absorbance of the rhabdom alone, after flushing the chamber with buffer solution, 100 mM KCl plus 90 mM phosphate buffer at pH 7.65. (B) Curve 1 (broken line), absorbance of phenol red in free solution. Replot of spectrum 1 from (A). Curve 2–3 (continuous line), absorbance of phenol red associated with isolated rhabdoms. Difference spectrum obtained by subtracting spectrum 3 in (A) from spectrum 2 in (A). Curve 4, baseline recorded off the rhabdom and after flushing the phenol red from the chamber.

TABLE 1.   pH INDICATED BY PHENOL RED IN THE PRESENCE OF CRAYFISH RHABDOM

Average pH of Phenol Red in Solution	Average pH of Phenol Red in Rhabdom	Number of Measurements	Chamber Thickness
7.65 ± 0.01	7.59 ± 0.03	15	30 μm
7.62 ± 0.01	7.60 ± 0.01	14	220 μm
7.11 ± 0.01	7.07 ± 0.02	11	220 μm

Comparison of the values of pH ± *SEM* calculated from the spectra of phenol red measured in free solution near isolated crayfish rhabdoms and from phenol red that is in association with rhabdoms. The method described in the legend to Fig. 5 was used throughout. pH was calculated from the optical absorbances at 480 and 560 nm.

absorbance of the rhabdoms before exposure to phenol red and after removal of the dye from the bath showed that there was no irreversible binding of the dye.

The pH indicated by phenol red was calculated from the absorbances at 480 and 560 nm. The results of 40 experiments are summarized in Table 1. The pH calculated from the spectrum of phenol red in the rhabdom was usually within 0.1 pH unit of the nearby solution. On average the difference was 0.044 pH units, with the rhabdom appearing more acid (p <0.01, t-test). In summary, when phenol red comes in contact with the isolated photoreceptor membranes of crayfish rhabdoms, it continues to indicate the pH of the bathing solution to an accuracy of 0.1 pH unit.

3.6.   Light-Induced Change in Absorbance of Intracellular Phenol Red

The absorbance of phenol red inside a *Limulus* ventral photoreceptor cell changed during illumination (Fig. 6A). The absorbance of intracellular phenol red at 560 nm decreased in two phases—a first phase that lasted 1–2 s and a second, slow phase that continued while illumination was maintained (Fig. 6C). In 7 experiments the measurement of absorbance began immediately at the onset of illumination after 5 min darkness. No additional rapid component of light-induced absorbance change during the first 30 ms of illumination was measured in these experiments (Fig. 8D). The light-induced difference spectrum was very similar to that elicited by acidification of the cytosol by acetate–ASW and to the behavior of the dye in vitro (Fig. 2b).

The light-induced acidification of the cytosol calculated from the absorbance change at 560 nm (in Fig. 6A) is $\Delta pH_i = -0.076$. The absorbance change at 430 nm was difficult to estimate because the spectrum was noisy; however, the calculated $\Delta pH_i = -0.063$. The light-induced ΔpH_i illustrated in Fig. 6 was greater than we usually observed. For all cells injected with phenol red, a bright flash given after at least 10 min of dark adaptation caused ΔpH_i of -0.029 ± 0.002 ($n = 46$). The light-induced pH change did not vary with light intensity at the light intensities used in this study (4.6×10^{-4} to 7.3×10^{-3} W/cm^{-2}).

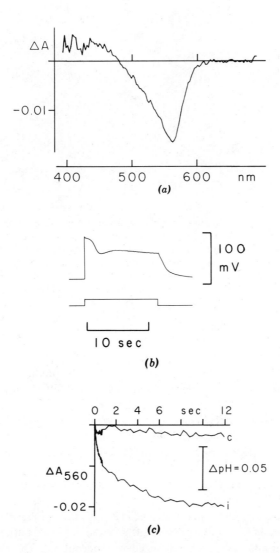

Figure 6. Light-induced changes in absorbance of phenol red. After 10 min in darkness, the nerve was illuminated with white light for 12 s, and the absorbance spectrum was recorded using the scanning spectrophotometer. Same cell as Figs. 1 and 2. (A) The light-induced difference spectrum of intracellular dye was determined as follows. Before impalement of the cell with a pipette, the absorbance changes of the cell that occurred during a 12-s illumination were measured (curve ($a + c$) of Fig. 1). The cell was then impaled, and phenol red was injected to an absorbance of 0.14 at 480 nm. After 10 min in darkness, the absorbance change in the cell that occurred during a second 12-s illumination was measured. The absorbance change measured before electrode impalement was subtracted from the absorbance change measured after phenol red injection to give the curve plotted in (A). (B) Upper trace: the receptor potential measured simultaneously with absorbance after injection of phenol red. Lower trace: light monitor. (C) The absorbance change at 560 nm is plotted as a function of time from the onset of illumination. c: Control, before impalement of the cell by a pipette; i: after injection of phenol red.

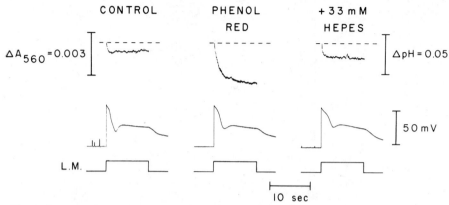

Figure 7. Effect of HEPES injection on the light-induced change of phenol red absorbance. A photoreceptor cell was impaled with two pipettes. One pipette contained phenol red solution and was used to measure the membrane voltage (middle row of traces). The second pipette contained 1 M HEPES buffer at pH 7.4. The upper traces show the change in absorbance measured during 10-s flashes of 560-nm light presented after 15 min in darkness. A horizontal dashed line is drawn through the beginning of each trace. Left-hand records: control, before injection of either phenol red or HEPES. A significant absorbance change was measured, probably due to phenol red that had leaked out of the pipette. Center: after pressure-injection of phenol red to an absorbance of 0.025 at 480 nm. Right: after injection of HEPES to an estimated intracellular concentration of 33 mM. L.M.: light monitor.

A change in absorbance of intracellular dye might be produced by factors other than a change in pH: For instance, a light-induced fall in the absorbance of intracellular phenol red could be due to dye bleaching or a light-induced change in the binding of dye to some intracellular component. We tested these possibilities by using bromphenol blue in place of phenol red. For bromphenol blue the maximum of the difference spectrum caused by a change in pH is at 615 nm. Cells were injected with bromphenol blue to a final dye absorbance of 0.060 ± 0.004 at 610 nm ($n = 7$). No light-induced change in the absorbance at 610 nm was subsequently detected by the single wavelength photometer.

To test whether the light-induced change in absorbance of intracellular phenol red is due to a change in intracellular pH, a 1 M solution of the pH buffer HEPES at pH 7.4 was injected into cells already injected with phenol red. The HEPES solution contained $^{35}SO_4^{2-}$ at high specific activity; this tracer allowed the final intracellular concentration of HEPES to be estimated (see Sect. 2). HEPES injection (to an average of 33 ± 5 mM) reduced the light-induced change of dye absorbance to 44 ± 9% ($n = 9$) of the value before injection (Fig. 7).

3.7. Absorbance Changes of Intracellular Phenol Red During Dark Adaptation

To examine the aftereffects of bright illumination on pH_i, photoreceptor cells previously injected with phenol red were dark-adapted for 15 min and then illuminated by two flashes (Fig. 8A). The second flash (the "test flash") was given at

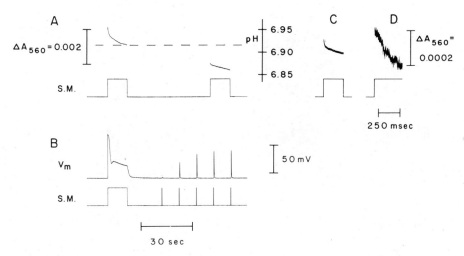

Figure 8. (*A*) Aftereffects of bright illumination on pH$_i$, studied with paired flashes. A photoreceptor cell containing phenol red (dye absorbance 0.025 at 480 nm) was dark-adapted for 15 min and then illuminated with two flashes (560 nm, 10-s duration). In this example the onset of the test flash occurred 1 min after the onset of the adapting flash. Top trace: absorbance. A horizontal dashed line is drawn through the value of the absorbance at the end of the adapting flash. S.M.: stimulus monitor. (*B*) Photoreceptor responsiveness recovers monotonically after bright illumination. The same photoreceptor was dark-adapted for 15 min and then illuminated with a single adapting flash (560 nm, 10-s duration) equal in intensity to the flashes in (*A*). After the adapting flash a series of 100-ms flashes (3.3 log units lower intensity) was given at 10-s intervals. V_m: membrane voltage. S.M.: stimulus monitor. The time scale below this record refers to (*A*), (*B*), and (*C*). (*C*) The same photoreceptor was dark-adapted for 15 min and then illuminated with two flashes (560 nm, 10-s duration). The onset of the test flash occurred 5 min after the onset of the adapting flash; both pH$_i$ and the light-induced change in pH$_i$ had nearly recovered their dark-adapted values. Top trace: absorbance measured during the test flash. The absorbance measurement began at the onset of illumination; that is, there was no 40-ms delay before measurement began. The bandwidth for this record was 0–625 Hz. Lower trace: Stimulus monitor. Scales as in (*A*) and (*B*). (*D*) There is no detectable extra component of absorbance change in the first 30 ms of illumination. Top trace: the absorbance change measured during the first 300 ms of the flash of (*C*) is plotted at 10× vertical gain on an expanded time scale. Bandwidth 0–625 Hz. Lower trace: Stimulus monitor. The time and absorbance scales below and to the right of this record refer to (*D*) only.

a variable time after the first flash (the "adapting flash"). The test flash served two purposes. First, the absorbance at the onset of the test flash was compared to the absorbance at the end of the adapting flash; the difference in absorbance was a measure of the change in pH$_i$ that occurred during the intervening dark period. Second, the absorbance change during the test flash itself was compared with the absorbance change during the adapting flash; this comparison showed the effect of prior illumination on the light-induced change in pH$_i$.

During the first 40 s of darkness after the adapting flash, the pH$_i$ indicated by phenol red continued to fall significantly (Fig. 8*A*, 9*A*), whereas the responsiveness of the cell began to recover (Fig. 8*B*). The pH$_i$ then gradually increased, returning toward the dark-adapted level during 15 min (Fig. 8*C*, 9*A*).

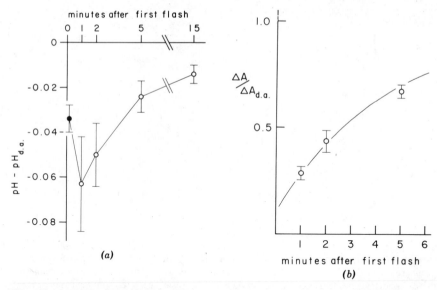

Figure 9. (A) Recovery of pH$_i$ after illumination. The cells ($n = 5$) were dark-adapted for 15 min and then illuminated with pairs of flashes (560 nm, 10-s duration). The adapting flash caused a fall in intracellular pH. The test flash was used to measure the intracellular pH either 1, 2, or 5 min after the onset of the adapting flash. The symbol ● denotes the pH at the end of the adapting flash relative to the pH at the onset of the adapting flash ($n = 33$). The symbol ○ denotes the pH at the onset of the test flash relative to the pH at the onset of the adapting flash (1 min: $n = 14$ trials; 2 min: n = 15 trials; 5 min: $n = 13$ trials). The symbol at 15 min represents the pH at the onset of an adapting flash (i.e., after 15 min darkness) relative to the pH at the onset of the preceding adapting flash ($N = 37$ trials). The error bars show ± standard error. (B) Recovery of ΔpH$_i$ after illumination. Protocol as (A). Results from 7 cells are combined. The absorbance change during the test flash is plotted as a fraction of the absorbance change during the adapting flash. The symbols represent mean and standard error (1 min: $n = 19$ trials; 2 min: $n = 19$ trials; 5 min: $n = 18$ trials). The continuous line is the best-fitting exponential, $\tau = 4.52$ min.

The absorbance change during the test flash was smaller than the change during the adapting flash (Fig. 8A). The absorbance change during the test flash became greater as the dark period between the adapting flash and the test flash was made longer (Fig. 9B). The data can be fitted by a single exponential (time constant = 4.52 min).

3.8. Spatial Scanning Experiments

To assess spatial inhomogeneities of pH indicated by intracellular phenol red, single photoreceptors that appeared isolated from neighboring photoreceptor cell bodies were examined with the spatial scanning photometer. An example of such an experiment is illustrated in Fig. 10. After injection, phenol red was distributed relatively uniformly along the cell (Fig. 10B). The apparent pH$_i$ indicated by phenol red in dark-adapted cells was not uniform over the whole cell. Dye in the portion of the cell opposite the axon indicated a pH 0.12 ± 0.02 ($n = 7$) units more acid than dye in the portion of the cell adjacent to the axon (Fig. 10C). A systematic error

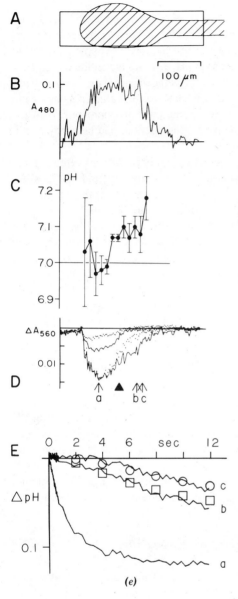

Figure 10. Spatial inhomogeneity of pH_i and light-induced ΔpH_i. (*A*) Tracings of the cell and of the rectangular measurement aperture taken through the measurement optics. Light passing through the rectangular aperture fell on the photodiode array. (*B*) Spatial distribution of dye absorbance. The dye absorbance at 480 nm is plotted as a function of distance along the aperture. (*C*) Apparent intracellular pH at the onset of illumination after 15 min in darkness. pH values were calculated from the absorbances at 560 and 480 nm. The points represent the mean ($\pm$ SEM) over each group of 5 pixels. (*D*) Light-induced change in absorbance. After 15 min in darkness, the cell was illuminated with 560 nm light for 12 s (4.3 mW/cm^2). The difference between the absorbance measured at various times after flash onset (260 ms, 700 ms, 5 s and 12 s and the absorbance at flash onset was plotted as a function of distance along the aperture. (*E*) Time course of the absorbance change. The absorbance changes at three individual points indicated as arrows *a*, *b*, and *c* in (*D*) are plotted as a function of time after flash onset. The absorbance changes are scaled by a factor proportional to the reciprocal of the total dye absorbance at each point so that the ordinate can be directly calibrated as a pH change. The hollow symbols ($\square$, $x = 42$ μm; $\circ$, $x = 55$ μm) are solutions of the semiinfinite diffusion equation given in the text using $\Delta pH^* = -0.12$; $D = 1.02 \times 10^{-10}$ m^2/s.

arising in the optical measurement system can be ruled out; for four cells the image of the axon was projected onto one end of the diode array, and for three cells the image of the axon was projected onto the opposite end of the array. Light induced a decrease in the absorbance of phenol red at 560 nm in the portion of the cell opposite the axon, whereas no change in absorbance occurred in the portion of the cell adjacent to the axon (Fig. 10*D*). This local decrease in absorbance indicated a rapid light-induced fall of pH_i within one region of the cell (the "active region") (Fig. 10*E*). During prolonged illumination there was a delayed fall of pH in the region of the cell adjacent to the axon. At progressively further distances from the active region there were both increased delays before detectable pH changes and decreased rates of pH change (Fig. 10E). During a 12-s illumination, the decrease of pH_i in the active region was $\Delta pH_i = -0.11 \pm 0.03$ ($n = 7$), whereas the pH decrease averaged across the whole cell was $\Delta pH_i = -0.043 \pm 0.004$. This latter value is close to the values obtained using either the single wavelength photometer or the scanning spectrophotometer ($\Delta pH_i = -0.029 \pm 0.002$, $n = 46$) in which the measuring aperture had been adjusted to cover the image of the whole cell.

In two experiments a cell was illuminated with a pair of flashes delivered at 1-min intervals to test whether a rapid component of pH_i recovery occurred in a localized region of the cell during dark-adaptation. In both experiments pH_i continued to fall in all regions of the cell during the dark period following the first flash.

We studied two photoreceptor cell bodies that were not isolated but that lay in contact with other photoreceptor cell bodies. In each cell an active region was found at the zone of contact with a second cell. In one of the two cells there was a second, distinct active region at the end of the cell opposite the zone of contact. Although the geometry of the active regions was more complex in these two cells, both the intracellular pH and light-induced change in pH were similar to those in isolated cells.

4. DISCUSSION

4.1. Absorbance Changes in the Absence of Dye

The optical transmission of undyed axons in the *Limulus* ventral eye rose during 12 s of illumination. Although this optical change is treated formally as an absorbance change, it probably results from a change in light scattering. The steep monotonic decrease of the signal with wavelength is characteristic of light scattering (Rayleigh, 1871). The additional change in optical transmission recorded from an unstained photoreceptor cell body was maximal at 530 nm and possibly resulted from visual pigment bleaching (Murray, 1966).

4.2. Phenol Red as an Indicator of Intracellular pH

Phenol red performed well as an indicator of changes of intracellular pH. When the intracellular pH was changed by bathing the preparation in ASW containing weak acids or bases, the resultant difference spectra closely resembled those of phenol red in vitro. We therefore believe that phenol red can be used to detect changes of intracellular pH that occur during normal physiological events.

Two possible sources of error must be considered in order to estimate the absolute value of intracellular pH from the absorbance of intracellular phenol red: (1) The absorbance spectra of the basic and acidic forms of intracellular phenol red may be different from the spectra measured *in vitro*, so the absorbance spectrum of intracellular dye cannot be matched to a unique *in vitro* absorbance spectrum. (2) The pK of intracellular phenol red may differ from the pK measured *in vitro* because of association with some intracellular component; that is, there may be a protein error (Clark, 1928).

We have found no evidence for a change in the absorbance spectra of the basic and acidic forms of phenol red either upon association with isolated crayfish photoreceptor membranes or upon injection into *Limulus* ventral photoreceptor cells. The absorbance spectrum of intracellular dye was very similar to one of the family of *in vitro* curves; moreover, the indicated pH (and therefore the calculated concentration of the basic and acidic forms) is the same whether the peak at 430 nm or the peak at 560 nm is considered. These results confirm the findings of Lisman and Strong (1979) and are in contrast to the findings of Baylor et al. (1982) in frog muscle cells.

The intracellular pressure-injection of pH buffers suggests that some intracellular component causes a significant protein error. Injection of MOPS at an estimated pH (after dilution) of 7.1 caused the pH indicated by intracellular phenol red to fall, although the value of pH indicated by the phenol red before injection of buffer was 6.9. If the true bulk intracellular pH were 6.9, then injection of buffer at 7.1 should increase rather than decrease the intracellular pH. The pH indicated by phenol red appeared to approach a limit during prolonged injections of buffer. The limit may represent the intracellular pH that equals the pH of the diluted injection solution, so further injection causes no further pH change. If this assumption is valid, then the difference of 0.4 unit between the pH of the buffer solution and the limiting pH indicated by phenol red during injection can be attributed to a protein error of 0.4 pH unit.

The intracellular pH (of all the dark-adapted cells in this paper) determined by comparison of absorbance spectra of intracellular phenol red with spectra measured in vitro is about 6.9 ($n = 65$). This value is similar to the value of 7.0 obtained by Lisman and Strong (1979) using phenol red. If the protein error estimate of 0.4 pH unit is correct, then the true intracellular pH of dark-adapted *Limulus* ventral photoreceptor cells would be approximately 7.3. This value is in agreement with previous electrophysiological estimates of pH$_i$ (Coles and Brown, 1976).

We investigated the possibility that rhabdomeric components including rhodopsin were responsible for the protein error by the experiments on isolated crayfish rhabdoms. The pH indicated by phenol red in the presence of the rhabdomeric membranes was 0.04 pH unit more acid than the bulk pH. This result indicates that the rhabdomeric membrane is probably not the principal source of protein error in our measurement.

4.3. Spatial Gradient of pH$_i$ in Dark-Adapted Cells

We consistently observed that the pH$_i$ indicated by phenol red within dark-adapted cells was lower by approximately 0.1 pH unit in the region of the cell opposite the axon than in the region of the cell adjacent to the axon. This pattern may indicate a real gradient of pH$_i$, but could also reflect a spatial difference in the apparent pK of intracellular dye. Consistent with the latter suggestion, the apparent pK of phenol red changes by approximately 0.04 unit when the dye is associated with crayfish rhabdoms.

4.4.   Intracellular pH Buffering Capacity

The pH indicated by intracellular phenol red fell rapidly by an average of 0.27 unit when the bath was changed to ASW in equilibrium with 5% CO_2. pH_i then continued to fall slowly. The rapid component of pH_i change probably results from diffusion of CO_2 into the cell, and the slow component probably results from a subsequent outward diffusion of bicarbonate ions (Boron and DeWeer, 1976). The rapid change in pH can be used to calculate the intracellular pH buffering capacity (by the method of Boron, 1977). Using a value for CO_2 solubility of 0.0365 mM/mm Hg (Boron, 1977) and correcting the intracellular pH values by 0.4 pH unit to account for the protein error, the estimated pH buffering capacity is 43 ± 8 mM/pH unit. For comparison, the buffering capacity of the lateral eye photoreceptor of *Balanus eburneus* was determined to be 15 mM/pH unit using pH-sensitive microelectrodes (Brown and Meech, 1979).

4.5.   Light-Induced Change in Intracellular pH Indicated by Phenol Red

Illumination of dark-adapted photoreceptor cells induced an absorbance change of intracellular phenol red that had two phases—a fast phase that lasted 1–2 s, and a slow phase that lasted as long as illumination was maintained. There was no additional rapid component of light-induced absorbance change in the first 30 ms of illumination. Therefore, because the main phase of the absorbance change took 1–2 s, only a small error will be introduced by taking the absorbance measured approximately 30 ms after the onset of light to be the absorbance of the dark-adapted cell. We believe that the light-induced absorbance change of intracellular phenol red is due to a light-induced fall of intracellular pH. The evidence is fivefold:

1.   The difference spectrum of the light-induced absorbance change is very similar to the difference spectrum for phenol red acidification both in vitro and in vivo. In particular, light induces an absorbance increase from 400 to 480 nm; this demonstrates that the signal could not be produced solely by photobleaching of dye.

2.   No light-induced absorbance changes were measured in cells injected with bromphenol blue. Bromphenol blue has a pK of 4.0, and its absorbance will not respond to pH changes in the physiological range. However, bromphenol blue is a structural analogue of phenol red and might be expected to show similar bleaching and binding behavior.

3.   Injection of the pH buffer HEPES reduced the light-induced absorbance change of intracellular phenol red.

4.   During illumination the absorbance of intracellular phenol red changed first in a spatially restricted active region. Light-induced dye bleaching and other photochemical reactions would be expected to occur uniformly over the cell volume.

Subsequent changes of absorbance in regions of the cell progressively farther from the active region occurred with both increased delays and decreased rates. Light-induced changes in pK of the dye would not be expected to undergo this apparent diffusion.

5.   Although Ross et al. (1977) found no significant voltage-dependent change in the absorbance of squid axons stained with phenol red, Salzberg and Obaid (personal communication, 1983) have found that phenol red is a weak, voltage-sensitive dye in squid axons. However, the light-induced absorbance change of phenol red in *Limulus* ventral photoreceptors was observed when the membrane voltage was held constant by a voltage clamp.

Therefore, we believe that the intracellular pH of dark-adapted *Limulus* ventral photoreceptor cells, averaged over the whole cell, fell by about 0.03 unit during 10–12 s of bright illumination. A characteristic distortion of the receptor potential, distinct from nonspecific damage, was sometimes observed after injection of phenol red. The values of pH_i and the changes of pH_i induced by illumination that we report were pooled from all cells injected with phenol red that appeared normal in all respects other than this specific distortion. However, if data are pooled from cells in which the receptor potential was normal for all conditions and which were injected with dye to less than 0.05 absorbance unit at 480 nm, none of the results is significantly changed. For this subset of the data, pH_i indicated at the onset of illumination after at least 10 min in the dark was 6.86 ± 0.03 ($n = 16$), and the change in pH_i during 10–12 s of illumination was -0.035 ± 0.005 unit ($n = 13$).

The light-induced decrease of pH was recorded in a spatially restricted active region during the first 2 s of illumination. *Limulus* ventral photoreceptor cells often have two lobes (Calman and Chamberlain, 1982)—an arhabdomeric lobe and a rhabdomeric lobe specialized for photon capture (Stern et al., 1982). The rhabdomeric lobe is usually located opposite the axon in photoreceptor cell bodies that are isolated from neighboring photoreceptor cell bodies (Calman and Chamberlain, 1982). We believe that the active region, in which the early light-induced pH change occurs, is identical with or is located within the rhabdomeric lobe. In parts of the cell remote from the active region, the pH_i began to fall only when illumination was prolonged. At progressively further distance from the active region there was both an increased initial delay before a detectable fall in pH_i and a decreased rate of that pH_i fall. We presume that this pattern is due to diffusion of acid from the active region into the rest of the cell.

The data from 7 cells (e.g., Fig. 10*E*) were fitted to a one-dimensional, semi-infinite diffusion model as follows. We assume that the pH changed in a single instantaneous step of amplitude ΔpH^* at the boundary between the active region and the rest of the cell. In the 7 cells analyzed, the pH change in the active region was almost complete before the pH in remote regions began to change. Thus the approximation of an instantaneous change at the boundary is reasonable for regions remote from the active region. The pH change at a distance x remote from the boundary is given by

$$\Delta pH_x = \Delta pH^*/\mathrm{erfc}(x/2\sqrt{Dt})$$

where D is the diffusion constant, t is the time after the onset of illumination, and erfc is the error function complement (Crank, 1975). This model gives a reasonable fit to the experimental data (hollow symbols on Fig. 10E). The average value of D calculated using this model was $6 \pm 2 \times 10^{-11}$ m^2/s ($n = 7$). This value is an order of magnitude smaller than the value of D for a small molecule in free solution and suggests that diffusion of acid within the cytosol is slowed by an unknown combination of path tortuosity and reactions with buffers.

4.6.   Origin of the Light-Induced Change in pH$_i$

H$^+$ Release from Rhodopsin.   A light-induced release of H$^+$ from rhodpsin might occur as a result of a conformational change in the rhodopsin molecule that unmasked a titratable proton binding site (Hubbard et al., 1965) or changed the extent to which rhodopsin is phosphorylated (Paulsen and Hoppe, 1978). The maximum pH change that could result from a light-induced release of one H$^+$ per rhodopsin can be estimated. There are as many as 2×10^9 rhodopsin molecules per *Limulus* cell (Lisman and Bering, 1977) of average volume 10^{-9} L, so the maximum total [H$^+$] released is 3 μM. If the intracellular buffering capacity is 43 mM/pH unit then the pH change will be 7×10^{-5} units. The measured light-induced pH change is more than 2 orders of magnitude greater than this and therefore cannot be produced solely by the release of the order of one H$^+$ per rhodopsin.

Calcium Buffering.   The intracellular calcium concentration rises rapidly after the onset of illumination and then declines to a steady level after about 2 s (Brown et al., 1977). This rise in calcium concentration probably occurs in the rhabdomeric lobe (Harary and Brown, 1984), which also is probably where the early fall of pH occurs. The change in free calcium concentration elicited by a bright stimulus is as large as 0.4 mM (Harary, 1983). Many calcium-sequestering mechanisms that have been studied release at least one H$^+$ for each calcium ion sequestered (Chance, 1965; Carvalho, 1972; Meech and Thomas, 1980). Thus during the first 2 s of illumination at least 0.4 mM of H$^+$ might be released into the cytosol. With an intracellular pH buffering capacity of 43 mM/pH unit, the pH change produced by calcium sequestration could be at least 0.4/43 = 0.009 units. 0.4 mM may be an underestimate of the true calcium load because of the presence of time-independent calcium buffering mechanisms. Nevertheless, this calculation shows that calcium sequestration could cause a pH change of the same order of magnitude as that measured.

4.7.   Does a Light-Induced Fall in Intracellular pH Play a Role in Light Adaptation?

Our results indicate that the illumination of *Limulus* ventral photoreceptors induces a fall in pH$_i$ of approximately 0.03 pH unit averaged over the entire profile of the cell body. The pH change in the light-sensitive region, the rhabdomeric lobe, is greater (ΔpH$_i$ is approximately -0.1 unit). It is very unlikely that pH changes of this magnitude could affect the sensitivity of a cell to light. We have found that injections of pH buffers that decrease the measured intracellular pH by 0.15 unit do not significantly affect the light response. Moreover, Coles and Brown (1976) injected pH buffers into *Limulus* ventral photoreceptor cells to a final concentration of 40–200 mM; they found no change in sensitivity after injection of buffer solutions at pH 6.3, a value 0.6 pH unit more acid than the intracellular pH indicated by phenol red. Injection of buffers at pH 6.0 and 5.6 reduced the sensitivity slightly (0.3 log unit; unpublished result of Coles and Brown). Therefore a pH change of 0.1 in the rhabdomeric lobe probably will have a negligible effect on sensitivity.

The time course of the change in pH_i induced by light provides further evidence against a major role for pH_i in light adaptation. The intracellular pH continued to decrease during the first 50 s of darkness after bright illumination, whereas the responsiveness of the cell to light increased during this period. On the basis of these results and those we have presented previously (Bolsover and Brown, 1982), we believe that pH plays little or no role in controlling light adaptation in *Limulus* photoreceptors.

ACKNOWLEDGMENTS

We thank Dr. Paul DeWeer and Dr. John E. Lisman for much helpful criticism. This work was supported by NIH grants EY-05166, EY-00222 and EY-05168.

REFERENCES

Baylor, S. M., W. K. Chandler, and M. W. Marshall (1982) Optical measurements of intracellular pH and magnesium in frog skeletal muscle fibres. *J. Physiol. (Lond.)*, **331**, 105–137.

Bolsover, S. R., and J. E. Brown (1982) Light adaptation of invertebrate photoreceptors: Influence of intracellular pH buffering capacity. *J. Physiol. (Lond.)*, **330**, 297–305.

Bolsover, S. R., J. E. Brown, and T. H. Goldsmith (1983) Intracellular pH of *Limulus* photoreceptors studied with phenol red. *Biophys. J.*, **41**, 26a.

Boron, W. F. (1977) Intracellular pH transients in giant barnacle muscle fibers. *Am. J. Physiol.* **233**, C61–C73.

Boron, W. F., and P. DeWeer (1976) Intracellular pH transients in squid giant axons caused by CO_2, NH_3 and metabolic inhibitors. *J. Gen. Physiol.* **67**, 91–112.

Brown, H. M., and R. W. Meech, (1979) Light induced changes of internal pH in a barnacle photoreceptor and the effect of internal pH on the receptor potential. *J. Physiol. (Lond.)*, **297**, 73-93.

Brown, J. E., P. K. Brown and L. H. Pinto (1977) Detection of light induced changes of intracellular ionized calcium concentration in *Limulus* ventral photoreceptors using arsenazo III. *J. Physiol. (Lond.)*, **267**, 299–320.

Calman, B. G., and S. C. Chamberlain (1982) Distinct lobes of *Limulus* ventral photoreceptors. II: Structure and ultrastructure. *J. Gen. Physiol.*, **80**, 839–862.

Carvalho, A. P. (1972) Binding and release of cations by sarcoplasmic reticulum before and after removal of lipid. *Eur. J. Biochem.*, **27**, 491–502.

Chance, B. (1965) The energy-linked reaction of calcium with mitochondria. *J. Biol. Chem.*, **240**, 2729–2748.

Clark, W. M. (1928) *The Determination of Hydrogen Ions*. Williams and Wilkins, Baltimore.

Coles, J. A., and J. E. Brown (1976) Effects of increased intracellular pH–buffering capacity on the light response of *Limulus* ventral photoreceptros. *Biochim. Biophys. Acta*, **436**, 140–153.

Crank, J. (1975) *The Mathematics of Diffusion*, 2nd ed. Oxford University Press, London.

Goldsmith, T. H. (1978) The spectral absorption of crayfish rhabdoms: Pigment, photoproduct and pH sensitivity. *Vision Res.*, **18**, 463–473.

Good, N. E., G. D. Winget, W. Winter, T. N. Connolly, S. Izawa, and R. M. M. Singh (1966) Hydrogen ion buffers for biological research. *Biochemistry*, **5**, 467–477.

Harary, H. H. (1983) Optical probes of the physiology of *Limulus* ventral photoreceptor. Ph.D. thesis, Harvard University, Cambridge, MA.

Harary, H. H., and J. E. Brown (1984) Spatially non-uniform changes in intracellular calcium ion concentrations. *Science*, **224**, 292–294.

Hubbard, R., D. Bownds, and T. Yoshizawa (1965) The chemistry of visual photoreception. *Cold Spring Harbor Symp. Quant. Biol.*, **30**, 301–315.

Lisman, J. E., and H. Bering (1977) Electrophysiological measurement of the number of rhodopsin molecules in single *Limulus* photoreceptors. *J. Gen. Physiol.*, **70**, 621–633.

Lisman, J. E., and J. A. Strong (1979) The initiation of excitation and light adaptation in *Limulus* ventral photoreceptors. *J. Gen. Physiol.*, **73**, 219–243.

Meech, R. W., and H. M. Brown (1976) Invertebrate photoreceptors: A survey of recent experiments on photoreceptors from *Balanus* and *Limulus. In P. S. Davies, Ed., Perspectives in Experimental Biology*, Vol. 1, *Zoology;* Pergamon, New York, pp. 331–351.

Meech, R. W., and R. C. Thomas (1980) Effect of measured calcium chloride injections on the membrane potential and internal pH of snail neurones. *J. Physiol. (Lond.)*, **298**, 111–129.

Millecchia, R., and A. Mauro (1969) The ventral photoreceptor cells of *Limulus*. II: The basic photoresponse. *J. Gen. Physiol.*, **54**, 310–330.

Murray, G. C. (1966) Intracellular absorption difference spectrum of *Limulus* extra-ocular photolabile pigment. *Science*, **154**, 1182–1183.

Paulsen, R., and I. Hoppe (1978) Light activated phosphorylation of cephalopod rhodopsin. *FEBS Lett.*, **96**, 55–58.

Rayleigh, J. W. S. (1871) On the light from the sky, its polarization and colour. Philos. *Mag.*, **41**, 107–120.

Roos, A., and W. F. Boron (1981) Intracellular pH. *Physiol. Rev.*, **61**, 296–434.

Ross, W. N., B. M. Salzberg, L. B. Cohen, A. Grinvald, H. V. Davila, A. S. Waggoner, and C. H. Wang (1977) Changes in absorption, fluorescence, dichroism and birefringence in stained giant axons: Optical measurement of membrane potential. *J. Membr. Biol.*, **33**, 141–183.

Stern, J., K. Chinn, J. Bacigalupo, and J. Lisman (1982) Distinct lobes of *Limulus* ventral photoreceptors. I: Functional and anatomical properties of lobes revealed by removal of glial cells. *J. Gen. Physiol.*, **80**, 825–837.

CHAPTER 18

INTRACELLULARLY TRAPPED pH INDICATORS

JOHN A. THOMAS

Department of Biochemistry
University of South Dakota School of Medicine
Vermillion, South Dakota

1.	INTRODUCTION	312
2.	LOADING TECHNIQUE	313
3.	MONITORING INDICATOR RESPONSES	314
4.	CALIBRATION OF RESPONSES	316
5.	THE COMPARTMENT SPECIFICITY OF CARBOXY-FLUORESCEIN AND FLUORESCEIN	317
6.	MONITORING MITOCHONDRIAL pH WITH CARBOXYFLUORESCEIN DIACETATE	321
7.	CALIBRATION OF MITOCHONDRIAL pH IN INTACT HEPATOCYTES	321
8.	OTHER USES OF TRAPPED INDICATORS	323
9.	SUMMARY	324
	REFERENCES	324

1.   INTRODUCTION

The pH sensitivity of enzymes and the importance of pH to cell viability have been appreciated for many years. Less well appreciated has been the ability of cells actively to regulate their internal pH and to maintain different pH values within various subcellular compartments. Much of the interest of the past 10 years in measuring transmembrane pH differences can be attributed to the pioneering work of Peter Mitchell and the subsequent acceptance of the importance of pH gradients in mitochondrial ATP synthesis and other energy transduction processes. The concomitant development of a variety of methods to measure proton concentrations in membrane-bound compartments has led to an increased awareness of the universality of proton pumps and the potential role of pH in metabolic regulation.

Any alteration of cytoplasmic pH could affect a variety of metabolic processes. Glycolysis (Tomoda et al., 1977) and gluconeogenesis (Kashiwagura et al., 1984), for example, are both markedly affected by small changes in pH. Intracellular pH has been implicated as a regulatory factor for such diverse processes as cell division, activation of eggs and sperm, and the mediation of the response to extracellular signals. The activation of many cell types from a quiescent state to an active metabolic state has been reported to be accompanied by an elevation in cytosolic pH at a time when the cells are stimulated to produce metabolic acids (Nuccitelli and Deamer, 1982). In mammalian cells, recent evidence indicates that this intracellular alkalinization is mediated by activation of protein kinase C, which in turn stimulates the plasma membrane Na^+-H^+ exchanger (Moolenaar et al., 1984).

The most popular methods now available for measuring intracellular pH include ^{31}P-NMR, microelectrodes, the distribution ratio of weak acids and bases, and spectroscopic methods utilizing pH indicator dyes. These various methods and their uses in the study of biological systems have been recently reviewed (Roos and Boron, 1981; Nuccitelli and Deamer, 1982).

The dye indicator method offers a combination of useful features that make it attractive from an experimental standpoint as compared to the other methods. These advantages include relatively simple equipment requirements, sufficient sensitivity for single cell measurements (Slavik and Kotyk, 1984), and the ability to monitor pH continuously. In this contribution we describe a pH indicator method possessing these attributes as well as the capacity to distinguish pH changes occurring in the cytosolic and mitochondrial compartments.

We have been using fluorescein (F) and 5(6)-carboxyfluorescein (CF) to monitor intracellular pH in hepatocytes and Ehrlich ascites tumor cells. Although these two pH indicators are structurally and spectroscopically very similar, CF reports pH within the cytosolic compartment, whereas F is primarily an indicator for the mitochondrial matrix. We will present the methodology for utilizing these indicators in cell suspensions to monitor qualitatively the pH in the cytosolic and mitochondrial compartments and summarize the evidence supporting their specificity for these compartments. In addition, an overview of the techniques for quantitating pH in both the cytosolic and mitochondrial compartments in whole cells will be presented.

2. LOADING TECHNIQUE

The method for introducing fluorescein into cells was first established by Rotman and Papermaster (1966). The uncharged, colorless molecule fluorescein diacetate (FA$_2$) diffuses into cells where intracellular esterases release the charged chromophore fluorescein (Fig. 1). The high negative charge (p$K_a \simeq 4$) of the released chromophore retards its diffusion back out of the cell. Rotman and Papermaster used the intracellular retention of fluorescein as a means of monitoring the intactness of cell membranes; any disruption of the plasma membrane resulted in the rapid release of the dye.

Since we were aware of the pH sensitivity of the fluorescein chromophore from our previous work (Thomas and Johnson, 1975), it seemed a natural extension of these experiments to take advantage of the trapped indicator to measure intracellular pH. Later we synthesized the diacetate derivative of CF (CFA$_2$), which has a lower rate of leakage from most cells on account of its additional charged carboxyl group (Thomas et al., 1979). More recently Tsien (see Rink et al., 1982) has synthesized a derivative with four additional carboxyl groups; Tsien is also responsible for the introduction of the intracellular calcium indicator quin2 (Tsien et al., 1981), which is generated from a permeant precursor by intracellular esterases. All these indicators are commercially available.

To load the indicator intracellularly, cell suspensions are incubated with the appropriate diacetate derivative until a detectable yellow color develops, signifying that the precursor has entered the cell and the intracellular esterases have released the indicator. Cells are then diluted with incubation buffer and separated from any external indicator, generally by repeated centrifugations until the supernatant is colorless. The specific procedure for loading cells with these indicators varies somewhat depending on the cell type and on the relative entry and leakage rates. CFA$_2$ generally enters cells more slowly than does FA$_2$ because of its charged group (p$K_a \simeq 4$). Lowering the pH of the incubation medium usually increases the entry rate of CFA$_2$ as a result of protonation of this group. The leakage rate for CF varies markedly with cell type: relatively low in neurons, $t\frac{1}{2} \sim 1$ day (Cohan et al., 1983);

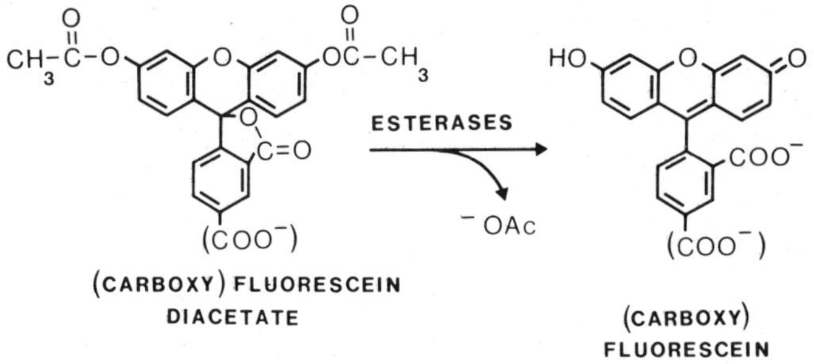

Figure 1. Structures of the fluorescein pH indicators and their diacetate precursors. Pictured is 6-carboxyfluorescein; fluorescein lacks the carboxyl group in parentheses.

intermediate in hepatocytes (Strzelecki et al., 1984), sperm cells (Babcock, 1983), and Ehrlich ascites tumor cells (Thomas et al., 1979), $t\frac{1}{2} \sim 1$ h; and rapid from kidney tubules (Boron, 1982), $t\frac{1}{2} \sim 13$ min, and certain acidophilic bacteria (Thomas et al., 1976), $t\frac{1}{2} \sim 30$ s. The leakage rates are quite temperature dependent, with Q_{10} values often above 3. Spectral measurements should be corrected for leakage by subtraction of supernatant absorbances after cell removal. Internal concentrations higher than 0.3 mM should be avoided because of the nonideal spectral behavior of the indicator (Babcock, 1983). For specific loading procedures, the reader is referred to the cited references.

3. MONITORING INDICATOR RESPONSES

Spectral responses may be monitored by either fluorescence or absorbance. Most of our experience is with the absorbance technique. For absorbance measurements, it is advisable to use an instrument designed for turbid solutions. We use an Aminco DW-2 spectrophotometer equipped with a magnetic stirrer and thermostated cell compartment. We have settled on the following procedure, which permits us to follow pH changes qualitatively (i.e., without correction for dye leakage) as a function of time but which provides sufficient information to calculate absolute pH values later if desired.

A spectrum of loaded cells is taken from 420 nm to 570 nm, using sham-loaded cells in the reference cuvette. This spectrum is examined for artifacts due to light scattering: it should extend to the baseline at 570 nm, and the general shape should

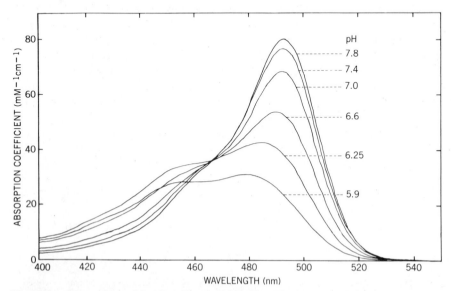

Figure 2. pH dependence of 6-carboxyfluorescein in buffers. The spectra of 6-carboxyfluorescein were measured in 100 mM MOPS buffers at the indicated pH values. Adapted from Thomas et al. (1979).

conform to one of the spectra seen in Fig. 2. An aliquot of the cell suspension is then centrifuged in a microfuge, and the supernatant is saved to evaluate dye leakage. The instrument is then switched to the dual wavelength mode (490–465 nm) to monitor pH changes continuously versus time. Any additions made to the sample cuvette are also made to the sham-loaded cells. At the conclusion of the experiment, another spectrum is taken (split-beam mode) versus the reference cuvette, and a second aliquot is centrifuged, saving the supernatant as before.

For absolute pH calibration, it is necessary to correct the spectra of the cell suspensions for dye leakage by subtracting the corresponding spectra of the supernatant fractions. Actually, it is only necessary to subtract the corresponding absorbance values at 490 (pH-sensitive wavelength) and 465 nm (the isosbestic wavelength; see Fig. 2). The ratio of the *corrected* absorbances at 490 and 465 nm (A_{490}/A_{465}) gives a concentration-independent parameter that is related to pH (Fig. 3; see Sect. 4). To calculate pH for any intermediate time period, prorated corrections are applied to the original 490-nm absorbance value. Dye leakage is assumed to be linear between the initial and final leakage values, so the appropriate correction factors are determined by extrapolation.

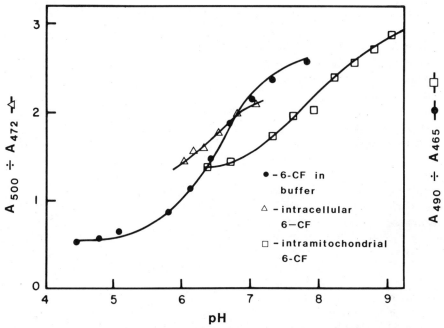

Figure 3.   Calibration curves for 6-carboxyfluorescein relating spectral shape to pH. For 6-CF in buffers, the data are taken from Fig. 2. For intracellular 6-CF, Ehrlich ascites tumor cells were incubated with 6-CFA$_2$ and suspended in 30 mM buffers containing 130 mM KCl and 1 mM MgCl$_2$. Internal and external pH were equilibrated with 10 μg/mL of nigericin. For intramitochondrial 6-CF, mitochondria incubated with 6-CHA$_2$ were placed in 50 mM MOPS buffers containing 0.15 M KCl. Internal and external pH were equilibrated with 10 μg/mL nigericin and 3 μM 1799. All spectral measurements were made versus the controls at the same concentration but lacking indicator.

If light-scattering artifacts are suspected, the reference wavelength may be varied on either side of the measuring wavelength in separate experiments (e.g., 490 nm versus 465 nm, or 490 nm versus 510 nm). Since light scattering varies as the inverse square of the wavelength, the scattering artifact will be in opposite directions in the two experiments, whereas the direction of the pH responses will be unaffected.

For fluorescence measurements, two different excitation wavelengths are chosen, such as 490 and 450 nm, with 520 nm a suitable emission wavelength. The ratio of fluorescence intensities observed at these two excitation wavelengths gives a concentration-independent parameter related to pH. Again, appropriate corrections should be made for leakage. Since the quantum yield varies with wavelength, the apparent pK of the indicator will depend on the particular wavelength pair chosen (Babcock, 1983). In our hands, the probe exhibits a pK_a of $\sim$ 6.7 at 25°C.

4. CALIBRATION OF RESPONSES

Two methods have been utilized to quantitate absolute internal pH: (1) the "null" titration proposed by Babcock for single-point determinations, and (2) the ionophore–equilibration and calibration method (Thomas et al., 1979), which allows continuous monitoring of pH. In the null method, one determines the external pH at which no spectral response is reported by the internal indicator when the plasma membrane is permeabilized with digitonin. At low concentrations of digitonin, the plasma membrane becomes permeable to small molecules but not to proteins. When the internal and external pH's are identical, no spectral change is registered by the internal indicator upon permeabilization of the plasma membrane. The actual null point pH is determined by graphical interpolation of the spectral responses obtained at various external pH values.

In the ionophore–equilibration method a calibration curve, relating internal pH to spectral shape, is first constructed (Fig. 3). The internal pH of cells loaded with the indicator is varied by equilibrating the external and internal pH in various buffers by means of ionophores. The addition of nigericin (a cation/H$^+$ exchanger) or of valinomycin plus a protonophore, will equilibrate K$^+$ and H$^+$ across the cell membrane so that

$$\frac{[H^+]_i}{[H^+]_o} = \frac{[K^+]_i}{[K^+]_o}$$

A convenient approach (Thomas et al., 1979) is to bathe the cells in a medium whose potassium concentration equals that of the cell interior (130 mM), so that intra- and extracellular pH's become equal. A spectrum is taken at each pH$_o$ value, using sham-loaded cells in the reference cuvette. The absorbances at 490 and 465 nm must be corrected for dye leakage, as described above (see Sect. 3). The ratio of the two corrected absorbances (490/465 nm) is then plotted versus external pH.

It is important to ascertain that pH$_i$ and pH$_o$ are indeed equilibrated by the particular conditions used; some buffer constituents such as serum albumin can decrease the effectiveness of the ionophores. By this calibration method the intracellular environment of the indicator is probably conserved better than with the null technique, which requires compromising the integrity of the plasma membrane.

Figure 3 shows several absorbance calibration curves obtained by the ionophore technique in which the 490/465 nm ratio is plotted versus pH. It is obvious from this figure that the spectral properties of the intracellular indicator are altered (especially in mitochondria) as compared to buffer solutions, with respect to both the pK_a of the indicator and its sensitivity to pH. Although these indicators are relatively insensitive to their environment and to the ionic composition, they will bind to proteins, especially when the protein concentration is high. In model studies, the addition of bovine serum albumin caused similar effects on pK_a and pH sensitivity (Udoff and Norman, 1979). According to these authors, protein binding is associated with a red shift of the absorbance maximum of up to 20 nm with serum albumin, but shifts observed in cells were generally less than 3 nm (7 nm for isolated mitochondria).

5. THE COMPARTMENT SPECIFICITY OF CARBOXYFLUORESCEIN AND FLUORESCEIN

There are several lines of evidence indicating that CF reports cytoplasmic pH, while F mainly reports mitochondrial pH. The two dyes respond differently to the addition of mitochondrial stimulators or inhibitors in whole cells. Figure 4 shows the spectral

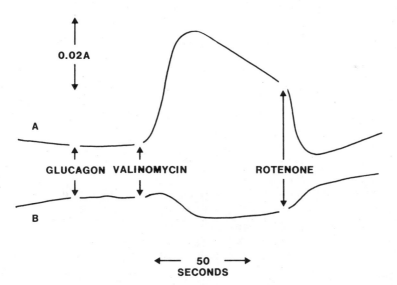

Figure 4. Absorbance responses of hepatocytes loaded with (A) fluorescein diacetate or (B) carboxyfluorescein diacetate. Hepatocytes were prepared by the collagenase perfusion method and loaded with the pH indicator as previously described (Strzelecki et al., 1984). Cells were incubated in 50 mM HEPES buffer, pH 7.4, containing 110 mM NaCl, 4.4 mM KCl, and 1.1 mM MgCl$_2$ at a concentration of 25 mg/mL (wet weight). Absorbance was monitored at 490–465 nm. Final concentrations: glucagon, 3 µg/mL; valinomycin, 0.07 µg/mL; rotenone, 3 µg/mL.

responses observed in hepatocytes loaded with either CF or F to the addition of valinomycin and rotenone. Valinomycin collapses the mitochondrial membrane potential, causing the pH gradient to become transiently larger (mitochondria alkaline), and rotenone inhibits the mitochondrial electron transport chain, collapsing any remaining pH gradient. The spectral responses reported by F conform to those expected for the mitochondrial space, whereas CF responses are in the opposite direction, consistent with a cytoplasmic location. Direct observations by fluorescence microscopy are compatible with this interpretation. CF-loaded cells always show a uniform fluorescence distribution within the cell, regardless of incubation conditions. In contrast, F-loaded cells display a highly localized fluorescence consistent with a mitochondrial location when mitochondrial pH gradients are present, but not when the mitochondrial pH gradient is collapsed (Thomas et al., 1982).

Our interpretation of these results is that F is able to cross the inner mitochondrial membrane, but CF is not. Cell fractionation studies indicate that most of the esterase activity is cytoplasmic in the Ehrlich ascites cell (Thomas et al., 1979) and probably for other cell types as well. Thus, as the diacetate derivative diffuses into the cell, cytoplasmic esterases immediately release the charged chromophore in the cytoplasm

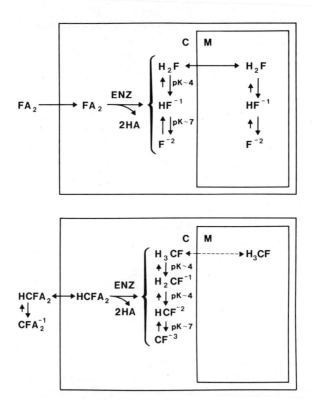

Figure 5.   Proposed cellular compartmentation of fluorescein (F, top) and carboxyfluorescein (CF, bottom) in whole cells. C, cytoplasmic compartment; M, mitochondrial compartment.

(Fig. 5). However, F is permeant across the mitochondrial membrane, whereas CF, on account of its extra charge ($pK \sim 4$), is not. This enables F, like other weak acids, to concentrate in the relatively basic mitochondrial matrix compartment. Since F has two dissociable groups (with pK_a's of ~ 4 and ~ 7, respectively), a mitochondrial transmembrane pH gradient of 1.0 causes the F concentration in the matrix to be about 80- to 100-fold higher than in the cytoplasmic compartment.

These conclusions are supported by experiments with isolated mitochondria. When CF is added externally to isolated mitochondria in buffered solutions, it is unresponsive to mitochondrial pH transitions caused by substrates and inhibitors (Fig. 6); in the same experiments, F faithfully reports the matrix pH transitions.

Movement of these indicators can be monitored by fluorescence polarization (Table 1). When mitochondrial pH gradients are large, F moves into the

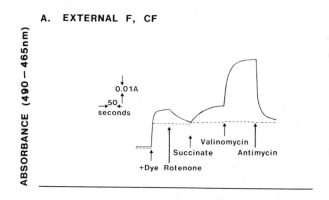

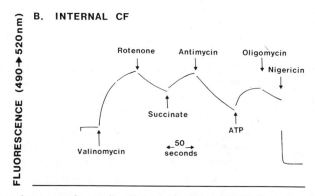

Figure 6. Spectral responses of externally added or internally trapped pH indicators to mitochondrial pH transitions. (*A*) Beef heart mitochondria were suspended at 1 mg protein/mL in 50 mM MES buffer, pH 6.1, containing 110 mM KCl and 1 mM MgCl$_2$. CF (dashed line) and F (solid line) at 3 μM. (*B*) Same conditions as (*A*), except mitochondria were pretreated with carboxyfluorescein diacetate. Final concentrations for both (*A*) and (*B*) succinate, 10 mM; valinomycin, 0.5 μM; antimycin, 10 μM; rotenone, 12 μM; ATP, 1 mM; oligomycin, 2 μg/mL; nigericin, 1 μM.

TABLE 1.   FLUORESCENCE POLARIZATION[a]

	Additions					
	(1) Rotenone	(2) Succinate	(3) Valinomycin	(4) Antimycin	(5) ATP	(6) 1799
CF_{int}	0.20	0.19	0.12	0.20	0.14	0.21
CF_{ext}	0.03	0.03	0.03	0.03	0.03	0.05
F_{ext}	0.05	0.05	0.16	0.03	0.08	0.03

[a]Effect of mitochondrial energization on the fluorescence polarization of externally added CF and F and internally trapped CF. Experimental conditions as in Fig. 6. Fluorescence polarization was determined in an Aminco Bowman spectrophotofluorometer equipped with Glan polarizers. Additions (1–6) were consecutively made to the cuvette, with polarization values being determined after steady-state fluorescence values were reached.

mitochondrion, and its fluorescence polarization increases; collapsing the pH gradient returns the polarization to its initial value. CF is unresponsive in these experiments when added externally.

The movement of F into mitochondria in suspension in response to pH transitions can also be demonstrated by centrifugation experiments. If the mitochondrial pH gradient is large, the indicator will tend to accumulate in the mitochondria; if it is small, mitochondrial indicator concentration will be low. This relationship between mitochondrial pH gradients and the distribution ratio of F has been quantitated (Strzelecki et al., 1984). The transmembrane pH gradient of isolated mitochondria was varied by appropriate combinations of ionophores, K^+, and phosphate, the external pH being kept constant. The mitochondrial pH gradient was evaluated using the DMO method, as applied by Addanki et al. (1968). In duplicate experiments mitochondria were incubated with either CF or F, and the accumulation of indicator was determined by fluorescence assay after rapid separation of the mitochondria from the suspending media by centrifugation in a microfuge. The results (Strzelecki et al., 1984) support the notion that F is permeant across the mitochondrial membrane, whereas CF is not. As predicted, a pH gradient of 1 resulted in a distribution ratio of almost 100 for fluorescein.

The difference in permeability between the two indicators has a straightforward physicochemical basis. As in the case of other weak acids, it is presumed that only the fully protonated, neutral form is permeant across the mitochondrial membrane. Since CF has an added ionizable group of $pK \sim 4$, the relative concentration of the fully protonated species at pH 7 is only about $2.5 \times 10^{-5}\%$ of the total dye present; the corresponding value for F is $2.5 \times 10^{-2}\%$. Thus, fluorescein would be expected to permeate the mitochondrial membrane about 1000 times faster than CF under the same conditions, simply because of the relative concentrations of the permeant species. Thus, the half time for entry of CF would be predicted to be about 6 h (versus 20 s for F), much too long to be seen in the time scale of our experiment.

Other cell types in which carboxyfluorescein derivatives have been reported to be reliable indicators of cytosolic pH include bovine sperm (Babcock, 1983), Ehrlich ascites tumor cells (Thomas et al., 1979), hepatocytes (Strzelecki et al., 1984), gastric glands (Paradiso et al., 1984), platelets (Grinstein and Furuya, 1984), lymphocytes (Grinstein et al., 1984), and heart cells (Piwnica-Worms and Lieberman, 1983).

6. MONITORING MITOCHONDRIAL pH WITH CARBOXYFLUORESCEIN DIACETATE

The fact that CF is impermeant across the mitochondrial inner membrane allowed the mitochondrial calibration curve to be made (Fig. 3). Incubation of isolated mitochondria with CFA_2 resulted in the trapping of CF within the matrix space, owing to matrix esterase activity. Calibration curves were performed as with whole cells, equilibrating internal and external pH with ionophores in various buffers. Similar experiments with FA_2 failed because of the high degree of permeability of F to these membranes. Table 2 shows the internal pH, as determined from the mitochondrial calibration curve (Fig. 3), of isolated mitochondria incubated under a variety of conditions.

7. CALIBRATION OF MITOCHONDRIAL pH IN INTACT HEPATOCYTES

Although fluorescein can be used to follow mitochondrial pH transitions qualitatively in intact cells, its quantitative use has proved more difficult, primarily owing to its rapid rate of equilibration across the mitochondrial membrane. Since the concentration of fluorescein in the mitochondrial matrix is dependent on the magnitude of the mitochondrial pH gradient, any change in pH on either side of the

TABLE 2.   INTERNAL pH OF ISOLATED MITOCHONDRIA AS MEASURED BY TRAPPED CF.[a]

	Intramitrochondrial pH	
Conditions	(− Phosphate)	(+ Phosphate)
No additions	9.2	8.3
+ Valinomycin (6.7 ng/mL)	9.0	8.2
+ Ammonium Acetate (6.7 mM)	9.1	8.7
+ Valinomycin, 1799, and nigericin	7.5	7.4

[a]All experiments were performed in 10 mM MOPS buffer, pH 7.4, containing 150 mM KCl, 10 mM MgCl$_2$, 15 mM β-hydroxybutyrate, and 10 mM sodium phosphate where indicated. Mitochondrial protein was 1–2 mg/mL.

inner mitochondrial membrane results in redistribution of the dye. However, in the presence of a typical mitochondrial pH gradient of ~1, much of the indicator would be expected to be in the matrix space and could be assumed to be solely monitoring mitochondrial pH.

We were recently able to determine the magnitude of mitochondrial pH gradients in isolated hepatocytes by a combination of the indicator methods and the distribution ratio method (Strzelecki et al., 1984), using the weak acid ([14C])5,5'-dimethyloxazolidine-2,4-dione (DMO). DMO freely distributes across the plasma membrane and mitochondrial membrane in whole cells according to the pH gradient across each membrane. Knowledge of the cytoplasmic pH from CF measurements allows the expected distribution of DMO across the plasma membrane to be calculated. The actual measured distribution of DMO exceeds this calculated value, because the weak acid accumulates in the basic mitochondrial compartment. This excess accumulation of DMO by intact cells is then used to calculate the pH gradient across the mitochondrial inner membrane, according to the equation (Strzelecki et al., 1984):

$$pH_{mit} = \log \frac{[DMO]_{mit}}{[DMO]_{cyt}} \times (10^{pK_a} + 10^{pH_{cyt}}) - 10^{pK_a}$$

No disruption of membranes is required by this method, but it does require that the volumes of the cytoplasmic and mitochondrial compartments be known. The total cellular water space can be determined using tritiated water, with [14C]inulin being used to correct for extracellular water. The mitochondrial volume can be estimated from the amount of citrate synthase activity in cells (LaNoue et al., 1984) or from other literature values. For hepatocytes, the mitochondrial matrix space was assumed to be 16% of the total cell volume. The total amount of cellular [14C]DMO is determined by centrifuging the cells through silicone oil into a heavier layer of

TABLE 3. pH DETERMINATION OF CYTOSOLIC AND MITOCHONDRIAL COMPARTMENTS IN ISOLATED HEPATOCYTES UNDER DIFFERENT METABOLIC CONDITIONS.[a]

	pH		
Metabolic Conditions	Medium	Cytosol	Mitochondria
Control	7.23 ± 0.03	6.88 ± 0.4	7.72 ± 0.06
Uncoupled 1799 150 μM	7.23 ± 0.05	6.98 ± 0.7	7.08 ± 0.13
Valinomycin 50 nM	7.28 ± 0.02	6.50 ± 0.6	8.03 ± 0.10
Valinomycin 5 nM	7.28 ± 0.04	6.90 ± 0.5	7.73 ± 0.05
Glucagon 1.5 μg mL	7.22 ± 0.03	6.87 ± 0.06	7.62 ± 0.05
Glucagon + valinomycin 5 nM	7.26 ± 0.03	6.84 ± 0.09	7.90 ± 0.08
c-AMP 0.75 mM	7.19 ± 0.02	6.80 ± 0.04	7.74 ± 0.08

[a]Experimental techniques are described in the text. Adapted from Strzelecki et al. (1984).

perchloric acid. Table 3 shows the cytoplasmic and mitochondrial pH values in hepatocytes under various conditions, as determined by this method.

This technique for determining mitochondrial pH gradients is not without drawbacks. It assumes that the cell consists of only two compartments, cytosolic and mitochondrial. Presumably there is no permeability barrier to protons between the nucleus and the cytoplasm, so these are considered as one compartment. The lysosomal compartment represents a small fraction of the total cell volume. Since the lysosomes possess an acidic interior, the weak acid DMO would tend to be excluded. In addition, the spectral contribution from any trapped CF would be quenched in this acidic compartment. Other cellular compartments are also believed to contribute little to these measurements.

This represents one of the few documented methods available to measure mitochondrial matrix pH in whole cells. Hoek et al. (1980) published a method to measure mitochondrial pH in hepatocytes; however, the method requires rapid cellular fractionation, assuming no change in the distribution of DMO occurs during the fractionation time. Although some reports using NMR had indicated that mitochondrial pH could be monitored using the ^{31}P-phosphate signal, subsequent results have shown that phosphate in the mitochondrial compartment is not detectable by NMR (Gadian et al., 1982).

8.   OTHER USES OF TRAPPED INDICATORS

Besides intracellular pH measurements, Babcock (1983) has used the null titration technique with CF to determine the intracellular concentration of Na^+ and K^+. The method is based on the electroneutral exchange of cations and protons catalyzed by certain ionophores. CF-loaded cells are placed in buffer adjusted to the same pH value as the internal pH, as determined by independent measurements. If the internal and external pH are equal but the internal and external K^+ concentrations differ, the addition of nigericin will cause an exchange of K^+ for protons. This pH change is reported by the internal pH indicator. The internal K^+ concentration is equal to the external K^+ when no spectral response is observed. Cytoplasmic Na^+ has been determined similarly, using the ionophore monensin instead of nigericin. The accuracy of the method depends on the absolute specificity of these ionophores for their respective ions, which is probably an oversimplification.

Recently Grinstein et al. (1983) used a double-null point titration technique to measure the free concentrations of Mg^{2+} and Ca^{2+} within platelet alpha granules. The dye 9-aminoacridine was found to accumulate within these relatively acidic compartments via the pH gradient, and the accumulated indicator could be used to monitor the intragranular pH. The authors circumvented the requirement for an ion-specific ionophore by using two ionophores having different selectivities for Ca^{2+} and Mg^{2+}. The external concentrations of the divalent metals were varied until no response was observed from the internal indicator. At the null point, the pH gradient across the granule membrane is balanced by the divalent cation gradient. The pH gradient was determined from [^{14}C]methylamine distribution measurements.

Calculation of the actual concentrations of Ca^{2+} and Mg^{2+} required solving two simultaneous equations.

9. SUMMARY

The trapped indicator technique has several advantages for monitoring intracellular pH. It can be performed with equipment available in most laboratories, using either fluorescence or absorbance measurements. It is nondestructive to cells, and fluorescein dyes seem to have little effect on cell metabolism (however, see Spray et al., 1984). Continuous real-time monitoring of intracellular pH is possible, and the method can detect changes of 0.01 pH. It is sensitive enough for studies with monolayers (Thomas et al., 1982) or individual cells (Slavik and Kotyk, 1984; Udkoff and Norman, 1979). In addition, pH changes in the cytoplasmic and mitochondrial compartments can be distinguished by judicious use of CF and F.

On the negative side, one of the main problems encountered is that of leakage, especially for F. Spectral measurements must be corrected for leakage in order to assess pH accurately. Three ways of minimizing leakage are as follows: (1) Use a less leaky indicator, such as BCECF (Rink et al., 1982); (2) lower the incubation temperature; (3) continuously remove external indicator by perfusion technique (Boron, 1982).

Although CFA_2 specifically monitors cytoplasmic pH in several different cell types, this may not necessarily be a general phenomenon. As shown in this chapter, it can report mitochondrial pH transitions if it is first hydrolyzed by mitochondrial esterases. Thus, this specificity for the cytoplasm should be established for any new cell type studied.

As a final note, Spray et al. (1984) have recently reported that the intracellular hydrolysis of certain membrane-permeant esters causes an acidification of the cytoplasm in several cell types. The acidification was considerably in excess of that expected from the small amount of acid generation caused by the esterase reaction. How general this phenomenon is remains to be established.

REFERENCES

Addanki, S., F. D. Cahill, and J. F. Sotos (1968) Reliability of the quantitation of intramitochondrial pH and pH gradient of heart mitochondria. *Anal. Biochem.*, **25**, 17–29.

Babcock, D. F. (1983) Examination of the intracellular ionic environment and of ionophore action by null point measurements employing the fluorescein chromophore. *J. Biol. Chem.*, **258**, 6380–6389.

Boron, W. F. (1982) Optical measurement of intracellular pH in isolated, perfused renal proximal tubules of the salamander. *Fed. Proc.*, **41**, 4335 (abstract).

Cohan, C. S., R. D. Hadley, and S. B. Kater (1983) Zap axotomy: Localized fluorescent excitation of single dye-filled neurons induces growth by selective axotomy. *Brain Res.*, **270**, 93–101.

Gadian, D. G., G. K. Radda, M. J. Dawson, and D. R. Wilkie (1982) "pH measurements of cardiac and skeletal muscle using ^{31}P-NMR," in R. Nuccitelli and D. W. Deamer, Eds., *Intracellular pH: Its Measurement, Regulation, and Utilization in Cellular Functions*, Liss, New York, pp. 61–77.

Grinstein, S., and W. Furuya (1984) Intracellular distribution of acridine derivatives in platelets and their suitability for cytoplasmic pH measurements. *Biochim. Biophys. Acta,* **803**, 221–228.

Grinstein, S., W. Furuya, J. VanderMeulen, and R. G. V. Hancock (1983) The total and free concentrations of Ca^{2+} and Mg^{2+} inside platelet secretory granules. *J. Biol. Chem.,* **258**, 14774–14777.

Grinstein, S., S. Cohen, H. M. Lederman, and E. W. Gelfand (1984) The intracellular pH of quiescent and proliferating human and rat thymic lymphocytes. *J. Cell. Physiol.,* **121**, 87–95.

Hoek, J. B., D. G. Nicholls, and J. R. Williamson (1980) Determination of the mitochondrial protonmotive force in isolated hepatocytes. *J. Biol. Chem.,* **256**, 1458–1464.

Kashiwagura, T., C. J. Deutsch, J. Taylor, M. Erecinska, and D. F. Wilson (1984) Dependence of gluconeogenesis, urea synthesis, and energy metabolism of hepatocytes on intracellular pH. *J. Biol. Chem.,* **259**, 237–243.

LaNoue, K. F., T. Strzelecki, and F. Finch (1984) The effect of glucagon on hepatic respiratory capacity. *J. Biol. Chem.,* **259**, 2116–2121.

Moolenaar, W. H., L. G. J. Tertoolen, and S. W. de Laat (1984) Phorbol ester and diacylglycerol mimic growth factors in raising cytoplasmic pH. *Nature,* **312**, 371–374.

Nuccitelli, R., and D. W. Deamer (Eds.) (1982) *Intracellular pH: Its Measurement, Regulation and Utilization in Cellular Function.* Liss, New York.

Paradiso, A. M., R. Y. Tsien, and T. E. Machen. (1984) $Na^+–H^+$ exchange in gastric glands as measured with a cytoplasmic-trapped, fluorescent pH indicator. *Proc. Natl. Acad. Sci. USA,* **81**, 7436–7440.

Piwnica-Worms, D., and M. Lieberman (1983) Microfluorometric monitoring of pH_i in cultured heart cells: $Na^+–H^+$ exchange. *Am. J. Physiol.,* **244**, C422–C428.

Rink, R. J., R. Y. Tsien, and T. Pozzan (1982) Cytoplasmic pH and free Mg^{++} in lymphocytes. *J. Cell Biol.,* **95**, 189–196.

Roos, A., and W. F. Boron (1981) Intracellular pH. *Physiol. Rev.,* **61**, 296–434.

Rotman, B., and B. W. Papermaster (1966) Membrane properties of living mammalian cells as studied by enzymatic hydrolysis of fluorogenic esters. *Proc. Natl. Acad. Sci. USA,* **55**, 134–141.

Slavik, J., and A. Kotyk (1984) Intracellular pH distribution and transmembrane pH profile of yeast cells. *Biochim. Biophys. Acta,* **766**, 679–684.

Spray, D. C., J. Nerbonne, A. C. DeCarvalho, A. L. Harris, and M. V. L. Bennett (1984) Substituted benzyl acetates: A new class of compounds that reduce gap junctional conductance by cytoplasmic acidification. *J. Cell Biol.,* **99**, 174–179.

Strzelecki, T., J. A. Thomas, C. D. Koch, and K. F. LaNoue (1984) The effect of hormones on proton compartmentation in hepatocytes. *J. Biol. Chem.,* **259**, 4122–4129.

Thomas, J. A., and D. L. Johnson (1975) Fluorescein conjugates of cytochrome *c* as internal pH probes in submitochondrial particles. *Biochem. Biophys. Res. Commun.,* **65**, 931–939.

Thomas, J. A., P. C. Kolbeck, and T. A. Langworthy (1982) "Spectroscopic determination of cytoplasmic and mitochondrial pH transitions using trapped pH indicators," in R. Nuccitelli and D. W. Deamer, Eds., *Intracellular pH: Its Measurement, Regulation, and Utilization in Cellular Functions,* Liss, New York, pp. 105–123.

Tomoda, A., S. Tsuda-Hirota, and S. Minakami (1977) Glycolysis of red cells suspended in solutions of impermeable solutes. *J. Biochem.,* **81**, 697–701.

Tsien, R. Y., T. Pozzan, and T. J. Rink (1981) Calcium homeostasis in intact lymphocytes: Cytoplasmic free calcium monitored with a new intracellular trapped fluorescent indicator. *J. Cell Biol.,* **94**, 325–334.

Udkoff, R., and A. Norman (1979) Polarization of fluorescein fluorescence in single cells. *J. Histochem. Cytochem.,* **27**, 49–55.

Wagel, S. A., and Ingebritsen, W. R., Jr. (1975) Isolation, purification, and metabolic characteristics of rat liver hepatocytes. *Methods Enzymol.,* **35**, 579–594.

CHAPTER **19**

NEW TETRACARBOXYLATE CHELATORS FOR FLUORESCENCE MEASUREMENT AND PHOTOCHEMICAL MANIPULATION OF CYTOSOLIC FREE CALCIUM CONCENTRATIONS

ROGER Y. TSIEN

Department of Physiology–Anatomy
University of California
Berkeley, California

---

1. INTRODUCTION: FLUORESCENT Ca^{2+} INDICATORS COMPARED WITH OTHER METHODS FOR MEASURING CYTOSOLIC FREE Ca^{2+}

    1.1 MEASUREMENT AND MANIPULATION OF CYTOSOLIC FREE Ca^{2+} 328

    1.2 CALCIUM-SENSITIVE ELECTRODES 328

    1.3 NUCLEAR MAGNETIC RESONANCE 330

    1.4 ABSORBANCE INDICATORS 330

    1.5 CHEMILUMINESCENCE 331

    1.6 FLUORESCENT PROBES: LIMITATIONS 332

2. NEW PROBES FOR MEASURING AND MANIPULATING Ca^{2+}

    2.1 "FURA-2" AND "INDO-1" 332

    2.2 FLUORESCENCE MICROSCOPY 335

2.3   FLOW CYTOMETRY   339

2.4   "CAGED CALCIUM"   343

REFERENCES   344

1.   INTRODUCTION: FLUORESCENT Ca^{2+} INDICATORS COMPARED WITH OTHER METHODS FOR MEASURING CYTOSOLIC FREE Ca^{2+}

1.1   Measurement and Manipulation of Cytosolic Free Ca^{2+}

One of the prime applications for optical indicators in biology is the study of how calcium ions act as intracellular signals. Fluctuations in cytosolic free calcium concentrations, $[Ca^{2+}]_i$, are hypothesized to be crucial in the triggering and control of a wide variety of cellular responses. How can one test such hypotheses? Ideally, one would first verify that during a physiological response, $[Ca^{2+}]_i$ does change with an amplitude and time course consistent with a triggering role. Optical indicators are often the best tools available for making such measurements, as will be discussed below. Assuming a rise in $[Ca^{2+}]_i$ is detected, a second important test is to raise cytosolic Ca^{2+} by artificial experimental means and see whether the physiological response is elicited. This test also requires $[Ca^{2+}]_i$ measurement to check whether the range of $[Ca^{2+}]_i$ required for the response is the same as that generated by the physiological stimulus. Traditional methods for raising $[Ca^{2+}]_i$ include ionophore administration or direct microinjection or iontophoresis of Ca^{2+}. A promising new approach is to use optical probes in reverse. As described below, photochemically sensitive Ca^{2+} chelators can be synthesized which change irreversibly from high to low Ca^{2+} affinity upon illumination. Light thereby releases the "caged" calcium and generates a jump in $[Ca^{2+}]_i$ that should eventually be better controllable in amplitude, time course, and spatial extent than achievable by classical techniques. The final test of the importance of $[Ca^{2+}]_i$ is to find means to prevent the natural stimulus from raising $[Ca^{2+}]_i$ as it normally would. If the physiological response is also prevented, the $[Ca^{2+}]_i$ rise is shown to be actually necessary, not just an auxiliary side effect. Here too, quantitative measurement of $[Ca^{2+}]_i$ is needed to show how effectively the $[Ca^{2+}]_i$ was clamped.

1.2   Calcium-Sensitive Electrodes

The techniques for measuring cytosolic free Ca^{2+} have been extensively reviewed (2, 3, 15, 20, 21, 24, 25) with comparisons between the relative advantages of ion-selective microelectrodes, nuclear magnetic resonance (NMR), optical indicators, and null points for plasma membrane permeabilization. Of these the most widely used are the Ca^{2+}-selective microelectrodes and optical indicators including photoproteins, bisazo dyes, and tetracarboxylate chelators. My personal view, based on several years' experience with microelectrodes before moving back to indicator dyes, is that the latter generally offer a much higher ratio of meaningful new results to man-hours expended. Depending on luck, skill, and requirements, days to years can be spent trying to make Ca^{2+}-selective electrodes, most of which may prove insufficiently sharp, selective, or stable. Most of the published intracellular measurements with Ca^{2+}-selective microelectrodes are essentially confirmatory in that they merely show by direct means what was already widely believed on other evidence. Examples from our own work include demonstrations that $[Ca^{2+}]_i$ does indeed rise during muscle contraction, that β-adrenergic agents can affect cardiac Ca^{2+} sequestration and myofilament sensitivity (11), or that neuronal $[Ca^{2+}]_i$ can rise after a train of Ca currents during action potentials (1).

In retrospect, it is not easy to think of significant unexpected results that have become accepted mainly on the basis of Ca^{2+} microelectrode studies, though we were fairly satisfied with our own efforts at the time. All too often, penetrations seem to show remarkably little change of $[Ca^{2+}]_i$ during drastic cell responses. Why have electrode studies been mainly confirmatory in their conclusions? Probably because the variability from electrode to electrode, penetration to penetration, and cell to cell makes for enough experimental scatter to obscure any but preconceived notions. By contrast, indicator dye molecules, once properly synthesized, are relatively reproducible sensors. They can now be loaded into millions of cells without microinjection and used to read an average value of $[Ca^{2+}]_i$ from the whole population. This ensemble averaging favors reproducibility and means that the $[Ca^{2+}]_i$ response can be measured from the very same population as the biochemical or physiological response is measured, thus minimizing sampling error.

Mammalian cells of great clinical and pathological interest can be studied directly with just a commercially available spectrofluorometer without recourse to skilled micromanipulations. Although reproducibility is no protection against systematic errors, dye experiments do tend to start giving believable results more quickly than Ca^{2+}-selective microelectrode projects do. The first and most widely used fluorescent indicator of cytosolic $[Ca^{2+}]_i$ has been quin2 (15, 24); although it has often been used to demonstrate $[Ca^{2+}]_i$ rises long suspected on indirect evidence, there have been enough surprises or answers to controversial questions to avoid total boredom. Examples include demonstrations that (a) mitochondria in healthy cells do not hold large amounts of Ca^{2+} (23); (b) abnormally low $[Ca^{2+}]_i$ can raise the Ca^{2+} permeability of the plasma membrane (10, 23); (c) secretory exocytosis from many cells can occur without any rise in $[Ca^{2+}]_i$ (5, 12, 17) and is actually inhibited in the parathyroid by raised $[Ca^{2+}]_i$ (18); (d) transmission of signals to release intracellular Ca stores is highly sensitive to ATP depletion (14); (e) phorbol ester tumor promoters interact with $[Ca^{2+}]_i$ signals not by raising $[Ca^{2+}]_i$ but rather by lowering the $[Ca^{2+}]_i$ thresholds of certain processes (5, 9, 17, 22); (f) inositol phospholipid breakdown can occur without any rise in $[Ca^{2+}]_i$ (27); (g) cyclic AMP antagonism of $[Ca^{2+}]_i$ can occur both by direct lowering of $[Ca^{2+}]_i$ and by reduction in sensitivity to $[Ca^{2+}]_i$ (6,

8); and (h) lymphocyte regulatory volume decrease (16), capping of surface antigens (14), and stimulation by interleukin-2 (28) can all proceed without detectable rises in $[Ca^{2+}]_i$.

The above criticisms of Ca^{2+} electrodes apply to intracellular measurements, not to extracellular applications or to permeabilized cell experiments. In these latter uses, electrode tip size rarely needs to be minimized, penetration damage is irrelevant, and electrical isopotentiality with the reference electrode is usually assured. The ion-selective tip can generally be steered to an optimal or at least well-defined location, unlike intracellular electrode whose position is uncertain with respect to spatial gradients. $[Ca^{2+}]_i$ levels to be detected are rarely as low as intracellular levels. There is less disparity between the speed of extracellular signals and the response time of the electrodes. For these reasons, electrodes are often the best and most generally applicable method for measuring extracellular Ca^{2+} levels.

1.3   Nuclear Magnetic Resonance

NMR of ^{19}F-labeled chelators is a new, nonoptical technique with considerable promise in some applications. Smith et al. (19) have synthesized tetracarboxylate chelators bearing fluorine substituents whose NMR chemical shifts are influenced by cation binding. Simple analysis of the ^{19}F NMR spectrum reveals the proportion of Ca^{2+} complex relative to free chelator. This ratio multiplied by the Ca^{2+} dissociation constant indicates the free $[Ca^{2+}]_i$. The major advantage of NMR is its noninvasiveness. If the known methods for loading chelators into disaggregated cells can be extended to work on bulk tissues in vivo, the NMR method might be able to show $[Ca^{2+}]_i$ in organs within the intact organism, a feat impossible with any existing technique. Also some of the fluorinated chelators can simultaneously signal concentrations of other ions such as Zn^{2+} or Fe^{2+}. Against these unique advantages must be set the extreme expense of the instrumentation and the difficulties often found in handling isolated tissues in the cavity of the superconductive magnet. The inherent low sensitivity of NMR detection even for a favorable isotope like ^{19}F demands some combination of high probe concentrations, large amounts of tissue, or long periods for signal accumulation. Fast responses on small amounts of tissue (e.g., single cells) are likely to be beyond NMR detection for a long time.

1.4   Absorbance Indicators

Many workers have made extensive use of dyes such as arsenazo III and antipyrylazo III, whose absorption spectra are sensitive to Ca^{2+}. This work is reviewed in this volume and in previous articles (2, 3, 20, 21, 25). Absorbance measurements of $[Ca^{2+}]_i$ are most suitable on single large excitable cells. Because intrinsic light scattering contributes significantly to the resting optical density, measures of static baseline $[Ca^{2+}]_i$ are much harder to obtain than amplitudes of fast transient elevations of Ca^{2+}. Any movement of the tissue tends to contribute large artifacts quite difficult to compensate for, even if multiwavelength measurements are made. The actual transmission changes are on the order of 10^{-2} to 10^{-5}, so that highly stabilized light sources, vibration-free mountings, and silicon photodiode detectors are used. The difficulty of detection increases sharply as the size and path length of the cell decrease. By contrast, fluorescence measurements are applicable to both small and large cells, singly or in populations. With decent fluorophores it is not difficult to get the dye fluorescence high enough to swamp cell autofluorescence and leakage of scattered light so that static $[Ca^{2+}]_i$ levels can be calibrated. Fluorescence changes with $[Ca^{2+}]_i$ are relatively large, of the order of 10^{-1} to 10^1 relative to the baseline fluorescence, so the mechanical stability of the instrument and the tissue is less critical. Though sensitivity to low absolute light levels is required, readily available photomultipliers and intensified television cameras are adequate and are readily interfaced to digital processing. Small cells can be studied not only by fluorescence microscopy but also by flow cytometry, the foremost technique for analyzing heterogeneity in a cell population.

1.5   Chemiluminescence

Fluorescence and chemiluminescence measurements share inherent advantages of photon detection against a low background, though the chemiluminescence background tends to be much easier to reduce to very low photon count rates. This drawback of fluorescence is more than compensated for by the enormously greater photon flux emitted from a fluorophore in comparison to a chemiluminescent emitter. Because the fluorophore taps the excitation beam for its energy, each molecule can and does emit thousands of photons during an experiment. By contrast a photoprotein can only emit a photon once, and in practice the useful photon output is kept even lower by the requirement that the photoprotein not be significantly exhausted during the physiological recording. For this reason, normal aequorin signals from individual cardiac myocytes (volume about 40 pL) are on the order of 1–10 photon counts per sec with an extremely efficient light-collecting setup (4), whereas the fluorescent Ca^{2+} indicator fura-2 readily gives 10^4–10^5 photon counts per second from a single thymocyte (volume 0.1–0.2 pL) viewed with an ordinary microscope that is far less efficient in gathering light. Having more photons not only greatly improves the photon statistics but also means that one need not be fanatical about reducing stray light or photomultiplier dark count.

The above discussion has focused on the most fundamental advantages and disadvantages of ion-selective microelectrodes, NMR, absorbance, and fluorescent and chemiluminescent indicators. It will have been obvious that I favor fluorescence as the readout mode with the greatest long-term promise, especially for fairly small mammalian cells that are of more direct clinical relevance than the classical giant cell model systems from invertebrates. Fluorescence readings can be obtained from cell populations en masse in a cuvette, from single cells on a microscope stage, or from populations dissected by flow cytometry. The first mode requires the least specialized equipment and is ideal for correlating $[Ca^{2+}]_i$ with biochemical responses that are also measured on cell populations. Microscopic viewing of single cells is the simplest way to see how $[Ca^{2+}]_i$ evolves in individual cells or localized regions of a cell. It is particularly appropriate when the cell population is heterogeneous in visible morphology. Flow cytometry is the quickest general method for objectively analyzing the heterogeneity of $[Ca^{2+}]_i$ in a cell population. In principle it should be able to correlate $[Ca^{2+}]_i$ with other parameters, such as surface marker phenotype, membrane potential, and cytosolic pH, that can also be read out by fluorescence. Also flow cytometry should be able to sort cells on the basis of their $[Ca^{2+}]_i$ and thereby serve in a preparative not just analytical mode.

1.6 Fluorescent Probes: Limitations

Despite the above advantages of fluorescence, it does have important limitations. Chief among these is the need to get light into and out of the tissue. Excessive pigmentation, light scattering, thickness, or actinic sensitivity can easily rule out any optical technique. However, epifluorescence observation is often more tolerant of these interferences than absorbance is, because optical access is required from only one side instead of two, and because light scattering has a less direct effect on fluorescence than on absorbance. Fiber optics represents a promising but relatively unexplored technique for exciting and collecting fluorescence from nonreadily accessible locations. Another potential stumbling block arises if the tissue has particularly strong autofluorescence relative to the achievable or permissible dye loadings. Since autofluorescence generally decreases as excitation and emission wavelengths increase, this problem is one of the main reasons we are trying to extend the wavelengths of fluorescent Ca^{2+} indicators. A third problem is that fluorescence can be artifactually perturbed by aspects of the intracellular environment such as viscosity, temperature, and quenching agents. All methods for assessing cytosolic ion activities could in principle have systematic errors due to differences in the behavior of the sensor in cytoplasm as opposed to calibration solutions. However, the specific parameters listed above are more likely to affect fluorophores than to perturb electrodes, NMR, or absorbance probes. Fortunately, the microviscosity seen by the fluorophore can be assessed by fluorescence polarization experiments and mimicked by adding thickening agents to the calibration solutions. Matching the calibration temperature to the tissue temperature is usually trivial. Simple forms of quenching are largely nulled out by the dual-wavelength fluorescence method described below.

The final major problem with fluorescence is that it is hard to design novel fluorophores. Only a small fraction of known chromophores fluoresce well in aqueous solution, and our ability to predict fluorescence properties is very limited. These difficulties retard development of fluorescent indicators that have all the right properties combined into a single structure.

2. NEW PROBES FOR MEASURING AND MANIPULATING Ca^{2+}

2.1 "Fura-2" and "Indo-1"

Though ideal molecules are still a long way off, we have made some recent progress toward better fluorescent indicators for measuring Ca^{2+} and photolabile chelators for manipulating Ca^{2+}. The first group comprises what we believe to be the best dyes now available for measurement of [Ca^{2+}]$_i$ in most intact cells. The most attractive members of this family, "fura-2" and "indo-1" (Fig. 1), have three main and two minor advantages over their predecessor quin2: (a) Molecule for molecule, the new dyes are about 30 times more fluorescent than quin2, an improvement due to a sixfold increase in extinction coefficient and about fivefold enhancement in quantum efficiency. This huge increase in brightness can be used either to reduce dye loading, thus minimizing exogenous buffering of [Ca^{2+}]$_i$ and the possibility of toxic side effects, or to make signal detection feasible from single cells, as will be shown below. (b) Both the Ca^{2+} complexes and the free species are highly fluorescent but at different wavelengths. Fura-2 changes mainly its peak excitation wavelength with Ca^{2+}, whereas indo-1 shifts both its excitation and emission maxima. Detecting Ca^{2+} by means of the ratio between intensities at two wavelengths is far more stable and reliable than relying on intensity at just one wavelength as was necessary with quin2. (c) Fura-2 and indo-1 can be used with wavelengths of 350 nm and above. While this may seem a small difference from the 339-nm excitation for quin2, those few nanometers make possible the use of this dye in conventional microscopes with glass optics and in flow cytometry systems using argon or krypton lasers. (d) The new dyes have about twofold greater effective dissociation constants for Ca^{2+} than quin2 has. Resolution of [Ca^{2+}]$_i$ levels near or above 10^{-6} M is therefore improved. (e) The new dyes have more selectivity than quin2 has for Ca^{2+} over competing

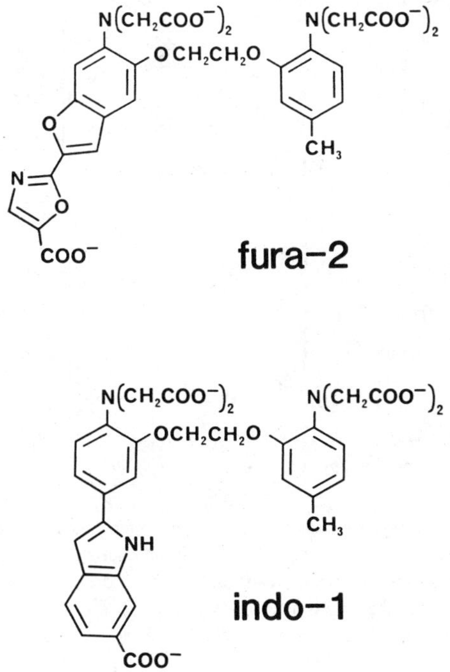

Figure 1.   Structures of the new fluorescent indicators fura-2 and indo-1.

divalents such as Mg^{2+}, Mn^{2+}, and Zn^{2+}. The improvements are factors of 4, 12, and 40, respectively. A full description of the new dyes' chemical synthesis and properties is in Grynkiewicz et al. (7). It should be noted that the syntheses are not trivial: To make the intracellularly hydrolyzable esters of fura-2 and indo-1 from commercially available chemicals requires 15 and 13 steps, respectively, including several that require experience in organic synthesis. Fortunately the new dyes have recently become commercially available from Molecular Probes Inc. and Calbiochem-Behring.

Fura-2 is preferred over indo-1 for cuvette and microscope experiments because its emission wavelengths (500–520 nm) are better separated from cell

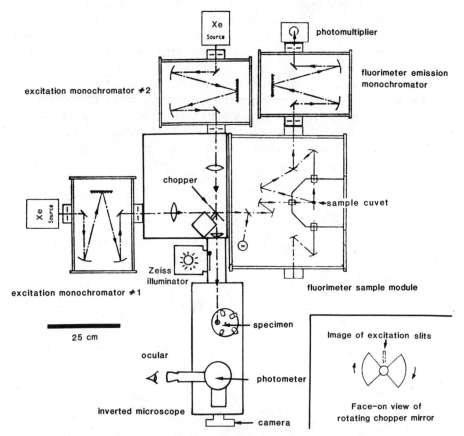

Figure 2. Optical layout of a dual-wavelength spectrofluorometer/microscope system constructed from a Spex Industries Fluorolog 2 system plus a Zeiss IM35 inverted microscope plus components made in our workshops. When the chopper is in the position shown in the insert at lower right, excitation monochromator #1 illuminates the fluorometer cuvette and #2 illuminates the microscope specimen. When the chopper has rotated by 90°, #1 feeds the microscope and #2 illuminates the cuvette. This cycle repeats 10–30 times a second. The Spex Industries microcomputer counts photons from either the fluorometer photomultiplier or the microscope photometer. In either case it keeps separate track of the signals at the two wavelengths.

autofluorescence than are the emission wavelengths of indo-1 (405 or 485 nm, depending on Ca^{2+}). (Indo-1 is preferred in flow cytometry because its large emission shift permits $[Ca^{2+}]_i$ measurement with only 1 exciting laser beam.) Preliminary trials showed that manual alternation of excitation wavelengths worked in principle but was far too laborious and slow in practice. Therefore a fluorometer system was constructed to chop rapidly between two sets of excitation lamps and monochromators preset to the two desired wavelengths. This system, highly modified from a Spex Fluorolog 2, feeds either a conventional cuvette or an inverted fluorescence microscope (Fig. 2). Light emitted from the cuvette passes through a monochromator, whereas light from a microscope image is selected by pinholes of adjustable size and passes through an interference filter. Either signal can be photon counted and accumulated in the Spex computer, which keeps separate track of the signals from the two excitation wavelengths. We prefer to view both of the traces and subtract background fluorescences if significant before dividing the 340–350 nm record by the 380–385 nm, so that we are aware of any dye loss by bleaching or leakage or drift of the cell away from the field of view.

Fura-2 and indo-1 can generally be loaded into small mammalian cells by incubating the cells with the intracellularly hydrolyzable, membrane-permeant acetoxymethyl esters. In suspensions of cells in a cuvette, much lower loadings of fura-2 are needed to get usable signals than were required with quin2. Typically 10–20 μM intracellular fura-2 gives a fluorescence more than matching cell autofluorescence, whereas 200–400 μM quin2 would be needed for the same increment over background. Lower loadings mean less buffering of Ca^{2+}, so measured $[Ca^{2+}]_i$ should more closely resemble those of truly unperturbed cells. So far in platelets and thymocytes, 10–30 μM dye loadings show much the same behavior of $[Ca^{2+}]_i$ as 100–300 μM loadings do. Over this range, exogenous dye buffering seems to be minor in comparison to the cells' endogenous buffering and homeostatic mechanisms. Other cells with faster transients and less surface-to-volume ratio might have less tolerance for added buffering and really benefit from the brightness of the new dyes. Lower loadings also are easier to attain by either ester hydrolysis or microinjection, and they reduce the potentiality for toxic side effects.

2.2 Fluorescence Microscopy

The full advantages of fura-2 over previous Ca^{2+} indicator dyes are most applicable to the study of single cells by fluorescence microscopy. Both the brightness of fluorescence and the dual wavelength sensitivity prove invaluable in giving calibratable signals. Figure 3 shows a fluorescence micrograph of fura-2-loaded thymocytes in comparison with the same field viewed by Nomarski differential interference contrast. Almost all the cells are fluorescent, though there is significant variation in brightness that seems to be more than explicable by cell size. The fluorescence within each cell is mostly diffuse without consistent signs of localization. With such visual brightness one would expect to have plenty of photons for a photomultiplier to count. Indeed, even using a monochromator and < 5 nm

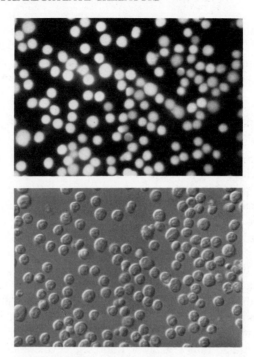

Figure 3. A field of mouse thymocytes photographed by epifluorescence (above) or Nomarski differential interference contrast optics (below). The cells were loaded with about 0.78 mM intracellular fura-2 by incubation with fura2/AM. The fluorescence photograph was taken with broad-band UV excitation and emission > 450 nm, using Ektachrome 400 film, a 20-s exposure, and ordinary processing. Note that nearly all the cells contain some dye and that the fluorescence is mostly diffuse. There is considerable variation in fluorescence from cell to cell, only part of which is explainable by nonuniform illumination.

excitation band pass, the light from a simple thymocyte gives $3 \times 10^4 - 10^5$ counts per second. The response of single thymocytes to mitogenic lectins is discussed briefly in Tsien et al. (26).

Sea urchin embryos are the preparation of single cells that we have most fully studied. This work, done mainly by Martin Poenie in collaboration with Janet Alderton and Richard Steinhardt of the Department of Zoology, University of California, Berkeley, is described in Poenie et al. (13). In brief, fura-1 is microinjected rather than loaded by acetoxymethyl esters because metabolically quiescent eggs from cold-water species of sea urchins do not hydrolyze the ester fast enough. A large $[Ca^{2+}]_i$ spike to 1.7 μM is observed as expected at fertilization. Though previous methods such as aequorin and arsenazo III had qualitatively detected this transient, its magnitude was heretofore poorly calibrated, and the eggs were generally too damaged to continue into the cell cycle; attempts with quin2 had failed because its weak fluorescence was drowned by changes in pyridine nucleotide autofluorescence. In one or two trials with a borrowed TV camera, the image excited at 380 nm, dimmed uniformly across the surface of the egg during fertilization, then

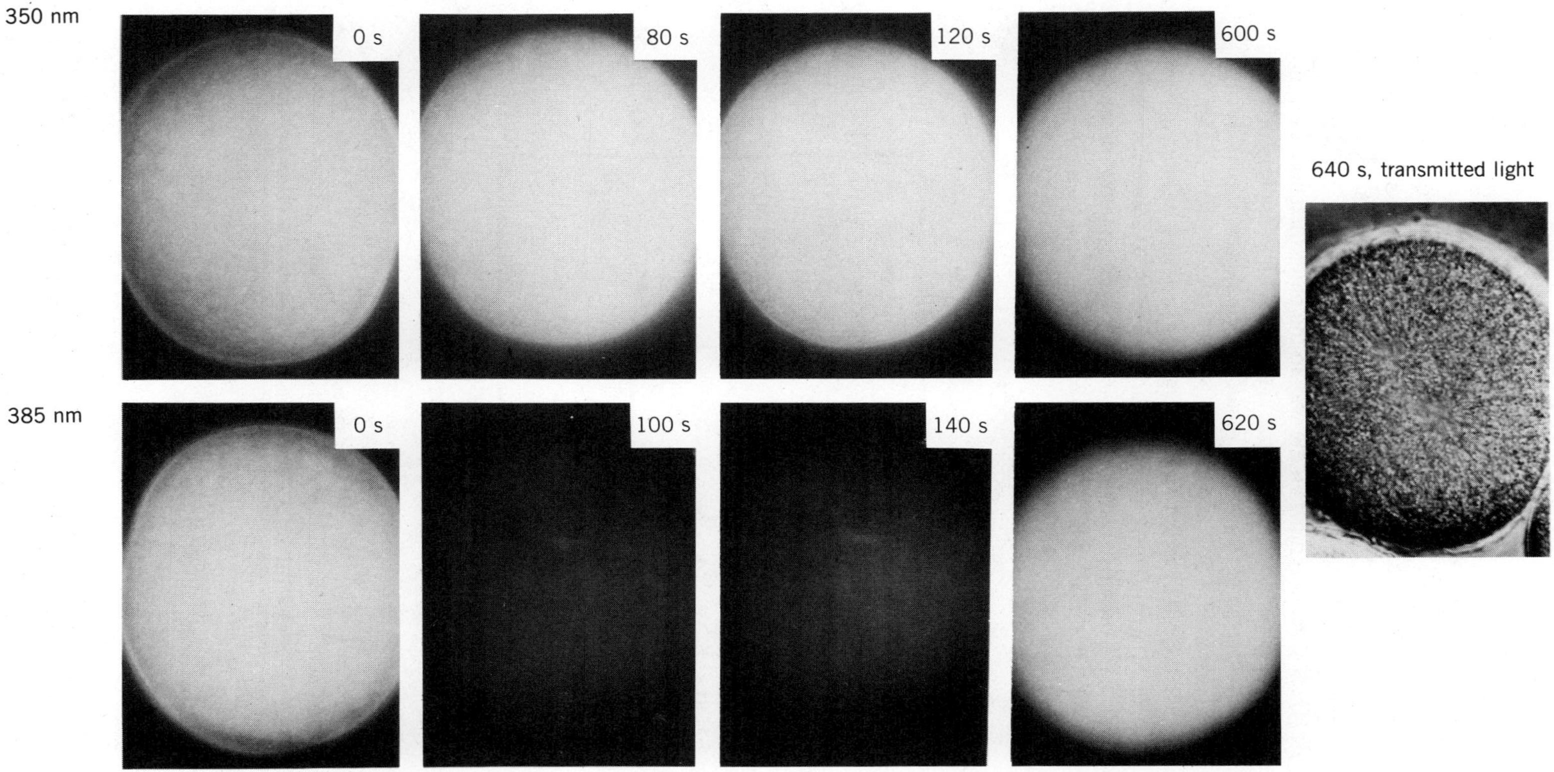

350 nm

385 nm

Figure 4. A montage of fluorescence micrographs of a single egg of the sea urchin *Lytechinus pictus*, previously microinjected with fura-2, undergoing fertilization. All eight fluorescence pictures were taken with identical 15-s exposures on Ilford XP1 film and subsequently processed identically. The upper four exposures used 350-nm fluorescence excitation, the lower four 385 nm with emission > 480 nm. Before fertilization (photos labeled 0 s), the 350-nm picture is slightly dimmer than the 385 nm. After fertilization (80, 100, 120, 140 s), the 350-nm images brighten slightly whereas the 385-nm images dim so much that they are nearly invisible. Eventually (600-620 s), the 385-nm brightness returns, though the ratio of 350:385 nm still remains somewhat greater (higher [Ca^{2+}]$_i$) than before fertilization. The images are now somewhat blurred because elevation of the fertilization membrane has lifted the cell away from the original plane of focus., A transmitted light view (764-s) shows the two pronuclei. Also, portions of two uninjected cells are visible; their absence from the fluorescence micrographs shows that autofluorescence is negligible.

337

gradually brightened again; there was no obvious sign of a traveling wave of $[Ca^{2+}]_i$. A similar view can be seen in still photographs (Fig. 4). In these improvised images, there was no way explicitly to calculate the ratio of 350-nm to 385-nm excitation. Proper spatial imaging of $[Ca^{2+}]_i$ gradients awaits acquisition of video image processing equipment.* Fura-2-injected eggs do progress into the cell cycle. In normal seawater, $[Ca^{2+}]_i$ never returns all the way to the prefertilization level, and small transient increases are often observed around the times of pronuclear migration and fusion, nuclear envelope breakdown, the metaphase-to-anaphase transition, and cytokinesis. Even if external Ca^{2+} is chelated by EGTA after fertilization, the $[Ca^{2+}]_i$ fluctuations in the cell cycle can persist (Fig. 5), showing that internal stores are sufficient to generate them. Parthenogenetic activation by ammonia or phorbol ester treatment, which are known to cause pH_i to rise, also cause an upward creep of $[Ca^{2+}]_i$ toward levels similar to normally fertilized eggs at corresponding stages.

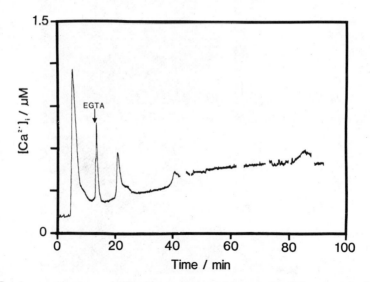

Figure 5. A sea urchin egg was fertilized in seawater containing only 1 mM Ca, one-tenth of normal. Subsequently, calcium-free seawater containing 3 mM EGTA, pH 8.0, was added as indicated in equal volume to reduce the free $[Ca^{2+}]$ to 10^{-7} M. Following the addition, the solution was exchanged with additional calcium-free seawater. Eggs that are fertilized and allowed to develop in 1 mM $[Ca^{2+}]$ seawater show a normal fertilization transient, the first large spike. The $[Ca^{2+}]_i$ peaks seen in normal development are little affected by the EGTA, provided the EGTA is added after the onset of the $[Ca^{2+}]_i$ rise associated with pronuclear movement. Here this transient peaked just after the arrow marking the EGTA addition. The third peak near 1300 s generally correlates with the streak stage of embryonic development. The step at 2500 s correlates with nuclear envelope breakdown; the bump at the end of the recording accompanies mitotic cleavage.

*We now have image-processing equipment that clearly shows a traveling wave of high $[Ca^{2+}]_i$ during fertilization of sea urchin eggs, as well as subsequent $[Ca^{2+}]_i$ gradients during development.

2.3 Flow Cytometry

Though fluorescence microscopy is uniquely suited to studying the time course of individual cells' responses, flow cytometry is obviously the ideal technique for statistically analyzing which cell populations do and do not respond. I have been fortunate to be able to collaborate with Drs. H. Alexander Wilson and Thomas M. Chused (Laboratory of Microbial Immunity, NIAID, National Institutes of Health), who are renowned for applying sophisticated multiwavelength flow cytometry to immunological questions. In preliminary experiments we have found the following: (a) Red cells do not become significantly fluorescent, so they need not be removed nor gated out from the lymphocyte preparations. (b) Lymphocytes show a wide range of brightness, much greater than can be explained just by cell size. However, in unmanipulated cells, the brightness at 405 nm is highly correlated with that at 485 nm (Fig. 6), so that the ratio of 405 to 485 nm, our measure of $[Ca^{2+}]_i$, forms quite a narrow distribution, as one might expect for a physiological variable subject to tight regulation. The superiority of ratio measurement over intensity at any one wavelength

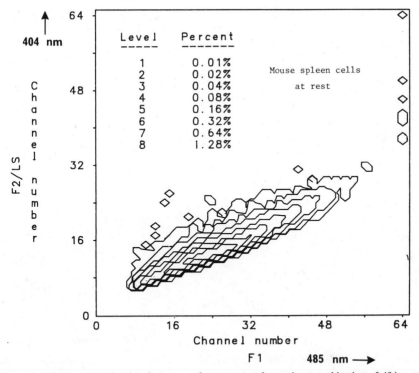

Figure 6. Contour plot showing frequency of occurrence of any given combination of 404-nm and 485-nm fluorescence from mouse spleen lymphocytes loaded with indo-1. Excitation was from a krypton laser at 356 nm; 50,000 cells were analyzed. Though there is considerable variability in the brightness at each wavelength, the two parameters are highly correlated, so that the ratio of 404 nm to 485 nm emission is narrowly distributed.

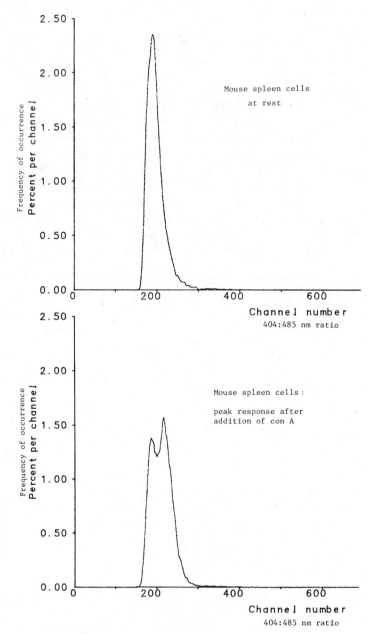

Figure 7. Frequency distribution of $[Ca^{2+}]_i$ values in mouse spleen cells before stimulation and after addition of the mitogenic lectin Con A, showing the heterogeneity of the response. Inceasing $[Ca^{2+}]_i$ corresponds to increasing channel number and ratio of 404-nm to 485-nm emission from indo-1.

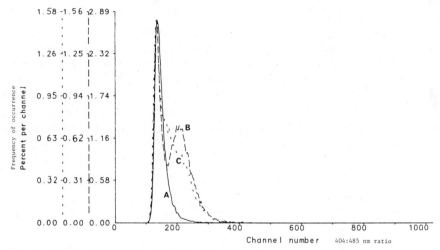

Figure 8. Frequency distribution of $[Ca^{2+}]_i$ values in mouse spleen cells before stimulation (trace *A*) and at two successive intervals (*B* and *C*) after addition of 40 μg/mL rabbit antimouse immunoglobulin. The time interval between successive traces was 1–2 min.

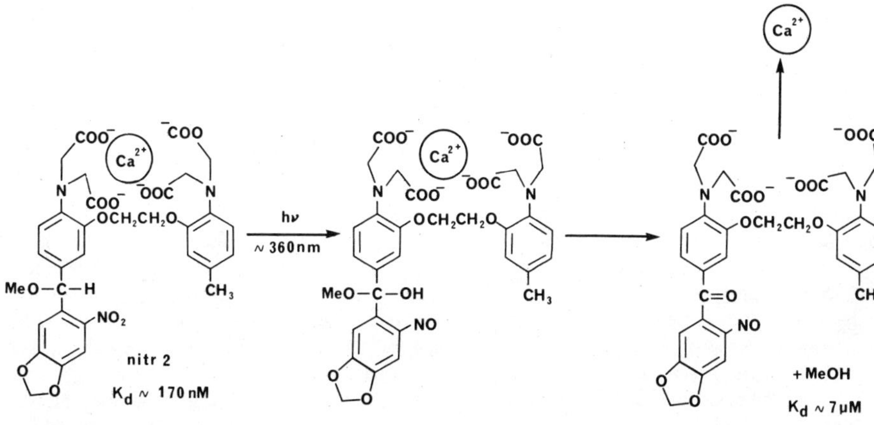

Figure 9. Mechanism by which nitr-2 changes Ca^{2+} affinity upon photolysis. Before photolysis, nitr-2 has a high affinity because the saturated carbon bridge on the left-hand side of the molecule does not transmit electron demand from the nitroaryl ring. Photolysis has the effect of inserting an oxygen from the nitro ring into the benzylic C–H bond, giving a hemiketal, which spontaneously eliminates MeOH to give a ketone. Now the chelating amino group is in direct conjugation with a strongly electron-withdrawing group, so that the Ca^{2+} affinity declines drastically, tending to release the Ca^{2+}.

341

is impressive. (c) Concanavalin A added to mouse spleen cells causes about half the cells to raise their $[Ca^{2+}]_i$, while about half seem unaffected (Fig. 7). This rise occurs by gradual growth of a new peak at a discretely higher 405:485-nm ratio, rather than an entire subpopulation gradually moving to higher and higher ratios. (d) Antiimmunoglobulin also raises $[Ca^{2+}]_i$ in about half the splenocytes (Fig. 8), though the rise occurs more quickly and decays sooner than with con A. (e) Con A also affects only a portion of thymocytes, though the distribution does not become as cleanly bimodal as with splenocytes. (f) Dual analysis of $[Ca^{2+}]_i$ and either plasma membrane potential (with potentiometric dyes) or lymphocyte surface markers (with

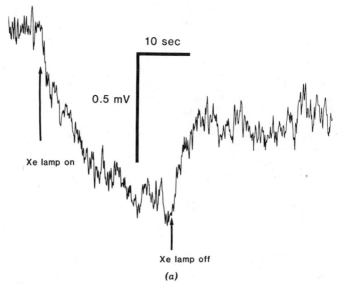

(a)

Figure 10a.   Illumination of nitr-2 can raise $[Ca^{2+}]_i$ inside neuron L2 of the abdominal ganglion of *Aplysia californica*. The neuron received a microinjection of nitr-2 solution amounting to about 2% of the cell's volume. This nitr-2 solution contained 400 m*M* nitr-2 tetraanion, 300 m*M* Ca^{2+}, with the ionic balance made up with K^+. The cell therefore received about 8 m*M* nitr-2 and 6 m*M* Ca^{2+}. It was also impaled with a separate microelectrode to record its membrane potential. At the time marked by the first arrow, a shutter was opened, permitting a 75-W xenon arc lamp to illuminate the cell through a quartz focusing lens. The membrane potential hyperpolarized as expected for a rise in $[Ca^{2+}]_i$ which would open a Ca^{2+}-activated potassium conductance in the membrane. This effect partly reversed when the light was blocked (second arrow). Development of this K^+ conductance is one of the best-understood hallmarks of elevating $[Ca^{2+}]_i$ in this cell type. A control experiment (not shown) verified that this cell was not sensitive to illumination before it was microinjected with nitr-2. There are at least four reasons why the hyperpolarizing effect was small: (1) Nitr-2 is still far from ideal in extinction coefficient and quantum efficiency. (2) The xenon lamp intensity could be increased. (3) The particular cell used contains a large amount of yellow pigment that absorbs UV light. Therefore only the small fraction of the cell volume most directly facing the lens was effectively irradiated. (4) The high ionic strength (300–500 m*M*) inside a marine invertebrate like *Aplysia* further weakens the binding of Ca^{2+} to unphotolyzed nitr-2. The lower the fraction of nitr-2 molecules bearing Ca^{2+} before photolysis, the fewer Ca^{2+} ions there are to be released by light and the less the quantum efficiency of isomerization of nitr-2 to the nitrosobenzophenone. Nitr-2 would be expected to be much more effective at lower ionic strengths.

fluorescent Ig reagents) is possible by argon excitation of indo-1 and krypton excitation of the second probe. In one such set of experiments, the restriction of Ca^{2+}-induced hyperpolarization to T cells was demonstrated.

2.4 "Caged Calcium"

The final group of new Ca^{2+} chelators uses light not to signal $[Ca^{2+}]_i$ but to release Ca^{2+}. These molecules are the first realizations of the much sought after "caged calcium." The structure and mode of operation of the best present version, nitr-2, are shown in Fig. 9. Before photolysis nitr-2 binds Ca^{2+} with an effective dissociation constant near 170 nM; afterward, the binding weakens to a K_d of 7 μM. My colleague, Dr. Robert Zucker of the Department of Physiology–Anatomy, University of California, Berkeley, has microinjected nitr-2 into *Aplysia* neurons and verified that illumination can now trigger membrane currents already known to be Ca^{2+} activated (Fig. 10a, b) However, nitr-2 could use considerable improvement in extinction coefficient, quantum efficiency of photolysis, preillumination affinity for Ca^{2+}, and speed of Ca^{2+} release (currently ∼0.2 s exponential time constant).

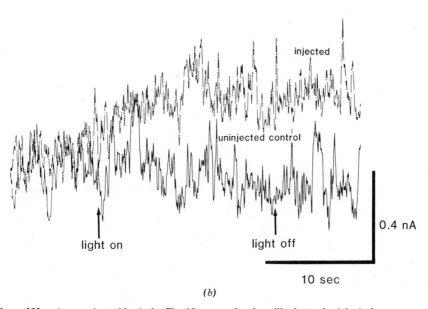

injected

uninjected control

light on   light off

0.4 nA

10 sec

(*b*)

Figure 10*b*.   An experiment identical to Fig. 10*a* except that the cell's electrophysiological response was measured under voltage clamp at -35 mV. Before the nitr-2 was introduced, illumination had no effect on the membrane current (trace marked uninjected control). After the nitr-2 was injected as in Fig. 10*a*, illumination caused an increased outward current, another manifestation of a Ca^{2+}-activated K^+ conductance. Further experiments (not shown) have shown that the outward current has a reversal potential between -65 and -75 mV and can be blocked by tetraethylammonium ions in the bath.

ACKNOWLEDGMENTS

This work was supported by NIH grants GM31004 and EY04372 and an award from the Searle Scholars Program.

REFERENCES

1. Alvarez-Leefmans, F. J., T. J. Rink, and R. Y. Tsien (1981) Free calcium in neurones of *Helix aspersa* measured with ion-selective microelectrodes. *J. Physiol. (Lond.)*, **315**, 531–548.
2. Blinks, J. R., W. G. Wier, P. Hess, and F. G. Prendergast (1982) Measurement of Ca^{2+} concentrations in living cells. *Prog. Biophys. Mol. Biol.*, **40**, 1–114.
3. Campbell, A. K. (1983) *Intracellular Calcium: Its Universal Role as Regulator*. Wiley, New York.
4. Cobbold, P. H., and P. K. Bourne (1984) Aequorin measurements of free calcium in single heart cells. *Nature (Lond.)*, **312**, 444–446.
5. DiVirgilio, F., P. D. Lew, and T. Pozzan (1984) Protein kinase C activation of physiological processes in human neutrophils at vanishingly small cytosolic Ca^{2+} levels. *Nature (Lond.)*, **310**, 691–693.
6. Feinstein, M. B., J. J. Egan, R. I. Sha'afi, and J. White (1983) The cytoplasmic concentration of free calcium in platelets is controlled by stimulators of cyclic AMP production. *Biochem. Biophys. Res. Commun.*, **113**, 598–604.
7. Grynkiewicz, G., M. Poenie, and R. Y. Tsien (1985) A new generation of Ca^{2+} indicators with greatly improved fluorescence properties. *J. Biol. Chem.*, **260**, 3440-3450.
8. Hallam, T. J., A. Sanchez, and T. J. Rink (1985) "Effect of excitatory and inhibitory icosanoids on cytoplasmic free calcium directly measured with the fluorescent indicator quin2," in P. Braquet, Ed., *Prostaglandins and Membrane Ion Transport,* Raven Press, New York, pp. 157–163.
9. Lagast, H., T. Pozzan, F. A. Waldvogel, and P. D. Lew (1984) Phorbol myristate acetate stimulates ATP-dependent calcium transport by the plasma membrane of neutrophils. *J. Clin. Invest.*, **73**, 878–883.
10. Lew, V. L., R. Y. Tsien, C. Miner, and R. M. Bookchin (1982) Physiological $[Ca^{2+}]_i$ level and pump-leak turnover in intact red cells measured using an incorporated Ca chelator. *Nature (Lond.)*, **298**, 478–481.
11. Marban, E., T. J. Rink, R. W. Tsien, and R. Y. Tsien (1980) Free calcium in heart muscle at rest and during contraction, measured with Ca^{2+}-sensitive microelectrodes. *Nature (Lond.)*, **286**, 845–850.
12. Meldolesi, J., W. B. Huttner, R. Y. Tsien, and T. Pozzan (1984) Free cytoplasmic Ca^{2+} and neurotransmitter release: Studies on PC12 cells and synaptosomes exposed to alpha-latrotoxin. *Proc. Natl. Acad. Sci. USA*, **81**, 620–624.
13. Poenie, M., J. Alderton, R. Y. Tsien, and R. Steinhardt (1985) Changes in free calcium levels with stages of the cell division cycle. *Nature (Lond.)*, **315**, 147-149.
14. Pozzan, T., P. Arslan, R. Y. Tsien, and T. J. Rink (1982) Anti-immunoglobulin, cytoplasmic free Ca^{2+}, and capping in B lymphocytes. *J. Cell Biol.*, **94**, 335–340.
15. Rink, T. J., and T. Pozzan (1985) *Cell Calcium*, **6**, 133–144.
16. Rink, T. J., A. Sanchez, S. Grinstein, and A. Rothstein (1983) Volume restoration in osmotically swollen lymphocytes does not involve changes in free Ca^{2+} concentration. *Biochim. Biophys. Acta*, **762**, 593–596.
17. Rink, T. J., A. Sanchez, and T. J. Hallam (1983) Diacylglycerol and phorbol ester stimulate secretion without raising cytoplasmic free calcium in human platelets. *Nature (Lond.)*, **305**, 317–319.

18.   Shoback, D., J. Thatcher, R. Leombruno, and E. Brown (1984) Relationship between parathyroid hormone secretion and cytosolic calcium concentration in dispersed bovine parathyroid cells. *Proc. Natl. Acad. Sci. USA*, **81**, 3113–3117.

19.   Smith, G. A., T. R. Hesketh, J. C. Metcalfe, J. Feeney, and P. G. Morris (1983) Intracellular calcium measurements by ^{19}F NMR of fluorine-labelled chelators. *Proc. Natl. Acad. Sci. USA*, **80**, 7178–7182.

20.   Thomas, M. V. (1982) *Techniques in Calcium Research*. Academic, London.

21.   Tsien, R. Y. (1983) Intracellular measurements of ion activities. *Annu. Rev. Biophys. Bioeng.*, **12**, 94–116.

22.   Tsien, R. Y., T. Pozzan, and T. J. Rink (1982) T-cell mitogens cause early changes in cytoplasmic free Ca^{2+} and membrane potential in lymphocytes. *Nature (Lond.)*, **295**, 68–71.

23.   Tsien, R. Y., T. Pozzan, and T. J. Rink (1982) Calcium homeostasis in intact lymphocytes: Cytoplasmic free Ca^{2+} monitored with a new, intracellularly trapped fluorescent indicator. *J. Cell Biol.*, **94**, 325–334.

24.   Tsien, R. Y., T. Pozzan, and T. J. Rink (1984) Measuring and manipulating cytosolic Ca^{2+} with trapped indicators. *Trends Biochem. Sci.*, **9**, 263–266.

25.   Tsien, R. Y., and T. J. Rink (1983) "Measurement of cytoplasmic free Ca^{2+}," in J. Barker and J. F. McKelvy, Eds., *Current Methods of Cellular Neurobiology*, Vol. III, Wiley, New York, pp. 249–312.

26.   Tsien, R. Y., T. J. Rink, and M. Poenie (1985) Measurement of cytosolic free Ca^{2+} in individual small cells using fluorescence microscopy with dual excitation wavelengths. *Cell Calcium*, **6**, 145–157.

27.   Vicentini, L. M., A. Ambrosini, F. DiVirgilio, T. Pozzan, and J. Meldolesi (1985) Muscarinic receptor-induced phosphoinositide hydrolysis at resting cytosolic Ca^{2+} concentration in PC12 cells. *J. Cell Biol.*, **100**, 1330–1333.

28.   Weiss, M. J., J. F. Daley, J. C. Hodgdon, and E. L. Reinherz (1984) Calcium dependency of antigen-specific (T3-Ti) and alternative (T11) pathways of human T-cell activation. *Proc. Natl. Acad. Sci. USA*, **81**, 6836–6840.

CHAPTER 20

THE USE OF TETRACARBOXYLATE FLUORESCENT INDICATORS IN THE MEASUREMENT AND CONTROL OF INTRACELLULAR FREE CALCIUM IONS

J. I. KORENBROT
D. L. OCHS
J. A. WILLIAMS
D. L. MILLER

Departments of Physiology, Biochemistry and Medicine
University of California Medical School
San Francisco, California

J. E. BROWN
Department of Ophthalmology
Washington University School of Medicine
St. Louis, Missouri

1.	INTRODUCTION	348
2.	LOADING THE INDICATOR DYE	350
3.	ASSESSMENT OF POSSIBLE TOXICITY CAUSED BY DYE LOADING	355
4.	ASSESSMENT OF PHYSIOLOGICAL EFFECTS OF DYE xLOADING	356
5.	PHOTOMETRIC STUDIES	359
	REFERENCES	362

1. INTRODUCTION

The ability to measure experimentally the cytoplasmic free calcium ion concentration in biological cells has provided evidence of the critical role this ion plays both as an internal messenger and as a regulator in cell physiology (reviewed by Rasmussen and Barret, 1984). Several methods, both optical and electrical, now exist to measure intracellular Ca^{2+} concentration (see reviews by Tsien, 1982; Blinks et al., 1982; Campbell, 1983; Thomas, 1984), and some of these methods are discussed elsewhere in this volume. Each method has advantages and limitations, and the selection of the most adequate method for a given experimental situation should be based on the sensitivity and reliability of the assay probe and the availability of appropriate instrumentation. It is our intention in this chapter to describe and discuss the experimental procedures, the advantages, and the perils of measurements of intracellular free Ca^{2+} that utilize the fluorescent Ca^{2+} indicators recently designed and synthesized by Tsien (1980, 1983).

The fluorescent Ca^{2+} indicators developed by Tsien are tetracarboxylic acids that bind Ca^{2+} with high affinity (usually with a K_d of about $10^{-7}M$) and high selectivity for Ca^{2+} over protons and other cations of biological importance. Typically, these dyes bind Ca^{2+} $10^3–10^4$ times better than Mg^{2+} and $10^6–10^8$ times better than monovalent cations such as Na^+ or K^+ (Tsien 1980, 1983). These indicator dyes

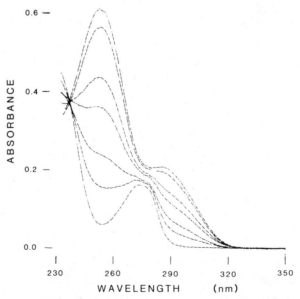

Figure 1. Absorption spectra of bapta in solutions of varying free Ca^{2+} concentrations. Solutions contained 40 μM bapta, 100 mM KCl, and 10 mM MOPS buffers, pH 7.0. The concentration of Ca^{2+} was varied by the use of appropriate mixtures of $CaCl_2$ and EGTA. The curve with highest absorbance at 250 nm was obtained in the presence of excess EGTA; the one with least absorbance at 250 nm was obtained at 1 mM free Ca^{2+}. From these curves, the K_d of Ca^{2+} binding to bapta was calculated to be 10^{-7} M.

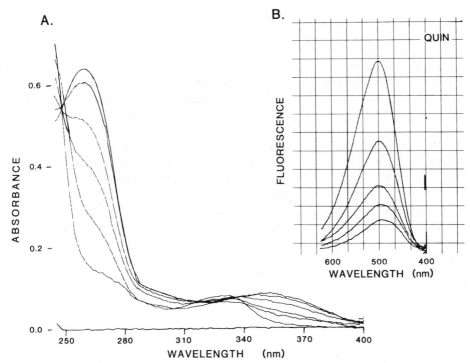

Figure 2. (*A*) Absorption spectra of a solution of quin in the presence of varying free Ca²⁺ concentrations. Solutions contained 18 μ*M* quin, 100 m*M* KCl, and 10 m*M* MOPS buffer, pH 7.0. The concentration of free Ca²⁺ was varied by the use of appropriate mixtures of CaCl₂ and EGTA. The curve with highest absorbance at 260 nm was obtained in the presence of excess EGTA; the curve with the least absorbance at 260 nm was obtained at 1 m*M* free Ca²⁺. From these curves the K_d of Ca²⁺ binding to quin was calculated to be 9×10^{-8}. (*B*) Uncorrected fluorescence emission spectra of the solutions whose absorbance spectra are shown in (*A*). Fluorescence excitation was at 339 ± 5 nm. The corrected wavelength of maximum emission is 517 nm.

have complex absorption spectra in the UV and visible ranges, and many of them also fluoresce. Of special significance is that these dyes change their optical properties upon binding Ca²⁺. Some of the dyes change their absorbance spectrum without changing fluorescence emission, for example, BAPTA (bis[*o*-amino-phenoxy]ethane-*N*, *N*, *N'*, *N'*-tetracetate), whereas others change both their absorbance spectrum and their fluorescence emission, for example, quin (2-[[2-[bis[(ethoxycarbonyl)methyl]amino]-5-methylphenoxy]methyl]-6-methoxy-8-[bis[(ethoxycarbonyl)methyl]amino]quinoline) (see Figs. 1, 2).

The tetracarboxylate dyes are useful in studies of cell physiology in part because of the relative ease with which they can be loaded into the cell's cytoplasm. This ease arises from a chemical trick designed by Tsien (1981). The carboxylic acid moieties in the dyes can be synthetically esterified with small hydrophobic groups to yield new hydrophobic compounds. An acetoxy–methoxy group is the most frequently used hydrophobic addition. Biological membranes are readily permeated by the tetracarboxylate dye esters because of their hydrophobicity. Therefore, an ester placed in the incubation medium can be introduced into the cytoplasm of cells. In the cell's interior, nonspecific esterases act on the ester bonds and catalyze the hydrolysis of the added small hydrophobic groups; thus the parent tetracarboxylate dye can be accumulated within the cytoplasm. This loading trick is entirely analogous to the one first used by Rotman and Papermaster (1966) and is used extensively to load the intracellular pH indicator dyes fluorescein and carboxyfluorescein (Thomas et al., 1979; Slavik, 1982).

The use of tetracarboxylate dyes in biological systems can be described as a procedure of four sequential steps: (1) loading of the dye; (2) assessment of possible cell toxicity caused by dye loading; (3) assessment of the physiological effects of dye loading and (4) photometric studies of changes in ion concentration. We will discuss here each of these procedural steps and exemplify them with experimental data from each of three different cell types in which transients in intracellular Ca^{2+} concentration are either known or presumed to be involved in cell function. The cells we have studied are pancreatic acinar cells, ventral eye photoreceptor cells from *Limulus polyphemus,* and rod photoreceptor cells from the toad retina.

Intracellular Ca^{2+} transients have long been suspected to play a role coupling chemical stimulation to enzyme secretion in pancreatic acinar cells. Specifically, binding of a chemical agonist to a membrane receptor is believed to lead to an increase in the level of cytosolic Ca^{2+}, which in turn initiates a process of vesicle exocytosis (Douglas, 1968; Williams, 1980; Rubin, 1982). In ventral eye photoreceptor cells of *Limulus polyphemus* it has already been demonstrated, through the use of optical probes (Brown and Blinks, 1974; Brown et al., 1977), that light stimulation leads to an increase in cytoplasmic Ca^{2+} that participates in the process of light adaptation (Lisman and Brown, 1975). Finally, in rod photoreceptors from the vertebrate retina, it has been hypothesized that an increase in cytoplasmic Ca^{2+} couples the excitation of the photopigment by light to the changes in the ionic permeability of the cell's plasma membrane. These permeability changes underlie the electrical response of the rods that is elicited by light (Hagins, 1972; reviewed in Korenbrot, 1984, 1985). Our discussion is limited to the use of two dyes currently at hand: BAPTA and quin; however, we hope it will prove of general use as other, improved dyes become available (Poeni et al., 1984).

2. LOADING THE INDICATOR DYE

The success of intracellular loading of the tetracarboxylate indicators can be assayed either photometrically or electrochemically. The conversion of the hydrophobic form of dye to the free acid form can be monitored photometrically (see Fig. 3) based on the finding that the λ_{max} of the fluorescence emission spectrum of the two forms differs. The electrochemical assay is based on the fact that the hydrolysis of the acetoxy–methoxy esters yields acetic acid and formaldehyde in addition to the free tetracarboxylic acid. Both formaldehyde and acetic acid rapidly cross biological membranes (Diamond and Wright, 1969), and therefore their rate of appearance can be monitored by measuring the extracellular pH (Tsien, 1981). The latter method is

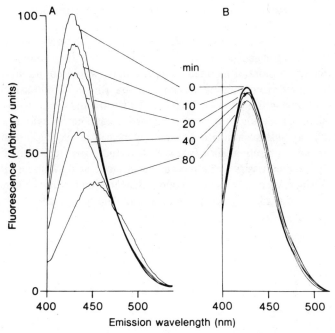

Figure 3. (*A*) Fluorescence emission spectra of a suspension of pancreatic acini measured at various times after the addition of 20 μ*M* AM-quin. Conditions of incubation are described in Ochs et al. (1983); times of incubation are indicated for each trace. Fluorescence was excited at 340 ± 5 nm. The time-dependent shifts in amplitude and wavelength of maximum emission reflect the conversion of AM-quin into quin. (*B*) Spectra measured in solutions identical to those used for cell incubation but in the absence of suspended acini.

probably less reliable than the former because extracellular deesterification, whether spontaneous or catalyzed by extracellular esterases (acetylcholine esterase, for example), cannot be distinguished from intracellular hydrolysis.

To achieve intracellular loading of the fluorescent dyes it is necessary to find the experimental conditions under which cytoplasmic nonspecific esterases remain sufficiently active to deesterify the hydrophobic form of the dyes in reasonably short times. Many types of cells from warm-blooded animals can be cultured in suspensions maintained at 37°C in media optimized for maximum cell viability. Such cultured cells have been commonly found to be competent to load quin when incubated with AM-quin. The concentration of AM-quin in the incubation medium is typically 20–50 μ*M*; these concentrations are above saturation of this hydrophobic compound in water. Tsien and collaborators (1984) have reviewed recent reports on many successful experiments of this type that have utilized AM-quin. For example, Fig. 3 shows fluorescence excitation spectra of suspensions of mouse pancreatic acini maintained at 37°C and incubated with AM-quin. The shift in both λ_{max} and emission intensity of the spectra indicate the intracellular accumulation of quin. These same cells, however, are much less effective in accumulating quin when incubated in

AM-quin at 25°C. Thus temperature, as is reasonable to expect, is an important parameter that must be controlled to optimize loading.

Perhaps other aspects of the metabolic well-being of the cells are relevant to successful dye loading in addition to incubation temperature. For example, we found that the rod photoreceptors of the toad (*Bufo marinus*) would not accumulate quin when incubated at room temperature in a conventional Ringer's solution (see Fig. 4); this Ringer's solution is similar to that first used by Brown and Pinto (1974) in which rods remain electrophysiologically competent for several hours (Baylor et al. 1979). We assayed the possible accumulation of quin by measuring the fluorescence of the rod outer segments in a retinal slice (see Fig. 4 legend). We studied the metabolic state of the toad retinal tissue by its production of lactic acid (Winkler, 1981) and found that the Ringer's solution is not an optimal incubation medium. Therefore, we tested the ability of toad photoreceptors to accumulate quin while incubating the

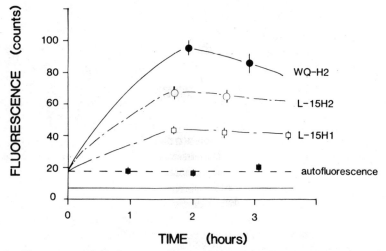

Figure 4.   Fluorescence emission from outer segments of toad rods incubated in various media containing AM-quin. Fluorescence emission was measured in a home-built microfluorometer. 300-μm-thick slices of isolated retinas were placed on their side in a nylon chamber on a microscope stage. The slices were viewed at high magnification with a high numerical aperture objective and the aid of an IR-sensitive TV camera. Adjustable slits placed in the image plane of the objective were used to restrict the measured light emission to that coming from the rod outer segments. Light passing through the adjustable slits was measured by a photomultiplier tube operated in a photon-counting mode. Light from a xenon arc lamp was passed through narrow band interference filters at 339 nm and shuttered to produce 1-s flashes that excited fluorescence. Isolated, thoroughly dark-adapted toad retinas were incubated in complete darkness at room temperature. The various media contained 50 μM AM-quin, were gently agitated, and were equilibrated with 95% O_2 plus 5% CO_2 or 100% O_2 depending on the bicarbonate content of the solution. At times indicated on the abscissa, retinas were withdrawn from the incubation medium, sliced, and prepared for fluorometry. Retinas incubated in a conventional Ringer's solution (Brown and Pinto, 1974) showed no more fluorescence than those measured in the absence of AM-quin (i.e., autofluorescencel) (■). Enriched media, such as Leibowitz-15 (Grand Island Biochemical), additionally buffered with either bicarbonate (L15-H1, □) or HEPES (L15-H2, ○) were effective in supporting the conversion of AM-quin to quin. The most effective medium was Wolf-Quimby medium (Grand Island Biochemical) enriched with bicarbonate (Besharse et al., 1980) (WQ-H2, ●).

retina in various culture media in which the metabolic state of the tissue was improved. As is shown in Fig. 4, enriched culture media do support a more efficient production of quin from AM-quin by rods. We found that the medium first described by Wolf and Quimby (1964), as modified by Besharse et al., (1980), sustains the most effective accumulation of quin. We also found that it was necessary to incubate the retinas with a vast excess of medium volume (about 10 mL/retina), and with continuous agitation.

It is also possible to fail completely to obtain cytoplasmic loading of a particular tetracarboxylate dye by incubating cells with the esterified form of that dye. We have been unable to produce either BAPTA or quin from AM-BAPTA or AM-quin respectively in halophilic bacteria (Korenbrot and Bogomolni, unpublished observation). Apparently loading has also failed in other bacteria, in some plant cells (Tsien, et al., 1984), and in cells of some invertebrates. For example, *Limulus* photoreceptors that were incubated for up to 12 h at room temperature in *Limulus* serum containing AM-quin failed to accumulate detectable intracellular quin. Interestingly, such a failure is not due to the lack of the required intracellular esterases because the photoreceptors did accumulate intracellular BAPTA in the presence of AM-BAPTA. After incubation in AM-BAPTA for 8 h in *Limulus* serum, the receptor potentials change in wave shape. The changes in wave shape illustrated in Fig. 5 are

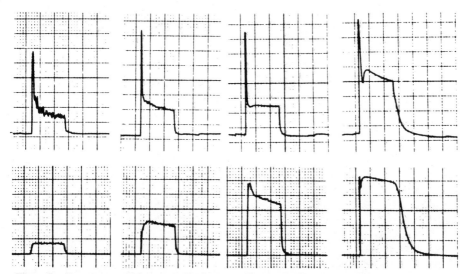

Figure 5. Receptor potentials measured in single *Limulus* ventral photoreceptors before and after loading with BAPTA. Electrophysiological methods are described in Lisman and Brown (1975). The top panels from left to right show signals measured from an untreated cell in response to 5-s flashes of white light of progressively increasing intensity. Light is doubled in intensity between panels. The receptor potentials elicited by the flashes have an early transient peak that declines to a plateau. In the bottom panels are shown responses elicited by the same light intensities from the same cell after it was incubated for approximately 8 h at room temperature in *Limulus* serum that contained 50 μM AM-BAPTA. The responses are more "square," and the plateau amplitudes elicited by the brighter stimuli are larger. The same effects can be obtained by intracellular injection of Ca^{2+} buffers such as EGTA or BAPTA.

characteristic of the effects of intracellular Ca^{2+} buffers in these cells (Lisman and Brown, 1975) and are evidence for the accumulation of BAPTA in the cytoplasm. These results suggest that the ester bonds in the esterified tetracarboxylate dyes may be inaccessible to the cytoplasmic esterases owing to steric hindrance and that such hindrance may vary from compound to compound. Thus, failure to accumulate a particular dye should not be taken to indicate certain failure of other dyes.

An essential requirement for the use of quin as a monitor of intracellular Ca^{2+} is that its cytoplasmic concentration remain constant throughout the measurements. That this experimental condition holds true must be tested by measuring the intracellular concentration of quin (Tsien et al., 1982) at various times after the desired load has first been achieved. In pancreatic acini, for example, the intracellular quin concentration remains stable only for approximately 50 min after the end of incubation in AM-quin. Moreover, careless handling of the cells can shorten this period considerably (Ochs et al., 1983, 1985). In extreme cases, cells can rapidly lose

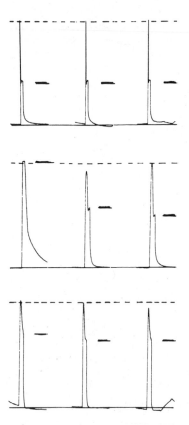

QUIN Pressure Injection

in Limulus

Figure 6. The effects of pressure injection of quin into *Limulus* ventral eye photoreceptors. Top row: Control responses. Middle row: Receptor potentials were recorded immediately after intracellular pressure injection of quin out of the recording micropipette. The peak to plateau transitions were lost immediately after injections, indicating that the intracellular Ca^{2+} was buffered by the injected quin. The effect was transient and the cell recovered within 5 min. Bottom row: After 5 min, another injection of quin caused the same transient changes in the receptor potentials.

quin, which is apparently the case in *Limulus* photoreceptors. Figure 6 illustrates the results of experiments in which quin was pressure injected into individual photoreceptor cells while their membrane voltage was being measured. The data show that the cell is loaded with quin immediately after the injection. The receptor potentials lack transients, as is characteristic of cells loaded with Ca^{2+} buffers (Lisman and Brown, 1975; also see Fig. 4). Surprisingly, the cell rapidly recovers, and the receptor potentials regain their transients, suggesting that the dye has disappeared from the cytoplasm. This cycle, as shown in Fig. 5, can be repeated several times. Parallel photometric measurements (not shown) indicate that the fluorescence of the cell increases during injection of quin and decreases after injection in parallel with the physiological changes. The fate of quin in these cells is undetermined, but the experimental data suggest that long-term stability of intracellular quin concentration should not be assumed and should be tested in each new experimental system.

3. ASSESSMENT OF POSSIBLE TOXICITY CAUSED BY DYE LOADING

The hydrolysis of the acetoxy–methoxy groups from the carboxylic groups in the fluorescent dyes yields stoichiometric amounts of free dye, acetic acid, and formaldehyde within the cytoplasm of the cell. When cells are loaded with AM-quin or AM-BAPTA, it is imperative to assess the possible toxic effects of the acid and aldehyde by-products. To date, the reported toxicity is surprisingly low (Tsien et al., 1984). In the vast majority of the successful experiments, cells were incubated with a large excess of medium volume over cell volume (typically 100:1 to 1000:1) and were continuously agitated. These experimental conditions probably promote a rapid diffusion of the potentially toxic by-products out of the cells and their effective dilution in the suspending medium. These incubation conditions should be observed whenever possible.

Cell toxicity should be assayed through studies of sensitive indicators of cell viability and cell integrity. One such study is the measurement of the leakage of soluble cytoplasmic markers out of the cells. For example, in acinar cells we have found that both intracellular K^+ and the enzyme lactic dehydrogenase are retained by cells loaded with quin or bapta to the same extent as unloaded cells for about 70 min after the loading begins. However, after this time, loaded cells lose these cytoplasmic markers at much higher rates than unloaded cells (Ochs et al., 1983, 1985). Similarly, rod photoreceptors remain able to generate electrical responses to light after loading with quin; however, the proportion of functioning cells in the population drops as the time in quin is prolonged. In general, cells loaded with dyes appear to remain viable for shorter times than normal cells.

To investigate cytotoxicity, there are two important control substances, AM-APDA(AM-*o*-aminophenoxymethane-*N*-diacetate) and methane–diol–diacetate (Hesketh et al., 1983). The first compound is hydrophobic and is hydrolyzed in the cytoplasm to yield acetic acid, formaldehyde, and a free dicarboxylic acid that does not bind Ca^{2+}. The second compound also enters the cell's cytoplasm and is hydrolyzed to yield only acetic acid and formaldehyde. These compounds thus provide excellent controls to ascertain the possible cytotoxic or physiological effects of the undesired by-products of hydrolysis of esterified tetracarboxylate dyes.

4. ASSESSMENT OF THE PHYSIOLOGICAL EFFECTS OF DYE LOADING

There are several optical properties of quin that conspire to limit its usefulness: (1) It has a low extinction coefficient at the optimum excitation wavelength, approximately $5 \times 10^3 \, \text{molar}^{-1}\text{cm}^{-1}$ at 339 nm; (2) it has a low quantum yield of fluorescence, about 0.03 for the Ca-free and 0.14 for the Ca^{2+}-bound forms (Tsien, 1980); and (3) the optimum excitation wavelength also excites autofluorescence from endogenous compounds such as reduced pyridine nucleotides. For these reasons, a high intracellular concentration of quin is necessary to obtain reliable fluorescence signals. This necessity seriously limits the usefulness of method in its present form. In typical experiments, quin is loaded to a cytoplasmic concentration between 0.5 and 5 mM and hence acts as an effective intracellular Ca^{2+} buffer. In such experiments the introduction of quin is very likely to alter and modify the very physiological process being measured.

Available evidence suggests that cytoplasmic quin in the range between 0.5 and 5 mM need not significantly alter the levels of resting free Ca^{2+}, as long as

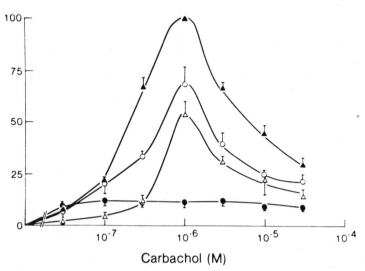

Figure 7. Effects of quin and BAPTA on the amylase release from pancreatic acini stimulated with carbachol choline. Quin and BAPTA alone or together were loaded by incubation with AM-quin and AM-BAPTA, respectively, as described (Ochs et al., 1983, 1984). Amylase release is plotted against carbachol concentration for untreated cells (), quin-loaded cells (○) BAPTA-loaded cells () or cells loaded with both quin and BAPTA (●). The amount of stimulated amylase release is normalized to the maximum release observed in response to 10^{-6} M carbachol choline in untreated cells.

extracellular Ca^{2+} is maintained near its normal value (Tsien et al. 1982; Hesketh, et al., 1983). That is, the homeostatic balance between active and passive Ca^{2+} transport processes that maintain free resting Ca^{2+} can apparently compensate for the presence of the buffer and make available enough Ca^{2+} to drive the buffer system to equilibrium at the normal resting Ca^{2+} concentration. However, the buffer can dampen the amplitude and alter the time course of Ca^{2+} transients that may occur in the cytoplasm. The quin fluorescent signal, therefore, is most likely to provide an underestimate of a true transient in Ca^{2+} concentration that might occur in unloaded cells. The seriousness of this complication can be illustrated with pancreatic acinar cells. Figure 7 shows the rate of amylase release by a suspension of mouse acini in response to various concentrations of carbachol choline. The rate of enzyme release is reduced in the presence of quin, and higher doses of carbachol choline are necessary to achieve a given fraction of the total release; that is, the sensitivity of the release process is decreased. Figure 7 also shows that BAPTA (loaded by incubating acini in AM-BAPTA) is even more effective than quin in reducing the secretory response and also shows that enzyme release can be totally prevented when quin and BAPTA are loaded simultaneously. The transient change in quin fluorescence induced by a fixed dose of carbachol choline is smaller in cells loaded with both quin and BAPTA than in cells loaded with quin alone (Ochs et al., 1985). This finding indicates that the reduction in enzyme release in the presence of these dyes is due to buffering of intracellular Ca^{2+}.

The usefulness of quin as an indicator dye is limited by its effectiveness as an intracellular Ca^{2+} buffer. However, the use of quin as a Ca^{2+} buffer has permitted us to explore the role of Ca^{2+} in phototransduction in vertebrate rods. Figure 8 illustrates the effects of intracellular quin on the ionic currents of the outer segments of toad rods. In the dark, there is a steady inward positive current across the outer segment membrane. Flash illumination transiently suppresses this current to an extent that depends on light intensity (Baylor et al., 1979). The time course of the current suppression can be modeled as the impulse response of a sequence of n first-order linear filters, each of time constant τ. As Fig. 8 shows, intracellular quin changes the kinetics of the photocurrent. However, the maximum peak amplitude for quin-loaded cells (range 12–17 pA) is approximately equal to that for untreated cells (range 14–17 pA). The photocurrents are slower in the presence of quin than in its absence. The altered wave shape can be described by the same model that describes the normal photocurrents by adjusting the value of τ appropriately. The dependence of the peak photocurrent on light intensity is also illustrated in Fig. 8 for both untreated and quin loaded cells. The intracellular Ca^{2+} buffer does not change the photosensitivity of the cells. This experimental fact is unexpected if changes in Ca ion concentration act as a simple message to couple rhodopsin photoexcitation to the control of the dark-current. If such simple mechanism operated in photoreceptors, then the dependence of photocurrent amplitude on light intensity should be shifted to higher intensities in quin loaded cells when compared with untreated photoreceptors. This prediction is analogous to the effects of quin on the dose-response curve in acinar cells (see Fig. 7). Rod outer segments transiently release Ca^{2+} into the extracellular medium in response to flash illumination (Gold and Korenbrot, 1980, 1981;

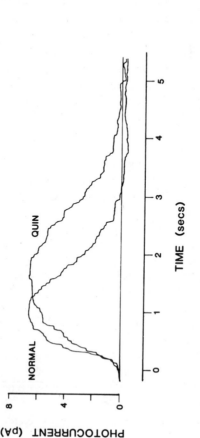

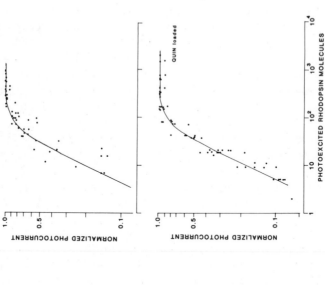

Figure 8. Shown on the left are membrane photocurrents measured from the outer segment of toad rods either from untreated cells (normal) or after incubation of the retina in AM-quin under the conditions described in the text. Photocurrents were measured with suction electrodes as described by Baylor et al. (1979) (Miller and Korenbrot, 1984). The responses shown were elicited by 25-ms flashes of 500-nm light that bleached 28 rhodopsin molecules in the normal cells and 30 molecules in the quin-loaded cells. The time course of the photoresponse is slower in the presence of quin; however, for dim flashes (less than 10 rhodopsin bleached per cell) this time course in both normal and treated cells can be fit by the same function $I = Ek(\frac{1}{\tau})^{n-1}(e^{-t/\tau})$ by keeping n constant (at a value of 4) and varying τ. In this expression I is photocurrent, E is light energy, and n and τ are adjustable parameters. Although the time to reach peak is slower in the presence of quin, the light sensitivity of the peak amplitude of the photocurrent is not affected by the intracellular buffer. Illustrated on the right are the normalized amplitudes of the peak of the photoresponses elicited by varying light intensity in normal and quin-loaded cells. Light intensity is expressed as the number of rhodopsin molecules bleached by a 25-ms flash of 500-nm light. Each symbol represents a different cell.

Yoshikami et al., 1980; Miller and Korenbrot, 1984). The time course and light sensitivity of this Ca^{2+} release are similar to those of the photocurrent. It has been argued, without direct proof, that (1) light induces a transient increase in intracellular free Ca^{2+}, and (2) there is a Na^+/Ca^{2+} exchanger in the plasmalemma that acts to restore the intracellular free Ca^{2+} to its resting or dark value by releasing Ca^{2+} extracellularly. The action of quin as an intracellular buffer provides a tool to test this hypothesis. For a given nonsaturating stimulus intensity, the amount of Ca^{2+} released extracellularly is greater in the presence of quin than in its absence (see Fig. 9). These data suggest that the light-dependent extracellular release of Ca^{2+} reflects changes in intracellular free Ca^{2+}. However, the data suggest that light-induced transient increases in intracellular Ca^{2+} in the rod have neither a stereotyped time course nor a fixed stoichiometry (relating the amplitude of the intracellular Ca^{2+} transient to the light intensity).

5. PHOTOMETRIC STUDIES

The use of quin as a fluorescent indicator of changes in cytoplasmic Ca^{2+} concentration is relatively simple when large populations of small cells are studied. The dye has been used most extensively in this type of experiment. In typical

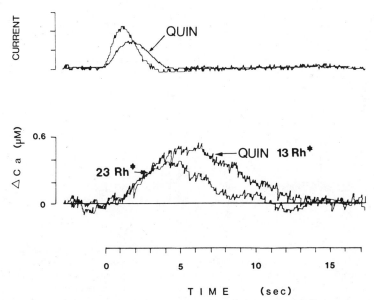

Figure 9. Simultaneous measurement of photocurrent and light-dependent Ca^{2+} release from the same toad rod outer segment (Miller and Korenbrot, 1984). These data were collected in the presence of 0.1 mM extracellular Ca concentration. Note that the time course of the photocurrent is slower after quin was loaded (traces indicated by arrows). Comparable total Ca^{2+} release requires much less intense stimulation of quin-loaded cells than untreated cells. That is, cells loaded with quin release more Ca^{2+} per rhodopsin molecule bleached (denoted Rh*) than untreated cells.

experiments, the fluorescence of several milliliters of a cell suspension (at a density of 10^6–10^8 cells/mL) can readily be measured in simple commercial instruments (review in Tsien et al., 1984). For example, Fig. 10 illustrates recordings of fluorescent transients in response to chemical stimulation of mouse pancreatic acini loaded with quin. The recordings were made in a commercial fluorometer; the cells were stirred gently during the measurement and were maintained at 37°C.

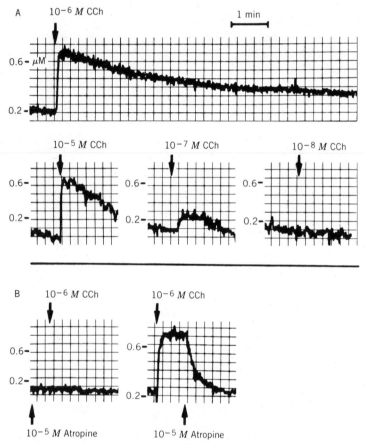

Figure 10.   Effects of chemical stimulation on the fluorescence of suspensions of pancreatic acini loaded with quin. (A) The panels show the change in fluorescence evoked by addition of increasing concentrations of carbachol choline. Each record was obtained from a different aliquot of loaded acini. Note that both the rate and amplitude of the fluorescence transient depend on the carbachol concentration. The fluorescence changes are transient, even though the rate of amylase release is steady for tens of minutes (Ochs et al., 1985). (B) Atropine, a blocker for the carbachol–choline receptor, prevents the change in fluorescence if added before the carbachol and reduces the fluorescence if added after the carbachol.

The change in fluorescence can be calibrated in units of Ca^{2+} concentration by either of two methods. The first method (Tsien et al., 1982) depends on the disruption of the cells with a detergent to release intracellular quin; that quin is titrated with excess Ca^{2+} and excess EGTA to determine the maximum and minimum possible fluorescence. Knowing these values, the K_d for Ca^{2+}, and that the binding of Ca^{2+} to quin involves a single site with 1:1 stoichiometry, it is possible to calibrate the fluorescence in units of Ca^{2+} concentration. In the second method (Heskett et al., 1983), the fluorescence end points are titrated in cells made permeable to divalent cations by ionomycin. In this second method, the quin fluorescence is taken to its end points with excess Ca^{2+} (for the maximum) and excess Mn^{2+} (for the minimum). Mn^{2+} is known to bind to quin with higher affinity than Ca^{2+} and completely quenches the quin fluorescence. In comparing the two methods it must be kept in mind that excess EGTA, unlike excess Mn^{2+}, does not quench quin fluorescence but reduces it to its value for the unbound state of quin (Rink et al., 1983).

Data such as those in Fig. 10 are collected in the first few seconds after cell stimulation. The cells must be maintained in suspension to allow measurements of changes in intracellular Ca^{2+} over tens of minutes. Continuous stirring of the fluorometric cuvette has been found to be deleterious to the cells (Ochs et al., 1985). As an alternative, Knight and Kestevin (1983) have maintained cells in suspensions in media that contain Dextran or Ficoll; these media have high density and low osmotic pressure. Also, we have studied separate aliquots removed at successive time intervals from a large stirred vessel that contained stimulated cells (Ochs et al., 1985).

Measurements made with large populations of cells (as discussed above) have serious limitations. Several assumptions must be made to interpret the photometric data. These assumptions cannot be rigorously verified. Among these assumptions are (1) that all cells present in the suspension are functionally intact and that their function is homogenous; (2) that the subcellular distribution of quin is homogeneous and limited to the cytoplasmic space. This is particularly troublesome because even if quin does exist only in the cytoplasm, Ca^{2+} transients themselves may not be, and probably are not, spatially homogeneous (Harary and Brown, 1984); and (3) that the K_d of quin in the cytoplasm is the same as that determined in vitro and remains unaltered during the physiological response. In addition, it must be assumed that all fluorescence changes that are measured arise from changes in quin fluorescence and not from other sources. This last assumption can be examined by making multiwavelength measurements and determining difference spectra.

The limitations in the use of quin might be reduced if measurements on individual cells or small cell assemblies were possible. Such measurements would require high photometric sensitivity with preservation of spatial resolution. For example, Fay and Tucker (1984) have reported quin fluorescence measurements in single cells with the use of a SIT camera and digital averaging techniques. In the future, the development of other fluorescent probes with higher absorbance and greater quantum yield of fluorescent emission (Poeni et al., 1984) should improve the information that can be collected with this type of probe. In the meantime, much can be learned with the present generation of probes if their limitations and advantages are recognized.

REFERENCES

Baylor, D. A., T. D. Lamb, and K. W. Yau (1979) The membrane current of single rod outer segments. *J. Physiol. (Lond.)*, **288**, 589–611.

Besharse, C., R. O. Terrill, and D. A. Dunis (1980) Light-evoked disc shedding by rod photoreceptors in vitro: Relationship to medium bicarbonate concentration. *Invest. Ophthalmol. Vis. Sci.*, **19**, 1512–1517.

Blinks, J. R., W. G. Wier, P. Hess, and F. G. Prendergast (1982) Measurement of Ca concentration in living cells. *Prog. Biophys. Mol. Biol.*, **40**, 1–114.

Brown, J. E., and J. R. Blinks (1974) Changes in intracellular free calcium concentration during illumination of invertebrate photoreceptors: detection with aequorin. *J. Gen. Physiol.*, **64**, 643–665.

Brown, J. E., and L. H. Pinto (1974) Ionic mechanism for the photoreceptor potential of the retina of *Bufo marinus*. *J. Physiol. (Lond.)*, **267**, 575–591.

Brown, J. E., P. K. Brown, and L. H. Pinto (1977) Detection of light-induced changes of intracellular ionized calcium concentration in *Limulus* ventral photoreceptor using arsenazo III. *J. Physiol. (Lond.)*, **236**, 299–320.

Campbell, A. K. (1983) *Intracellular Calcium.* Wiley, New York.

Diamond, J. A., and E. M. Wright (1969) Biological membranes: The physical basis of ion and non-electrolyte selectivity. *Annu. Rev. Physiol.*, **31**, 581–646.

Douglas, W. W. (1968) Stimulus-secretion coupling: The concept and clues from chromaffin and other cells. *Br. J. Pharmacol.*, **34**, 451–474.

Fay, F. S., and R. W. Tucker (1984) Measurement of free Ca distribution during mitogenesis using quin 2 in single BALB/c 3T3 cells. *J. Gen. Physiol.*, **84**, 15a.

Gold, G. H., and J. I. Korenbrot (1980) Light-induced Ca release by intact retinal rods. *Proc. Natl. Acad. Sci. USA*, **77**, 5557–5561.

Gold, G. H., and J. I. Korenbrot (1981) The regulation of calcium in the intact retinal rod: A study of light-induced calcium release by the outer segment. *Curr. Top. Membr. Transp.*, **15**, 307–328.

Hagins, W. A. (1972) The visual process: Excitatory mechanisms in the primary receptor cells. *Annu. Rev. Biophys. Bioengin.*, **1**, 131–158.

Harary, H. H., and J. E. Brown (1984) Spatially nonuniform changes in intracellular calcium ion concentrations. *Science*, **224**, 292–294.

Hesketh, T. R., G. S. Smith, J. P. Moore, M. V. Taylor, and J. C. Metcalfe (1983) Free cytoplasmic Ca concentration and the mitogenic stimulation of lymphocytes. *J. Biol. Chem.*, **258**, 4876–4882.

Knight, D. E., and N. T. Kestevin (1983) Evoked transient intracellular free Ca changes and secretion in isolated bovine adrenal medullary cells. *Proc. R. Soc. Lond.*, **B218**, 177–199.

Korenbrot, J. I. (1984) Role of intracellular messengers in signal transduction in retinal rods. *Prog. Retinal Res.*, **4**, 25–53.

Korenbrot, J. I. (1985) Signal mechanisms of phototransduction in retinal rods. *CRC Crit. Rev. Biochem.*, **17**, 223–256.

Lisman, J. E., and J. E. Brown (1975) Effects of intracellular injection of calcium buffers on light-adaptation in *Limulus* ventral photoreceptors. *J. Gen. Physiol.*, **66**, 489–506.

Miller, D. L., and J. I. Korenbrot (1984) Simultaneous current and calcium release measurement from a single rod photoreceptor. *Biophys. J.*, **45**, 341a.

Ochs, D. L., J. I. Korenbrot, and J. A. Williams (1983) Intracellular free calcium in isolated pancreatic acini: Effects of secretagogues. *Biochem. Biophys. Res. Commun.*, **117**, 122–128.

Ochs, D. L., J. I. Korenbrot, and J. A. Williams (1985) Relation between agonist induced changes in the concentration of free intracellular Ca and secretion of amylase by pancreatic acini. *Am. J. Physiol.*, **249**: G389–G398.

Poenie, M., J. Alderton, A. Steinhardt, and R. Y. Tsien (1984) Measurement of calcium during fertilization of sea urchin eggs using a new fluorescent indicator. *J. Gen. Physiol.*, **84**, 14a.

Rasmussen, H., and P. Q. Barrett (1984) Calcium messenger system: An integrated view. *Physiol. Rev.*, **64**, 938–984.

Rink, T. J., A. Sanchez, S. Grinstein, and A. Rothstein (1983) Volume regulation in lymphocytes does not involve a change in free cytoplasmic Ca. *Biochim. Biophys. Acta*, **762**, 593–596.

Rotman, B., and B. W. Papermaster (1966) Membrane properties of living mammalian cells as studied by enzymatic hydrolysis of fluorogenic esters. *Proc. Natl. Acad. Sci. USA*, **55**, 134–141.

Rubin, R. P. (1982) *Calcium and Cellular Secretion*. Plenum, New York.

Slavik, J. (1982) Intracellular pH of yeast cells measured with fluorescent probes. *FEBS Lett.*, **140**, 22–26.

Thomas, A. V. (1983) *Techniques in Calcium Research*. Academic, New York.

Thomas, J. A., R. N. Buchsbaum, A. Zimniak, and E. Racker (1979) Intracellular pH measurement in Erlich ascites tumor cells utilizing spectroscopic probes generated in situ. *Biochemistry*, **18**, 3315–3330.

Tsien, R. Y. (1980) New calcium indicators and buffers with high selectivity against magnesium and protons: Design synthesis and properties of prototype structures. *Biochemistry*, **19**, 2396–2404.

Tsien, R. Y. (1981) A non-disruptive technique for loading Ca buffers and indicators into cells. *Nature*, **290**, 527–528.

Tsien, R. Y. (1983) Intracellular measurement of ion activities. *Annu. Rev. Biophys. Bioeng.*, **12**, 91–116.

Tsien R. Y., and T. J. Rink, In J. Barker and J. McKelvy Eds., (1983) *Current Methods in Cellular Neurobiology*, Academic, New York, Vol. 4.

Tsien, R. Y., T. Pozzan, and T. J. Rink (1982) Calcium homeostasis in intact lymphocytes: Cytoplasmic free calcium monitored with a new, intracellularly trapped fluorescent indicator. *J. Cell Biol.*, **94**, 325–334.

Tsien, R. Y., T. Pozzan, and T. J. Rink (1984) Measuring and manipulating cytosolic Ca with trapped indicators. *Trends Biochem. Sci.*, **9**, 263–266.

Williams, J. A. (1980) Regulation of pancreatic acinar cell function of intracellular calcium. *Am. J. Physiol.*, **238**, 6269–6279.

Winkler, B. S. (1981) Glycolytic and oxydative metabolism in relation to retinal function. *J. Gen. Physiol.*, **77**, 667–692.

Wolf, K., and M. C. Quimby (1964) Amphibian cell culture: Permanent cell line from bullfrog (*Rana catesbiana*). *Science*, **144**, 1578–1580.

Yoshikami, S., J. S. George, and W. A. Hagins (1980) Light-induced calcium fluxes from outer segment layer of vertebrate retinas. *Nature*, **286**, 395–398.

PART 4

PHOTOBLEACHING AND PHOTOACTIVATION

CHAPTER 21

MEMBRANE DYNAMICS STUDIED BY FLUORESCENCE CORRELATION SPECTROSCOPY AND PHOTOBLEACHING RECOVERY

ELLIOT L. ELSON

Department of Biological Chemistry
Division of Biology and Biomedical Sciences
Washington University School of Medicine
St. Louis, Missouri

1. INTRODUCTION
   1.1 INFORMATION ABOUT MOLECULAR DYNAMICS IN
       CELLS PROVIDED BY FCS AND FPR   368
   1.2 SIMILAR CONCEPTUAL BASES OF FCS AND FPR   369
2. THEORY
   2.1 GENERAL CONSIDERATIONS   369
   2.2 TIME COURSE OF CONCENTRATION CHANGES IN
       TERMS OF PHENOMENOLOGICAL COEFFICIENTS   370
   2.3 DESCRIPTION OF FLUORESCENCE MEASUREMENTS
       IN TERMS OF CONCENTRATION CHANGES   372
3. EXPERIMENTAL IMPLEMENTATION   373
4. COMPARISON OF FCS AND FPR   374
5. ANALYSIS OF EXAMPLES
   5.1 SIMPLE REACTION SYSTEM UNDER SYSTEMATIC
       TRANSPORT   375

5.2   REVERSIBLE BIMOLECULAR REACTION   377

5.3   MULTIPLE BINDING REACTIONS   378

6. APPLICATIONS   379

REFERENCES   380

1.   INTRODUCTION

1.1   Information about Molecular Dynamics in Cells Provided by FCS and FPR

Fluorescence correlation spectroscopy (FCS) and fluorescence photobleaching recovery (FPR) are methods for measuring rates of molecular transport and the kinetics of chemical reactions coupled to transport in open systems. The transport could occur by diffusion or by systematic processes such as a convective flow or a drift driven by a force field. The two methods are closely related in their theoretical analysis and their experimental implementation. Both derive molecular rate coefficients from observations of the time course of changes in fluorescence excited in small open regions of the sample system. An important distinction between FCS and FPR, however, is that the former is used on systems in equilibrium. The observed changes in fluorescence result from the *microscopic* spontaneous fluctuations in the concentrations of the fluorescent molecules in the reaction volume that occur even in a system in equilibrium (1, 2). FPR measures the time course of fluorescence changes that occur during the relaxation back to equilibrium of an initial *macroscopic* concentration gradient which had been established by photobleaching a fraction of the fluorophores in a small subregion of the sample (3–7). Because of differences in the measurement of microscopic and macroscopic processes, the two methods differ in their range of application and in the information they are best suited to provide. Although applicable to a wider range of systems, both FCS and FPR can be used to study molecular dynamics in and on biological cells. This is because these methods can probe small subregions of cells and because they subject the sample to minimal perturbation. Therefore, they can provide information about cell structure, about the interactions of cellular components, and about the rates and mechanisms of physiologically important processes. For example, as will be discussed below, the lateral diffusion of cell surface proteins appears to be retarded by interactions with other cellular components that may reside in the cytoplasm, on the cell surface, or in the extracellular matrix. Hence measurements of the rates of diffusion of cell surface proteins and of the effects of structural perturbations on these rates can characterize the interactions of the surface proteins with each other and with other cellular components. Cellular responses to external ligands such as polypeptide hormones frequently require a limited aggregation of the initially randomly dispersed receptors.

Then a larger-scale aggregation and endocytosis of the ligand-receptor complexes typically follows (8). Similarly, intermolecular interactions among membrane-bound mechanisms of these processes, which could in principle be limited by the rates of intermolecular encounters and therefore by the rates of lateral diffusion on the membrane, can be investigated by FCS and FPR.

1.2   Similar Conceptual Bases of FCS and FPR

FCS and FPR both require the same kind of experimental data to determine phenomenological chemical rate constants and transport (e.g., diffusion) coefficients. Both methods center on measurements of the number of molecules of specified types in a defined open observation volume as a function of time. In an open chemical reaction system the reactant molecules can appear or disappear either by being generated or destroyed by chemical reaction or by being transported (e.g., by diffusion) into or out of the volume. Hence the rates of change of concentrations are determined by the chemical reaction rates and by the rates of transport (and the size of the observation region). Thus by measuring rates of concentration change in the observation region one can obtain both chemical rate constants and transport coefficients such as diffusion coefficients and flow or drift velocities. Although the coupling of chemical reaction and transport in an open system means that FCS and FPR can provide information simultaneously about both kinds of processes, the price of this advantage is an increase in the complexity of the theory needed to interpret the measurements. In the absence of reactions it is relatively simple to obtain transport coefficients. Chemical kinetics uncomplicated by transport can in principle be obtained from similar measurements on small *closed* systems. Photochemical interference and other problems, however, typically make this kind of measurement impractical.

Molecular concentrations are measured by fluorescence. Typically a focused laser beam traverses the sample; the volume (in 2-dimensional systems, the area) illuminated by the narrow beam constitutes the observation region. Molecules in this region that have perceptible fluorescence quantum yield and which are excited by the laser light emit fluorescence the intensity of which is a measure of their number in the region. Fluorescence is especially useful because of the sensitivity with which it can be detected, its chemical and spectroscopic selectivity, and its susceptibility to rapid and easy measurement with minimal perturbation of the experimental system, under physiological conditions if required. Another advantage is the high spatial resolution that can be obtained using fluorescence microscopy which permits measurements in small (~ 1 μm) subregions of biological cells.

2.   THEORY

2.1   General Considerations

For simplicity we shall suppose that the system to be observed is confined to a planar surface (the $z = 0$ plane), which might, for example, be the plasma membrane of a well-spread cell. No difference in principle but some increase in complexity is involved in extending these results to curved surfaces or three-dimensional volumes. Let us suppose that we have a chemical reaction system containing n components with the concentration of the jth component at time t and position $\mathbf{r}$ [$= r(x,y)$] denoted $c_j(\mathbf{r},t)$. The sample is traversed by a laser beam with intensity profile $I(\mathbf{r})$ in the x,y plane. Hence the fluorescence measured at time t, $F(t)$, will be given by

$$F(t) = \int I(\mathbf{r}) \sum {}_j Q_j \, c_j \, (\mathbf{r},t) \, d^2r$$

where Q_j accounts for the instrumental constants as well as the absorption coefficient and fluorescence quantum yield of the jth component and $d^2r = dx \, dy$.

The task of the theory is to relate the experimentally observed changes of the fluorescence, and hence of the concentrations of the fluorescent components, to the phenomenological diffusion coefficients and chemical rate constants and drift or flow velocities. This analysis must begin with a consideration of the time course of concentration changes.

2.2 Time Course of Concentration Changes in Terms of Phenomenological Coefficients

At the outset we shall suppose that concentrations in the observation region can change due to chemical reaction, diffusion, or systematic transport such as flow or drift. Hence

$$\frac{\partial C_j(\mathbf{r},t)}{\partial t} = D_j \nabla^2 C_j(\mathbf{r},t) - \nabla \cdot [C_j(\mathbf{r},t)\mathbf{V}_j] + \sum {}_l T_{jl} C_l(\mathbf{r},t) \tag{1}$$

where the terms on the right-hand side of the equation yield the rate of change of the concentration of the jth component due, respectively, to diffusion with diffusion coefficient D_j; to flow or drift with velocity $\mathbf{V}_j$ (in an electric field, e.g., each component with a different electrophoretic mobility could have a different drift velocity); and to chemical reaction. In the following we shall suppose that each $\mathbf{V}_j$ is constant and oriented in the x direction as would be true for a uniform electric or gravitational field along x or if the sample were translated with uniform velocity in the x direction relative to the laser beam. Then $\nabla \cdot (c_j \mathbf{V}_j) = V_j \, \partial c_j(\mathbf{r},t)/\partial x$. [The analysis is much more complex if $\mathbf{V}$ varies over space as, for example, it would in a liquid sample flowing through a tube (10).] We shall suppose that reaction rate expressions of higher than first order have been linearized (and hence that the concentrations of reactants participating in these reactions do not deviate too far from equilibrium). Then T_{jl}, the matrix of chemical kinetic coefficients, will be composed of reaction rate constants and equilibrium concentrations. Let us suppose that the observation area is small compared to the total area of the reaction system. Then Eq. (1) must be solved subject to the boundary conditions

$$\delta c_j(\mathbf{r},t) = 0 \quad \text{at } r = \infty$$

where $\delta c_j \, (\mathbf{r},t) = c_j \, (\mathbf{r},t) - \bar{c}_j$, the microscopic deviation of $c_j \, (\mathbf{r},t)$ from its equilibrium mean value $\bar{c}_j$.

In an FPR experiment the initial condition for Eq. (1) is set by the initial concentration gradient established by the photobleaching process. A brief, intense pulse of focused light irreversibly photolyzes a portion of the fluorophores in a small area. Although the photochemistry of fluorophores in cells is not well characterized and is likely to be complex, it is common to assume that the photolysis reaction occurs as an irreversible first-order process:

$$dc_j\,(\mathbf{r},t)/dt \;=\; -\alpha_j I'(\mathbf{r})c_j(\mathbf{r},t)$$

Then, for a bleach pulse which begins at $t = -T$ and ends at $t = 0$,

$$\Delta c_j(r,0) \;=\; \bar{c}_j[\exp(-\alpha_j I'(r)T) \;-\; 1]$$

where Δ denotes a macroscopic displacement from equilibrium, α_j is the photochemical rate constant for bleaching the jth component, and $I'(\mathbf{r})$ and T are, respectively, the intensity and duration of the bleaching pulse (6).

To express the initial condition for Eq. (1) appropriate for FCS measurements, we must take into account a fundamental distinction between methods based on macroscopic and microscopic concentration changes. The latter but not the former must, intrinsically, be analyzed statistically. After a macroscopic displacement the concentration relaxes deterministically to equilibrium governed by the conventional phenomenological chemical rate constants and transport coefficients. Therefore the values of these quantities can in principle be determined from single relaxation measurements with an accuracy limited only by the accuracy of the measurements. [Hence analysis of an FPR experiment requires the measured time course of fluorescence relaxation, $\Delta F(t)$, in terms of the rate of relaxation of the macroscopic concentration displacements $\Delta c_j(\mathbf{r},t)$ from their initial nonequilibrium values $\Delta c_j(\mathbf{r},0)$.] In contrast, the space–time behavior of spontaneous microscope fluctuations is related only probabilistically to the phenomenological coefficients. Measurement of a single fluctuation even with infinite accuracy would be insufficient to determine these coefficients accurately. A statistical analysis of many fluctuations is required. This is appropriately accomplished in terms of a correlation function $\phi_{jl}(\mathbf{r},\mathbf{r}',\tau)$, which describes the average correlation of a concentration fluctuation of the jth component at $\mathbf{r}$ and t with a fluctuation of the concentration of the lth component at $\mathbf{r}'$ and at a later time $t + \tau$.

$$\phi_{jl}(\mathbf{r},\mathbf{r}',\tau) \;=\; \langle\delta C_j(\mathbf{r},t)\delta C_l(\mathbf{r}',t+\tau)\rangle \;=\; \langle\delta C_j(\mathbf{r},0)\delta C_l(\mathbf{r}',\tau)\rangle$$

where $\langle\ \rangle$ denotes an ensemble or time average over many fluctuations. We have assumed that the system is stationary so that $\phi_{jl}(r,r',\tau)$ is independent of t. Therefore in the analysis of FCS experiments the initial conditions required for the solution of Eq. (1) are expressed in terms of the concentration correlation functions:

$$\phi_{jl}(r,r',0) \;=\; \langle\delta C_j(r,0)\delta C_l(r',0)\rangle \;=\; \bar{c}_l\delta_{jl}\delta(\mathbf{r} - \mathbf{r}')$$

This condition, based on an assumption of ideal behavior, asserts that at any instant of time a fluctuation in the concentration of one component at some position $\mathbf{r}$ is independent of a fluctuation of the same component at any other position $\mathbf{r}'$ or of any other component at any position. Furthermore, the number of molecules of a given species in any volume is governed by the Poisson distribution, and hence the mean-square fluctuation equals the mean number of molecules in the region.

It is convenient to proceed by carrying out a Fourier transformation of Eq. (1) to yield

$$\frac{d\tilde{C}_j(v,t)}{dt} = -[v^2D_j - iVv_x]\tilde{C}_j(v,t) + \sum_l T_{jl}\tilde{C}_l(v,t)$$

where $\tilde{c}_j(v,t)$ is the Fourier transform of $\delta c_j(\mathbf{r},t)$ with $v = (v_x,v_y)$ the transform variable and $v^2 = v_x^2 + v_y^2$. This may be rewritten as the matrix equation

$$\frac{d\tilde{c}_j(v,t)}{dt} = \sum_l M_{jl}\tilde{C}_l(v,t)$$

with matrix elements

$$M_{jl} = -[v^2D_l - i\,v_xV_l]\delta_{jl} + T_{jl}$$

This matrix equation is solved to yield a set of "normal modes" which combine the effects of the chemical reaction and the diffusion and flow or drift of each of the components. Associated with the sth mode there is a rate, $\lambda^{(s)}$, which is an eigenvalue of the matrix M_{jl}. There are also amplitude factors associated with the sth mode which are developed from the right eigenvector $X_l^{(s)}$ and the left eigenvector $(X^{-1})_j^{(s)}$ of M_{jl}, as previously described (1). Then for an FCS experiment

$$\phi_{jl}(r,r',\tau) = \frac{\tilde{c}_j}{4\pi^2} \int d^2v \exp[-iv \cdot (\mathbf{r}-\mathbf{r}')] \sum_s X_l^{(s)}(X^{-1})_j^{(s)} \exp(\lambda^{(s)}\tau)$$

and for an FPR experiment

$$\Delta c_j(\mathbf{r},t) = \frac{1}{4\pi^2} \int d^2v \exp[-iv \cdot (\mathbf{r}-\mathbf{r}')] \sum_s X_j^{(s)} \exp(\lambda^{(s)}t)$$

$$\cdot \sum_l (X^{-1})_l^{(s)}\tilde{c}_l \int d^2r' \exp(iv \cdot r)[\exp(-\alpha_l I'(r)T)-1]$$

2.3   Description of Fluorescence Measurements in Terms of Concentration Changes

In an FCS measurement the concentration correlation behavior is observed in terms of the fluorescence intensity autocorrelation function $G(\tau)$,

$$G(\tau) = \langle\delta F(o)\,\delta F(\tau)\rangle = \sum_{j,l} Q_jQ_l \int d^2r\,d^2r\, I(r)I(r')\,\phi_{jl}(r,r',\tau)$$

Experimentally $G(\tau)$ can be defined as

$$G(\tau) = \lim_{T\to\infty} \frac{1}{T}\int_o^T F(t)F(t+\tau)\,dt$$

In terms of the solution of Eq. (1), $G(\tau)$ may be expressed as

$$G(\tau) = (4\pi^2)^{-1} \sum_{jl} Q_jQ_l \int d^2v\,\Phi(v) \sum_s A_{jl}^{(s)} \exp(\lambda^{(s)}\tau)$$

where $A_{jl}^{(s)}(\nu) = X_j^{(s)}(X^{-1})_l^{(s)}$. The function $(4\pi^2)^{-1}\Phi(\nu)$ is simply the square magnitude of the Fourier transform of the beam profile $I(r)$ (cf. ref. 1).

In an FPR experiment the fluorescence relaxation is expressed as

$$\Delta F(t) = (4\pi^2)^{-1} \sum_j Q_j \int d^2\nu \sum_s \exp(\lambda^{(s)}\tau) \sum_l A_{jl}^{(s)}\bar{c}_l\Psi_l(\nu)$$

where $[\bar{c}_l/(4\pi^2)]\Psi_l(\nu)$ is the product of the Fourier transform of $I(r)$ and the Fourier transform of $\Delta c_l(\mathbf{r},0)$. There is a close similarity between $G(\tau)$ and $F(t)$. This is most clearly seen when Eq. (2) is expressed for very small extents of bleaching. Then it may be shown that for the bleaching mechanism postulated above, $G(\tau)$ and $F(t)$ differ only in the relative weight of the amplitude components (11). Thus apart from constants $G(\tau)$ and $F(t)$ at low levels of bleaching differ mainly in that Q_j in the former is replaced by α_j in the latter. As will be discussed below, however, there are systems for which realistic bleaching mechanisms will yield large differences between FCS and FPR observations on the same system (12, 13).

Koppel (14) has pointed out that FPR experiments that use certain beam profiles or sample geometries permit the interpretation of the fluorescence recovery directly in terms of the Fourier components of the solution of Eq. (1). One example is the pattern photobleaching approach introduced by McConnell and co-workers (15, 16). A periodic pattern of parallel stripes or a rectangular grid is bleached into the cell, and the rate of disappearance of the pattern by fluorescence recovery is monitored. This immediately provides the Fourier components of the recovery process. Another example developed by Koppel and co-workers is a normal mode analysis of diffusion on a spherical surface (17). In this case the concentration is expressed in terms of a series of Legendre polynomials rather than as a Fourier transform.

3. EXPERIMENTAL IMPLEMENTATION

The basic experimental function in both FCS and FPR experiments is the measurement of the time-dependent fluorescence emitted from a well-defined observation volume. The observation volume is usually defined as the volume or area illuminated by a laser beam which excites the fluorescence in the sample. The simplest and most commonly used approach bleaches and observes a circularly symmetrical spot in two dimensions or a cylindrically symmetrical volume in three. The lateral intensity profile is typically Gaussian. Hence, for spot photobleaching $I(r) = I(0)\exp(-2r^2/w^2)$, where w is the radius at which the intensity falls to $\exp(-2)$ of its central maximum intensity. Then the laser power is $P = \int I(r)\,d^2r = I_0\pi w^2/2$. The bleaching pulse is taken to have the same profile at higher intensity. Thus $I'(\mathbf{r})/I(\mathbf{r}) = A$, a constant. The laser beam is focused on the sample, and the fluorescence emitted is collected by a fluorescence microscope fitted with a sensitive photomultiplier typically operated in a photon-counting mode. The photon pulses are discriminated and amplified. Then in an FPR measurement the photocounts that characterize the fluorescence recovery are registered by a transient recorder or signal averager. Additional electronics are needed to initiate and terminate the bleaching pulse and to protect the photomultiplier from the intense fluorescence generated during the brief bleaching pulse. The bleaching pulse is most commonly generated using an approach developed by Koppel (18). A beam splitter divides the incident laser intensity into a weak beam, appropriate for exciting fluorescence with minimal photolysis of the sample fluorophore during the measuring phases of the experiment, and an intense beam (typically 1000- to 10,000-fold more intense than the monitoring beam) for photobleaching. The two beams are recombined by a second beam splitter. Between the splitters used to separate and recombine the beams, the intense beam is blocked by a shutter during the measurement phase of the experiment. The shutter is opened for a brief but well-defined interval to permit a bleaching pulse coaxial with the measuring beam to strike the target. A description of the optics, electronics, and data analysis procedures for simple spot photobleaching measurements has been published (19). More elaborate arrangements involving pattern photobleaching, point-by-point scanning (15, 18), and combinations of scanning and pattern photobleaching (20, 21) have also been introduced.

The optical arrangement for FCS experiments on cells can be very similar to that used for FPR measurements except that the beam splitters and shutter are not needed. In contrast to FPR, which requires the recording of only the relatively brief transient fluorescence recovery, however, an FCS measurement requires the collection of an extended record of many fluctuations and the computation of a fluorescence fluctuation autocorrelation function. This can be done by a general-purpose computer or by a special-purpose correlator. More detailed comparisons of FCS and FPR and their technical details have been published (7, 11).

4. COMPARISON OF FCS AND FPR

FPR permits the determination of phenomenological coefficients from a single transient relaxation of the system back to equilibrium from its macroscopically displaced initial state. In contrast FCS requires the observation of many spontaneous fluctuations the duration of each of which is comparable to the macroscopic relaxation process. [An analysis of the signal-to-noise properties of FCS measurements has been published (22).] Hence data must be acquired over a much longer period for an FCS than for an FPR experiment. Moreover, the microscopic concentration and fluorescence changes observed in an FCS measurement are typically much smaller than the macroscopic changes seen in FPR and so could be submerged even by small artifactual fluorescence fluctuations caused by small movements of the sample which would not interfere with FPR measurements. In general the long duration and stringent stability requirements of FCS measurements have made them unsuitable for studies of living biological systems. In contrast FPR has been applied extensively to measurements on living cells in culture.

A disadvantage of FPR is that it requires a photobleaching process, which is usually poorly understood, to generate the initital concentration gradient. Two aspects of the photochemical response of the system should be examined to interpret a photobleaching measurement. One is the kinetic mechanism of the photochemical destruction of the fluorophores. In our development we have assumed the simplest possible irreversible first-order processes. Both kinetic complexity and chemical reversibility of the photolysis reactions can considerably complicate interpretation of the results. Second, it is necessary to know the effect of photolysis on the chemical reactions characterized by T_{jl}. In the development outlined above we have assumed that photolysis of a fluorescent component both renders it invisible and removes it from participation in the chemical reaction system. This assumption may not correctly describe the behavior of the system, however. Consider, for example, the simple second-order reaction:

$$A \;+\; B \underset{k_b}{\overset{k_f}{\rightleftharpoons}} C \tag{3}$$

where B is a fluorescent ligand that binds to a nonfluorescent molecule A, to produce a fluorescent complex, C. We shall suppose that B and C, but not A, absorb the excitation radiation and are vulnerable to photolysis. The equations developed above assume that B and C are simply eliminated optically and chemically from the system by photobleaching. Then the initial conditions for a relaxation of this system would be

$$\Delta c_A(\mathbf{r},0) \;=\; 0$$

$$\Delta c_j(\mathbf{r},0) \;=\; \bar{c}_j[\exp(-\alpha_j I'(\mathbf{r})T) \;-\; 1]; \qquad j \;=\; B, C$$

Several other possibilities exist, however. For example, the photolysis pulse might destroy B whether free or bound with no effect on A whether free or bound. Then the effect of the pulse is to eliminate B and convert C to A. For this example, therefore, $\Delta c_B(\mathbf{r},0)$ and $\Delta c_C(\mathbf{r},0)$ are as described above, but $\Delta c_A(\mathbf{r},0) = -\Delta c_C(\mathbf{r},0)$. This difference in initial conditions does not influence the characteristic rates of the several relaxation modes of the two examples, but it does determine differences in their relative amplitudes. A much more profound effect of bleaching mechanism will be discussed below (12, 13). Further potential difficulties associated with photobleaching include the possibilities of thermal and photochemical artifacts introduced by the bleaching pulse. A variety of studies using many different approaches indicate that these potential problems do not have detectable effects on actual measurements of lateral diffusion of lipids and proteins on natural or synthetic membranes (23–28).

5. ANALYSIS OF EXAMPLES

5.1 Simple Reaction System under Systematic Transport

We shall consider a simple unimolecular isomerization in an open system subject to uniform translation at velocity V.

$$A \underset{k_b}{\overset{k_f}{\rightleftharpoons}} B$$

Then $K \;=\; k_f/k_b \;=\; \bar{c}_B/\bar{c}_A$ is the equilibrium constant and $R \;=\; k_f \;+\; k_b$, the conventional chemical relaxation rate. Although too simple for wide practical application in biological systems, this example will illustrate the characteristic contributions of reaction, diffusion, and systematic transport to the fluorescence recovery and correlation functions. For this system Eq. (1) becomes

$$\frac{\partial \delta C_A(r,t)}{\partial t} = D_A \nabla^2 \delta C_A - V \frac{\partial \delta C_A}{\partial x} - k_f \delta C_A + k_b \delta C_B$$

$$\frac{\partial \delta C_B(r,t)}{\partial t} = D_B \nabla^2 \delta C_B - V \frac{\partial \delta C_B}{\partial x} + k_f \delta C_A - k_b \delta C_A$$

We shall further simplify by assuming that $D_A = D_B = D$. Then the matrix $\mathbf{M}$ is

$$\begin{bmatrix} -[v^2 D - ivv_x + k_f] & k_b \\ k_f & -[v^2 D - ivv_x + k_b] \end{bmatrix}$$

Hence the characteristic relaxation rates (eigenvalues of M) are

$$\lambda^{(+)} = -(v^2 D - iVv_x)$$

$$\lambda^{(-)} = -(v^2 D - iVv_x + R)$$

and the right and left eigenvectors are

$$X^{(+)} = [X_A^{(+)}, X_B^{(+)}] = (1 + K^2)^{-1}[1, K]$$

$$X^{(-)} = 2^{-1/2}[1, -1]$$

$$(X^{-1})^{(+)} = \frac{\sqrt{1 + K^2}}{1 + K} [1, 1]$$

$$(X^{-1})^{(-)} = \frac{\sqrt{2} K}{1 + K} [1, -1]$$

Therefore the FCS correlation function is

$$G(\tau) = G_{AA} + G_{BB} + 2G_{AB}$$

$$= \frac{P^2}{4w^2(1 + K)} \frac{\exp[-(\tau/\tau_f)^2/(1 + \tau/\tau_D)]}{1 + \tau/\tau_D}$$

$$\cdot \{Q_A^2 \bar{c}_A[1 + K \exp(-R\tau)] + Q_B^2 \bar{c}_B[K + \exp(-R\tau)]$$

$$+ 2Q_A Q_B \bar{c}_B[1 - \exp(-R\tau)]\}$$

The characteristic time for displacement at uniform velocity is $\tau_f = w/V$; that for diffusion is $\tau_D = w^2/4D$. The FPR fluorescence recovery is

$$F(t) = \frac{2P^2}{4\pi^2 I_o(1 + R)} \sum_n \left\{ (Q_A + KQ_B) \left[\frac{\bar{c}_A(-\kappa_A)^n + \bar{c}_B(-\kappa_B)^n}{n!} \right] \right.$$

$$+ (Q_A - Q_B) \frac{K\bar{c}_A(-\kappa_A)^n - \bar{c}_B(-\kappa_B)^n}{n!} \exp(-Rt) \right\}$$

$$\cdot \exp \left[\frac{(t/\tau_f)^2 2n}{1 + n(1 + 2t/\tau_D)} \right] \left[(1 + n(1 + 2t/\tau_D)) \right]^{-1}$$

where $\kappa_j = \alpha_j I'_0 T$. Because this is a two-component system, there are two normal modes in the analysis each with a characteristic rate. In general diffusion, systematic transport (e.g., uniform translation) and chemical reaction would each contribute to each normal mode. In this simple example, however, we see that one mode is influenced only by diffusion and uniform translation whereas the other has contributions from all three processes that occur in the system. It is characteristic of FCS and spot photobleaching experiments that diffusion enters as a slow recovery process with a time dependence given by $1/(1 + t/\tau_D)$, uniform translation (for $\tau_f <$ τ_D) as a Gaussian function of the distance (in beam widths) translated, and chemical reaction as an exponential decay governed by a characteristic chemical relaxation time R^{-1}. We note that for this example $G(\tau)$ and $F(t)$ report on the chemical kinetics only if $Q_A \neq Q_B$ because, with $D_A = D_B$ and $V_A = V_B$, only the molecular optical properties can depend on the progress of the chemical reaction. If, however, $D_A \neq D_B$ or $V_A \neq V_B$, then the measurements could report on the chemical kinetics via changes in transport behavior even if there were no optical differences between A and B.

5.2 Reversible Bimolecular Reaction

The analysis of this example, embodied in Eq. (3) above, is useful in interpreting FCS and FPR measurements on interacting systems. It has been discussed in several different contexts (1, 14, 29, 30). For both simplicity and relevance to experimental systems of interest we shall suppose that A is a nonfluorescent molecule which is either immobile or slowly moving with diffusion coefficient D_A. (It might, for example, be a cytoskeletal component or immobile structure on the plasma membrane.) Component B (perhaps a labeled mobile cell surface protein) is fluorescent and has diffusion coefficient D_B. The apparent mobility of B is, however, reduced by its interaction with A to form the complex C. We shall furthermore assume that the interaction of A and B does not change the mobility of A. Hence $D_A = D_C << D_B$. A derivation of the FCS correlation function for this example has been presented in detail (1) accompanied by a discussion and experimental demonstration of the special case in which chemical reaction is fast compared to diffusion (31). There are three components in the system, so there are three normal modes denoted (0), (+), and (−). In the illustrated example for which reaction is taken to be fast compared to diffusion (31) the (0) mode refers only to the slow diffusion of A and so is of little interest. The second, (+), mode has the form of simple diffusion process $[\sim 1/(1 + \tau/\tau^{(+)}); \tau^{(+)} = w^2/4D^{(+)}]$, but with a compound diffusion coefficient $D(+) = f_B D_B + f_C D_C$ where $f_B = \bar{c}_B/(\bar{c}_B + \bar{c}_C) = K\bar{c}_A/(1 + K\bar{c}_A)$ and $f_C = 1 - f_B$. The third, (−), mode $[\sim \exp(-R\tau)/(1 + \tau/\tau^{(-)})]$ has both a diffusional contribution with compound diffusion coefficient $D(-) = f_B D_C + f_C D_B$ and an exponentially decaying chemical kinetic contribution governed by the conventional relaxation time for the linearized chemical reaction, $\tau^{-1}_{chem} = R = [k_f(\bar{c}_A + \bar{c}_B) + k_b]$. The relative amplitudes of the second and third components are governed by the equilibrium ratio of components B and C.

The kinetics of the interaction may appear in the experimental measurements in two different ways. If the fluorescence of B and C are substantially different, then the chemical reaction can be observed directly (31). The kinetics of the reaction may also be revealed by the retardation of the diffusion of B. This will depend on the relative magnitudes of the characteristic times for diffusion of B ($\tau_B = w^2/4D_B$) and chemical relaxation (29). If the chemical reaction is slow compared to diffusional recovery of B ($\tau_B \ll \tau_{chem}$), a given B molecule will be observed for the most part either free or complexed and only rarely in both states during the relaxation of a single fluctuation or macroscopic transient. Hence the B molecules will appear to reside in two distinct mobility classes: fast, owing to diffusion of free B, and slow, owing to diffusion of C. For D sufficiently small there may be negligible recovery of C during the measurement period so that the fraction of B in the complex will appear immobile. If, however, the reaction is fast compared to diffusional recovery ($\tau_{chem} \ll \tau_B$), each B will react many times with A during a single fluctuation or transient. Then all B molecules will be retarded to a comparable extent and will seem to diffuse with an effective diffusion coefficient $D(+)$.

5.3   Multiple Binding Reactions

As an example of the application of FCS and FPR to a more complex reaction system, we consider the following coupled system of binding reactions:

$$A + B \rightleftharpoons AB$$

$$AB + B \rightleftharpoons AB_2$$

$$\vdots$$

$$AB_{n-1} + B \rightleftharpoons AB_n$$

The set of coupled differential equations, accounting for both diffusion and binding reactions, which describes the behavior of this system, has not been solved in simple form. A detailed computer analysis and comparison with FPR and FCS measurements of the binding of ethidium bromide to DNA has been presented (12, 13). It was shown that the theoretical analysis accounted well for both FCS and FPR measurements. An interesting result of this study is that under some conditions the predicted and observed behaviors seen in an FCS measurement were quite different from those seen in an FPR measurement. When on the average several ligands were bound to each DNA molecule, the FCS relaxation behavior was dominated by the most highly bound members of the population. In simple terms this is a consequence of the dependence of the fluorescence fluctuation autocorrelation function on the square of the intensity emitted from the diffusing molecules. Hence when a molecule bearing 10 ligands diffuses into or out of the beam its contribution to $G(\tau)$ is 100-fold greater than the contribution of a molecule bearing a single ligand (cf. ref. 10). Therefore the FCS correlation function for a system in which there are several ligands per DNA molecule has quite different properties from one for a system in which no DNA molecule has more than one ligand and which therefore could be accounted for in terms of the simple bimolecular reaction—Eq. (3). In contrast, both theory and experimental measurements reveal that the FPR recovery curves both for systems in which there are many ligands per DNA and for those in which there are no more than one ligand can be accounted for in terms of the single bimolecular reaction—Eq. (3). This difference between the behavior of FCS and FPR measurements was interpreted to result from the bleaching mechanism. It was assumed that each ethidium on a given DNA molecule was bleached independently of the others. This is both reasonable a priori, because strong interactions among widely spaced ethidium molecules would not be expected, and sufficient to account for the experimental measurements. If, however, all the ethidium molecules on a given DNA molecule were photolyzed in a single cooperative process, then for systems in which there were several ligands per DNA the FPR recovery curves like the FCS correlation functions would behave quite differently from those in which there were no more than one ligand (12, 13). This example illustrates that uncertainties in the bleaching mechanism can introduce substantial ambiguities in the interpretation of FPR measurements.

6.   APPLICATIONS

FCS has found little application to studies of live cells because of its long data acquisition times and stringent requirements for sample stability. FPR, however, has been extensively used to study the mobility of cell surface proteins (32, 33) and, to a lesser extent, cytoplasmic constituents (34, 35). Unlike FCS, an FPR measurement conveniently yields the relative proportion of fluorophores that are mobile on the time scale of the measurement. In noninteracting systems, all the fluorophores are mobile. A substantial fraction of cell membrane proteins, however, is immobile on the experimental time scale. Hence their diffusion seems to be retarded by interactions with other structures. In some instances changes in the fraction of a specific membrane component that was mobile have provided information about changes in biologically significant interactions of that component and so have been more interesting than the diffusion coefficient of the component (36, 37). With few exceptions it seems generally to be true that cell surface proteins diffuse laterally at much slower rates than would be expected from the viscosity of the lipid bilayer (32, 33). It thus appears that there are forces in addition to bilayer viscosity that retard protein diffusion. Some of these forces seem to arise from interactions between the membrane proteins and the cytoskeleton (29, 38–41). In the erythrocyte the interaction of a major membrane protein, band 3, with cytoskeletal spectrin via the linker protein ankyrin has been extensively characterized at a biochemical level (42). In nucleated mammalian cells, however, the biochemical basis of the interactions of membrane proteins with the cytoskeleton is not yet well understood (cf. 43). Two models have been proposed to account for the effects of these interactions on the protein diffusion rate. One supposes that the principal retardation results from nonspecific steric interactions between the cytoplasmic portion of the membrane protein and a dense network of cytoskeletal filaments adjacent to the cytoplasmic surface of the membrane (44). The other model attributes the retardation to specific interactions between protein and cytoskeleton. As discussed above, interactions with lifetimes long compared to the fluorescence recovery time could account for the immobile fraction of molecules while more transient interactions could govern the slow but measurable diffusion rate of the mobile fraction (14, 29).

If encounters of membrane molecules are important for physiologically important reactions and cellular responses, it is interesting to ask whether the rates of these responses or reactions could be limited by the rate of membrane protein diffusion. It appears that the rate of aggregation of EGF–receptor complexes, although requiring diffusional encounters, is not limited by the rate of lateral diffusion (45). Similarly Gupte et al. (9) have shown that the diffusion rates of mitochondrial enzymes are sufficient to allow the redox reactions which they catalyze to occur at physiological rates by random collision of the membrane-bound enzyme molecules and cofactors. Hence they concluded that electron transport is coupled to the lateral diffusion of the redox components in the mitochondrial inner membrane but is not limited by the diffusion rates.

Although FCS is little used for studies of live cells, it has the ability to provide useful information that is not accessible by FPR. The amplitude of the correlation function, $G(0)$, is sensitive to the aggregation state of the fluorophores and therefore could in principle be used to observe aggregation processes on or in cells or in vitro systems (10, 46). Indeed measurement of the degree of polymerization of large DNA molecules to which ethidium bromide was bound was one of the first applications of fluorescence fluctuation measurements (46). Aggregation processes on cell surfaces are of interest in the activation of cellular responses (8) and in the assembly of specialized structures such as enveloped viruses or regions of high receptor density as in the neuromuscular junction. In principle FCS can be used to investigate these processes, although many technical difficulties remain to be overcome (47).

ACKNOWLEDGMENTS

The work carried out in the author's laboratory which has been cited in this paper was supported by NIH grants GM 21661 and GM 30299. I am grateful to D. Magde, W. Webb, R. Icenogle, and N. Petersen for their participation in various phases of the work discussed here.

REFERENCES

1. Elson, E. L., and D. Magde (1974) Fluorescence correlation spectroscopy. I. Conceptual basis and theory. *Biopolymers*, **13**, 1–27.
2. Elson, E. L., and W. W. Webb (1975) Concentration correlation spectroscopy: A new biophysical probe based on occupation number fluctuations. *Annu. Rev. Biophys. Bioeng.*, **4**, 311–334.
3. Peters, R., J. Peters, K. H. Tews, and W. Bahr (1974) A microfluorimetric study of translation diffusion in erythrocyte membranes. *Biochim. Biophys. Acta*, **367**, 282–294.
4. Edidin, M., Y. Zagyanski, and T. J. Lardner (1976) Measurement of membrane lateral diffusion on single cells. *Science*, **191**, 466–468.

5. Jacobson, K., Z. Derzko, E.-S. Wu, Y. Hou, and G. Poste (1976) Measurement of the lateral mobility of cell surface components in single living cells by fluorescence recovery after photobleaching. *J. Supramol. Struct.*, **5**, 565–576.

6. Axelrod, D., D. E. Koppel, J. Schlessinger, E. Elson, and W. W. Webb (1976) Mobility measurement by analysis of fluorescence photobleaching recovery. *Biophys. J.*, **16**, 1055–1069.

7. Koppel, D. E., D. Axelrod, J. Schlessinger, E. L. Elson, and W. W. Webb (1976) Dynamics of fluorescence marker concentration as a probe of mobility. *Biophys. J.*, **16**, 1315–1329.

8. Schlessinger, J., A. B. Schreiber, T. A. Libermann, I. Lax, A. Avivi, and Y. Yarden (1983) "Polypeptide-hormone-induced receptor clustering and internalization," in E. Elson, W. Frazier, and L. Glaser, Eds., *Cell Membranes. Methods and Reviews*, Vol. 1, Plenum, New York, pp. 117–149.

9. Gupte, S., E.-S. Wu, L. Hoechli, M. Hoechli, K. Jacobson, A. E. Sowers, and C. R. Hackenbrock (1984) Relationship between lateral diffusion, collision frequency, and electron transfer of mitochondrial inner membrane oxidation–reduction components. *Proc. Natl. Acad. Sci. USA*, **81**, 2602–2610.

10. Magde, D., W. W. Webb, and E. L. Elson (1978) Fluorescence correlation spectroscopy. III. Uniform translation and laminar flow. *Biopolymers*, **17**, 361–376.

11. Petersen, N. O., and E. L. Elson (1984) Measurement of diffusion and chemical kinetics by fluorescence photobleaching recovery and fluorescence correlation spectroscopy. *Meth. Enzym.*, (1985) in press.

12. Icenogle, R. D., and E. L. Elson (1983) Fluorescence correlation spectroscopy and photobleaching recovery of multiple binding reactions. I. Theory and FCS measurements. *Biopolymers*, **22**, 1919–1948.

13. Icenogle, R. D., and E. L. Elson (1983) Fluorescence correlation spectroscopy and photobleaching recovery of multiple binding reactions. II. FPR and FCS measurements at low and high DNA concentrations. *Biopolymers*, **22**, 1949–1966.

14. Koppel, D. E. (1981) Association dynamics and lateral transport in biological membranes. *J. Supramol. Struct.*, **17**, 61–67.

15. Smith, B. A., and H. M. McConnell (1978) Determination of molecular motion in membranes using periodic pattern photobleaching. *Proc. Natl. Acad. Sci. USA*, **75**, 2759–2763.

16. Smith, B. A., W. R. Clark, and H. M. McConnell (1979) Anisotropic molecular motion on cell surfaces. *Proc. Natl. Acad. Sci. USA*, **76**, 5641–5644.

17. Koppel, D. E., M. P. Sheetz, and M. Schindler (1980) Lateral diffusion in biological membranes—a normal-mode analysis of diffusion on a spherical surface. *Biophys. J.*, **30**, 187–192.

18. Koppel, D. E. (1979) Fluorescence redistribution after photobleaching. A new multipoint analysis of membrane translational dynamics. *Biophys. J.*, **28**, 281–292.

19. Petersen, N. O., S. Felder, and E. L. Elson (1985) "Measurement of lateral diffusion by fluorescence photobleaching recovery," in D. W. Weir, L. A. Herzenberg, C. G. Blackwell, and L. A. Herzenberg, Eds., *Handbook of Experimental Immunology*, 4th ed., Blackwell, London, in press.

20. Lanni, F., and b. R. Ware (1982) Modulation detection of fluorescence photobleaching recovery. *Rev. Sci. Instruments*, **53**, 905–908.

21. Davoust, J., P. F. Devoux, and L. Leger (1982) Fringe pattern photobleaching, a new method of the measurement of transport coefficients of biological macromolecules. *EMBO J.*, **1**, 1233–1238.

22. Koppel, K. E. (1974) Statistical accuracy in fluorescence correlation spectroscopy. *Physiol. Rev.*, **A10**, 1938–1945.

23. Axelrod, D. (1977) Cell surface heating during fluorescence photobleaching recovery experiments. *Biophys. J.*, **18**, 129–131.

24. Fahey, P. F., and W. W. Webb (1978) Lateral diffusion in phospholipid bilayer membranes and multilamellar liquid crystals. *Biochemistry*, **17**, 3046/3053.

25. Wolf, D. E., M. Edidin, and P. R. Dragsten (1980) Effect of bleaching light on measurements of lateral diffusion in cell membranes by the fluorescence photobleaching recovery method. *Proc. Natl. Acad. Sci. USA*, **77**, 2043–2045.

26. Wey, C.-L., R. A. Cone, and M. A. Edidin (1981) Lateral diffusion of rhodopsin in photoreceptor cells measured by fluorescence photobleaching and recovery. *Biophys. J.*, **33**, 225–232.

27. Koppel, D., and M. P. Sheetz (1981) Fluorescence photobleaching does not alter the lateral mobility of erythrocyte membrane glycoproteins. *Nature*, **293**, 159–161.

28. Jacobson, K., E. Elson, D. Koppel, and W. Webb (1983) International workshop on the application of fluorescence photobleaching techniques to problems in cell biology. *Fed. Proc.*, **42**, 72–79.

29. Elson, E. L., and J. A. Reidler (1979) Analysis of cell surface interactions by measurements of lateral mobility. *J. Supramol. Struct.*, **12**, 481–489.

30. Thompson, N. L., T. P. Burghardt, and D. Axelrod (1981) Measuring surface dynamics of biomolecules by total internal reflection fluorescence with photobleaching recovery or correlation spectroscopy. *Biophys. J.*, **33**, 435–454.

31. Magde, D., E. L. Elson, and W. W. Webb (1974) Fluorescence correlation spectroscopy. II. An experimental realization. *Biopolymers*, **13**, 29–61.

32. Cherry, R. J. (1979) Rotational and lateral diffusion of membrane proteins. *Biochim. Biophhys. Acta.*, **559**, 289–327.

33. Peters, R. (1981) Translational diffusion in the plasma membrane of single cells as studied by fluorescence microphotolysis. *Cell Biol. Int. Rep.*, **5**, 733–760.

34. Wojcieszyn, J. W., R. A. Schlegel, E.-S. Wu, and K. Jacobson (1981) Diffusion of injected macromolecules within the cytoplasm of living cells. *Proc. Natl. Acad. Sci. USA*, **78**, 4407–4410.

35. Kreis, T. E., B. Geiger, and J. Schlessinger (1982) Mobility of microinjected rhodamine actin within living chicken gizzard cells determined by fluorescence photobleaching recovery. *Cell*, **29**, 835–845.

36. Reidler, J. A., P. M. Keller, E. L. Elson, and J. Lenard (1981) A fluorescence photobleaching study of vesicular stomatitis virus injected BHK cells. Modulation of G protein mobility by M protein. *Biochemistry*, **20**, 1345–1349.

37. Johnson, D. C., M. J. Schlesinger, and E. L. Elson (1981) Fluorescence photobleaching recovery measurements reveal differences in envelopment of sindbis and vesicular stomatitis viruses. *Cell*, **23**, 423–431.

38. Golan, D. E., and W. Veatch (1980) Lateral mobility of band 3 in the human erythrocyte membrane studied by fluorescence photobleaching recovery: Evidence for control cytoskeletal interactions. *Proc. Natl. Acad. Sci. USA*, **77**, 2537–2541.

39. Sheetz, M. P., M. Schindler, and D. E. Koppel (1980) Lateral mobility of integral membrane proteins is increased in spherocytic erythrocytes. *Nature*, **285**, 510–512.

40. Tank, D. W., W.-S. Wu, and W. W. Webb (1982) Enhanced molecular diffusibility in muscle membrane blebs. Release of lateral constraints. *J. Cell Biol.*, **92**, 207–212.

41. Henis, Y. I., and E. L. Elson (1981) Inhibition of the mobility of mouse lymphocyte surface immunoglobulins by locally bound concanavalin A. *Proc. Natl. Acad. Sci. USA*, **78**, 1072–1076.

42. Bennett, V. (1982) The molecular basis for membrane–cytoskeleton association in human erythrocytes. *J. Cell. Biochem.*, **18**, 49–65.

43. Mangeat, P. H., and K. Burridge (1984) Immunoprecipitation of nonerythrocyte spectrin within live cells following microinjection of specific antibodies: Relation to cytoskeletal structures. *J. Cell Biol.*, **98**, 1363–1377.

44. Koppel, D. E., M. P. Sheetz, and M. Schindler (1981) Matrix control of protein diffusion in biological membranes. *Proc. Natl. Acad. Sci. USA*, **78**, 3576–3580.

45. Hillman, G. M., and J. Schlessinger (1982) Lateral diffusion of epidermal growth factor complexed to its surface receptors does not account for the thermal sensitivity of patch formation and endocytosis. *Biochemistry*, **21**, 1667–1672.

46. Weissman, M., H. Schindler, and J. Feher (1976) Determination of molecular weights by fluctuation spectroscopy: Application to DNA. *Proc. Natl. Acad. Sci. USA*, **73**, 2776–2780.

47. Petersen, N. O. (1984) Diffusion and aggregation in biological membranes. *Can. J. Biochem. Cell Biol.*, **62**, 1158–1166.

CHAPTER 22

CAGED ATP AS A TOOL IN ACTIVE TRANSPORT RESEARCH

JACK H. KAPLAN

Department of Physiology
University of Pennsylvania School of Medicine
Philadelphia, Pennsylvania

1. INTRODUCTION   385
2. CHEMISTRY AND PHOTOCHEMISTRY OF CAGED ATP   386
3. STUDIES ON THE SODIUM PUMP   390
4. STUDIES ON THE CALCIUM PUMP   393
5. CONCLUDING REMARKS   394
   REFERENCES   395

1. INTRODUCTION

The majority of the contributions in this volume describe experimental approaches and techniques where light absorption or emission is used to monitor changes in interesting physiological parameters. This chapter deals with the use of light to initiate biochemical or physiological processes where the incident illumination activates a chemical reaction, the consequence of which is to supply substrates for physiological processes.

Caged ATP (P^3-1-(2-nitro)phenylethyl adenosine triphosphate, Fig. 2, R = CH_3) is a derivative of ATP in which the terminal phosphate is esterified with a secondary 2-nitrobenzyl alcohol moiety. On illumination with ultraviolet light the phosphate ester bond is cleaved and free ATP is produced. The ability to "switch on" ATP-dependent reactions at will has enabled the investigator to study such reactions with hitherto unavailable kinetic control. Light activation has also enabled the initiation of ATP-dependent reactions in closed compartments which are inaccessible to rapid-mixing techniques (e.g., vesicles, resealed red cell ghosts, or muscle fibers). The synthesis, characterization, and photolability of caged ATP, caged ADP, and caged P_i were first described in 1978 by Kaplan et al. Since that time caged ATP has been used in two areas of research; in studies of the mechanism of muscle contraction (see Goldman et al., 1982a) and in studies of the mechanism of active cation transport (see below). In both of these areas ATP is the proximate energy source in an energy-transducing process. In muscle fibers, the free energy of hydrolysis of ATP is utilized in contraction to generate tension; in active transport systems, or ion pumps, the free energy of hydrolysis of ATP is utilized in transporting cations across biomembranes against steep electrochemical potential gradients. This chapter will focus on the results of studies on ion pumps. Some of the characteristics of the chemistry and photochemistry of caged ATP will be described, and mention will be made of the potential application of this approach to other biological substrates utilizing already established chemical strategies. The results of studies carried out on the Na, K-ATPase or sodium pump and the Ca-ATPase or calcium pump will then be discussed. As yet the number and range of reported studies has been small; however, an indication of some of the limitations and advantages of this approach has begun to emerge.

2.   CHEMISTRY AND PHOTOCHEMISTRY OF CAGED ATP

The photocleavage of organic molecules has been utilized for several years in synthetic organic chemistry where photolabile protecting groups have been employed. It was shown by Baltrop et al. (1966) that ultraviolet irradiation (for several hours) of solutions of the 2-nitrobenzyl esters of carboxylic acids in nonaqueous solvents led to the liberation of the free carboxylic acid (see Fig. 1). These observations, together with the reported synthesis and photolability of bis(2-nitrobenzyl)P_i (Rubenstein et al., 1975), led us to apply this strategy to the synthesis of a compound that would produce ATP on photolysis. Based on observations of Baltrop et al. (1966), the simplest molecule of this type would be the 2-nitrobenzyl ester of ATP (Fig. 3, R = H). This molecule was synthesized by

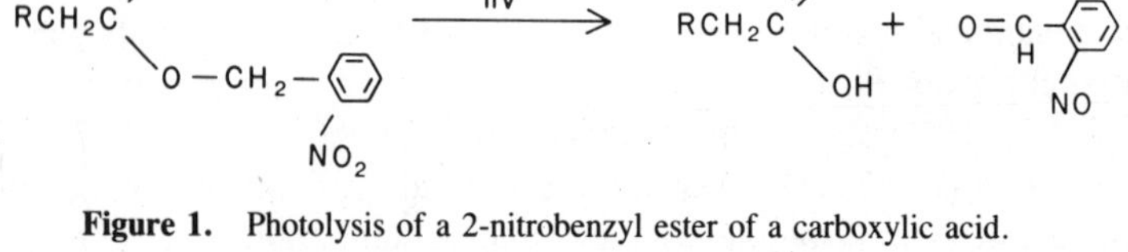

Figure 1.   Photolysis of a 2-nitrobenzyl ester of a carboxylic acid.

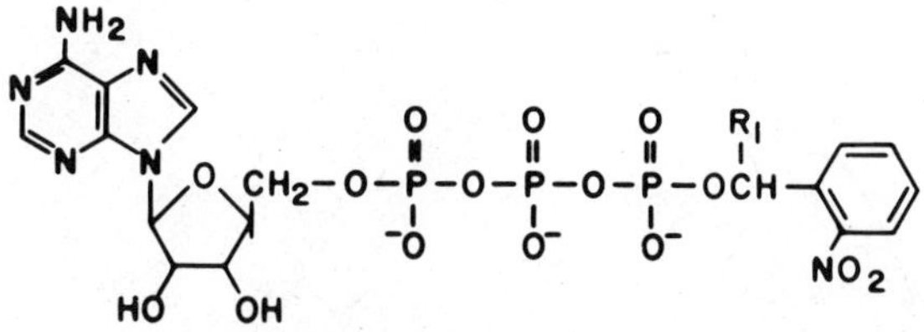

Figure 2. Structure of caged ATP.

condensing ADP-morpholidate with 2-nitrobenzyl phosphate using dicyclohexylcarbodiimide as the condensing agent; the yield of product was about 25%. One advantage of this synthetic route was that the photolability of this class of compounds could first be established using the caged phosphate where the photoreleased P_i (inorganic phosphate) could be easily assayed.

The same photochemical reaction seems to be involved in the phosphate esters as in the carboxylates, since the analogue of caged P_i bearing the nitro group para to the benzylic carbon was stable to ultraviolet illumination, whereas the ortho nitro compound was rapidly cleaved (Kaplan et al., 1978). Although photolysis of 2-nitrobenzyl phosphate yielded free P_i in good yield ($\sim$ 90%), photolysis of the ATP derivative (Fig. 2, R = H) gave only a low yield (25%) of free ATP when illuminated under identical conditions (see below). Although photolysis of the 2-nitrobenzyl ester of ATP was complete (assessed spectroscopically and chromatographically), the major product was not recognized as ATP by the luciferin–luciferase assay system. It seemed possible that an interaction occurred between a photoproduct and the adenosine moiety of ATP, yielding a structurally altered adenosine phosphate. Since the most likely photofragment was a nitroso aldehyde, we reasoned that the secondary alcohol derivative would yield the less reactive nitroso ketone on photolysis and such a reaction might be avoided. The same synthetic scheme was then employed using the secondary alcohol in the caged P_i component. The product of this condensation (Fig. 2, R = CH_3) yielded ATP in close to quantitative yield when illuminated (Kaplan et al., 1978).

The absorption spectrum of caged ATP is shown in Fig. 3. The spectrum is composed of absorption due to the adenosine moiety with a λ_{max} around 260 nm and a long tail extending beyond 340 nm owing to the absorbance of the 2-nitroaromatic residue. Appreciative absorption occurs at wavelengths greater than 320 nm; thus the irradiation of caged ATP can be performed in biological systems. Most of the naturally occurring chromophores (nucleoside bases, aromatic amino acids) absorb at wavelengths below 300 nm so that photodamage due to light absorption by endogenous chromophores may be avoided. In the initial studies, illumination was performed using the filtered output from a 1-kW Hg arc lamp so that light below 300 nm was removed. The filtered light gave a peak wavelength at 342 nm (73% transmission at peak) with a half bandwidth of 60 nm. Under the illumination conditions described, using a defocused light spot (2.5-cm diameter), ATP was released with a half-time of around 7–9 s (Kaplan et al., 1978). This time could be

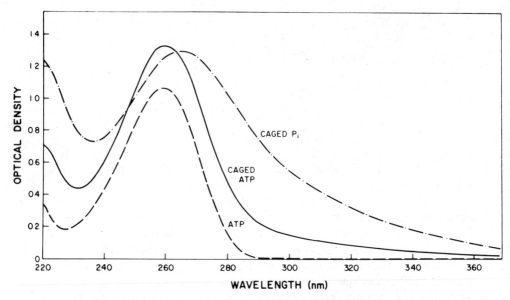

Figure 3.   Absorbance spectra of ATP and caged compounds.

reduced by focusing the beam to a smaller area. It was also shown that free intracellular ATP could be generated by irradiation of suspensions of resealed red cell ghosts containing caged ATP. Using a continuous light source, levels of about 100 μM ATP have been released during a 32-ms flash in a rotating shuttered system (Forbush, 1984a). More recent work has employed either photoflash lamps or frequency-doubled ruby lasers. The quantum yield for ATP release (measured using the Hg arc lamp) is relatively high (about 0.55), so that the photochemical reaction yields ATP with a rather high efficiency. Studies using a frequency-doubled ruby laser to photolyze caged ATP at 347 nm and ATP-induced dissociation of actomyosin to measure the release rate have shown that the kinetics of release of ATP are controlled by a dark reaction. The controlling process had a rate constant of about 2.2 $\times 10^9 \, [H^+] \, s^{-1}$ at 22°C at pH 5.8–9.5, which corresponds to 220 s^{-1} at pH 7 (McCray et al., 1980).

Initial studies using the purified renal Na,K-ATPase showed that the photofragment resulting from the release of ATP (probably the nitroso ketone) was a potent inhibitor of the enzyme (Kaplan et al., 1978). The inhibition was dependent upon photolysis of caged ATP and was prevented by including either glutathione or bisulphite in the reaction medium. These two agents are known to react with nitroso compounds, and they presumably trap out the photoproduct responsible for the enzymatic inhibition (see Fig. 4). This reaction between glutathione and the photofragment has subsequently been applied in studies aimed at elucidating the photochemical reaction mechanism and rate constants for steps involved in ATP release from caged ATP (Goldman et al., 1982b). The observation that amounts of

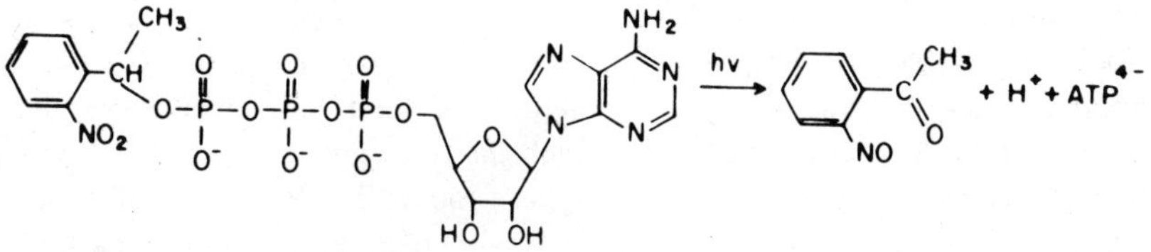

Figure 4.   Photolysis of caged ATP.

glutathione (or bisulfite) necessary to protect against inhibition were stoichiometrically related (1:1) to the amount of ATP released (and hence also the photofragment) suggested (1) that the photofragment was a potent inhibitor and (2) that glutathione reacted in a 1:1 manner to give a stable adduct. It was found necessary to include glutathione or bisulphite in studies using the purified Na,K-ATPase enzyme in order to prevent the photolysis-dependent inhibition. In studies on the sodium pump in resealed red cell ghosts, the addition of these protecting agents was not necessary (Kaplan et al., 1978). Presumably, when the system of interest is a minor component of the membrane, there are sufficient other pools of reactive sites available to mop up the reactive and probably not very specific photoproduced inhibitor.

The same experimental approach that has been employed for the adenosine phosphates can also be employed for the other nucleoside phosphates. A report describing the properties of a series of analogues of cyclic AMP contained the first reported synthesis and use of a caged cyclic AMP (Engels and Schlaeger, 1977). This compound, the 2-nitrobenzyl ester of the phosphate of cyclic AMP, was shown to enter cells of a neuronal cell line and after illumination initiate a differentiation response previously shown to be dependent upon intracellular cyclic AMP (Engels and Schlaeger, 1977). Presumably this approach can also be used for cyclic GMP, another hormone messenger of considerable interest to cell physiologists. The application of well-known photolability of 2-nitrobenzylic molecules is not the only illustration of a photochemical reaction used to rapidly initiate physiological processes. The interesting studies of Lester and co-workers using a photodependent isomerization to study cholinergic receptors (see Lester and Nerbonne, 1982) and the recently reported photoaromatization of calcium channel blockers leading to channel deblocking are examples of other applications of photochemical reactions employed to study cellular processes (Morad et al., 1983; Sanguinetti and Kass, 1984).

Precedents exist in the chemical literature for the photochemical manipulation of the concentration of a wide range of molecules of interest to the cellular physiologist. A list of such compounds is easily devised and might include the following: pH jumps by photolysis of esters of carboxylic or phosphoric acids, divalent cations via photoreleased or photolabile chelators, changes in hexose levels using caged glucose molecules, transport studies using caged amino acids where photodeblocking at the acidic or amino function is possible, initiation of ionic channel events using caged transmitters, peptide hormones where N-terminal or C-terminal blockage could be rendered photolabile, and so on. All of these approaches are plausible because of the photochemical reaction employed in the caged ATP strategy. Nitroaromatic compounds have a wide range of versatility in their photochemistry, and the particular lability of nitroaromatics in which the nitro group is ortho to a benzylic hydrogen has been long recognized. It was proposed in 1904 that "all aromatics which have a hydrogen ortho to a nitro group will be light sensitive" (see Morrison, 1969). The success of caged ATP as an ATP source is related to the effectiveness of ATP as a leaving group. A recent report employs another series of photochemical reactions of phosphate esters dependent on this type of property to generate efficient photochemical labelling of proteins with fluorescent reagents (see Givens and Matuszewski, 1984).

For new caged molecules to be successful, certain requirements should be met. The absorbance and excitation of the excited chromophore should be in an appropriate spectral range. Here caged ATP, with its simple 2-nitrobenzyl moiety, is probably close to the lowest wavelength limit. In some systems (e.g., visual tranduction), longer wavelength chromophores would be desirable. The caged molecules should be water soluble and not bind to the membrane proteins (ion pumps) or cytoplasmic proteins prior to photolysis. It has been observed that caged ATP will bind, prior to photolysis, to the ATP site of the sodium pump (see below); this may complicate the kinetics of effects subsequent to photoactivation. This requirement is in marked contrast to the desired properties of reagents for photoaffinity labeling, another light-activated process widely employed in cellular physiology studies. In photoaffinity labeling, the unphotolyzed reagent should bind reversibly to a specific site to which it becomes irreversibly attached after photolysis. The strategy for using photorelease techniques is to release the caged substrate after photolysis so as to increase its free concentration in a particular compartment. It should be pointed out that alterations in the chromophore to be excited may lead to changes in the action spectrum for release and the rate of release of the caged molecule. However, as yet the mechanistic details of the photorelease of caged ATP do not allow us to make such modifications and predict the consequences to the release rates for P_i from the modified caged P_i (the intermediate) or for ATP from the modified caged ATP. The complex photochemistry of 2-nitrobenzyl compounds ensures that many more possibilities are open following excitation than the single desired reaction (for examples, see Morrison, 1969).

3. STUDIES ON THE SODIUM PUMP

The sodium pump is responsible for maintaining high intracellular [K] and low intracellular [Na] in the face of low [K] and high [Na] in extracellular fluids. This is achieved by the expulsion of 3 Na ions and the coupled uptake of 2 K ions for each molecule of ATP hydrolyzed. The pump is also able to operate in several other modes including a Na:Na exchange (see Glynn and Karlish, 1975; Kaplan, 1983). This transport mode, which has been best characterized in human red blood cells, requires the presence of both ATP and ADP and involves a one-for-one exchange of intracellular and extracellular Na ions. Although ATP is required, there is no net hydrolysis. These observations led to the suggestion that the biochemical reaction sequence underlying this transport mode was a reversible part of the overall enzymatic hydrolysis cycle—namely the ATP:ADP exchange reaction. Most current ideas on the reaction mechanism of the sodium pump are derived from two separate bodies of information. The transport data come from studies on red cells, ghosts, or axons where sidedness has been maintained; the second body of information on enzymatic properties comes from studies on purified enzyme where the specific activities are high but all information on sidedness is indirect. To arrive at an understanding of the sodium pump reaction mechanism, the relationship between transport reactions and biochemical transformations must be better defined, especially with respect to the sidedness and affinity of activating cations for each type of process. The ATP:ADP exchange reaction, which had been assumed to accompany Na:Na exchange, involves the phosphorylation of the pump protein by ATP and the subsequent phosphorylation of ADP by the phosphoprotein. The way in which the reaction has been measured in enzyme preparations is to mix enzyme and the appropriate salts with [^{3}H]ADP and ATP and to measure the ouabain-sensitive (i.e., sodium pump mediated) rate of appearance of [^{3}H]ATP. The major problem in measuring this activity in a sided transporting system was to be able to initiate the reaction inside the resealed ghosts after their preparation was completed. If [^{3}H]ADP and ATP were trapped in the ghosts after hemolysis, several events would occur; ATP would break down during the 37°C incubation needed to reseal the ghosts, and the exchange reaction would be taking place continuously so that initial rates would be unobtainable. The use of caged ATP enabled these measurements to be made. Briefly, resealed ghosts were prepared to contain caged ATP, [^{3}H]ADP, and the appropriate salts; after the ghost suspensions were resealed and washed, the reaction was initiated by irradiation of the ghosts at 340 nm. This resulted in the release of ATP and the initiation of the ATP:ADP exchange reaction. By sampling and acid quenching the ghost suspension at 15-s intervals after photolysis (and release of ATP), the rate of the ATP:ADP exchange reaction could be followed (Kaplan and Hollis, 1980). The subsequent characterization of this partial reaction in a sided system enabled several conclusions to be drawn (Kaplan and Hollis, 1980; Kaplan, 1982). The complex triphasic curve relating [Na] to ATP:ADP exchange rate seen in unsided preparations could be accounted for on the basis of the separate effects (activating or inhibitory) of Na at either cytoplasmic or extracellular sites. These assignments could be made unambiguously once the intracellular and extracellular [Na] were independently manipulated and the ATP:ADP exchange rate was measured. The intracellular and extracellular affinities for Na in activating and inhibiting the ATP:ADP exchange rate correlated with the effects of internal and external Na on the Na:Na exchange transport process. The sided effects of K ions also correlated well with K effects on the transport reaction. It appears that the same cation sites that activate (or inhibit) the transport activate (or inhibit) the biochemical reaction, in keeping with the hypothesis that ATP:ADP exchange is the biochemical transformation taking place during Na:Na exchange. An additional piece of information was also obtained that was not available from studies on unsided (porous) systems. When Na:Na exchange could not take place (i.e., when there was no external Na), ATP:ADP exchange occurred at a significant rate (Kaplan and Hollis, 1980; Kaplan, 1982). Thus, although the transport and biochemical reactions are linked, they are not tightly coupled, and the biochemical reaction can go on in the absence of transport. These studies were made possible by the ability to release ATP rapidly inside the ghosts from a source that was resistant to endogenous ATPases.

Recent studies on Na,K-ATPase using microsomal vesicles from dog kidney outer medulla have utilized caged ATP to enable transport measurements to be made in right-side-out vesicles. In this system, following centrifugation of a kidney medulla homogenate, closed vesicles are obtained that form spontaneously with their membranes in the same orientation as in the renal cells. This places the ATP-hydrolyzing site of the sodium pump at the inner surface of the vesicle. Such vesicles contain high levels of Na,K-ATPase and thus have high rates of ATP hydrolysis. It is necessary then to protect the ATP until the time of the transport experiment. This was achieved using caged ATP. The membrane vesicles were loaded with caged ATP by osmotic shock and with radioisotopes (^{42}K or ^{22}Na) by equilibration. Photolysis of caged ATP in a 1-s flash (sufficient to photolyze about 80% of the 10 mM caged ATP) produced a rapid ^{22}Na efflux that was oubain-sensitive (Forbush, 1982). This approach will enable biochemical and transport studies to be performed in a sided oriented preparation where the sodium pump is a major fraction of the membrane protein. In a subsequent publication with the combined use of rapid filtration apparatus (see Forbush, 1984a) and photolysis of intravesicular caged ATP, it has been possible to measure the rates of Na movement through a single turnover of the Na pump and to resolve an efflux burst in the time range 20–1500 ms (Forbush, 1984b). The time course of ^{22}Na efflux at 15°C was found to be consistent with models of the Na pump in which extracellular Na release is an early step in the overall cycle.

In the course of these studies it was found that caged ATP binds to the catalytic site of the Na, K-ATPase. This possibility had not been previously investigated systematically. Forbush (1984b) was able to show that caged ATP binds to the Na, K-ATPase (in the absence of Mg^{2+}) with an affinity of ~43 μM; this compares with an affinity for ATP of ~ 0.5 μM. This means that for processes initiated by photolysis of caged ATP, where several rate constants are resolvable, the early steps may contain contributions from the dissociation rate of caged ATP from the catalytic site as well as ATP association. Whether the photochemical release rate is at all affected by the caged ATP being bound during photolysis has not yet been investigated. Indeed the effects of environment (ionic strength, dielectric constant) on the photolytic reaction are also unknown. As pointed out by Forbush (1984b), the consequences of the binding of caged ATP to a site prior to photolysis in such single turnover experiments are difficult to resolve when the photolysis is not complete and the medium is then composed of a mixture of ATP and caged ATP.

4.   STUDIES ON THE CALCIUM PUMP

The ATP-dependent active transport of calcium across the sarcoplasmic reticulum (SR) membrane in skeletal muscle causes the physiological relaxation of the myofibrils. In a recent series of studies of the structure and mechanism of the SR calcium pump, the photoactivated release of caged ATP has played a central role.

Oriented multilayers of stacked flattened disks of SR vesicles were obtained in a hydrated form with caged ATP between the disks. Ultraviolet flash photolysis of the caged ATP was used to initiate Ca^{2+} transport into the interiors of the disks. The transport was monitored spectrophotometrically (using metallochromic dyes), and the X-ray diffraction was simultaneously recorded within serial time windows of variable duration following the light flash (Herbette and Blasie, 1980). In these first studies lamellar diffractions were obtained from 5 s prior to a 100-ms light flash and compared to the diffraction obtained during 5 s immediately following the flash. Significant changes were evident in the ratio of second- and third-order lamellar reflections, which relaxed several minutes after the flash. The Ca^{2+} transport process in these multilayers was completed in 1–2 min. Herbette and Blasie (1980) showed that these diffraction differences resulted from a shift in electron density from the intravesicular surface of the single SR mechanism profile into the interior of the membrane's lipid core. The kinetics of ATP-induced Ca^{2+} uptake by dispersions of SR vesicles have been determined with a time resolution of about 10 ms using caged ATP (Pierce et al., 1983b). Ca uptake was monitored by using arsenazo III following flash photolysis of caged ATP with a frequency-doubled ruby laser. Two phases of uptake were distinguished in the temperature range $-2°C$ to $26°C$. The first phase of Ca uptake (with a specific rate constant of about 60 s^{-1} at 23–26°C and an activation energy of 16 kcal/mol) and the formation of phosphorylated enzyme (EP) occurred in the same time interval, and the stoichiometry was about two Ca ions per phosphoenzyme. This rapid uptake was not affected by the addition of a calcium ionophore and thus is probably not taken up into the free intravesicular volume. As pointed out by the authors, either these Ca^{2+} ions remain bound to the outside of the vesicles via an ATP-dependent increase in Ca affinity or they are not accessible from either side of the membrane and are "occluded" (Dupont, 1980). Such occlusion of ions has also been implicated as an important step in the transmembrane transport of ions by the sodium pump. The slow phase of Ca uptake had a specific rate constant of about 0.6 s^{-1} at 25–26°C and an activation energy of 22 kcal/mol. The ratio of ATP hydrolyzed to Ca ions transported during the slow phase was about 1:2; the uptake was abolished by the addition of a calcium ionophore. This slow phase then represents the translocation of Ca ions across the membrane. It was possible to time-resolve the fast phase of Ca uptake by SR because the specific rate was considerably slower than the release of ATP from caged ATP (220 s^{-1} at pH 7.0 and 22°C) and the diffusion of the photoreleased ATP to the ATPase site as well as the response of the metallochromic indicator arsenazo III. However, since the specific rates for the fast phase of calcium uptake and for ATP release are within a factor of 5, the kinetics of the fast phase may still be somewhat limited by the ATP release rate.

The most recent study in this series brought together transport studies in SR vesicle dispersions and in oriented multilayers. Since any advances in the time-resolved structural approach will be made using the oriented multilayer system, it was important to establish that the two systems behave kinetically in a similar way. In summary, both systems show the same biphasic kinetics of Ca^{2+} uptake, and both phases of uptake show essentially the same rates and activation energies. Not surprisingly, photolysis with either an ultraviolet lamp or frequency-doubled ruby laser gave comparable results; however, the laser offered the advantage of time resolution and synchronization of the Ca^{2+} transport cycles of the ensemble of Ca pump molecules. The time-resolved X-ray experiment, in which a structural change in the SR membrane profile was initiated by flash photolysis of caged ATP, can now be correlated with particular steps in of the Ca^{2+} transport process (see Pierce et al., 1983b).

5. CONCLUDING REMARKS

Once the properties of caged ATP are described, many of the potential applications made possible by the photorelease approach are obvious. Indeed the extension of this approach to other caged molecules is also straightforward, at least in principle. However, it is likely that before systematic investigations can be performed using such new compounds (even after successful synthesis), detailed studies on the photorelease rates and photorelease mechanisms will be necessary. These concluding comments will be directed only toward what we have learned from the application of caged ATP. As has been described, caged ATP can be utilized to initiate ATP-dependent processes in closed compartments (vesicles or ghosts) or in ordered structures (lamellae), where the direct addition of ATP is difficult or impossible or would disrupt the structure. Another advantage of this approach is that dilution considerations usually present in mixing experiments are avoided. This would be important in several areas. One example (as yet untried) might be in calorimetric studies on ATP-dependent enzymes. As with any experimental approach there are also limitations. The most obvious is kinetic. How fast is ATP released? Caged ATP is released rapidly (see above), but as in the case of the rapid phase of Ca^{2+} entry into SR vesicles, a molecule with a release rate 1 order of magnitude faster would be very useful. A second limitation, apparent in some sodium pump studies, is that caged ATP may bind to the ATP site prior to the photorelease of ATP, thus complicating the initial kinetics (see Forbush, 1984b). This type of effect has not been reported in muscle studies or in Ca^{2+} pump studies, but in the latter case it is possible that such an effect might have gone undetected. Obviously the extent to which the binding of caged ATP is a limitation will depend on the system being studied and the type of experiment. The potential problems due to the production of a photodependent inhibitor were solved in our initial studies by using glutathione or bisulfite (Kaplan et al., 1978); this approach has been successfully used where necessary in most subsequent studies on ion pumps or muscle. It is noteworthy that in all the studies reported till now in resealed ghosts, kidney vesicles, sodium pump enzyme, or SR vesicles, no untoward effects have been reported as a result of direct illumination in the 340-nm range.

It is clear from the published work with caged ATP that the photorelease approach has made possible (or more accessible) several different types of experiments on active transport systems. We can expect that in the near future the approach will be employed on a greater number of systems and that the application of the photorelease approach (of which caged ATP is but the first example) to other molecules will have an increasing impact on studies in cellular physiology.

ACKNOWLEDGEMENTS

The work described in this chapter that was performed in the author's laboratory was supported by NIH grant HL 30315. J.H.K. is a recipient of Research Career Development Award K04-HL-01092. I thank Linda J. Kenney for her comments on the manuscript and Dr. B. M. Salzberg for applying pressure.

REFERENCES

Baltrop, J. A., P. J. Plant, and P. Schofield (1966) Photosensitive protecting groups. *Chem. Commun.*, 822–823.

Dupont, Y. (1980) Occlusion of divalent cations in the phosphorylated calcium pump of sarcoplasmic reticulum. *Eur. J. Biochem.* **109**, 231–238.

Engels, J. and E. Schlaeger (1977) Synthesis, structure, and reactivity of adenosin cyclic 3',5'-phosphate benzyl triesters. *J. Med. Chem.* **20**, 907–911.

Forbush, B., III (1982). Characterization of right-side-out membrane vesicles rich in (Na,K)-ATPase and isolated from Dog kidney outer medulla. *J. Biol. Chem.*, **257**, 12678–12684.

Forbush, B., III (1984a) An apparatus for rapid kinetics analysis of isotopic efflux from membrane vesicles and of ligand dissociation from membrane proteins. *Anal. Biochem.*, **140**, 495–505.

Forbush, B., III, (1984b) Na$^+$ movement in a single turnover of the Na pump. *Proc. Nat. Acad. Sci. USA* **81**, 5310–5314.

Givens, R. S., and B. Matuszewski (1984) Photochemistry of phosphate esters: An efficient method for the generation of electrophiles. *J. Am. Chem. Soc.* **106**, 6860–6861.

Glynn, I. M., and S. J. D. Karlish (1975) The sodium pump. *Ann. Rev. Physiol.* **37**, 13–55.

Goldman, Y. E., M. G. Hibberd, J. A. McCray, and D. R. Trentham (1982a) Relaxation of muscle fibres by photolysis of caged ATP. *Nature,* **300**, 701–705.

Goldman, Y. E., H. Gutfreund, M. G. Hibberd, J. A. McCray, and D. R. Trentham, (1982b) The role of thiols in caged-ATP studies and their use in measurement of the photolysis kinetics of *o*-nitrobenzyl compounds. *Biophys. J.* **37**, 125a.

Herbette, L., and J. K. Blasie (1980) In F. L. Siegel, E. Carafoli, R. H. Kretsinger, D. H. McLennan, and R. H. Wasserman, Eds., *Calcium Binding Proteins: Structure and Function,* Elsevier North-Holland Amsterdam, pp. 115-120.

Kaplan, J. H. (1982) Sodium pump-mediated ATP:ADP exchange: The sided effects of sodium and potassium ions, *J. Gen. Physiol.* **80**, 915–937.

Kaplan, J. H. (1983) Sodium ions and the sodium pump: Transport and enzymatic activity, *Amer. J. Physiol.* **245**, G327–G333.

Kaplan, J. H., and R. J. Hollis (1980) The external Na dependence of ouabain-sensitive ATP:ADP exchange initiated by photolysis of intracellular caged-ATP in human red cell ghosts, *Nature* **288**, 587–589.

Kaplan, J. H., B Forbush, III., and J. F. Hoffman (1978) Rapid photolytic release of adenosine 5'-triphosphate from a protected analog: Utilization by the Na:K pump of human red blood cell ghosts, *Biochemistry* **17**, 1929–1935.

Lester, H. A., and J. M. Nerbonne (1982) Physiological and pharmacological manipulations with light flashes. *Ann. Rev. Biophys. Bioeng.* **11**, 151–175.

McCray, J. A., L. Herbette, T. Kihara, and D. R. Trentham (1980) A new approach to time-resolved studies of ATP-requiring biological systems: Laser flash photolysis of caged ATP. *Proc. Nat. Acad. Sci. USA* **77**, 7237–7241.

Morad, M., Y. E. Goldman, and D. R. Trentham (1983) Rapid photochemical inactivation of Ca^{2+}-antagonists shows that Ca^{2+} entry directly activates contraction in frog heart. *Nature* **304**, 635–638.

Morrison, H. A. (1969) The photochemistry of nitro and nitroso groups. In H. Fever, Ed., *Chemistry of the Nitro and Nitroso Groups.*, Wiley, New York, pp. 165–213.

Pierce, D. H., A. Scarpa, M. R. Topp, and J. K. Blasie (1983a) Kinetics of calcium uptake by isolated sarcoplasmic reticulum vesicles using flash photolysis of caged adenosine 5'-triphosphate. *Biochemistry* **22**, 5254–5261.

Pierce D. H., A. Scarpa, D. R. Trentham, M. R. Topp, and J. K. Blasie (1983b) Comparison of the kinetics of calcium transport in vesicular dispersions and oriented multilayers of isolated sarcoplasmic reticulum membranes *Biophys. J.* **44**, 365–373.

Rubenstein, M. B. Amit and A. Patchornick (1975) The use of a light-sensitive phosphate protecting group for some mononucleotide syntheses. Tetrahedron Lett., 1445–1448.

Sanguinetti, M. C., and R. S. Kass (1984) Photoalteration of calcium channel blockade in the cardiac Purkinje fiber. *Biophys. J.* **45**, 873–880.

CHAPTER 23

LASER PULSED RELEASE OF ATP AND OTHER OPTICAL METHODS IN THE STUDY OF MUSCLE CONTRACTION

Y. E. GOLDMAN

Department of Physiology
University of Pennsylvania School of Medicine
Philadelphia, Pennsylvania

1.	INTRODUCTION	398
2.	LASER PULSE PHOTOLYSIS TECHNIQUE	399
3.	ILLUMINATION SOURCES	401
4.	PHOTOLYSIS KINETICS OF CAGED ATP	403
5.	OPTICAL RULERS AND PROTRACTORS	405
6.	APPLICATIONS TO MUSCLE CONTRACTION	408
7.	CONCLUSIONS	411
	REFERENCES	412

Acknowledgments: I thank Drs. M. G. Hibberd and D. R. Trentham and Ms. J. A. Dantzig for allowing me to present their unpublished data and Drs. S. M. Baylor and D. D. Thomas for helpful discussions. The work was funded by the Muscular Dystrophy Associations of America and by NIH grants HL15835 to the Pennsylvania Muscle Institute and AM26846 and AM00745.

397

1. INTRODUCTION

The widely accepted view of energy transduction by muscle is that contraction occurs by relative sliding of two interdigitating sets of protein filaments in the sarcomere (A. F. Huxley, 1957; H. E. Huxley, 1969). A cyclic mechanical interaction between myosin heads in the thick filaments and actin in the thin filaments is driven by hydrolysis of adenosine triphosphate (ATP) to adenosine diphosphate (ADP) and inorganic phosphate (P_i) (A. F. Huxley, 1974). To relate the chemical and mechanical events, a tentative scheme was proposed by Lymn and Taylor (1971) from rapid reaction studies on isolated myosin and actomyosin. An updated version of their scheme is shown in Fig. 1.

The reaction cycle involves (a) mechanical attachment of a myosin head to actin to form a cross-bridge, (b and c) a power stroke, shown hypothetically as a rotation of the head, tending to cause filament sliding, and (d) detachment. These steps are supposed to correspond with (a) association of a myosin–ADP–P_i complex to actin, (b and c) release of P_i, then ADP and (d) ATP binding to the nucleotide-free (rigor) cross-bridge and actomyosin dissociation. Major goals of current research are to test this hypothesis for the mechanochemistry of muscle and to determine if the structural change corresponding to the power stroke is indeed a rotational motion of the myosin head.

Optical techniques have contributed to recent advances in both of these areas. A laser pulse photolysis technique for abruptly initiating the cross-bridge cycle has been developed in collaboration with D. R. Trentham and M. G. Hibberd and is discussed

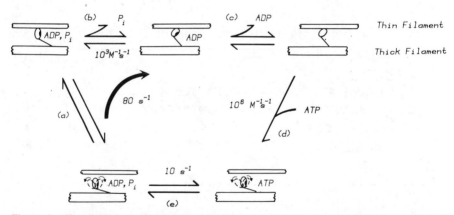

Figure 1.   Hypothetical reaction scheme to relate actomyosin biochemistry with the mechanical cross-bridge cycle. Each cartoon depicts a biochemical intermediate state of the actomyosin ATPase reaction. Myosin is considered to be either attached to actin in specific attitudes (upper row) or detached from actin and relatively mobile (lower row). The cross-bridge power stroke is correlated with release of product phosphate (reaction (b)), implying that an actomyosin•ADP state bears the force which causes active filament sliding. This assumption does not rule out force generation by actomyosin•ADP•P_i or actomyosin with no nucleotide. The rate constants listed with the reactions were measured in muscle fibers as discussed in the text.

in Sections 2, 3, and 4. Orientation-sensing and distance-sensing optical probes of myosin structure are discussed in Sections 5 and 6.

2. LASER PULSE PHOTOLYSIS TECHNIQUE

The object of the photolysis studies has been to produce a step change in concentration of ATP or other phosphate compounds within a muscle fiber in order to start the cross-bridge cycle abruptly and from a defined state. By monitoring the resulting mechanical events we can measure the rates of elementary cross-bridge reactions. We can determine which steps depend on mechanical strain and which are relatively slow and thus candidates for possible rate-limiting control points. By comparing these data with corresponding reaction rates measured with isolated actomyosin, we can determine how organization of the proteins into a filament lattice modifies the kinetics. Thus we hope to identify the reaction pathway during energy transduction.

The surface membranes of fibers from rabbit psoas, frog semitendinosus, scallop aductor, or *Lethocerus* (giant water bug) flight muscle were removed by glycerol extraction, detergent treatment, or mechanical dissection. The fibers were put into rigor by washing in ATP-free media. Caged ATP (P^3-1-(2-nitro)-phenylethyladenosine 5′-triphosphate (Kaplan et al., 1978), 0–20 mM, was then allowed to diffuse into the fiber interior. Steady concentrations of Ca^{2+}, P_i, and ADP or temperature and filament overlap were altered for specific experiments. Equilibration of the fiber with the caged ATP and variation of the other ionic constituents at constant total ionic strength did not markedly alter rigor tension. A pulse of 347-nm radiation from a laser, described in Section 3, photolyzed up to 50% of the caged ATP to ATP and 2-nitrosoacetophenone. Liberated ATP then bound to the rigor cross-bridges, initiating tension and stiffness transients as the cross-bridges detached, reattached, and generated active force.

Results from this type of experiment on a rabbit muscle fiber (Fig. 2) show that tension and stiffness decay rapidly after caged ATP photolysis in the absence of Ca^{2+} (Goldman et al., 1982b, 1984a). When Ca^{2+} is present, release of ATP switches the muscle fiber from rigor (moderate force, high stiffness) into an active contraction (high force, moderate stiffness) without an initial full relaxation (Goldman et al., 1984b). These transients indicate that cross-bridge reattachment occurs rapidly after detachment.

The results from a series of photolysis experiments can be summarized by the numerical values shown in Fig. 1 for the elementary reaction rates in muscle fibers. These values were measured by varying ATP, ADP, P_i, and Ca^{2+} concentrations in photolysis experiments. ATP binding and detachment of the rigor cross bridge (reaction d) is rapid ($\approx 10^6\ M^{-1}\ s^{-1}$; Goldman et al., 1984a), not Ca^{2+}-dependent (Goldman et al., 1984b), and similar in rate to ATP-induced dissociation of isolated actomyosin (White and Taylor, 1976). Reaction (d) is too fast at physiological ATP concentration (5 mM) to control overall cross-bridge cycling in a muscle fiber. The ATP hydrolysis step (reaction e) has been studied in muscle fibers by another method (oxygen exchange; Hibberd et al., 1984b, 1985b) and appears to follow a similar

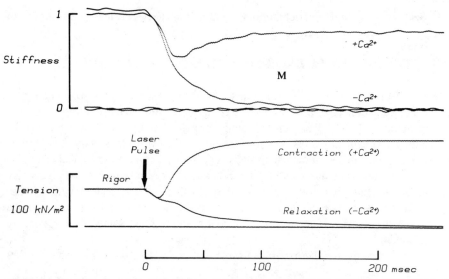

Figure 2.   Mechanical transients initiated by photolysis of caged ATP in a glycerol-extracted single fiber from rabbit psoas muscle. Records obtained in separate trials in the presence and absence of Ca^{2+} are superimposed. The bathing solution contained Caged ATP, 10 mM; free Mg^{2+}, 1 mM; TES buffer, 100 mM; pH, 7.1 at 21°C; total ionic strength, 200 mM and EGTA, 53 mM for the $-Ca^{2+}$ traces or EGTA, 20 mM; free Ca^{2+}, 30 μM; and HDTA, 33 mM for the $+Ca^{2+}$ traces. A laser pulse (arrow) liberated 0.38 mM ATP from caged ATP. Stiffness was detected by synchronous demodulation of the force required to alter the fiber length sinusoidally 1 μm at 500 Hz. The lowest stiffness and tension traces are baselines recorded after relaxation of the fiber. Release of ATP rapidly switches the fiber from rigor to relaxation or contraction.

chemical pathway with a comparable rate to that measured with the solubilized proteins (Webb and Trentham, 1981). Cross-bridge reattachment and force generation (reactions (*a*) and/or (*b*) occur at about 80 s^{-1} (Goldman et al., 1984b). This rate is independent of ATP concentration, as expected from Fig. 1. The step leading to force generation seems to be reversible; inorganic phosphate (P$_i$) in the millimolar range can apparently bind to the force-generating intermediate (an actomyosin–ADP complex) at approximately 10^3 M^{-1} s^{-1} and reverse the power stroke (Hibberd et al., 1984a, 1985a). The presence of 0.02–1 mM ADP slows detachment of rigor cross-bridges. This slowing may indicate a cooperative influence of the Ca^{2+}-dependent regulatory system and thus may serve a role in regulation of contraction (Dantzig et al., 1984).

These kinetics provide support for the scheme of Fig. 1. Several of the reactions—namely detachment of the rigor cross-bridge (*d*), ATP cleavage (*e*), and reattachment (*a*)—are rapid and not markedly different from the corresponding reactions in isolated and solubilized actomyosin. On the other hand, the steps involving release of product P$_i$ and ADP depend on mechanical cross-bridge strain, and there are differences in rates between fibers and the solublized proteins. Since the product release steps are reversible, an influence of mechanical strain on these chemical reactions thermodynamically implies a coupling to force generation.

Further investigations will focus on quantitating the relationship between mechanical stress and the chemical product release steps.

3. ILLUMINATION SOURCES

The components of the laser used for the muscle photolysis studies (Control Laser model 634) are shown in Fig. 3. A ruby rod (160 × 13 mm diameter) is water cooled and surrounded by a helical xenon flash lamp. The resonant cavity for laser oscillation is formed by the partially reflecting front end of the ruby rod and by a totally reflecting back mirror. When the flash lamp is triggered, absorption of ultraviolet light excites chromium ions in the ruby crystal to a high-energy state capable of optically stimulated emission of photons. Laser oscillation is initially prevented by a Pockels cell and Brewster angle stack-plate polarizer, which damp the resonant quality (Q) of the laser cavity. Near the peak of the xenon lamp pulse, high voltage on the Pockels cell (3 kV) is rapidly removed by a thyristor circuit. The combination of the Pockels cell and Brewster stack then become transparent, switching the Q of the laser cavity sufficiently high to promote laser oscillation. The laser produces 3–5 J of red light (694 nm) per pulse. The Q switching causes the pulse duration to be extremely short (50 ns) with high peak power ($\approx$ 100 MW) necessary for frequency doubling.

Ultraviolet light at 347 nm is generated from the 694-nm laser beam by a nonlinear optical element, a crystal of rubidium dihydrogen arsenate (RDA, Inrad 520-004). The crystal has a small, nonlinear coefficient leading to dielectric polarization proportional to the *square* of the oscillating electric field of the 694-nm light in addition to the normal component directly proportional to the electric field. This nonlinearity of polarization introduces harmonics of the primary radiation which propagate in the crystal. Components of 347-nm radiation generated at various depths in the crystal are out of phase and destructively interfere unless the propagation

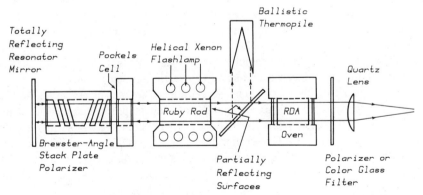

Figure 3.   Components of the frequency-doubled ruby laser used for photolysis of caged ATP in muscle fibers. The ruby rod and totally reflecting resonator mirror constitute the laser oscillator. The Pockels cell and intracavity polarizer serve to decrease the duration of the laser pulse. The ballistic thermopile monitors the ruby laser output pulse. The temperature-controlled RDA crystal doubles the frequency to provide 347-nm UV light. See text for further details.

velocity for the second harmonic (347 nm) matches the velocity at 694 nm. Propagation velocity is related to the reciprocal of the refractive index (n) and for RDA, $n_{347} = n_{694}$ when the crystal is heated to 101°C. The RDA "temperature tuning" curve for 347-nm output is narrow (full-width half-maximum ≈ 0.8°C), so the crystal is housed in a thermally insulated and servo-regulated oven.

At the exit of the frequency doubler, the 347-nm beam is polarized perpendicular to the 694-nm beam. Thus a second Brewster stack polarizer transmits the ultraviolet light and deflects the primary beam. Other methods for separating the two beams are by wavelength dispersion in a quartz prism or with a color glass filter (Oriel 5181). The final ultraviolet beam is a highly collimated, 12-mm diameter, 50-ns, 100- to 300-mJ pulse at 347 nm.

In a typical experiment, a trough containing the muscle fiber in a rigor buffer with 10 mM caged ATP is irradiated with a 100-mJ UV pulse from the laser. The final concentration of ATP released from caged ATP is 0.5–1 mM, determined by a luciferin–luciferase assay (Goldman et al., 1982b) or by high-performance liquid chromatography (Goldman et al., 1984a).

The frequency-doubled ruby laser provides a brief, high-energy pulse at a wavelength appropriate for photolysis of caged ATP. The UV radiation can be easily focused onto a small preparation, because the output beam is highly collimated. However, the instrument is costly to obtain and to operate, and therefore alternative sources of UV illumination should be considered for other applications.

In addition to the ruby laser used in our laboratory, several other laser sources are commercially available. The neodymium:YAG laser emits at 1064 nm, and this wavelength can be efficiently frequency-tripled to produce 355-nm pulses. The nonlinear crystals used in this system (KD*P) are not as prone to optical radiation damage as the RDA used in the ruby laser system. Excimer lasers provide a choice of several emission wavelengths depending on the gas fill. With XeF gas, collimated and uniform 350-nm pulses are available up to about 1 J in energy. Although this relatively new technology has a high initial cost, the choice of several wavelengths and the high expected reliability make excimer lasers an attractive possibility for photolysis experiments. The nitrogen laser emits pulses at 337 nm and probably is the least expensive commercial pulsed UV laser. The output energy is less than the other lasers discussed, but for small samples, such as single cells, a N$_2$ laser may suffice, since the light can be focused to a small (diffraction-limited) spot.

Arc lamps are relatively inexpensive and intense sources of UV radiation. The output of the continuous mercury arc lamp contains an intense spectral line at 356 nm, and this source is used in muscle fiber studies for pilot photolysis experiments and to produce ramp changes of ATP concentration (Goldman et al., 1984a). The radiation from a 200-W mercury arc lamp is collected by a high-aperture quartz condenser. The light is focused onto the muscle fiber by a f/2.0 quartz lens, and 300–400 nm radiation is selected by a water filter to remove heat and by a color glass filter. An electromechanical shutter times the exposure for photolysis. In a typical experiment a

1-s exposure from the filtered mercury arc lamp liberates 300 μM of ATP from 2 mM caged ATP in a 3 × 8 × 1 mm trough.

Illumination periods on the order of 1 s limit the experimental time resolution for kinetic experiments. This problem can be overcome by using a pulsed arc lamp. A commercial xenon arc lamp excited by a 200-μs, 200-J pulse, in an optical arrangement as described for the continuous lamp, liberated 250 μM ATP from 5 mM caged ATP.

A disadvantage of arc lamp systems compared to lasers is that the light is more difficult to direct into a small region because the source is several millimeters long. High-aperture lenses with concomitant reduction of working distance are required for focusing to a small spot. Since the arc size generally increases with lamp power or energy, a higher-power lamp does not ensure delivery of increased radiation to the experimental preparation. Thus geometrical, temporal, spectral, and economic considerations must be balanced in the selection of a photolysis radiation source.

4. PHOTOLYSIS KINETICS OF CAGED ATP

Experiments to elucidate the photochemical reaction scheme for caged ATP hydrolysis have indicated that upon illumination, a photochemical intermediate is formed immediately and then splits on the millisecond time scale (see Chapter 22, Fig. 4). The intermediate compound provides a chromophoric probe into the dark reactions leading to photolysis (McCray et al., 1980). It introduces an extra component of absorption at 406 nm and decays at the same rate as ATP is liberated (McCray et al., 1986). The ATP liberation rate is proportional to proton concentration and has an apparent first-order rate constant of 118 s^{-1} in the buffer used for physiological experiments at pH 7.1 (Goldman et al., 1984b).

The compound that separates from ATP undergoes further slow reactions to form the by-product of caged ATP photolysis, 2-nitrosoacetophenone (McCray et al., 1986). This by-product can react with the contractile proteins, and in fiber experiments it proved necessary to include reduced glutathione (10 mM) in the medium to prevent loss of sensitivity to ATP after photolysis trials (Goldman et al., 1982a, 1984a). The glutathione probably protects the fibers by competing with protein thiol groups for the 2-nitrosoacetophenone.

The practical result of these dark reactions is that after a brief illumination, ATP is liberated exponentially with a time constant of 8 ms. The reactions of the cross-bridge cycle occur with time constants in this same range (1–50 ms), so there is a reasonable concern whether the photolysis reactions limit or obscure the biological processes under study. In the muscle fiber experiments, the final ATP concentration liberated (0.5–1mM) is usually greater than the myosin head concentration (≈ 0.2 mM). In that situation, the first ATP molecules released from caged ATP can bind to most myosin sites before the first time constant of the dark reaction. Thus the photochemical kinetics do not severely hinder our conclusions.

A kinetic simulation demonstrates this point quantitatively. The following set of sequential reactions was solved numerically (Gear, 1971):

$$\text{Caged ATP} \xrightarrow{k_c} \text{ATP}$$

$$\text{Actomyosin} + \text{ATP} \xrightarrow{k_d} \text{myosin}$$

Typical initial experimental conditions would be 0 ATP, 0.6 mM photoexcited caged ATP immediately after the laser pulse, and 0.2 mM actin-bound myosin heads. The first-order rate constant for liberation of ATP (k_c) was set at 100 s^{-1}, and the second-order rate constant for ATP binding to actomyosin and detaching the cross bridges (k_d) was $10^6 M^{-1} s^{-1}$. Under these conditions the half-time for detachment of cross-bridges in the simulated reaction sequence is 6.3 ms—that is, within the first time constant of ATP release. Figure 4(A) shows simulated curves of cross-bridge detachment when the rate k_c of the photolysis dark reaction is varied. If k_c is increased or decreased by a factor of 3, the half-time of cross-bridge detachment is altered in the corresponding direction by a factor of about 2. Note that there is a delay phase sensitive to k_c. When the rate constant for ATP-induced cross-bridge detachment (k_d) is increased or decreased by a factor of 3 (Fig. 4B), the half-time for simulated cross-bridge detachment also changes by a factor of about 2. Thus the values of k_c and k_d both make substantial contributions to the observed kinetics for cross-bridge detachment. If k_c is known from other measurements, then k_d can be estimated by comparing the experimental data to these simulations. However, estimation of k_d requires an accurate value for k_c.

The effect of altering the rate of ATP release can be experimentally tested by varying pH, since k_c is proportional to the H$^+$ concentration. Figure 5 shows records of tension relaxation from rigor initiated by photolysis of caged ATP in the absence of free Ca^{2+} and at three different pH values. Each panel shows two superimposed trials in which the fiber was either held isometric (i) or stretched (s) 0.54% 1 s before the laser pulse. Cross-bridges detach and then reattach briefly in

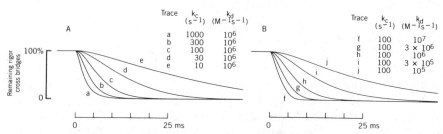

Figure 4.   Kinetic simulation of ATP liberation from caged ATP followed by ATP-induced detachment of rigor cross-bridges. Photoexcited caged ATP was considered to release 0.6 mM ATP exponentially with rate constant k_c. The liberated ATP then bound to rigor cross-bridges and detached them with a second-order rate constant k_d. (A) The expected effect of varying k_c. (B) The expected effect of varying k_d. Both rate constants significantly influence the simulated cross-bridge detachment.

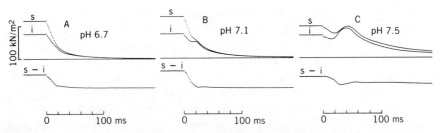

Figure 5.   Relaxation from rigor after photolysis of caged ATP in the absence of Ca^{2+} at three pH values. Each panel shows two superimposed tension recordings in an isometric trial (i) and in a subsequent trial (s) when the fiber was stretched by 0.54% 1 s before the laser pulse. The flat traces are baselines recorded in the photolysis solution after relaxation. The lowest traces are algebraic differences ($s-i$) between the prestretch and isometric tension recordings. Experimental conditions were as in Fig. 2 except for pH as indicated.

these conditions before the final relaxation (Goldman et al., 1982b, 1984a). To isolate the first detachment step from these subsequent steps, the two traces are subtracted on the assumption that the extra mechanical strain imposed by the prestretch for trace (s) is lost when the cross-bridge detaches. The algebraic differences ($s - i$) between the prestretch and isometric transients are shown beneath the original recordings.

The difference traces indicate the detachment of just the original rigor cross-bridges, so they decay much faster than tension that has contributions from reattachments. At pH 7.1 the difference trace decays to zero after a slight delay. This delay phase is increased in duration at pH 7.5. The change in half-time for cross-bridge detachment changes less than the rate constant for caged ATP release for either an increase or decrease of pH. Thus the rate of cross-bridge detachment is not solely limited by the photochemistry, as predicted in the simulations of Fig. 4. A direct effect of pH on cross-bridge detachment would also appear in the experiment of Fig. 5. However, with isolated actomyosin, changes of pH do not markedly alter ATP-induced dissociation (Finlayson et al., 1969; White and Taylor, 1976).

The experiments show that the photochemistry of caged ATP does not prevent quantitation of the rates of elementary cross-bridge reactions. Nevertheless, a faster caged ATP would still be desirable. Some compounds have been synthesized with this goal (Goldman et al., 1982a; Nerbonne and Lester, 1983; Nerbonne et al., 1984).

5.   OPTICAL RULERS AND PROTRACTORS

The high degree of macromolecular structural regularity of muscle allows application of many optical techniques to the problem of contraction. Sliding of fixed-length filaments was first deduced from light-microscopical observations (A. F. Huxley and Niedergerke, 1954; H. E. Huxley and Hanson, 1954). Birefringence (Taylor, 1976) and correlation spectroscopy of light scattered from muscle fibers (Haskell and Carlson, 1981) may provide structural information at the molecular level. A variety of techniques based on coherent light diffraction by the sarcomere periodicity or optical detection of fiber segment length have been used to signal filament sliding and to "clamp" the sarcomere length for mechanical studies (Cleworth and Edman, 1972; Pollack et al., 1977; McCarter, 1981; Goldman and Simmons, 1984; Gordon et al., 1966; A. F. Huxley et al., 1981).

Another group of techniques can detect angles and rotations of macromolecules and the distance between well-defined sites in a biological structure (Morales et al., 1982). These methods may identify the structural change corresponding to the "power stroke" of the cross-bridge cycle. Orientation- and distance-sensitive optical methods have also been applied to detect motions of biological membrane components, and several excellent reviews have been published (Stryer, 1978; Cherry, 1978; Jovin et al., 1981; Kinosita et al., 1984). However, they have not been widely applied to gated excitable membrane channels, the traditional territory of "cell physiologists."

Photons absorbed or emitted by chromophoric molecules are generally polarized with respect to specific molecular axes. Thus, orientation of a population of chromophores can be detected by the preferred polarization axis for light absorption or luminescence. The electronic levels of a chrompophore are illustrated in Fig. 6 (Cherry, 1978; Kinosita et al., 1984). S_0 is the ground state, and absorption of a photon of appropriate energy excites the molecule to higher energy levels S_1, S_2, and so on. Nonradiative processes such as thermal transfer of energy to surrounding molecules typically result in rapid decay of S_2 and S_3 states to the lowest directly excited state, S_1. This state may decay thermally, or, if the fluorescence quantum yield is appreciable, the molecule may return to S_0 by emitting a photon at a somewhat longer wavelength than the excitation wavelength. This so-called "prompt fluorescence" is emitted 5–50 ns after excitation, depending on the particular molecule and the conditions, such as solvent, temperature, and environment of the chromophore.

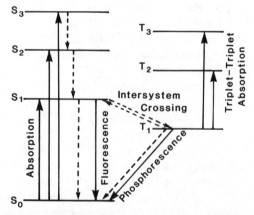

Figure 6. Schematic electronic energy level diagram of a typical chromophoric molecule. Transitions involving absorption or radiation of a photon are indicated by solid arrows. Thermal transitions are dashed. Vibrational states have been omitted for simplicity. Modified from Cherry (1978).

"Intersystem crossing" is a reversal of one of the electron spins of the excited molecule leading to state T_1, which is paramagnetic because it has two unpaired electrons. A magnetic field splits the energy of this state into three levels, and thus T_1 and a series of higher-energy paramagnetic states (T_2, T_3, etc.) are termed triplet states. The ground state (S_0) and states directly excited by photon absorption (S_1, S_2, etc.) are not split by a magnetic field and are thus termed singlet states. Transition from T_1 to S_0 is relatively slow (μs to ms time scale), because the required spin reversal is quantum-mechanically "forbidden." Radiative decay of T_1 directly to S_0 is termed phosphorescence and in certain molecules has an appreciable probability. Phosphorescence occurs at longer wavelengths than fluorescence.

In some cases, thermal activation or chemical reactions can allow T_1 to repopulate S_1; this is followed by radiative decay to S_0. This route is termed delayed fluorescence and occurs at the same wavelength as the prompt fluorescence but on the time scale of phosphorescence. Delayed fluorescence and phosphorescence emission occur up to milliseconds after excitation depending on the molecule and environmental factors. For instance, molecular oxygen quenches phosphorescence.

These absorption and emission properties can be used to detect orientation and rotation of the chromophore as follows. The probability (p) for absorption of a photon by a molecule depends on the angle θ between the polarization of the incident light and the absorption dipole of the molecule according to the relation

$$p \propto \cos^2(\theta) \tag{1}$$

Thus, polarized light preferentially "tags" molecules oriented with their absorption dipoles near the polarization axis. If the chromophores are oriented, the light absorption will have a preferential polarization, an effect termed linear dichroism. If the molecules are rigidly fixed in place, fluorescence or phosphorescence will also be polarized, although not necessarily along the same axis as the absorption dichroism. Polarization of the emitted light is termed fluorescence or phosphorescence "anisotropy." If the molecules are mobile on the time scale of the excited state, they will rotate between absorption and emission, thus decreasing the anisotropy. Decay of fluorescence or phosphorescence anisotropy can be time-resolved if the excitation is brief compared to the lifetime of the excited state. These signals provide a direct measure of the rate and extent of rotational motion of the chromophoric molecule.

Besides anisotropy of the emitted photons, there are several other ways to probe the orientations. Depletion of the population of S_0 states is detectable by decreased absorption or decreased fluorescence excited by a weak probe beam. A new dichroic absorption may develop corresponding to triplet–triplet transitions between energy levels T_1 and T_2 or T_3 (Jovin et al., 1981).

Another useful decay mode of the excited state is fluorescence energy transfer. A fluorescent molecule can directly excite a nearby molecule if the fluorescence emission spectrum of the first molecular species overlaps the absorption spectrum of the second. The two molecules are termed donor and acceptor. Energy transfer is highly sensitive to the distance between the two chromophores and to the angle between the donor emission and the acceptor absorption dipoles. Fluorescence energy transfer can be detected by decreased donor fluorescence, decreased lifetime of the donor excited state, or fluorescence at the acceptor emission wavelenth. At high acceptor concentrations, "trivial absorption" of photons emitted by the donor will occur, but this can be distinguished from fluorescence energy transfer which is direct interaction between the molecules. Quantitative distance measurements can be made by applying theoretical expressions (Stryer, 1978):

$$R = R_0(1/E - 1)^{1/6}$$

$$\tag{2}$$

$$R_0 = 9.7 \times 10^2 \ (JK^2\phi_D n^{-4}) \quad nm$$

$$\tag{3}$$

where R is the distance to be determined between the two chromophores, E is the fraction of energy transferred, R_0 is the distance at which $E = 1/2$, J is the integral of overlap between donor emission and acceptor absorption spectra, K^2 is an orientation and mobility factor, ϕ_D is the fluorescence quantum yield of the donor in the absence of the acceptor, and n is the refractive index of the medium. Since energy transfer depends on R^6, the estimate of R obtained by the method in the case of mobile chromophores is closer to a minimum distance than a mean separation (Thomas et al., 1978). K^2 is usually not known but can be set within limits (Stryer, 1978; Dale and Eisinger, 1975).

For protein systems the choices of chromophores for orientation and distance measurements are usually intrinsic tryptophan and tyrosine residues, fluorescent analogues of ligands, and reaction substrates or dye probe molecules covalently bound to a specific site on the protein such as a cysteine residue. Extrinsic probes have the advantage that the optical properties can be optimized for the experiment, but the specificity or heterogeneity of labeling and the perturbtion of the system under study by the probes are concerns. However, these factors have been controlled in many biological systems.

6.   APPLICATIONS TO MUSCLE CONTRACTION

Only a few experiments are described here, but comprehensive discussion is provided in Morales et al. (1982), Wilson and Mendelson (1983), Dos Remedios and Cooke (1984), and Eads et al. (1984). Figure 7 shows an experimental setup for studies of polarized fluorescence of a muscle fiber (Wilson and Mendelson, 1983). Brief (0.3-ns) pulses of polarized light from a N_2 laser pumped dye laser are directed at the fiber, and fluorescent emission is collected by an objective lens. A band-pass filter and Wollaston prism direct the collected light polarized perpendicular and parallel to the fiber on to photomultiplier tubes. Electronic gating of the photomultiplier output voltage is used to eliminate scattered laser light from the recorded signals. Small mathematical corrections can be applied to the data for the effect of the objective lens aperture (Borejdo et al., 1982; Wilson and Mendelson, 1983).

Several extrinsic probe molecules that have been useful in muscle contraction studies are shown in Fig. 8. 1,5-IAEDANS (N-iodoacetyl-N'-(1-sulfo-

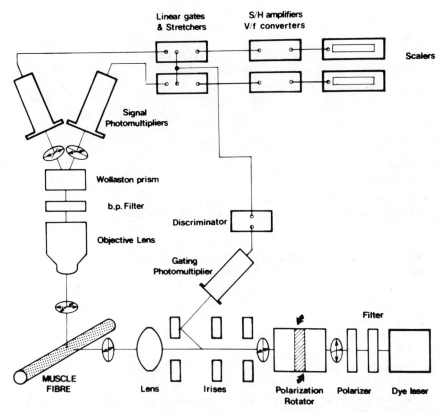

Figure 7. Microfluorometer used by Wilson and Mendelson (1983) to measure fluorescence anisotropy of AEDANS bound to myosin heads in muscle fibers. Pulses of 390-nm light were directed on to the muscle fiber at a polarization angle (indicated by the double-headed arrows) set by a polarization rotator. The partially polarized fluorescent emission from the bound AEDANS was collected by an objective lens, 420 to 435-nm band-pass (b.p.) filter, Wollaston prism, and two photomultiplier tubes. A third photomultiplier tube gated the recording electronics to reduce the unwanted signal from directly scattered light.

5-napthyl)ethylenediamine) forms a covalent bond to a particularly reactive cysteine residue in the myosin head, termed SH1. Under optimal conditions, iodoacetamide derivatives of dye molecules bind fairly specifically to this residue. Up to about 70% of the myosin heads of glycerol-extracted rabbit psoas fibers have been labeled with iodoacetamide probes with little alteration of active force production or shortening velocity (Nihei et al., 1974; Crowder and Cooke, 1984).

Time-resolved measurements of polarization anisotropy decay of labeled myosin heads in solution suggest that the probe molecule attaches at a relatively fixed angle to the protein (Mendelson et al., 1973). Polarized fluorescence from fibers labeled with 1,5-IAEDANS indicates the degree of order in the array of probe molecules and the angles of their absorption and emission dipoles with respect to the fiber axis. In rigor, the specifically bound probe molecules are well ordered and oriented at 30–40° to the

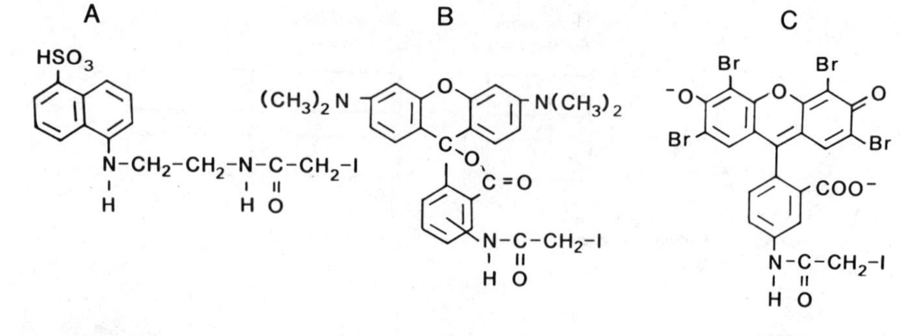

Figure 8.   Structures of fluorescent (*A*) and (*B*) and phosphorescent (*C*) probes. All three molecules bind covalently and specifically to the SH1 cysteine residue in the myosin head.

fiber axis (Wilson and Mendelson, 1983). This result reflects ordering of the cross-bridge attachment angle, but the numerical value need not represent the angle of the myosin head since the probe orientation with respect to the protein molecule is not accurately known. The probe molecules are less ordered in relaxed fibers, indicating either static disorder or thermal motions of myosin about its two flexible hinges.

The orientation of IATR (iodoacetamido tetramethyl rhodamine; Fig. 8*B*) bound to SH1 in fibers has been measured by the preferential polarization of light absorption (i.e., linear dichroism) which was detected from the magnitude of total fluorescence (Borejdo et al., 1982). No dichroism was observed in relaxed fibers, indicating disorder of the probe angles. The extent of head mobility in relaxed fibers or isolated thick filaments has not been determined conclusively (Thomas and Cooke, 1980; Borejdo et al., 1982; Mendelson and Wilson, 1982; Poulsen and Lowy, 1983; Wilson and Mendelson, 1983; Eads et al., 1985). Tortional mobility of the myosin head about its attachment to the thick filament could contribute substantially to the observed disorder if the probe axis is not parallel to the long axis of the head (Mendelson and Wilson, 1982).

In rigor fibers, IATR is oriented at a specific angle (80°) to the fiber axis (Borejdo et al., 1982). A striking finding with this dye is that addition of ADP to a fiber in rigor causes a dramatic change in the angle of the probe absorption dipole. However, when ADP binds to rigor cross-bridges, little change of force occurs, and no substantial orientation change has been observed with other probes. These puzzling results are probably not explained simply by a movement of IATR on the protein surface, since ADP does not alter the lifetime of the excited state or the initial anisotropy of IATR (Borejdo et al., 1982). The experiments might be compatible with differences of orientation among probe molecules, or they might indicate a segmental motion in the myosin head (Morales et al., 1982).

During active contraction, the fluorescent probes indicate a relatively wide angular distribution of cross bridges (Nihei et al., 1974; Borejdo et al., 1982). A correlation spectroscopy study of IATR fluorescence polarization indicated repetitive rotational motions of active cross-bridges at the approximate frequency $(1-5s^{-1})$ of the ATPase rate per myosin head (Borejdo et al., 1979). Since attached and detached cross-bridges contribute to the angular distribution, the range and motions for attached active cross-bridges remain to be unambiguously determined (Yanagida, 1981; Cooke et al., 1982; H. E. Huxley et al., 1983; Matsuda and Podolsky, 1984; Brenner et al., 1984).

1,5-IAEDANS and IATR have estremely short ($<$ 20 ns) fluorescence decay times compared to the millisecond time scale of the myosin head motions expected from mechanical studies (A. F. Huxley and Simmons, 1971). Therefore, time-resolved studies of the cross-bridge motions accompanying the power stroke require a probe with a longer excited state lifetime.

5-IAE (5-iodoacetamidoeosin; Fig. 8C) is a tetrabromofluorescein derivative with phosphorescence decay in the millisecond time scale. Decay of phosphorescence anisotropy in myosin and myosin fragments bound to 5-IAE indicates considerable mobility of the head about the hinge regions, in agreement with the fluorescence studies. However, when the myosin molecules are packed into thick filaments, both the rate and amplitude of motions are restricted (Eads et al., 1984). Preliminary application of the phosphorescence technique to muscle fibers has been reported (Eads et al., 1983).

Distance measurements by the fluorescence energy transfer method described in Section 5 have not yet been made in organized muscle fibers. However, distances have been measured between several pairs of well-defined sites in myosin and actomyosin (Morales et al., 1982; Moss and Trentham, 1983; Trayer and Trayer, 1983; Dos Remedios and Cooke, 1984). As an example, the fast-reacting cysteine, SH1, on myosin, and a cysteine residue, Cys-373, on actin were labeled with donor 1,5-BrAEDANS and acceptor 5-iodoacetamidofluorescein (IAF), respectively (Trayer and Trayer, 1983). Approximately 45% of the fluorescence of myosin-bound AEDANS was quenched by fluorescence energy transfer when IAF-labeled actin bound to the myosin. Relative mobilities of the probes on the proteins were determined by emission anisotropy in viscous solutions. The interprobe distance was calculated to be 4–6 nm. Since the myosin head is approximately 12 nm long (Mendelson and Kretschmar, 1980), the SH1 residue seems to be located away from the actin-binding site. The ATP-binding site has also been located distally from actin (Dos Remedios and Cooke, 1984).

7. CONCLUSIONS

To transform chemical to mechanical energy, contractile proteins are assembled in a complex but strikingly regular array. This organization makes dynamic structural measurements essential for elucidating the mechanism of contraction. The optical probes, briefly described above, complement electron microscopy and X-ray diffraction by providing real-time signals with relatively direct structural interpretations. The experiments have not yet proved or ruled out the rotating myosin head hypothesis for the cross-bridge power stroke.

The unit of contractile machinery most convenient for detailed mechanical studies is the single intact or skinned muscle fiber. It has macroscopic dimensions (20–100 nl) and contains a large number (10^{13}) of myosin heads. These factors limit the kinetic information that can be obtained by rapid mixing of solutions with the muscle preparations, since the results are kinetically dominated by solute diffusion. Recent progress in mechanical studies on myofibrils (Iwazumi, 1982) and on reconstituted "actomyosin motors" (Yano et al., 1982; Sheetz and Spudich, 1983) may directly decrease mixing times.

The photochemical method decreases the effective mixing time for nucleotides in muscle fibers by several orders of magnitude. Synthesis of new photolabile nucleotide anologues may improve time resolution further. The present techniques provide strong support for our working hypothesis on energy transduction in muscle and suggest a link between release of the products of ATP hydrolysis and physiological control of the cross-bridge cycle.

REFERENCES

Borejdo, J., S. Putnam, and M. F. Morales (1979) Fluctuations in polarized fluorescence: Evidence that muscle cross-bridges rotate repetitively during contraction. *Proc. Natl. Acad. Sci. USA,* **76,** 6346–6350.

Borejdo, J., O. Assulin, T. Ando, and S. Putnam (1982) Cross-bridge orientation in skeletal muscle measured by linear dichroism of an extrinsic chromophore. *J. Mol. Biol.,* **158,** 391–414.

Brenner, B., L. C. Yu, and R. J. Podolsky (1984) X-ray diffraction evidence for cross-bridge formation in relaxed muscle fibers at various ionic strengths. *Biophys. J.,* **46,** 299–306.

Cherry, R. J. (1978) Measurement of protein rotational diffusion in membranes by flash photolysis. *Methods Enzymol.,* **54,** 47–61.

Cleworth, D. R., and K. A. P. Edman (1972) Changes in sarcomere length during isometric tension development in frog skeletal muscle. *J. Physiol. (Lond.),* **227,** 1–17.

Cooke, R., M. S. Crowder, and D. D. Thomas (1982) Orientation of spin labels attached to cross-bridges in contracting muscle fibres. *Nature,* **300,** 776–778.

Crowder, M. S., and R. Cooke (1984) The effect of myosin sulphydryl modification on the mechanics of fibre contraction. *J. Musc. Res. Cell Motil.,* **5,** 131–146.

Dale, R. E., and J. Eisinger (1975) in R. F. Chen and Edelhoch, H., Eds., *Biochemical Fluorescence: Concepts,* Vol. 1, Dekker, New York, pp. 115–284.

Dantzig, J. A., M. G. Hibberd, Y. E. Goldman, and D. R. Trentham (1984) ADP slows cross-bridge detachment rate induced by photolysis of caged ATP in rabbit psoas muscle fibers. *Biophys. J.,* **45,** 8a.

Dos Remedios, C. G., and R. Cooke (1984) Fluorescence energy transfer between probes on actin and probes on myosin. *Biochim. Biophys. Acta,* **788,** 193–205.

Eads, T., R. Austin, B. Citak, D. Momont, and D. Thomas (1983) Phosphorescence anisotropy measurements of slow motions in eosin-labeled myosin. *Biophys. J.,* **41,** 146a.

Eads, T. M., D. D. Thomas and R. H. Austin, (1984) Microsecond rotational motions of eosin-labeled myosin measured by time-resolved anisotropy of absorption and phosphorescence. *J. Mol. Biol.,* **179,** 55–81.

Finlayson, B., R. W. Lymn, and E. W. Taylor (1969) Studies on the kinetics of formation and dissociation of the actomyosin complex. *Biochemistry,* **8,** 811–819.

Gear, C. W. (1971) *Numerical Initial Value Problems in Ordinary Differential Equations,* Ch. 11. Prentice-Hall, Englewood Cliffs, NJ.

Goldman, Y. E., and R. M. Simmons (1984) Control of sarcomere length in skinned muscle fibres of *Rana temporaria* during mechanical transients. *J. Physiol. (Lond.)*, **350**, 497–518.

Goldman, Y. E., H. Gutfreund, M. G. Hibberd, J. A. McCray, and D. R. Trentham (1982a) The role of thiols in caged ATP studies and their use in measurement of the photolysis kinetics of *o*-nitrobenzyl compounds. *Biophys, J.*, **37**, 125a.

Goldman, Y. E., M. G. Hibberd, J. A. McCray, and D. R. Trentham (1982b) Relaxation of muscle fibres by photolysis of caged ATP. *Nature*, **300**, 701–705.

Goldman, Y. E., M. G. Hibberd, and D. R. Trentham (1984a) Relaxation of rabbit psoas muscle fibres from rigor by photochemical generation of adenosine-5'-triphosphate. *J. Physiol. (Lond.)*, **354**, 577–604.

Goldman, Y. E., M. G. Hibberd, and D. R. Trentham (1984b) Initiation of active contraction by photogeneration of adenosine-5'-triphosphate in rabbit psoas muscle fibres. *J. Physiol. (Lond.)*, **354**, 605–624.

Gordon, A. M., A. F. Huxley, and F. J. Julian (1966) Tension development in highly stretched vertebrate muscle fibres. *J. Physiol. (Lond.)*, **184**, 143–169.

Haskell, R. C., and F. D. Carlson (1981) Quasi-elastic light-scattering studies of single skeletal muscle fibers. *Biophys. J.*, **33**, 39–62.

Hibberd, M. G., J. A. Dantzig, Y. E. Goldman, and D. R. Trentham (1984a) P_i speeds ATP-induced relaxation of skinned muscle fibers. *Proc. Int. Un. Pure Appl. Biophys. 8th Int. Biophys. Cong.*, 204.

Hibberd, M. G., M. R. Webb, Y. E. Goldman, and D. R. Trentham (1984b) Oxygen exchange accompanies ATP hydrolysis catalyzed by active and relaxed glycerinated rabbit psoas fibers. *Biophys. J.*, **45**, 152a.

Hibberd, M. G., J. A. Dantzig, D. R. Trentham, and Y. E. Goldman (1985a) Phosphate release and force generation in skeletal muscle fibers. *Science*, **228**, 1317–1319.

Hibberd, M. G., M. R. Webb, Y. E. Goldman, and D. R. Trentham (1985b) Oxygen exchange between phosphate and water accompanies calcium-regulated ATPase activity of skinned fibers from rabbit skeletal muscle. *J. Biol. Chem.*, **260**, 3496–3500.

Huxley, A. F. (1957) Muscle structure and theories of contraction. *Prog. Biophys.*, **7**, 255–318.

Huxley, A. F. (1974) Review lecture. Muscular contraction. *J. Physiol. (Lond.)*, **243**, 1–43.

Huxley, A. F., and R. Niedergerke (1954) Structural changes in muscle during contraction. Interference microscopy of living muscle fibres. *Nature*, **173**, 971–973.

Huxley, A. F., and R. M. Simmons (1971) Proposed mechanism of force generation in striated muscle. *Nature*, **233**, 533–538.

Huxley, A. F., V. Lombardi, and L. D. Peachey (1981) A system for fast recording of longitudinal displacement of a striated muscle fibre. *J. Physiol. (Lond.)*, **317**, 12–13P.

Huxley, H. E. (1969) The mechanism of muscular contraction. *Science*, **164**, 1356–1366.

Huxley, H. E., and J. Hanson (1954) Changes in the cross-striations of muscle during contraction and stretch and their structural interpretation. *Nature*, **173**, 973–976.

Huxley, H. E., R. M. Simmons, A. R. Faruqi, M. Kress, J. Bordas, and M. H. J. Koch (1983) Changes in the X-ray reflections from contracting muscle during rapid mechanical transients and their structural implications. *J. Mol. Biol.*, **169**, 469–506.

Iwazumi, T. (1982) High performance instrument for myofibrillar mechanics. *Biophys. J.*, **37**, 357a.

Jovin, T. M., M. Bartholdi, W. L. C. Vaz, and R. H. Austin (1981) Rotational diffusion of biological macromolecules by time-resolved delayed luminescence (phosphorescence, fluorescence) anisotropy. *Ann. N.Y. Acad. Sci.*, **366**, 176–196.

Kaplan, J. H., B. Forbush III, and J. F. Hoffman (1978) Rapid photolytic release of adenosine 5'-triphosphate from a protected analogue: Utilization by the Na:K pump of human red blood cell ghosts. *Biochemistry*, **17**, 1929–1935.

Kinosita, K., Jr., S. Kawato, and A. Ikegami (1984) Dynamic structure of biological and model membranes: Analysis by optical anisotropy decay measurement. *Adv. Biophys.*, **17**, 147–203.

Lymn, R. W., and E. W. Taylor (1971) Mechanism of adenosine triphosphate hydrolysis by actomyosin. *Biochemistry*, 10, 4617–4624.

Matsuda, T., and R. J. Podolsky (1984) X-ray evidence for two structural states of the actomyosin cross-bridge in muscle fibers. *Proc. Natl. Acad. Sci. USA*, **81**, 2364–2368.

McCarter, R. (1981) Studies of sarcomere length by optical diffraction. *Cell Musc. Motil.*, **1**, 35–62.

McCray, J. A., L. Herbette, T. Kihara, and D. R. Trentham (1980) A new approach to time-resolved studies of ATP-requiring biological systems: Laser flash photolysis of caged ATP. *Proc. Natl. Acad. Sci. USA*, **77**, 7237–7241.

McCray, J. A., H. Gutfreund, and D. R. Trentham (1986) The kinetics and mechanism of caged—ATP photolysis (In preparation.)

Mendelson, R. A., and K. M. Kretzschmar (1980) Structure of myosin subfragment 1 from low-angle X-ray scattering. *Biochemistry*, **19**, 4103–4108.

Mendelson, R. A., and M. G. A. Wilson (1982) Three-dimensional disorder of dipolar probes in a helical array. Application to muscle cross-bridges. *Biophys. J.*, **39**, 221–227.

Mendelson, R. A., M. F. Morales, and J. Botts (1973) Segmental flexibility of the S-1 moiety of myosin. *Biochemistry*, **12**, 2250–2255.

Morales, M. F., J. Borejdo, J. Botts, R. Cooke, R. A. Mendelson, and R. Takashi (1982) Some physical studies of the contractile mechanism in muscle. *Annu. Rev. Phys. Chem.*, **33**, 319–351.

Moss, D. J., and D. R. Trentham (1983) Distance measurement between the active site and cysteine-177 of the alkali one light chain of subfragment 1 from rabbit skeletal muscle. *Biochemistry*, **22**, 5261–5270.

Nerbonne, J. M., and H. A. Lester (1983) The design of photoactivated molecules: Chemical and structural control of reaction rates and efficiencies. *Biophys. J.*, **41**, 3a.

Nerbonne, J. M., S. Richard, J. Nargeot, and H. A. Lester (1984) New photoactivatable cyclic nucleotides produce intracellular jumps in cyclic AMP and cyclic GMP concentrations. *Nature*, **310**, 74–76.

Nihei, T., R. A. Mendelson, and J. Botts (1974) Use of fluorescence polarization to observe changes in attitude of S-1 moieties in muscle fibers. *Biophys. J.*, **14**, 236–242.

Pollack, G. H., T. Iwazumi, H. E. D. J. ter Keurs, and E. F. Shibata (1977) Sarcomere shortening in striated muscle occurs in stepwise fashion. *Nature*, **268**, 757–759.

Poulsen, F. R., and J. Lowy (1983) Small-angle X-ray scattering from myosin heads in relaxed and rigor frog skeletal muscles. *Nature*, **303**, 146–152.

Sheetz, M. P., and J. A. Spudich (1983) Movement of myosin measured in vitro. *Biophys. J.*, **41**, 155a.

Stryer, L. (1978) Fluorescence energy transfer as a spectroscopic ruler. *Annu. Rev. Biochem.*, **47**, 819–846.

Taylor, D. L. (1976) Quantitative studies on the polarization optical properties of striated muscle. *J. Cell Biol.*, **68**, 497–511.

Thomas, D. D., and R. Cooke (1980) Orientation of spin-labeled myosin heads in glycerinated muscle fibers. *Biophys. J.*, **32**, 891–906.

Thomas, D. D., W. F. Carlsen, and L. Stryer (1978) Fluorescence energy transfer in the rapid-diffusion limit. *Proc. Natl. Acad. Sci. USA*, **75**, 5746–5750.

Trayer, H. R., and I. P. Trayer (1983) Fluorescence energy transfer between the myosin subfragment-1 isoenzymes and F-actin in the absence and presence of nucleotides. *Eur. J. Biochem.*, **135**, 47–59.

Webb, M. R., and D. R. Trentham (1981) The mechanism of ATP hydrolysis catalyzed by myosin and actomyosin, using rapid reaction techniques to study oxygen exchange. *J. Biol. Chem.*, **256**, 10910–10916.

White, H. D., and E. W. Taylor (1976) Energetics and mechanism of actomyosin adenosine triphosphatase. *Biochemistry*, **15**, 5818–5826.

Wilson, M. G. A., and R. A. Mendelson (1983) A comparison of order and orientation of crossbridges in rigor and relaxed muscle fibres using fluorescence polarization. *J. Musc. Res. Cell Motil.* **4**, 671–693.

Yanagida, T. (1981) Angles of nucleotides bound to cross-bridges in glycerinated muscle fiber at various concentrations of ε-ATP, ε-ADP and ε-AMPPNP detected by polarized fluorescence. *J. Mol. Biol.*, **146**, 539–560.

Yano, M., Y. Yamamoto, and H. Shimizu (1982) An actomyosin motor. *Nature,* **299**, 557–558.

CHAPTER 24

DESIGN AND APPLICATION OF PHOTOLABILE INTRACELLULAR PROBES

JEANNE M. NERBONNE

Division of Biology
California Institute of Technology
Pasadena, California

1.	INTRODUCTION	418
2.	PHOTOLABILE INTRACELLULAR MESSENGERS	418
3.	OPTICAL METHODS FOR PHYSIOLOGICAL EXPERIMENTS	424
4.	"CONCENTRATION JUMPS" OF CYCLIC NUCLEOTIDES AND THE MODULATION OF THE Ca^{2+} CURRENT	427
5.	OTHER OBSERVATIONS WITH PHOTOLABILE CYCLIC NUCLEOTIDES	431
6.	KINETIC IMPLICATIONS OF cAMP EFFECTS	432
7.	PHOTOLABILE Ca^{2+} ANTAGONISTS	432
8.	Ca^{2+} ANTAGONISTS AND THE MODULATION OF I_{si}	433
9.	PHOTOLABILE Ca^{2+} ANTAGONISTS AND WHOLE-CELL RECORDING TECHNIQUES	436
10.	CONCLUSIONS	438
	REFERENCES	439

The author's present address is Department of Pharmacology, Washington University School of Medicine, 660 South Euclid Avenue, St. Louis, MO 63110.

1.   INTRODUCTION

It is generally believed that intracellular second messengers play key physiological roles in many systems by mediating the responses to neurotransmitters, neurohormones, and to other external and/or internal stimuli. The cyclic nucleotides cAMP and cGMP are perhaps the most widely discussed intracellular messengers (Robison et al., 1965; Robison and Sutherland, 1971; Sutherland and Robison, 1966; Sutherland, 1970) but, in a large variety of cell types, Ca^{2+} (Rasmussen and Goodman, 1977; Iwatsuki and Peterson, 1978) and H^+ (Iwatsuki and Peterson, 1978, 1979; Giaume et al., 1980; Turin and Warner, 1980; Spray et al., 1981, 1982; Reber and Weingart, 1982; Schuetze and Goodenough, 1982; Campos de Carvalho et al., 1984) are known or postulated to also be important second messengers. More recently, ATP (Noma, 1983; Trube and Heschler, 1983) and the phosphatidyl inositides and their metabolites (Fisher et al., 1984) have also been suggested to play functional roles as intracellular messengers. It clearly is quite likely that other ions and small molecules will join the ranks of these important chemical messengers in the future. In spite of the general acceptance of the intracellular second-messenger hypothesis (Sutherland, 1970; Robison and Sutherland, 1971; Drummond, 1973, 1983; Drummond and Severson, 1977; Kupferman, 1979; Nirenberg et al., 1983), numerous questions do remain about the detailed mechanisms (i.e., time courses, causal relationships, etc.) of events thought to depend directly or indirectly on increases or decreases in intracellular messenger concentrations. Quantitative descriptions of second-messenger-mediated processes are often difficult to obtain experimentally, most likely owing to the problems associated with controlling accurately the intracellular concentrations of these species. Conventional techniques (e.g., use of analogues, enzyme stimulation or inhibition, microinjection) provide the experimenter with little control over the species being manipulated, the effective concentrations provided, and/or the time course over which the changes are produced. We have, therefore, been interested in the development of methods for altering intracellular messenger concentrations through the use of light-activated precursors. A photochemical methodology was selected, as it was anticipated that this approach could provide both temporal and spatial resolution of a magnitude unmatched by the more conventional techniques. This methodology has the additional property that it should provide a means for altering intracellular messenger concentrations that is noninvasive to the preparation and therefore less likely to result in long-term (or irreversible) physiological damage. Here, I shall describe in some detail our efforts in this area with respect to the development, chemical characterization, and biological properties of these probes and their applications in physiological experiments.

2.   PHOTOLABILE INTRACELLULAR MESSENGERS

Several laboratories have been involved in recent years in efforts aimed at the development and characterization of photolabile intracellular probes—in particular, light-activatable intracellular messengers. The goal of this work is to design and prepare molecules which: (1) are physiologically inert prior to irradiation; (2) can be loaded into cells by diffusion, microinjection, or dialysis (e.g., when used in conjunction with whole-cell patch-clamp recording techniques); and (3) are photosensitive—absorption of light liberating the intracellular probe of interest (plus 1 or more physiologically inert fragments). This methodology involves the use of photoprotecting groups (Pillai, 1980) and, in most cases, we and others have used the o-nitrobenzyl moiety (Eq. (1)) originally designed by Patchornik and colleagues (Patchornik et al., 1970; Zehavi et al., 1972; Amit et al., 1974; Rubinstein et al., 1975) and now used extensively by organic chemists for protection during elaborate, multistep syntheses (Pillai, 1980).

$$\begin{array}{ccc} \underset{X_n}{\text{CHROCOCH}_3} & \xrightarrow{\lambda>300\,nm} & \underset{X_n}{\text{COR}} & + \; H^+ \; + \; CH_3COO^- \end{array} \qquad (1)$$

Irradiation of o-nitrobenzyl derivatives recovers the protected molecule (in Eq. (1), the acetate ion) and liberates the nitroso photoproduct as either an aldehyde or a ketone, depending on the substitution at the benzylic carbon; further photoreactions of these primary photoproducts are possible under some conditions (Patchornik et al., 1970; Amit et al., 1974). This approach was utilized by Kaplan and co-workers (Kaplan et al., 1978) in the development of "caged ATP" and by Engels and co-workers (Engels and Schlaeger, 1977; Engels and Reidys, 1978; Korth and Engels, 1979) in the preparation of membrane permeant (and photolabile) cyclic nucleotide analogues. Caged ATP has been exploited extensively in mechanistic studies of muscle contraction (McCray et al., 1980; Goldman et al., 1982a,b; Hibberd et al., 1983; Goldman et al., 1984a,b) and Na^+/K^+-ATPase activity (Kaplan et al., 1978; Kaplan and Hollis, 1980), whereas the photolabile cyclic nucleotides were principally used as membrane-permeant prodrug (rather than light-activated) analogues (Korth and Engels, 1979) that could be activated by the action of intracellular enzymes. Recently, Senter and co-workers (Senter et al., 1985) have also used o-nitrobenzyl protecting groups in the design and preparation of a new class of photocleavable protein cross-linking reagents. Most recently, Tsien (see chapter by R. Y. Tsien in this volume) has utilized this chemical methodology in the development of a photolabile, caged Ca^{2+} chelate.

In spite, however, of the widespread interest in the concept of "photoprotection," little was known of the detailed reaction mechanisms leading to photoremoval (Pillai, 1980), and it was therefore not well understood how the rates and efficiencies of these reactions were controlled. A possible exception to this statement is that it does seem clear that at least one hydrogen atom must be present on the benzylic carbon (ortho to the nitro group) for the reaction to proceed with any reasonable efficiency. Kinetic studies of ATP release have been reported (McCray et al., 1980; Goldman et al., 1982a), and some attempts have been made to clarify the structural requirements for efficient, fast reactions of caged ATP (McCray et al., 1980; Goldman et al., 1982a). In order to design useful photoprotecting groups for biological studies, it appeared

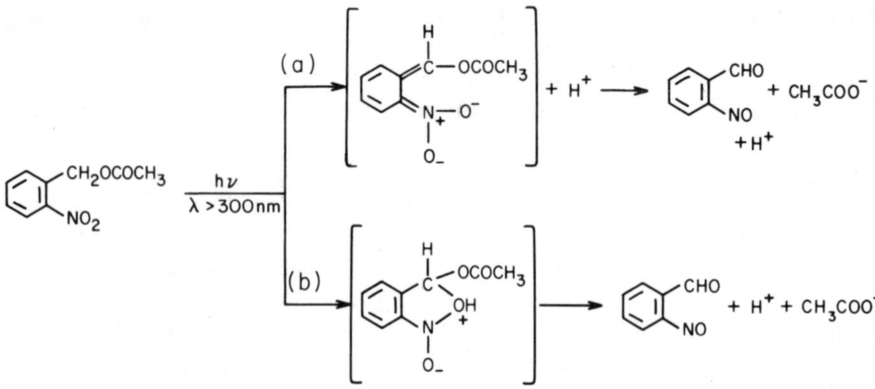

Figure 1. Photolysis of o-nitrobenzyl esters: Irradiation liberates the protected molecule and a nitrosoaldehyde (or ketone) photolysis by-product. Two reaction schemes, (a) and (b), can be proposed for the mechanism underlying conversion of the o-nitrobenzyl moiety and thus leading to photorelease. The two mechanisms differ in the structure of the intermediate formed on photolysis and in the disposition of the H^+ produced by irradiation. All results obtained to date suggest that (a) is the more likely of the two reaction mechanisms (see text).

necessary to study in some detail the reaction itself and the factors influencing the reaction. We therefore prepared a series of structurally diverse o-nitrobenzyl esters in order to: (1) study the photorelease process directly; (2) determine reaction rates and efficiencies; and (3) characterize the structural features that control and/or influence these reactions (rates and efficiencies).

There are (at least) two reaction schemes that could be proposed to explain the photolysis reactions of the o-nitrobenzyl moiety (Fig. 1). These mechanisms differ principally in two respects: (1) the disposition of the H^+ produced on photolysis—that is, in (a) the H^+ is dissociated from the acinitro intermediate (in brackets), whereas in (b) the H^+ remains associated; and (2) the actual chemical structure proposed for the acinitro intermediate—that is, in (a), an extended resonance structure is postulated, whereas in (b), a cyclic intermediate is shown. In flash photolysis experiments, we find that, although the decay rates of the acinitro intermediates produced on irradiation are highly variable, all compounds release H^+ (as assayed by the pH-sensitive dye phenol red) within the 5-ms resolution of the photolysis apparatus (Fig. 2, lower panels) and, therefore, simultaneous with the production of the intermediate. In addition, we observe (Fig. 2, upper panels) that the photolysis that led to the production of the acinitro intermediate was, in many cases, accompanied by large, red spectral shifts; these results are consistent with the formation of an intermediate(s) with extended resonance possibilities relative to the starting material(s) and product(s). These findings taken together reveal that (a) is the more likely reaction mechanism. In agreement with the previous suggestion of McCray et al. (1980), we propose, therefore, that the rate of breakdown of the acinitro intermediate is equal to the rate of photorelease of the protected moiety (e.g., the acetate ion in Eq. (1)). As noted above, in some cases, intermediate (i.e., acinitro)

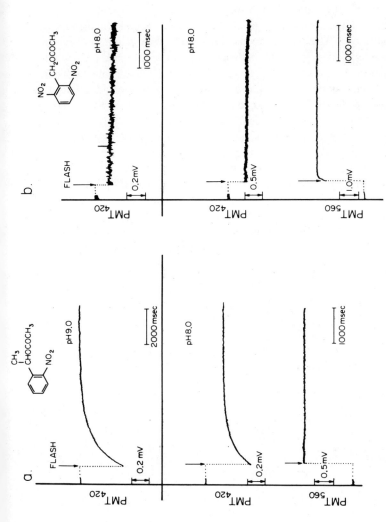

Figure 2. Flash photolysis studies of the rates and efficiencies of photorelease from *o*-nitrobenzyl esters. Single flashes were presented to the test samples at the time indicated by the arrows, and the resulting spectral changes were detected using a photomultiplier. The photomultiplier tube (PMT) output at each of the observation wavelengths, 420 and 560 nm, is plotted as a function of time. The rate of photorelease is assumed to be equal to the rate of decay of the acinitro intermediate, observed and monitored at 420 nm (top panels). H^+ production, used as an estimate of reaction efficiency, was measured using the pH-sensitive dye phenol red (lower panels). The dye spectrum was evaluated at 560 nm, a pH-sensitive wavelength, and the amplitude of the pH "jump" produced by a flash was determined. Although acinitro decay could be measured for some compounds (*a*), this was not the case with all *o*-nitrobenzyl derivatives prepared to date. In the photolysis of 2,6-dinitrobenzyl acetate, for example, rather than a transient absorption change, a step change was observed (*b*), a result we interpret to mean that the reactions are complete within the 5-ms recording resolution of the photolysis apparatus (limited by the mechanical properties of the shutter). In contrast, all compounds, regardless of structure, were observed to liberate H^+ within 5 ms of irradiation (*a*,*b*); proton release and acinitro decay are not, therefore, coincident.

421

decay can be monitored optically in flash photolysis experiments (Fig. 2a). With some derivatives, however, no flash-induced long wavelength transients can be observed (Fig. 2b), implying that the reactions are complete within the recording resolution (5 ms) of the photolysis setup.

If these assumptions are all correct, then, one can conclude that the rates of release vary markedly with the structures of both the protecting group (Table 1) and the leaving group (Table 2), and further that, in some cases, the reaction is controlled by the protecting group alone. At least two groups, the 4,5-dimethoxy-2-nitrobenzyl and the 2,6-dinitrobenzyl moieties, yield photorelease that is apparently complete within the 5-ms resolution of the measuring apparatus (see: Fig. 2b and Table 1). These results suggest the conclusion that these protecting groups could be generally useful for the protection of a wide variety of biologically interesting molecules and that they would yield fast, efficient release independent of the nature of the biological molecule that one is attempting to protect. It remains, however, to be demonstrated that: (1) the reaction can (in all cases) be controlled independent of the protected moiety; and, more significantly, (2) the rate of breakdown of the acinitro intermediate does indeed reflect the rate of photorelease. In support of the latter postulate (2), is the observation that release of ATP from caged ATP proceeds at a rate equal to the rate of acinitro breakdown (Goldman et al., 1984a), a finding that is also in agreement with the original suggestion of McCray et al. (1980). The efficiency of photorelease from all analogues prepared to date, as measured by the amplitudes of the pH "jumps"

Table 1.   Substituent Effects on Decay Rates ($k(s^{-1})$)

COMPOUND	pH 7.2	pH 8.0	pH 9.0	pH 12.0
(2-nitrobenzyl) CH$_2$OCOCH$_3$ / NO$_2$	1.4 ± 0.1	1.0 ± 0.1	0.6 ± 0.1	
CH$_3$ / CHOCOCH$_3$ / NO$_2$	3.1 ± 0.1	1.1 ± 0.1	0.8 ± 0.1	
CH(OCOCH$_3$)$_2$ / NO$_2$	—	—	> 200	
(methylenedioxy) CH$_2$OCOCH$_3$ / NO$_2$	—	> 200		1.7 ± 0.1
CH$_3$O, CH$_3$O (dimethoxy) CH$_2$OCOCH$_3$ / NO$_2$	—	> 200		12 ± 1.0
NO$_2$ / CH$_2$OCOCH$_3$ / NO$_2$	—	> 200		> 200

produced in response to flashes, is high (Table 3) and apparently influenced only very slightly by structural modifications. Although we have not yet measured the absolute quantum yields for H^+ release, a value of 0.50 has been reported for the quantum yield for ATP release from caged ATP (Kaplan et al., 1978). Proton release has been determined for all derivatives and compared here, therefore, to that measured for ATP in our experimental set-up. The values relative to o-nitrobenzyl acetate are given in Table 3; relative quantum yields would be obtained by multiplying the values in the third column of Table 3 by 0.3. We find, in addition, that these reactions are relatively insensitive to the pH of the solvent (over the pH range 4.5–8.5) and are similarly not influenced by the dielectric constant or the ionic strength of the medium.

We have also designed, synthesized, and characterized the 4,5-dimethoxy-2-nitrobenzyl esters of cAMP and cGMP (Fig. 3); we find (Nerbonne et al., 1984) that photorelease from these analogues is complete within at most 5 ms and proceeds with a relativelty high quantum yield (Table 3). These reactions are uncomplicated by competing secondary side reactions such as those seen in the photolysis of o-nitrobenzyl cGMP (Nargeot et al., 1983).

Table 2.   Leaving Group Effects on Decay Rates ($k(s^{-1})$)

COMPOUND	pH7.2	pH8.0	pH9.0	pH12.0
(benzyl, CH₂OCOCH₃, NO₂)	1.4 ± 0.1	1.0 ± 0.1	0.6 ± 0.1	
(benzyl, CH₃ CHOCOCH₃, NO₂)	3.1 ± 0.1	1.1 ± 0.1	0.8 ± 0.1	
(benzyl, CH₂OCOOCH₃, NO₂)	1.6 ± 0.1	1.2 ± 0.1	1.0 ± 0.1	
(benzyl, CH₂Cl, NO₂)	—	14.0 ± 1.0	10.6 ± 0.2	4.5 ± 0.2
(benzyl, CH₂O—cAMP, NO₂)	1.5 ± 0.1	1.2 ± 0.1	1.2 ± 0.1	
(benzyl, CH₂O—cGMP, NO₂)	—	3.0 ± 0.3	2.2 ± 0.1	
(benzyl, CH₃ CHO—P$_i$, NO₂)	$(200)^*$			
(benzyl, CH₃ CHO—ATP, NO₂)	30.0 ± 2.0 $(200)^*$	12.0 ± 1.0 $(30)^*$	6.0 ± 2.0 $(4)^*$	

Table 3. Reaction Efficiencies

COMPOUND	ESTIMATED ΔpH	RELATIVE ΔpH = RELATIVE RX. EFFICIENCY
CH$_2$OCOCH$_3$ / NO$_2$	0.20	1.0
CH$_3$ / CHOCOCH$_3$ / NO$_2$	0.25	1.25
CH(OCOCH$_3$)$_2$ / NO$_2$	0.20	1.0
NO$_2$ / CH$_2$OCOCH$_3$ / NO$_2$	0.50	2.5
CH$_2$OCOCH$_3$ / NO$_2$	0.20	1.0
CH$_3$O / CH$_2$OCOCH$_3$ / CH$_3$O / NO$_2$	0.20	1.0
CH$_2$Cl / NO$_2$	0.25	1.25
CH$_2$OCOOCH$_3$ / NO$_2$	0.25	1.25
CH$_2$—O—cAMP / NO$_2$	0.20	1.0
CH$_3$O / CH$_2$—O—cAMP / CH$_3$O / NO$_2$	0.30	1.5
CH$_2$—O—cGMP / NO$_2$	0.12	0.6
CH$_3$ / CH$_2$—O—ATP / NO$_2$	0.30	1.5

3. OPTICAL METHODS FOR PHYSIOLOGICAL EXPERIMENTS

In the flash photolysis setup noted above and in our physiological experiments, light flashes are provided by a xenon flash lamp system, modified to yield high-energy, low-repitition (about 1 every 10 s) flashes (Lester and Nerbonne, 1982; Nargeot et

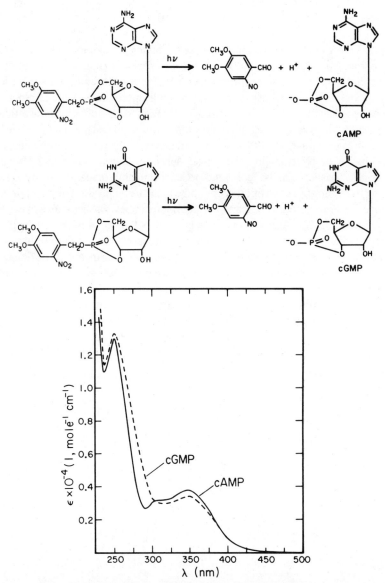

Figure 3. Structures, photoreactions, and optical absorption spectra of the 4,5-dimethoxy-2-nitrobenzyl esters of cAMP (——) and cGMP (– – –). Irradiation at wavelengths >300 nm yields the ionized acids of cAMP or cGMP. The esters were synthesized, purified, and characterized as described (Nerbonne et al., 1984), using modifications of procedures employed previously (Engels and Schlaeger, 1977; Engels and Reidys, 1978; Davies and Schwarz, 1965). The esters are obtained as diastereomeric mixtures; isomer assignments and product ratios were determined by spectral analysis, and separation and quantification of the isomers were afforded by high-performance liquid chromatography (Nerbonne et al., 1984). The axial and equatorial isomers of 4,5-dimethoxy-2-nitrobenzyl cAMP and cGMP display identical absorption spectra and, in addition, react with equal efficiencies and rates to yield free cAMP or cGMP upon irradiation. In all experiments, therefore, diastereomeric mixtures are employed (Nerbonne et al., 1984). The esters are stable in crystalline form indefinitely. In aqueous solutions, they slowly undergo thermal hydrolysis to yield free cAMP and cGMP; frozen solutions of both compounds are stable for at least 30 days. The esters are only partially soluble in water, but, when dissolved first in dimethyl sulphoxide (DMSO), then diluted into the appropriate Ringer, final concentrations of several hundred micromolar can be achieved.

425

al., 1982; Lester et al., this volume). UV-grade optics are employed to increase the energy output from the lamp in the 300 to 400 nm range, the most useful wavelengths to effect photoconversion of the *o*-nitrobenzyl analogues. Recently, we succeeded in designing and building a flash lamp system with approximately threefold higher output energy than those previously employed through the use of ellipsoidal reflectors (unpublished observations). Flash duration and/or intensity can be varied by changing the configuration of the energy storage capacitors (4 capacitors; each: 550 μF at 450 V) or by changing the power supply voltage. When using inverted microscopes, flashes can be delivered to experimental preparations by directly mounting the lamp above the microscope stage in the usual position of the illuminator and condenser. For both inverted and upright microscopes, flashes can also easily be delivered through the microscope objectives. Although delivery through the microscope objective has the advantage of a small spot size, standard optics do not transmit well below approximately 350 nm, and, in our experiments, the efficiency of photoconversion is lowered at wavelengths greater than 400 nm. Direct delivery of flashes, on the other hand, permits a wider range of irradiation wavelengths to be selected; the area illuminated is, however, large (> 3.5 mm). The experimental arrangement of choice is generally best determined by the preparation under study and the particular experimental paradigms to be employed (also, see: Lester et al., this volume). It is also experimentally possible to produce high-energy flashes and, as

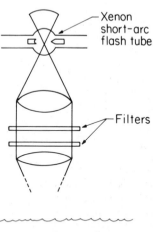

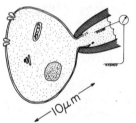

Figure 4. Schematic view of the experimental setup (shown here for recordings using the whole-cell patch-clamp technique). Light from a xenon short-arc flash lamp is filtered and focused on the preparation; the arrangement is similar when used in conjunction with other physiological recording techniques. Alternatively, flashes can be delivered through the microscope objective (see text).

a result, high per-flash conversions, using lasers (McCray et al., 1980; Goldman et al., 1982a,b; 1984a,b; Hibberd et al., 1983).

In the experiments described here and, indeed, most routinely, the flash lamp is mounted and light is delivered to the preparation from above (as shown in Fig. 4). The light is filtered to remove wavelengths below 300 nm and is focused on the preparation through the use of a quartz, secondary focusing lens; additional optical filters can be placed in the light path as needed. The capacitors are arranged in parallel, and the power supply is operated at or near 450 V. In this configuration, each flash delivers approximately 1 J (output energy, as measured with a bolometer) to the preparation; the flash duration is approximately 1 ms. This arrangement has proved satisfactory with a variety of cell types and experimental paradigms. Most interesting, however, are the consistent observations that flashes (300–400 nm), in the presence and in the absence of light-sensitive molecules, do not result in measurable physiological artifacts or damage on the time scale of the electrophysiological experiments. There is also no evidence for photodynamic damage in preparations exposed to multiple flashes (each 1-ms duration) on the time scale (minutes to hours) of our physiological experiments. It is certainly possible that there are long-term, damaging effects of irradiation on such preparations; to date, however, we have not examined this possibility. The flash lamp trigger produces a large, transient electrical artifact that can be reduced and, in many cases, almost eliminated by shielding (also see: Lester et al., this volume). Photoelectric artifacts are similarly avoided by appropriately shielding the preparation and the electrode(s).

4. "CONCENTRATION JUMPS" OF CYCLIC NUCLEOTIDES AND THE MODULATION OF THE Ca^{2+} CURRENT

We have utilized the photolabile cyclic nucleotides to study the modulation of the slow inward Ca^{2+} current (I_{si}) and tension in cardiac muscle (Nargeot et al., 1983; Nerbonne et al., 1984; Richard et al., 1985). Both the slow inward current (as well as other currents in cardiac tissue) and the contractile properties of heart muscle have been well characterized in a variety of species (for reviews, see: Reuter, 1979, 1983, 1984; Chapman, 1979; Fabiato and Fabiato, 1979; Fozzard, 1977; Tsien, 1983). Experiments were performed on isolated atrial trabeculae from bullfrog heart using the double sucrose-gap voltage-clamp method and both the o-nitrobenzyl and the 4,5-dimethoxy-2-nitrobenzyl esters of cAMP and cGMP. In the presence of the cAMP nitrobenzyl esters, we find that flashes increase the amplitude and duration of the action potential (Nargeot et al., 1983; Nerbonne et al., 1984; Richard et al., 1985) and increase the amplitudes of I_{si} and tension (Fig. 5a); flashes alone have no measurable effects. cAMP jumps have, in contrast, little effect on the kinetics (Fig. 5a) or the voltage dependences of the current and tension (Fig. 5b) even under conditions where the amplitudes have increased severalfold (Nargeot et al. 1983; Nerbonne et al. 1984; Richard et al. 1985). The time course of the increase in tension parallels that seen in the modulation of I_{si} (Richard et al., 1985); the time to peak

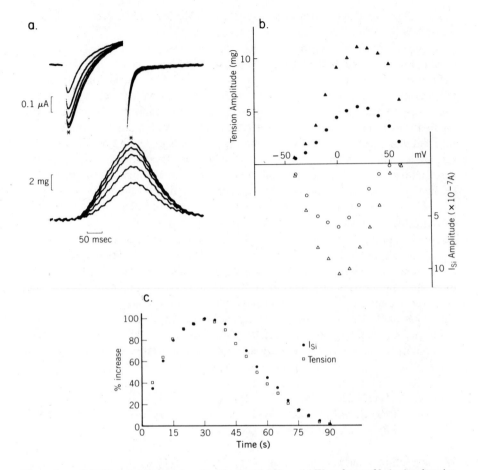

Figure 5. cAMP "jump" effects on I_{si} and tension in atrial fibers. (*a*) Wave forms of I_{si} (top) and tension (bottom) in a voltage-clamped (double sucrose gap) frog atrial trabecula. The Ringer contained 1 μM tetrodotoxin (TTX) to supress the fast, inward Na^+ current, and 50 μM 4,5-dimethoxy-2-nitrobenzyl cAMP. The voltage was held at the resting potential and depolarized 80 mV at 5-s intervals; the lowest amplitude traces (first sweep) show the steady-state levels of I_{si} and tension before any flashes. One second prior to the second sweep, a single flash was delivered: I_{si} and tension amplitudes increased, continued to increase over the next several sweeps, leveled off (maximum reponses are indicated by the asterisks), and finally returned to the preflash values (recovery not shown). (*b*) Voltage dependence of I_{si} and twitch tension. The fiber was held at the resting potential (-70 mV) and depolarized to various test potentials (-50 to $+50$ mV) at 5-s intervals. Trials were recorded before (○, ●) and 30 sec after (△, ▲) a single flash in the presence of 60 μM 4,5-dimethoxy-2-nitrobenzyl cAMP; the peak amplitudes of I_{si} and tension after flashes are plotted. (*c*) Time course of the increase and recovery in I_{si} and tension amplitudes after a flash. The fiber was held at rest (-80 mV) and depolarized 80 mV (step duration 200 ms) at 5-s intervals. The Ringer contained 1 μM TTX and 50 μM 4,5-dimethoxy-2-nitrobenzyl cAMP; the flash was delivered at time zero (Nargeot et al., 1983; Nerbonne et al., 1984; Richard et al., 1985).

response and the time course of recovery after flashes superimpose (Fig. 5c). It therefore appears that all of the effects of cAMP on phasic tension can be ascribed to changes in the amplitude of I_{si} and that cAMP, in the concentration range used here (see below), does not directly alter any process in the development or recovery of phasic tension; only tension amplitude is increased as a direct result of increased Ca^{2+} influx. As it is clear that phosphorylation of contractile proteins and the Ca^{2+} pump can be observed in response to β stimulation (Tada et al., 1975; Katz, 1979; Perry, 1979; Hartzell and Titus, 1982; Opie, 1982), it is interesting that cAMP jumps increase the amplitude without measurably altering the waveform of the phasic contraction. It is possible that cAMP influences the properties of contractile proteins in a manner not revealed by kinetic analysis of the tension waveforms or that the cAMP concentrations necessary to alter tension kinetics are higher than those produced by cAMP jumps in the experiments described here. Although we cannot eliminate either of these possibilities, we can conclude that it is not necessary to invoke a role for cAMP in the modulation of phasic tension other than its effect on the slow inward current.

It has been demonstrated that the sarcolemmal protein calciductin is the major target of cAMP-dependent protein kinase and that voltage-dependent Ca^{2+} uptake by isolated sarcolemmal vesicles is linearly correlated with the phosphorylation of calciductin (Rinaldi et al., 1981, 1982). These results have led to the hypotheses that calciductin might be the slow channel protein itself (or a closely associated regulatory component of the slow channel protein) and that β-agonists increase I_{si} by stimulating, directly, calciductin phosphorylation (Rinaldi et al., 1982). Although our results cannot be taken as direct evidence in support of these mechanisms, they are consistent with the hypothesis.

Responses to flashes lasted longer at higher drug concentrations (Fig. 6a), at lower temperatures, and in the presence of phosphodiesterase inhibitors (Nargeot et al., 1983; Nerbonne et al., 1984). The first increases in I_{si} were detected 150–200 ms after flashes (Fig. 6b) in all experiments regardless of which cAMP precursor was employed (Nargeot et al., 1983; Nerbonne et al., 1984). These findings imply that the time courses of the observed increases in I_{si} reflect the biochemistry of events thought to underlie cAMP-induced modulation of I_{si} (Reuter, 1979, 1983; Osterrieder et al., 1982; Trautwein et al., 1982; Bean et al., 1983; Brum et al., 1983; Cachelin et al., 1983; Reuter et al., 1983; Tsien, 1983) rather than the chemical properties of the photolabile probes. In addition, we find that I_{si} amplitude increases linearly in time with no detectable delay when extrapolated to zero response (Nargeot et al., 1983; Nerbonne et al., 1984), implying the presence of a single rate-limiting step in I_{si} modulation.

We have been unable to measure any effects of flashes in the presence of 4,5-dimethoxy-2-nitrobenzyl cGMP (Nerbonne et al., 1984) on the action potential, I_{si}, or tension at concentrations up to 250 μM. In addition, when the dimethoxy

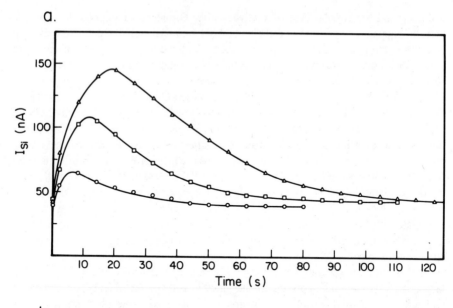

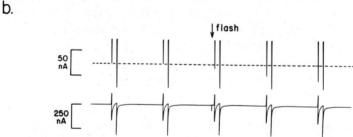

Figure 6.   Time course of the increase in I_{si} following concentration jumps of cAMP. (*a*) Increase and recovery of I_{si} amplitude following a single flash in the presence of 10 μM (●), 25 μM (□), and 100 μM (△) 4,5-dimethoxy-2-nitrobenzyl cAMP. Flashes were delivered at time zero, and the fiber was depolarized every 6 s; all other conditions were identical to those described in the legend to Fig. 5. The peak response amplitudes, times-to-peak responses, and recovery times following flashes all increased with the starting concentration of the *o*-nitrobenzyl cAMP esters employed and therefore with the amplitude of the cAMP concentration jump produced (see: Nargeot et al., 1983; Nerbonne et al., 1984; Richard et al., 1985). (*b*) Rapid response to a flash in the presence of 30 μM *o*-nitrobenzyl cAMP. The fiber was held at the resting potential and depolarized 80 mV at 2-s intervals. The upper trace, displayed with an offset, is a higher-gain record of the lower trace and is off scale except at the peak of I_{si}. A single flash was delivered at the time indicated and, within 150 msec, a small, but reproducible, increase in I_{si} was seen (see: Nargeot et al., 1983).

nitrobenzyl esters of cAMP and cGMP are both present in the bath, there is neither attenuation nor augmentation of cAMP's effect on I_{si}, action potential, or tension following flashes.

How large are the cAMP concentration jumps produced in these experiments? For technical reasons, we have not measured the amounts of cAMP and cGMP actually produced intracellularly following flashes. However, since the esters are lipophilic and presumably equilibrate rapidly inside cells, and since the reaction efficiency is relatively insensitive to environmental effects, we assume that the yields of free cyclic nucleotides produced intracellularly equal those produced extracellularly. In solution, we have determined that a single flash converts 5% of the molecules. From starting ester concentrations of 2–100 μM, therefore, a single flash would produce 0.1- to 5-μM cAMP (or cGMP) intracellularly. If the esters partition preferentially into membranes, nonuniform distributions and concentration jumps would certainly exist, and, as a result, it would not be possible to establish the precise concentration dependence of events that are regulated by cAMP (or cGMP). Quantitative studies might, therefore, be better done with membrane-impermeant analogues.

5. OTHER OBSERVATIONS WITH PHOTOLABILE CYCLIC NUCLEOTIDES

In contrast to its effect on phasic tension, cAMP decreases the amplitude of the tonic tension while increasing the delayed K^+ current; the kinetics of tension development after a voltage step and the recovery after repolarization do not appear to be altered by cAMP (Richard et al., 1985). It is possible, therefore, that cAMP directly influences the amount of Ca^{2+} entering via the Na^+/Ca^{2+} exchange mechanism during prolonged depolarizations. The rate of Na^+/Ca^{2+} exchange does not appear, however, to be altered by cAMP, a conclusion that is supported by the observation (Richard et al., 1985) that cAMP jumps do not change the rate of relaxation of contraction; relaxation is attributed to Ca^{2+} transport (e.g., extrusion via Na^+/Ca^{2+} exchange) (Goto et al., 1972; Kavaler and Anderson, 1978; Klitzner and Morad, 1983). It is not possible to conclude the mechanism whereby cAMP reduces tonic tension, although it is clear that the kinetics and concentration dependences of cAMP-induced modulation of tonic tension and the delayed K^+ current are quite different from those measured for cAMP-stimulated increases in I_{si} and phasic tension. It appears unlikely, therefore, that a single mechanism can explain all of the physiological effects of increasing intracellular cAMP. In these experiments, we also observe no measurable effects of cGMP concentration jumps on the tonic tension or on the delayed K^+ conductance either alone or associated with cAMP increases (Richard et al., 1985).

In spontaneously beating sinus venosus fibers, we find that cAMP jumps increase the action potential frequency (Garnier et al., 1983). All of these results, taken together, therefore, reveal that cAMP concentration jumps mimic all of the effects of β-adrenergic stimulation in frog atrial fibers and, most probably, in frog heart in general. This conclusion is perhaps not surprising, as it is consistent with the notion that, in most tissues, β agonists act via stimulation of adenylate cyclase (Minneman and Molinoff, 1980). Our results, in addition, however, suggest strongly that cAMP exerts its effects through the involvement and the modulation of multiple processes, and it appears likely that each of these various mechanisms is sensitive to and regulated by different intracellular cAMP concentrations. This particular conclusion is obtained as a direct result of exploiting the photochemical methodology for increasing intracellular cAMP and, moreover, suggests that detailed mechanistic studies of the multiple effects of cAMP (or cGMP) should be possible using this approach.

6.   KINETIC IMPLICATIONS OF cAMP EFFECTS

As noted above, we consistently observe that I_{si} and tension amplitudes are increased following flashes (in the presence of the photolabile cAMP analogues) within a few hundred milliseconds and that maximum responses require tens of seconds (Nargeot et al., 1983; Nerbonne et al., 1984; Richard et al., 1985); similar results are obtained with both analogues in spite of the dramatically different rates of photorelease (see Table 1). We interpret the measured time courses to reflect the biochemical events that are generally assumed (Reuter, 1979, 1983; Osterrieder et al., 1982; Trautwein et al., 1982; Bean et al., 1983; Brum et al., 1983; Cachelin et al., 1983; Reuter et al., 1983; Tsien, 1983) to mediate current and tension modulation by cAMP. The observation that, following a single intracellular cAMP concentration jump, I_{si} increases linearly in time with no detectable delay and extrapolates to zero (Nargeot et al., 1983; Nerbonne et al., 1984) suggests in addition that a single rate-limiting step is involved between the production of cAMP and the increase in the current (and tension). However, it is not possible to conclude from our work which biochemical step (e.g., kinase activation, kinase dissociation, etc.) might be rate limiting. Although all of the observed effects of cAMP jumps are consistent with these biochemical postulates, we were interested to determine whether indeed Ca^{2+} channels (or any membrane channels, for that matter) could be manipulated on the time scale of a few milliseconds with a photolabile probe. To explore this possibility, we initiated studies aimed at determining the time scale over which channels could be manipulated using photolabile Ca^{2+} antagonists.

7.   PHOTOLABILE Ca^{2+} ANTAGONISTS

Nifedipine and other "calcium antagonists" (Fleckenstein, 1977, 1983; Nayler, 1983; Janis and Triggle, 1983; Triggle and Swamy, 1983; Kass, 1983) inhibit Ca^{2+} influx in cardiac and smooth muscle and are thought to act via a "channel blockade" mechanism (Adams, 1976, 1977; Armstrong, 1966; Strichartz, 1973; Hille, 1977; Hondeghem, 1977, 1984; Neher and Steinbach, 1978; Lester et al., 1979; Starmer et al., 1984; Uehara and Hume, 1984). Although this concept is supported experimentally in many cases, numerous questions remain about the detailed relationship between binding and blockade in the case of the structurally diverse organic calcium channel blockers (DePover et al., 1982; Glossmann et al., 1982, 1983; Fleckenstein, 1983; Holck et al., 1983; Janis and Triggle, 1983; Kanaya et al., 1983; Lee and Tsien, 1983; Murphy et al., 1983; Morgan et al., 1983; Nayler, 1983; Pang and Sperelakis, 1983; Hess et al., 1984; Miller and Freedman, 1984). For example: (1) Do drugs bind preferentially to open, closed and/or inactivated channels? (2) Are there multiple binding sites? (3) Do these drugs act at extracellular and/or intracellular binding sites? (4) Does blocking or unblocking depend on membrane potential or its history? (5) Do these compounds preferentially alter channel gating?

Nifedipine (Eq. (2)) contains an o-nitrobenzyl moiety and is photolabile (Ebel et al., 1978); irradiation in solution yields a molecule devoid of Ca^{2+} channel-blocking activity (Ebel et al., 1978; Morad et al., 1983; Sanguinetti and Kass, 1984) and the reactions leading to photodestruction are complete within 100 μs (Morad et al., 1983).

$$ (2) $$

It can be speculated that these properties make nifedipine, as well as other photolabile Ca^{2+} antagonists (Morad et al., 1983; Sanguinetti and Kass, 1984), not only a potentially reversible Ca^{2+} antagonist (Sanguinetti and Kass, 1984) but also a useful compound for mechanistic studies of Ca^{2+} channels and their regulation.

8.   Ca^{2+} ANTAGONISTS AND THE MODULATION OF I_{si}

Exploiting the photolabile properties of nifedipine to permit determination of the time scale over which Ca^{2+} channels could be modulated and to attempt mechanistic studies of Ca^{2+} antagonist action, we examined the waveforms of I_{si} in isolated, voltage-clamped frog atrial trabeculae in the presence of nifedipine and during flash-induced removal of nifedipine (Nerbonne et al., 1985).

Nifedipine decreases the amplitudes of I_{si} and tension in isolated, voltage-clamped (double sucrose gap) atrial fibers (ID$_{50}$ = 0.5 μM) without altering the kinetics or the voltage dependence; no "use-dependent" block is observed (see Ehara and Kaufman, 1978; Lee and Tsien, 1983). Current and tension suppression by nifedipine are reversed by light; in the absence of nifedipine, light has no effects. Preirradiated solutions of nifedipine (to 20 μM) in the presence and in the absence of light similarly do not alter the waveforms of I_{si} or tension. To study the time course and kinetics of I_{si} reactivation following photoremoval of nifedipine, single flashes were delivered prior to, or during, depolarizing voltage steps. When a flash was presented at the start of (Fig. 7a) or during (Fig. 7c) a step, current amplitude is increased, and the rate of reactivation of I_{si} equals the normal, voltage-dependent rate of current activation (time-to-peak current 6 ± 1 ms) (Fig. 7a,c). The increase in I_{si} proceeds, therefore, approximately 10^4 times faster than with cAMP jumps. An additional effect of the flash is seen: the apparent rate of decay of I_{si} is increased (Fig. 7a,c) immediately after the flash, returning to its preflash value in the subsequent trials (Fig. 7b,d). Peak amplitudes, in contrast, are the same in the two trials after flashes. I_{si} decay, however, is not accelerated when flashes are delivered 500–800 ms

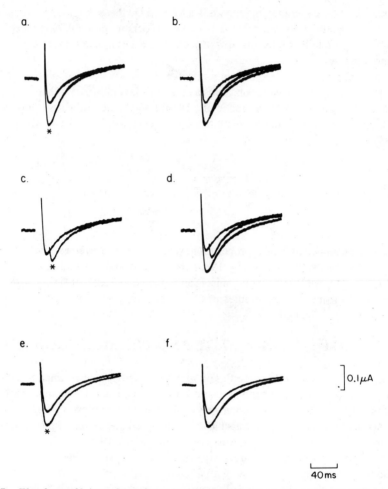

a.

b.

c.

d.

e.

f.

0.1 μA

|_____|
40 ms

Figure 7. Waveforms of I_{si} in a voltage-clamped atrial trabecula in the presence of 0.5 μM nifedipine and 1 μM TTX; the fiber was held 20 mV depolarized from the resting potential and depolarized 70 mV at 5-s intervals. Each pair of panels (a,b; c,d; e,f) presents two or three successive trials. The preflash steady-state level of I_{si} was lowest in the first trial of each panel; flashes were delivered prior to or during the second trials. The first two trials are superimposed in panels on the left (a,c,e); flash trials are indicated by the asterisks. The third trials are included in b, d, and f. (a) A single flash was delivered simultaneously with the start of the depolarizing step: I_{si} amplitude increased twofold. Although the rate of I_{si} activation was the same before and after the flash, the rate of I_{si} decay increased after the flash. (b) The accelerated decay rate was a transient effect of the flash: although peak amplitudes were identical in the two trials following the flash, the wave forms of I_{si} were different. When the flash was delivered 20 ms after the start of the step (c), similar results were obtained: I_{si} amplitude was increased, the rate of reactivation of I_{si} paralleled the normal rate of current activation, and the rate of decay of I_{si} was again increased transiently after the flash (d). When the flash was delivered > 500 ms before the start of the step as in (e), I_{si} amplitude recovered with little or no change in decay kinetics (f). Similar effects were seen when the action potential and tension or I_{si} and tension were monitored simultaneously (see: Nerbonne et al., 1985).

or more before trials (Fig. 7e); the wave forms of I_{si} in trials 2 and 3 superimpose (Fig. 7f). The acceleration of I_{si} decay is therefore observed transiently in the first few hundred milliseconds after a flash, and the amplitude of the effect varies with flash timing. After a flash, tension amplitude is recovered independent of the timing of the flash and with no measurable changes in kinetics.

Similar results, including the transient flash effects just described, are obtained in experiments in which action potentials and tension are monitored simultaneously, suggesting that similar mechanisms are involved (Nerbonne et al., 1985). Additional experiments revealed that the transient flash effect on I_{si} inactivation could not be attributed to blockade of channels other than the slow channels (Kass and Tsien, 1973; Nishi et al., 1983) or to increased Ca^{2+}-dependent outward currents, and it has therefore been attributed to a direct effect on I_{si} inactivation (Nerbonne et al., 1985). Although the underlying mechanism responsible for the time dependence is not completely clear, it does seem likely that it is this effect and not a requirement for membrane repolarization , as previously suggested (Morad et al., 1983), that leads to incomplete recovery of action potentials and tension seen in some experiments (Morad et al., 1983; Nerbonne et al., 1985). This conclusion is supported by the finding that, under voltage clamp, the amplitude of I_{si} is recovered fully within a few milliseconds at all membrane potentials after flash-induced removal of nifedipine. The experimental difficulties associated with adequate control of the ionic compositions of the intra- and extracellular media in this preparation complicate further mechanistic analysis of the observed inactivation rate enhancement.

Nonetheless, these results do demonstrate that: (1) Ca^{2+} channels are unblocked within at most a few milliseconds after photoremoval of nifedipine; (2) the rate of reactivation of I_{si} parallels the normal, voltage-dependent rate of I_{si} activation; and (3) membrane repolarization is not required for reversal of the nifedipine-induced suppression of I_{si}. The speed of I_{si} recovery (approximately 10^4 times faster than that observed following cAMP jumps) seen after photoremoval of nifedipine reveals that the slow Ca^{2+} channels can indeed be manipulated on the time scale of milliseconds and support the previous conclusion that, after cAMP jumps, I_{si} increases over a time course that reflects the biochemical steps involved in the modulation of I_{si} by cAMP.

In addition, findings (1) and (2) imply that nifedipine was bound to and stabilized Ca^{2+} channels in a closed, resting state. If the drug were bound to and blocked open channels, then one would expect photoremoval, when effected during a voltage step, to reveal open channels directly; that is, unblocking would have proceeded at a rate equal to the photolysis rate (i.e., < 1 ms) rather than at the rate associated with channel gating. Such an effect was not observed here. It might be argued, however, that the limitations (Ramon et al., 1975; Poindessault et al., 1976; Attwell and Cohen, 1977; Reuter and Scholz, 1977a; Beeler and McGuigan, 1978) of the sucrose-gap voltage-clamp technique prevent the resolution of events faster than the rise time of I_{si}. If this effect were significant in these experiments, then relief from open channel blockade would likely go undetected. As a result, conclusions regarding the mechanism of nifedipine-induced suppression of I_{si} would be drawn in error. We therefore decided to extend these studies using patch-clamp methods.

9.   PHOTOLABILE Ca²⁺ ANTAGONISTS AND WHOLE-CELL RECORDING TECHNIQUES

In order to better quantitate these effects, similar experiments were performed (Gurney et al., 1985) on single heart cells (cultured, neonatal rat ventricular myocytes) using the whole-cell patch-clamp method (Hamill et al., 1981; Sakmann and Neher, 1983, 1984). Using this approach, I_{si} could be isolated more efficiently from contaminating outward currents, and current kinetics could be resolved more accurately than is possible with the sucrose-gap method. In this preparation, nifedipine also suppresses Ca^{2+} current amplitude (ID_{50} = 0.5 μM) and similarly does not alter the activation kinetics or voltage dependence of the current and does not display "use dependence." In many experiments, Ba^{2+} is substituted for Ca^{2+} to carry inward current, as under these conditions current amplitude is increased (Reuter and Scholz, 1977a; Nishi et al., 1983) and the rates of current activation and inactivation are greatly reduced (Lee and Tsien, 1983; Bean et al., 1983). As in other

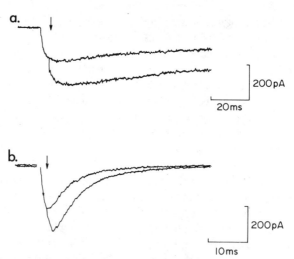

Figure 8.   Reversal of nifedipine blockade of whole-cell Ba^{2+} and Ca^{2+} currents in cultured rat ventricular myocytes. Currents were recorded in response to step depolarizations to 10 mV from holding potentials of -50 mV in the whole-cell patch-clamp configuration. The bathing solution contained 20 μM TTX and 0.5 μM nifedipine and either 10 mM Ba^{2+} (a) or 5 mM Ca^{2+} to carry inward current; the pipette solution contained 140 mM Cs^{+} to block outward K^{+} currents. The steady-state, partially suppressed currents in the presence of nifedipine are the lower amplitude records in (a) and (b). In each case, a single flash was delivered at the times indicated by the arrows: currents were increased, and the rate of current activation paralleled the normal (voltage-dependent) rate of current activation. The similarities in the kinetics of current activation before and after a flash imply that nifedipine blocks closed Ca^{2+} channels (see Gurney et al., 1985).

preparations, nifedipine does not alter the voltage dependence or the kinetics of activation of Ba^{2+} currents; the rates of current inactivation are, however, increased (see below).

We find that the nifedipine-induced suppression of Ba^{2+} and Ca^{2+} currents can be reversed by single flashes (Fig. 8a,b) and that the rates of current reactivation equal the normal, voltage-dependent rates of activation (Gurney et al., 1985). These

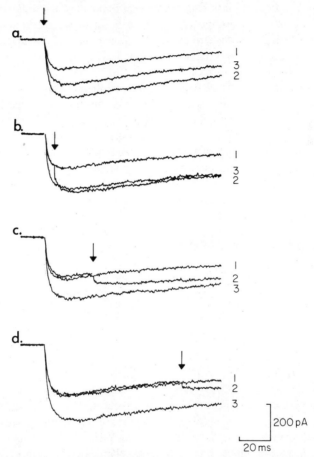

Figure 9. Recovery from nifedipine blockade depends on flash timing. Whole-cell Ba^{2+} currents recorded from a cultured (rat) ventricular myocyte; all experimental conditions were identical to those described in the legend to Fig. 8. In each experiment, a single flash was delivered to the preparation at various times (as indicated by the arrows) during the 125-ms depolarizing voltage-clamp step. The fractional recovery is dependent on when, during the step, the flash occurred: flashes delivered early provide complete (a) or nearly complete (b) recovery; flashes presented late yield very little recovery (d), results which imply that nifedipine binds effectively to inactivated Ca^{2+} channels. In experiments similar to these, we have demonstrated that nifedipine blocks "closed," "open," and "inactivated" channels, findings consistent with the modulated receptor model (Hille, 1977) of drug action (see Gurney et al., 1985).

findings are consistent with the previous conclusion (Nerbonne et al., 1985) that nifedipine binds to and stabilizes closed, resting Ca^{2+} channels. In addition, we observe that Ba^{2+} currents inactivate faster in the presence of nifedipine, suggesting the additional presence of open-channel blockade. Recovery of Ba^{2+} current amplitude, which can be effected by a flash, is also dependent on when the flash is presented during a depolarizing voltage-clamp step (Fig. 9a–d). Flashes delivered early in a trial (Fig. 9a,b) yield complete (or nearly complete) recovery, whereas those presented late reveal little recovery (Fig. 9c,d). Nifedipine, therefore, apparently also binds to inactivated channels; removal of the drug from inactivated channels does not directly open these channels (or increase current amplitude).

It is interesting to note here that neither open-channel nor inactivated-channel block is revealed in experiments performed with Ca^{2+} as the current carrier presumably because inactivation, which is Ca^{2+}- and voltage-dependent (Tsien, 1983; Reuter, 1984; Lux and Brown, 1984), is already very fast (Fig. 8b). These results therefore suggest that under normal conditions (i.e., when used clinically) the predominant effect of nifedipine is most likely to "block" closed Ca^{2+} channels, thus preventing them from opening. The results obtained with Ba^{2+} as the current carrier, however, also imply that nifedipine can bind to closed, open, and inactivated Ca^{2+} channels, a suggestion that is consistent with the "modulated receptor" model for drug action (Hille, 1977; Hondeghem and Katzung, 1977, 1984; Murphy et al., 1983; Starmer et al., 1984; Uehara and Hume, 1984).

10. CONCLUSIONS

These results provide an overview of our progress to date in the development and utilization of photolabile intracellular probes. The 4,5-dimethoxy-2-nitrobenzyl esters of cAMP and cGMP are anticipated to be generally useful molecules for the production of intracellular "concentration jumps" in virtually any tissue or preparation that is reasonably transparent and lacks an endogenous physiological reponse to light. In pigmented systems—for example, in cultured bag cells of the marine mollusk, *Aplysia californica*—we do not obtain reproducible (i.e., quantitative) effects of flashes (Gurney and Nerbonne, 1984; Nerbonne and Gurney, 1985). For further quantification of the flash-induced cyclic nucleotide concentration jumps, it will be interesting to prepare and study the effects of membrane-impermeant, light-sensitive cyclic nucleotide analogues. Although we have, in addition, developed a number of photolabile proton donors for physiological studies, we have not yet successfully exploited these molecules to produce reliable H^+ jumps. The principal difficulty encountered to date (Spray et al., 1984) is that these compounds themselves are apparently hydrolyzed efficiently and reduce intracellular pH in the absence of light. These molecules may, however, provide a useful noninvasive means for chronically altering intracellular pH (Spray et al., 1984). We shall, however, continue our efforts in the development of photolabile proton donors that are thermally and enzymatically nonlabile.

It is anticipated that the chemical development and further characterization of these probes will continue, as it appears clear that an understanding of the chemical parameters controlling rates and efficiencies is crucial to the development of new photolabile intracellular (and extracellular) probes and, in particular, to the design of the optimal, generally useful photolabile protecting group. It will also certainly be of great interest to search for other biologically relevant molecules (like nifedipine) which are inherently light-sensitive and to exploit the photolabile properties of these molecules in mechanistic studies such as those described above. It seems clear that the combination of photochemical and electrophysiological methodologies such as detailed here will be extremely powerful in the future, providing mechanistic insights into ion channel regulation and modulation by intracellular messengers and other internal and external stimuli. This combination of approaches, particularly when patch-clamp techniques (Hamill et al., 1981; Sakmann and Neher, 1983, 1984) are employed, should enable experiments to be performed and analyses to be conducted that could not have been undertaken using more conventional physiological and pharmacological techniques.

ACNOWLEDGMENTS

I acknowledge and thank my colleagues and collaborators, J. Nargeot, S. Richard, D. Lo, and A. M. Gurney, who have been involved in many aspects of this work over the past several years and, in particular, H. A. Lester in whose laboratory most of this work was performed. I also thank the Centre d'Etudes et de Recherches Servier, Neuilly/Seine (France) for providing a travel grant. This research has been supported by the American Heart Association, Los Angeles Affiliate (Postdoctoral Fellowship), and the National Office (Established Investigatorship), the National Institutes of Health (GM-29836), INSERM, France (835012), and the CNRS.

REFERENCES

Adams, P. R. (1976) Drug blockade of open end-plate channels. *J. Physiol. (Lond.)*, **260**, 531–552.

Adams, P. R. (1977) Voltage jump analysis of procaine action at frog end-plate. *J. Physiol. (Lond.)*, **268**, 291–318.

Amit, B., U. Zehavi, and A. Patchornik (1974) Photosensitive protecting groups of amino sugars and their use in glycoside synthesis. 2-Nitrobenzyloxycarbonylamino and 6-nitroveratryloxycarbonylamino derivatives. *J. Org. Chem.*, **39**, 192–196.

Armstrong, C. M. (1966) Time course of TEA$^+$-induced anomalous rectification in squid giant axons. *J. Gen. Physiol.*, **50**, 491–503.

Attwell, D., and I. Cohen (1977) The voltage-clamp of multicellular preparations. *Prog. Biophys. Mol. Biol.*, **31**, 201–245.

Bean, B. P., M. C. Nowycky, and R. W. Tsien (1983) β-Adrenergic modulation of the number of functional calcium channels in frog ventricular heart cells. *Nature (Lond.)*, **30**, 371–375.

Beeler, G. W., and J. A. S. McGuigan (1978) Voltage clamping of multicellular myocardial preparations: Capabilities and limitations of existing methods. *Prog. Biophys. Mol. Biol.*, **34**, 219–254.

Brum, G., V. Flockerzi, F. Hofmann, W. Osterrieder, and W. Trautwein (1983) Injection of the catalytic subunit of cAMP-dependent protein kinase into isolated cardiac myocytes. *Pflügers Arch.*, **398**, 147–154.

Cachelin, A. B., J. E. dePeyer, S. Kokubun, and H. Reuter (1983) Calcium channel modulation by 8-bromo-cyclic AMP in cultured heart cells. *Nature (Lond.)*, **304**, 462–464.

Campos de Carvalho, A., D. C. Spray, and M. V. L. Bennett (1984) pH dependence of transmission at electrotonic synapses of the crayfish septate axon. *Brain Res.*, **321**, 279–286.

Chapman, R. A. (1979) Excitation–contraction coupling in cardiac muscle. *Prog. Biophys. Mol. Biol.*, **35**, 1–52.

Davies, H. W., and M. Schwarz (1965) The effects of hydrogen bonding on the absorption spectra of some substituted benzaldehyde tosylhydrazone anions. *J. Org. Chem.*, **30**, 1242–1244.

DePover, A., M. A. Matlib, S. W. Lee, G. P. Dube, G. Grupp, and A. Schwartz (1982) Specific binding of [^{3}H]-nitrendipine to membranes from coronary arteries and heart in relation to pharmacological effects. Paradoxical stimulation by diltiazem. *Biochem. Biophys. Res. Commun.*, **108**, 110–117.

Drummond, G. I. (1973) Metabolism and functions of cyclic AMP in nerve. *Prog. Neurobiol.*, **2**, 119–176.

Drummond, G. I. (1983) Cyclic nucleotides in the nervous system. *Adv. Cyclic Nucleotide Res.*, **15**, 373–494.

Drummond, G. I., and D. L. Severson (1977) Cyclic nucleotides and cardiac function. *Circ. Res.*, **44**, 145–153.

Ebel, V. S., H. Schutz, and A. Hornitschek (1978) Untersuchungen zur Analytik von Nifedipin unter besonderer Berucksichtigung der bei Lichtexposition entstehenden Umwandlungsprodukte. *Arzneim. Forsch.*, **28**, 2188–2193.

Ehara, T., and R. Kaufmann (1978) The voltage- and time-dependent effects of ($-l$)-verapamil on the slow inward current in isolated cat ventricular myocardium. *J. Pharmacol. Exp. Ther.*, **207**, 49–55.

Engels, J., and R. Reidys (1978) Synthesis and application of the photolabile guanosine 3′,5′-phoshoric-o-nitrobenzyl ester. *Experientia*, **34**, 14–15.

Engels, J., and E. J. Schlaeger (1977) Synthesis, structure, and reactivity of adenosine cyclic 3′,5′-phosphate benzyl triesters. *J. Med. Chem.*, **20**, 907–911.

Fabiato, A., and F. Fabiato (1979) Calcium and cardiac excitation–contraction coupling. *Annu. Rev. Physiol.*, **41**, 473–484.

Fisher, S. K., L. A. A. Van Rooijen, and B. W. Agranoff (1984) Renewed interest in the polyphosphoinositides. *Trends Biochem. Sci.*, **9**, 53–56.

Fleckenstein, A. (1977) Specific pharmacology of calcium in myocardium, cardiac pacemakers, and vascular smooth muscle. *Annu. Rev. Pharmacol. Toxicol.*, **17**, 149–166.

Fleckenstein, A. (1983) History of calcium antagonists. *Circ. Res.* (Suppl. #1), **52**, 3–16.

Fozzard, H. A. (1977) Heart: Excitation–contraction coupling. *Annu. Rev. Physiol.*, **39**, 201–220.

Garnier, D., H. A. Lester, J. Nargeot, J. M. Nerbonne, and S. Richard (1983) Excitation–contraction coupling in frog atrial muscle during intracellular jumps of cAMP. *J. Physiol. (Lond.)*, **344**, 38P.

Giaume, C., M. E. Spira, and H. Korn (1980) Uncoupling of invertebrate electrotonic synapses by carbon dioxide. *Neurosci. Lett.*, **17**, 197–202.

Glossmann, H., D. R. Ferry, F. Luebbecke, R. Mewes, and F. Hofmann 1982. Calcium channels: Direct identification with radioligand binding studies. *Trends Pharm. Sci.*, **39**, 431–437.

Glossman, H., D. R. Ferry, F. Lübbecke, R. Mewes, and F. Hofmann (1983) Identification of voltage operated calcium channels by binding studies: Differentiation of subclasses of calcium antagonist drugs with ^{3}H-nimodipine radioligand binding. *J. Recept. Res.*, **3**, 177–190.

Goldman, Y. E., H. Gutfreund, M. G. Hibberd, J. A. McCray, and D. R. Trentham (1982a) The role of thiols in caged-ATP studies and their use in measurement of the photolysis kinetics of o-nitrobenzyl compounds. *Biophys. J.*, **37**, 125a.

Goldman, Y. E., M. G. Hibberd, J. A. McCray, and D. R. Trentham (1982b) Relaxation of muscle fibres by photolysis of caged ATP. *Nature (Lond.)*, **300**, 701–705.

Goldman, Y. E., M. G. Hibberd, and D. R. Trentham (1984a) Relaxation of rabbit psoas muscle fibers from rigor by photochemical generation of adenosine-5'-triphosphate. *J. Physiol. (Lond.)*, **354**, 577–604.

Goldman, Y. E., M. G. Hibberd, and D. R. Trentham (1984b) Initiation of contraction by generation of adenosine-5'-triphosphate in rabbit psoas muscle. *J. Physiol. (Lond.)*, **354**, 605–624.

Goto, M., Y. Kimoto, M. Saito, and Y. Wada (1972) Tension fall after contraction of bullfrog atrial muscle examined with the voltage clamp technique. *Jpn. J. Physiol.*, **22**, 637–650.

Gurney, A. M., and J. M. Nerbonne (1984) Nifedipine blockade of neuronal calcium currents. *Soc. Neurosci., Abstracts* **10**: 868.

Gurney, A. M., J. M. Nerbonne, and H. A. Lester (1984) Photoremoval of nifedipine reveals the likely mechanism of action of this Ca^{++} antagonist. *Biophys. J.*, **45**, 35a.

Gurney, A. M., J. M. Nerbonne, and H. A. Lester (1985) Photoinduced removal of nifedipine reveals mechanisms of calcium antagonist action on single heart cells. J. Gen. Physiol. **86**, 353–379.

Hamill, O. P., A. Marty, E. Neher, B. Sakmann, and F. J. Sigworth (1981) Improved patch-clamp techniques for high resolution current recording from cells and cell-free membrane patches. *Pflügers Arch.*, **391**, 85–100.

Hartzell, H. C., and L. Titus (1982) Effects of cholinergic and adrenergic agonists on phosphorylation of a 165,000-dalton myofibrillar protein in intact cardiac muscle. *J. Biol. Chem.*, **257**, 2111–2120.

Hess, P., J. B. Lansman, K. S. Lee, and R. W. Tsien (1984) Cardiac calcium current: Regulation by Ca antagonists and Ca agonists. *J. Physiol. (Lond.)*, **356**, 9P.

Hibberd, M. G., Y. E. Goldman, and D. R. Trentham (1983) The mechanical response of muscle fibers to a rapid increase in concentration of ATP. A. Oplatka and Balaban, M., Eds., *Biological Structures and Coupled Flows*, Academic Press, New York, pp. 223–238.

Hille, B. (1977) Local anesthetics: Hydrophilic and hydrophobic pathways for the drug–receptor reaction. *J. Gen. Physiol.*, **69**, 497–515.

Holck, M., S. Thorens, and G. Hausler (1983) Does [^{3}H]nifedipine label the calcium channel in rat myocardium? *J. Recept. Res.*, **3**, 191–198.

Hondeghem, L. M., and B. G. Katzung (1977) Time- and voltage-dependent interactions of antiarrhythmic drugs with cardiac sodium channels. *Biochim. Biophys. Acta*, **472**, 373–398.

Hondeghem, L. M., and B. G. Katzung 1984. Antiarrhythmic agents: The modulated receptor mechanism of action of sodium and calcium channel-blocking drugs. *Annu. Rev. Pharmacol. Toxicol.*, **24**, 387–423.

Iwatsuki, N., and O. H. Peterson (1978) Pancreatic acinar cells: Acetylcholine-evoked electrical uncoupling and its ionic dependency. *J. Physiol. (Lond.)*, **274**, 81–96.

Iwatsuki, N., and O. H. Peterson (1979) Pancreatic acinar cells: The effect of carbon dioxide, ammonium chloride and acetylcholine on intercellular communication. *J. Physiol. (Lond.)*, **291**, 317–326.

Janis, R. A., and D. J. Triggle (1983) New developments in Ca^{++} channel antagonists. *J. Med. Chem.*, **26**, 775–785.

Kanaya, S., P. Arlock, B. G. Katzung, and L. M. Hondeghem (1983) Diltiazem and verapamil preferentially block inactivated cardiac calcium channels. *J. Mol. Cell. Cardiol.*, **15**, 145–148.

Kaplan, J. H., and R. J. Hollis (1980) External Na dependence of ouabain-sensitive ATP:ADP exchange initiated by photolysis of intracellular caged-ATP in human red cell ghosts. *Nature (Lond.)*, **288**, 587–589.

Kaplan, J. H., B. Forbush, and J. F. Hoffmann (1978) Rapid photolytic release of adenosine 5'-triphosphate from a protected analogue: utilization by the Na:K pump of human red blood cell ghosts. *Biochemistry*, **17**, 1929–1935.

Kass, R. S. (1983) Measurement and block of voltage-dependent calcium current in heart. A comparison of the actions of D600 and nisoldipine. G. F. Merrill and H. R. Weiss, Eds., *Calcium Entry Blockers, Adenosine and Nuerohumors*, Urban and Schwartzenberg, Baltimore, MD. pp. 1–19.

Kass, R. S., and R. W. Tsien (1975) Multiple effects of calcium antagonists on plateau currents on cardiac Purkinje fibers. *J. Gen Physiol.*, **66**, 169–192.

Katz, A. M. (1979) Role of contractile proteins and sarcoplasmic reticulum in the response of the heart to catecholamines: A historical review. *Adv. Cyclic Nucleotide Res.*, **11**, 303–343.

Kavaler, F., and T. W. Anderson (1978) Indirect evidence that calcium extrusion causes relaxation of frog ventricular muscle. *Fed. Proc.*, **37**, 300.

Klitzner, T., and M. Morad (1983) Excitation–contraction coupling in frog ventricle: possible Ca^{++} transport mechanisms. *Pflügers Arch.*, **398**, 274–283.

Korth, M., and J. Engels (1979) The effects of adenosine- and guanosine 3',5'-phosphoric acid benzyl esters on guinea pig ventricular myocardium. *Naunyn-Schmied. Arch. Pharmacol.*, **310**, 103–111.

Kupferman, I. (1979) Modulatory actions of neurotransmitters. *Annu. Rev. Neurosci.*, **2**, 447–465.

Lee, K. S., and R. W. Tsien (1983) Mechanism of calcium channel blockade by verapamil, D600, diltiazem and nitrendipine in single dialysed heart cells. *Nature (Lond.)*, **302**, 790–794.

Lester, H. A., and J. M. Nerbonne (1982) Physiological and pharmacological manipulations with light flashes. *Annu. Rev. Biophys. Bioeng.*, **11**, 151–175.

Lester H. A., M. E. Krouse, M. M. Nass, N. H. Wassermann, and B. F. Erlanger (1979) Light-activated drug confirms a mechanism of ion channel blockade. *Nature (Lond.)*, **280**, 509–510.

McCray, J. A., L. Herbette, T. Kihara, and D. R. Trentham (1980) A new approach to time-resolved studies of ATP-requiring biological systems: Laser flash photolysis of caged ATP. *Proc. Natl. Acad. Sci. USA*, **77**, 7327–7241.

Miller, R. J., and S. B. Freedman (1984) Are dihydropyridine binding sites voltage sensitive calcium channels? *Life Sci.*, **34**, 1205–1221.

Minneman, K. P., and P. Molinoff (1980) Classification and quantitation of β-adrenergic receptor subtypes. *Biochem. Pharmacol.*, **29**, 1317–1323.

Morad, M., Y. E. Goldman, and D. R. Trentham (1983) Rapid photochemical inactivation of Ca^{++}-antagonists shows that Ca^{++} entry directly activates contraction in frog heart. *Nature*, **304**, 635–638.

Morgan, J. P., W. G. Wier, P. Hess, and J. R. Blinks (1983) Influence of Ca^{++}-channel blocking agents on calcium transients and tension development in isolated mammalian heart muscle. *Circ. Res.* (Suppl. 1), **52**, 47–52.

Murphy, K. M. M., R. J. Gould, B. L. Largent, and S. H. Snyder (1983) A unitary mechanism of calcium antagonist drug action. *Proc. Natl. Acad. Sci. USA*, **80**, 860–864.

Nargeot, J., H. A. Lester, N. J. M. Birdsall, J. Stockton, N. H. Wassermann, and B. F. Erlanger (1982) A photoisomerizable muscarinic antagonist: Studies of binding and conductance relaxations in frog heart. *J. Gen. Physiol.*, **79**, 657–678.

Nargeot, J., J. M. Nerbonne, J. Engels, and H. A. Lester (1983) Time course of the increase in the myocardial slow inward current after a photochemically generated concentration jump of intracellular cAMP. *Proc. Natl. Acad. Sci. USA*, **80**, 2395–2399.

Nayler, W. G. (1983) The heterogeneity of the slow channel blockers (calcium antagonists). *Int. J. Cardiol.*, **3**, 391–400.

Neher, E., and J. H. Steinbach (1978) Local anaesthetics transiently block currents through single acetylcholine receptor channels. *J. Physiol. (Lond.)*, **277**, 153–176.

Nerbonne, J. M. and A. M. Gurney (1985) Blockade of Ca^{++} and K^+ currents in bag cell neurons of *Aplysia Califirnica* by dihydrophyridine Ca^{++} antagonists. (Submitted).

Nerbonne, J. M., S. Richard, J. Nargeot, and H. A. Lester (1984) New photoactivatable cyclic nucleotides produce intracellular jumps in cyclic AMP and cyclic GMP concentrations. *Nature (Lond.)*, **310**, 74–76.

Nerbonne, J. M., S. Richard, and J. Nargeot (1985) Calcium channels are unblocked within a few msec after photoconversion of nifedipine. J. Mol. Cell. Cardiol. **17**, 511–515.

Nirenberg, M., S. Wilson, H. Higashida, A. Rotter, K. Krueger, N. Busis, R. Ray, J. G. Kenimer, and M. Adler (1983) Modulation of synapse formation by cyclic adenosine monophosphate. *Science*, **222**, 794–799.

Nishi, K., N. Akaike, Y. Oyama, and H. Ito (1983) Actions of calcium antagonists on calcium currents in helix neurons, specificity and potency. *Circ. Res.* (Suppl. 1), **52**, 53–59.

Noma, A. (1983) ATP-regulated K^+ channels in cardiac muscle. *Nature (Lond.)*, **305**, 147–148.

Opie, L. (1982) Role of cyclic nucleotides in heart metabolism. *Cardiovasc. Res.*, **16**, 483–507.

Osterrieder, W., G. Brum, J. Hescheler, W. Trautwein, F. Hofmann, and V. Flockerzi (1982) Injection of subunits of cyclic AMP-dependent protein kinase into cardiac myocytes modulates Ca^{++} current. *Nature (Lond.)*, **298**, 576–578.

Ouedraogo, C. O., D. Garnier, J. Nargeot, and B. Pourrias (1982) Electrophysiological and pharmacological study of the inotropic effects of adrenaline, dopamine and tryptamine on frog atrial fibers. *J. Mol. Cell. Cardiol.*, **14**, 111–121.

Pang, D. C., and N. Sperelakis (1983) Nifedipine, diltiazem, bepridil and verapamil uptakes into cardiac and smooth muscles. *Eur. J. Pharmacol.*, **87**, 199–207.

Patchornik, A., B. Amit, and R. B. Woodward (1970) Photosensitive protecting groups. *J. Am. Chem. Soc.*, **92**, 6333–6335.

Perry, S. (1979) The regulation of contractile activity in muscle. *Biochem. Soc. Trans.*, **7**, 593–617.

Pillai, V. N. (1980) Photoremovable protecting groups in organic synthesis. *Synthesis*, 1–26.

Poindessault, J. P., A. Duval, and C. Leoty (1976) Voltage clamp with double sucrose gap technique: External series resistance compensation. *Biophys. J.*, **16**, 105–120.

Ramon, F., N. Anderson, R. W. Joyner, and J. W. Moore (1975) Axon voltage-clamp simulations. IV. A multicellular preparation. *Biophys. J.*, **15**, 55–69.

Rasmussen, H., and D. B. P. Goodman (1977) Relationships between calcium and cyclic nucleotides in cell activation. *Physiol. Rev.*, **57**, 421–509.

Reber, W. R., and R. Weingart (1982) Ungulate cardiac Purkinje fibres: The influence of intracellular pH on the electrical cell–cell coupling. *J. Physiol. (Lond.)*, **328**, 87–104.

Reuter, H. (1979) Properties of two inward membrane currents in the heart. *Annu. Rev. Physiol.*, **41**, 413–424.

Reuter, H. (1983) Calcium channel modulation by neurotransmitters, enzymes and drugs. *Nature (Lond.)*, **301**, 569–574.

Reuter, H. (1984) Ion channels in cardiac cell membranes. *Annu. Rev. Physiol.*, **46**, 473–484.

Reuter, H., and H. Scholz (1977a) A study of the ion selectivity and the kinetic properties of the calcium dependent slow inward current in mammalian cardiac muscle. *J. Physiol. (Lond.)*, **264**, 17–47.

Reuter, H., and H. Scholz (1977b) The regulation of the calcium conductance of cardiac muscle by adrenaline. *J. Physiol. (Lond.)*, **264**, 49–62.

Reuter, H., A. B. Cachelin, J. E. dePeyer, and S. Kokubun 1983. Modulation of calcium channels in cultured cardiac cells by isoproterenol and 8-bromo-cAMP. *Cold Spring Harbor. Symp. Quant. Biol.*, **48**, 193–200.

Richard, S., J. M. Nerbonne, J. Nargeot, H. A. Lester, and D. Garnier (1985) Photochemically produced intracellular concentration jumps of cAMP mimic the effects of catecholamines on excitation contraction coupling in frog atrial fibers. Pflügers Archiv. **403**, 312-317.

Rinaldi, M. L., C. J. LePeuch, and J. G. Demaille (1981) The epinephrine-induced activation of the cardiac slow Ca^{++} channel is mediated by the cAMP-dependent phosphorylation of calciductin, a 23000 M_r sarcolemmal protein. *FEBS Lett.*, **129**, 277-281.

Rinaldi, M. L., J.-P. Capony, and J. G. Demaille (1982) The cyclic AMP dependent modulation of cardiac sarcolemmal slow calcium channels. *J. Mol. Cell. Cardiol.*, **14**, 277-289.

Robison, G. A., and E. W. Sutherland (1971) Cyclic AMP and the function of eukaryotic cells: An introduction. *Ann. N.Y. Acad. Sci.*, **185**, 5-9.

Robison, G. A., R. W. Butcher, I. Oye, H. E. Morgan, and E. W. Sutherland (1965) The effect of epinephrine on adenosine-3',5'-phosphate levels in the isolated, perfused rat heart. *Mol. Pharmacol.*, **1**, 168-177.

Rubinstein, M., B. Amit, and A. Patchornik (1975) The use of a light-sensitive phosphate protecting group for some mononucleotide syntheses. *Tetrahedron Lett.*, **17**, 1445-1448.

Sakmann, B., and E. Neher, Eds. (1983) *Single Channel Recording*. Plenum, New York.

Sakmann, B., and E. Neher (1984) Patch clamp techniques for studying ionic channels in excitable membranes. *Annu. Rev. Physiol.*, **46**, 455-472.

Sanguinetti, M. C., and R. S. Kass (1984) Photoalteration of calcium channel blockade in the cardiac Purkinje fiber. *Biophys. J.*, **45**, 873-880.

Schuetze, S. M., and D. A. Goodenough (1982) Dye transfer between cells of embryonic chick becomes less sensitive to CO_2 treatment with development. *J. Cell. Biol.*, **92**, 694-705.

Senter, P. D., M. J. Tansey, J. M. Lambert, and W. A. Blattler (1985) Novel photocleavable protein crosslinking reagents and their use in the preparation of antibody–toxin conjugates. (Submitted.)

Spray, D. C., A. L. Harris, and M. V. L. Bennett (1981) Gap junctional conductance is a simple and sensitive function of intracellular pH. *Science*, **211**, 712-715.

Spray, D. C., A. L. Harris, and M. V. L. Bennett (1982) Comparison of pH and Ca dependence of gap junctional conductance. R. Nuccitelli and D. Deamer, Eds., *Intracellular pH*, Alan R. Liss, New York.

Spray, D. C., J. M. Nerbonne, A. Campos de Carvalho, A. L. Harris, and M. V. L. Bennett (1984) Substituted benzyl acetates: A new class of compounds that reduce gap junctional conductance by cytoplasmic acidification. *J. Cell Biol.*, **99**, 174-179.

Starmer, C. F., A. O. Grant, and H. C. Strauss (1984) Mechanisms of use-dependent block of sodium channels in excitable membranes by local anesthetics. *Biophys. J.*, **46**, 15-27.

Strichartz, G. R. (1973) The inhibition of sodium currents in myelinated nerve by quaternary derivatives of lidocaine. *J. Gen. Physiol.*, **62**, 37-57.

Sutherland, E. W. (1970) On the biological role of cyclic AMP. *JAMA*, **214**, 1281-1288.

Sutherland, E. W., and G. A. Robison (1966) The role of cyclic-3',5'-AMP in responses to catecholamines and other hormones. *Pharmacol. Rev.*, **18**, 145-161.

Tada, M., M. A. Kirchberger, and A. M. Katz (1975) Phosphorylation of a 22,000 dalton component of the cardiac sarcoplasmic reticulum by adenosine-3',5'-monophosphate dependent protein kinase. *J. Biol. Chem.*, **250**, 2640-2647.

Trautwein, W., J. Taniguchi, and A. Noma (1982) The effect of intracellular cyclic nucleotides and calcium on the action potential and acetylcholine response of isolated cardiac cells. *Pflügers Arch.*, **392**, 307-314.

Triggle, D. J., and V. C. Swamy (1983) Calcium antagonists: Some chemical–pharmacological aspects. *Circ. Res.* (Suppl. 1), **52**, 17-28.

Trube, G., and J. Heschler (1983) Potassium channels in isolated patches of cardiac cell membranes. *Naunyn-Schmiedebergs Arch. Pharmacol.*, **322**, R64, No. 255.

Tsien, R. W. (1983) Calcium channels in excitable membranes. *Annu. Rev. Physiol.*, **45**, 341–358.

Turin, L., and A. E. Warner (1980) Intracellular pH in early *Xenopus* embryos: Its effect on current flow between blastomeres. *J. Physiol. (Lond.)*, **300**, 489–504.

Uehara, A., and J. R. Hume (1984) Interactions of organic Ca channel antagonists with Ca channels in isolated frog atrial cells: Test of a modulated receptor hypothesis. *Biophys. J.*, **45**, 50a.

Zehavi, U., B. Amit, and A. Patchornik (1972) Light sensitive glycosides. I. 6-Nitroveratryl β-D-glucopyranoside and 2-nitrobenzyl β-D-glucopyranoside. *J. Org. Chem.*, **37**, 2281–2288.

CHAPTER 25

EXPERIMENTS WITH PHOTOISOMERIZABLE MOLECULES AT NICOTINIC ACETYLCHOLINE RECEPTORS IN CELLS AND MEMBRANE PATCHES FROM RAT MUSCLE

HENRY A. LESTER, LEE D. CHABALA,
ALISON M. GURNEY, ROBERT E. SHERIDAN

Division of Biology
California Institute of Technology
Pasadena, California

1.	INTRODUCTION	448
2.	OPTICAL SYSTEM	448
3.	PHOTOCHEMISTRY	451
4.	THE PHOTOISOMERIZABLE AGONISTS Bis-Q AND QBr	451
5.	REVERSIBLY BOUND Bis-Q	451
6.	TETHERED QBr	453
7.	SINGLE-CHANNEL RECORDINGS	455
8.	PHOTOISOMERIZATION OF BOUND AGONIST MOLECULES	457

Dr. Chabala's present address is Department of Physiology and Biophysics, Cornell University Medical College, New York, NY 10021.

Dr. Sheridan's present address is Department of Pharmacology, Georgetown University School of Medicine and Dentistry, Washington, DC 20007.

9. THE PHOTOISOMERIZABLE OPEN-CHANNEL BLOCKER
    EW-1   459
10. CONCLUSIONS   460
    REFERENCES   461

1. INTRODUCTION

In our experiments, electrophysiological techniques are employed to monitor the number of open channels in a biological membrane while photochemical manipulations are used (1) to perturb the concentration of drug molecules near receptors or (2) to change the structure of a drug–receptor complex (Lester and Nerbonne, 1982).

Over the past 3 years, our group has been combining high-resolution whole-cell and single-channel recording methods (Hamill et al., 1981; Sakmann and Neher, 1983; see Fig. 1) with light-flash techniques in order to gain more information about the mechanisms of action of intracellular and extracellular messengers (see also chapter by Nerbonne in this volume). For our applications, the gigaohm seal technique offers several advantages: Currents up to a few nanoamperes can be passed without serious series resistance errors, and problems associated with photoelectric or mechanical artifacts are minimized. For these reasons, we have found it simple and convenient to incorporate gigaohm seal recording methods into our light-flash experiments.

2. OPTICAL SYSTEM

The whole-cell method is ideally applied to cells smaller than those traditionally studied with intracellular electrodes; a typical experiment can be carried out on a cell 5–30 μm in diameter. In addition, membrane patches a few square micrometers in area can be studied either while attached to a cell or excised. We therefore adapted our optical system to produce flashes of comparable dimensions imaged onto cells or membrane patches. As other chapters in this volume show, it is possible to employ lasers to deliver light flashes. However, we wanted to select the excitation wavelength from a broad spectrum of available light and we sought an inexpensive setup that requires minimum maintenance.

We therefore chose to make simple modifications on our existing optical system, which employs xenon short-arc flash lamps (Nass et al., 1978; Nargeot et al., 1982).

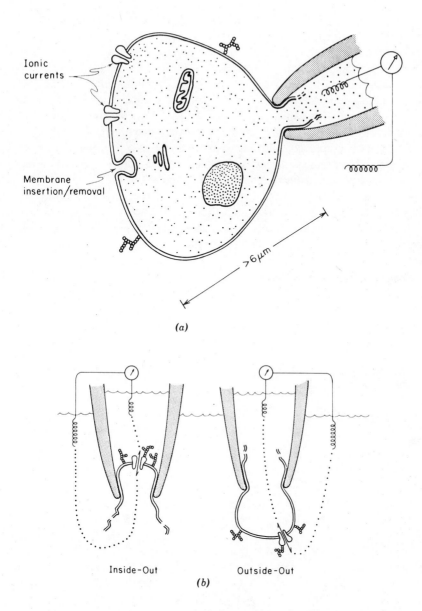

Ionic
currents

Membrane
insertion/removal

>6 μm

(a)

Inside-Out   Outside-Out

(b)

Figure 1. Schematic diagram of patch recordings with gigaohm seals. (*a*) Whole-cell clamp. In addition to ion channels, the diagram also shows membrane insertion/removal. This process can be monitored electrically (Marty and Neher, 1982) and might be stimulated by appropriate flash-activated intracellular molecules. Thus the light-flash technique might also be applied to study endocytotic or exocytotic events. (*b*) Excised patches. Dotted lines show the paths of current flow through open channels.

The flash lamp is placed in a standard housing, complete with rear reflector and condenser. A similar housing and power supply are available from the Chadwick-Helmuth Corp., El Monte, CA, although our power supply has a slightly higher energy storage capacity (2200 μF at 450 V) and a greater discharge frequency ($0.1\ \mathrm{s}^{-1}$ at this charge). The condenser is placed near an adjustable, movable aperture at the intermediate image plane of a compound microscope (Fig. 2; in the modified Leitz Diavert and Dialux microscopes that we use, it is convenient to employ the camera port for this purpose). A "complete closure" diaphragm iris (Rolyn Optics, Arcadia, CA) is employed, allowing spots as small as a few micrometers at the electrode tip. The beam splitter, which enables the operator to observe the image of the aperture during adjustment, can be removed to allow 100% of the flash lamp output to reach the preparation. Contrast elements, polarizers, and other optical components are removed from the light path, so that the beam passes only through the objective lens (we routinely use a 32 ×, N.A. 0.4 lens). This system would, if necessary, allow even smaller test spots and objectives with higher numerical aperture.

The system requires careful construction in two areas. (1) The trigger pulse (14 kV) and the discharge current (2000 A) must both be shielded from the

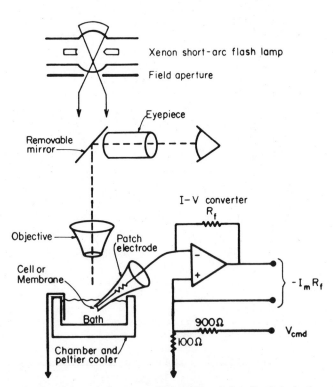

Figure 2. Schematic view of the optical apparatus for recording whole-cell and single-channel currents during photochemical manipulations.

current-to-voltage (*I*-to-*V*) converter in the patch-clamp headstage, so that the discharge artifact is on the order of 1 pA—a shielding factor of 10^{15} over a distance < 1 m! (2) The flash lamp thumps its housing when it discharges, again disturbing the headstage. Both these problems are solved by placing the flash lamp housing in its own shielded cage, mechanically separate from the table holding the *I*-to-*V* converter.

3. PHOTOCHEMISTRY

This chapter describes experiments with a series of photoisomerizable azobenzene derivatives, synthesized at Columbia University (Bartels et al., 1971), that have interesting properties at nicotinic acetylcholine receptors. As previously reviewed in greater detail (Lester and Nerbonne, 1982), the *cis* and *trans* isomers have differing pharmacological potencies. The photoisomerization is complete within 1 μs after the absorption of a photon (Sheridan and Lester, 1982), so the time course of the photochemistry depends primarily on that of the flash and can be made essentially instantaneous compared with the time constants that describe channel gating. The individual isomers are metastable in the dark. The concentrations of *cis* and *trans* isomers can be monitored with optical absorption measurements, although such measurements cannot usually be made simultaneously with the electrophysiological ones. For exact measures of the photoisomerization potency of a given flash delivered through the compound microscope, the beam is collimated and expanded to fill a special cuvette used for absorbance measurements. Corrections are then made for vignetting effects and for light loss in the lenses (Chabala et al., 1985a,b).

4. THE PHOTOISOMERIZABLE AGONISTS Bis-Q AND QBr

Two azobenzene derivatives, Bis-Q and QBr, have previously been used to study acetylcholine receptors at synapses from fishes (Lester and Nerbonne, 1982; Sheridan and Lester, 1982; Weinstock, 1983). We have found that a similar pharmacological profile is displayed in cultured myoballs from rat and chick and in the cloned cell line, BC3H-1. The *trans* isomers are agonists, effective at submicromolar concentrations; the *cis* isomers have little or no potency as agonists (Chabala et al., 1982, 1984, 1985a,b; Chabala and Lester, 1984, 1985, 1986; Lester and Chabala, 1984).

5. REVERSIBLY BOUND Bis-Q

In the experiment of Fig. 3, these properties are exploited to produce concentration jumps of agonist molecules. Rat myoballs are bathed in a solution of *cis*-Bis-Q, and an individual myoball at the center of the microscope field is voltage clamped. In the

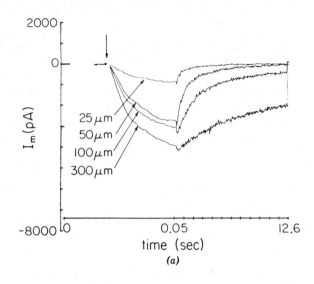

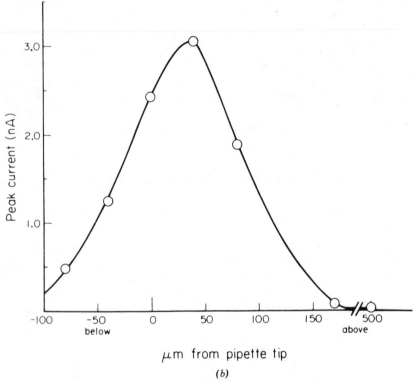

Figure 3. Effects of spot diameter and plane of focus. (*a*) Wave forms of agonist-induced currents during flash-activated "concentration-jump" experiments with *cis*-Bis-Q (1.5 μ*M*). At vertical arrow, the flash increases [*trans*-Bis-Q] from nearly zero to 800 n*M*. The diameter of the light spot at the myoball is varied as noted for each trace; this is accomplished by changing the iris diaphragm (Fig. 2). Cell diameter is 30 μm. Note that the time scale is compressed by a factor of 250 for the last half of the trace. (*b*) Changes in the peak response amplitude when the microscope objective is moved in the vertical direction. Voltage, − 100 mV; temperature, 15°C.

absence of light, there is essentially no agonist-induced current. Flashes of 1-ms duration are imaged onto the cell; these produce a local concentration jump of *trans*-Bis-Q which in turn produces a transient increase in the conductance of the membrane. For light spots larger than the cell, the amplitude of the conductance increase does not depend strongly on this diameter; the peak conductance varies by less than 50% for a 36-fold change in the spot area. In agreement with previous studies from our laboratory using other preparations, the time course of this increase depends on the molecular properties of the agonist–receptor interaction and is much slower than the concentration jump. On the other hand, the decay time of the conductance does depend strongly on this diameter and seems to reflect primarily the diffusive dissipation of the bolus of agonist. With spots on the order of 50 μm in diameter, the agonist diffuses away within a few seconds. The exact localization of the image is critical both in the focal plane and along the focusing axis (Fig. 3*b*).

Because the bolus of agonist occupies a volume at least 4 orders of magnitude smaller than the bath, such flash applications can be repeated hundreds of times with no local accumulation of *trans*-Bis-Q. Because the cell is exposed to agonist for such a short time, cumulative physiological effects are also minimal; we generally observe no long-term drifts after dozens of flashes. Finally, the size of the concentration jump can readily be varied by filtering the flash. This situation is well suited for dose–response studies, especially since the *cis* isomer of Bis-Q has recently been isolated (Nerbonne et al., 1983). We were able to confirm that the dose–response relation for Bis-Q has a sigmoid start; at low concentrations, the relation has a slope very near 2 when plotted on double-logarithmic coordinates (Fig. 4). The "Hill coefficient" remains near 2 over the range of voltages between − 100 mV and + 80 mV and even after receptors are reduced by exposure to 2 m*M* dithiothreitol. These results confirm and extend the idea that the open state of the acetytcholine receptor channel is much more likely to be associated with the presence of two reversibly bound agonist molecules than with a single one. The agonist concentration-jump technique in the presence of *cis*-Bis-Q is also useful for studies of the voltage dependence of the macroscopic transition rates (cf. Chabala and Lester, 1984, 1985).

6. TETHERED QBr

There are interesting exceptions to this conclusion about functional stoichiometry. After the dithiothreitol treatment described above, several drugs can irreversibly bind to receptors by alkylating the newly exposed sulfhydryl groups. Although most of these "tethered" ligands are blockers, at least three tethered agonist molecules are known (Silman and Karlin, 1969; Bartels et al., 1971; Lester et al., 1980). Many details are known about the reaction between these molecules and the acetylcholine receptor. The particular cysteine residue that is derivatized on the α subunit of the receptor has recently been identified (Kao et al., 1984). Of particular interest is the possibility that tethered agonists can activate the channel even if only one of the two α subunits of a receptor oligomer is alkylated. This suggestion was tested and supported in previous studies on *Electrophorus* electroplaques using the photoisomerizable

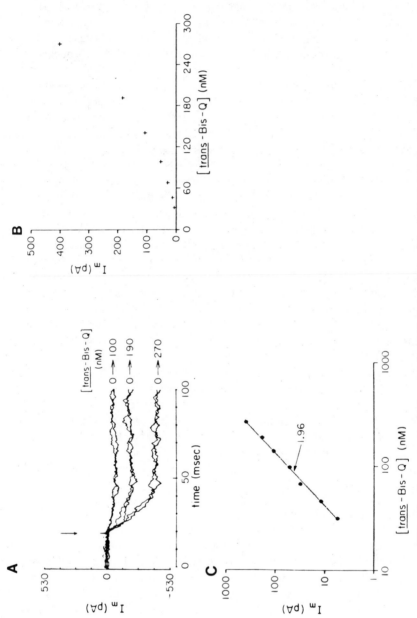

Figure 4. Dose–response relations measured with flash activation. The *cis*-Bis-Q concentration is 500 n*M*. (*a*) Traces showing the response to three different concentration jumps, produced by filtering the flash; two episodes at each flash intensity are shown to illustrate the reproducibility of the technique. (*b*) Plot of the data from the entire experiment, on linear coordinates. (*c*) Plot of the same data on double-logarithmic coordinates. The slope (least-squares fit) is 1.96; −100 mV; 15°C.

tethered agonist, QBr (Lester et al., 1980; see also Walker et al., 1984). The present experiments with myoballs provide further support.

Myoballs were treated for 10 min at room temperature with 2 mM dithiothreitol, washed, exposed for a further 10 min to 2 μM QBr, then washed again. Although QBr is a very poor agonist before reduction, this treatment produces an irreversible conductance with all the hallmarks of the cholinergic response produced by reversibly bound agonists and by other tethered agonists such as bromoacetylcholine (BrACh): The conductance increases at more negative voltages, the mean channel lifetime is a few milliseconds, and the single-channel conductance is about 30 pS at 15°C (there are also a few channels with a conductance of 45 pS).

Of present interest, tethered QBr remains photoisomerizable, and only the *trans* isomer is an agonist. There is an important difference with soluble Bis-Q, however: Once the *trans* isomer of tethered QBr is produced with light, it does not diffuse away. Instead, UV light must be applied to reisomerize the tethered QBr population to the inactive *cis* state. With a mercury arc lamp and a 341-nm interference filter, the reisomerization requires about 1 min and results in a *cis*-photostationary state that contains only 3–4% *trans* isomer. These procedures can be used to produce the "tethered" analogue of the dose–reponse curve.

In order to measure functional stoichiometry with tethered QBr, we asked whether the agonist-induced current depends on the first or second power of the mole fraction of bound *trans*-QBr molecules (Chabala and Lester, 1986). The experiments are less precise than those with Bis-Q, because (1) there is no convenient way to find zero agonist-induced current, and (2) the prolonged exposures to agonist do cause a gradual decline in the agonist-induced current. Nonetheless, the results do allow a decision among the two hypotheses (Fig. 5). The good fit to a linear relation suggests that most—if not all—of the conductance is contributed by receptors whose open state is controlled by the configuration of a *single* tethered QBr molecule. The time constant of the light-flash and voltage-jump relaxations is about 5 ms at -100 mV and 15°C; this is presumably the lifetime of the monoliganded open state.

7. SINGLE-CHANNEL RECORDINGS

With these concepts established, it was possible to examine the single-channel recordings with a view to resolving the basis for a recently discovered phenomenon. In addition to the normal population of ACh channels activated by reversibly bound agonist molecules (lifetime $\sim$ 10 ms under the conditions of our experiments), there is a population of briefer openings ($<$ 1 ms) (Colquhoun and Sakmann, 1981; Jackson

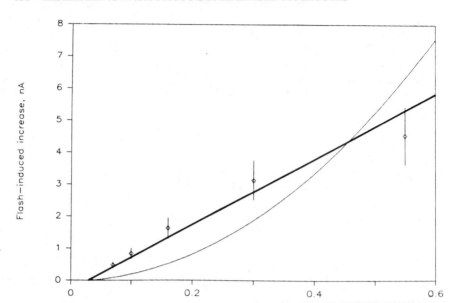

Figure 5.   Functional stoichiometry for the tethered agonist QBr. Light-flash relaxations are measured as in Figs. 2 and 3. The plot shows the relaxation amplitudes for flashes that produce varying mole fractions of *trans*-QBr. Between flashes, the tethered QBr is returned to a nearly pure *cis* state by exposing the preparation to steady UV light as described in the text. Each data point is the average of 4–8 observations ± SEM. The lines show least-squares fits of the data to the first or second power of the mole fraction of *trans*-QBr. The maximum current, corresponding to pure *trans*-QBr, is 0.4 nA for $n = 1$ and 21 nA for $n = 2\%$ -100 mV; 15°C.

et al., 1983; Labarca et al., 1984; Sine and Steinbach, 1984). It has been hypothesized that these brief openings represent rather inefficient channel activation by monoliganded receptors (Colquhoun and Sakmann, 1981). The integrated conductance due to these openings is so small that it would not be expected to change the Hill coefficient of 2 determined for macroscopic currents with reversibly bound agonists.

In our laboratory, single-channel recordings also revealed this brief component for Bis-Q, carbachol, and even for *d*-tubocurarine (see also Morris et al., 1983), extending the generality of this finding. The chief point, however, is that very similar distributions are also seen with the two tethered agonists, QBr and BrACh (Fig. 6) . As seen for reversibly bound agonists, most of the conductance is contributed by the longer component of open times; also as seen for reversibly bound agonists, the time constant of this component, 5 ms for QBr, equals the value expected from macroscopic voltage-jump and light-flash relaxations. Because the active state is already monoliganded with tethered agonists, this finding suggests that the two distributions of open times do not arise separately from monoliganded and biliganded receptors.

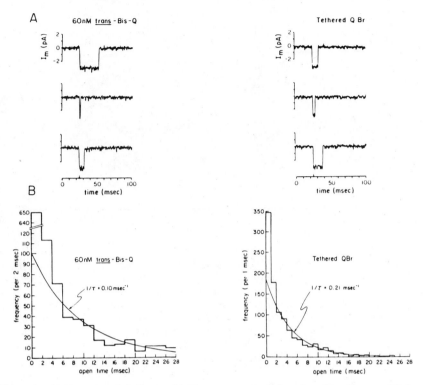

Figure 6. (*a*) Examples of single acetylcholine–receptor channels activated by the reversibly bound agonist Bis-Q (left), and the tethered agonist QBr (right). (*b*) Histograms of channel lifetimes (burst durations) for the two agonists. Each histogram shows a large, fast component (note the break on the vertical axis for Bis-Q) and a slower component with the indicated time constant.

Therefore, the origin of the briefer openings remains unresolved. One possibility is that this brief component with tethered QBr actually arises from biliganded channels. The situation for tethered agonists would therefore be opposite to that suggested for reversibly bound agonists (Colquhoun and Sakmann, 1981), and the dose–response relation would show negative cooperativity ($n < 1$ in experiments like that of Fig. 5). We are currently examining this hypothesis. Another possibility is that the brief openings represent partially desensitized receptors, although preliminary data (Lester and Chabala, 1984) do not support this idea.

8. PHOTOISOMERIZATION OF BOUND AGONIST MOLECULES

The experiments with QBr are not, of course, bona fide concentration jumps; instead, they exemplify an experiment unique to the light-flash technique. The drug–receptor

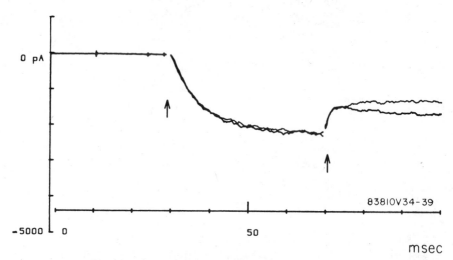

Figure 7. Responses to photoisomerization of bound *trans*-Bis-Q molecules, as described in the text. The *cis*-Bis-Q concentration is 300 n*M*. Two superimposed traces are shown; each is the average of three individual episodes. In both traces, the first flash is unfiltered and produces a jump from nearly zero to 180 n*M trans*-Bis-Q. A second flash then produces *transcis* photoisomerizations of bound Bis-Q molecules; as a result, channels close (phase 1; see text). For the heavy trace, the second flash is predominantly UV (UG-11 filter, Schott glass), so that a net *transcis* photoisomerization is produced in the bulk solution. This leads to a sustained decrease in current. For the light trace, the second flash is unfiltered and produces little or no change in the *trans*-Bis-Q concentration. As a result, phase 1 is transient.

complex is manipulated directly. Another example of such a manipulation occurs with reversibly bound Bis-Q (Fig. 7). The experiment begins in the presence of *cis*-Bis-Q. A flash produces a concentration jump of the agonist, *trans*-Bis-Q, and the conductance increases, as in the experiment of Fig. 3*a*. A second flash is then delivered; it produces a much more rapid decrease in conductance. This decrease is called phase 1 (Nass et al., 1978). (It is important to note that the time scale remains the same for the entire trace.) This rapid decrease occurs because the bound Bis-Q population is almost purely *trans;* as a result, the flash produces a net *transcis* photoisomerization of bound Bis-Q. Some receptors are left with one or more *cis*-Bis-Q molecules; because *cis*-Bis-Q is a poor agonist, channels close rapidly, and the *cis*-Bis-Q molecules dissociate (Nass et al., 1978; Sheridan and Lester, 1982; Chabala et al., 1985a). The rate constant for the rapid decrease in Fig. 7 is at least 10 times greater than that for the concentration-jump increase earlier in the trace. This difference thus represents the difference in channel lifetimes between the active *trans* isomer and the inactive *cis* isomer. A marked contrast between the channel lifetimes for the two isomers was also found in earlier studies on *Electrophorus* electroplaques (Nass et al., 1978; Sheridan and Lester, 1982).

We are attempting to view phase 1 at the single-channel level with instrumentation described more fully by Kegel et al. (1985). The experiment is performed with an excised outside-out patch continually exposed to *trans*-Bis-Q. Channels open at random intervals. The instruments continually monitor the signal from the

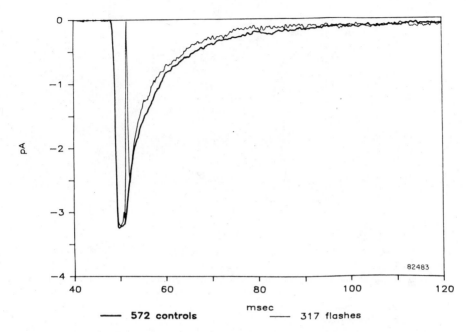

Figure 8. Data from an experiment to "catch" single open acetylcholine receptor channels. The instrumentation is described in the text. The wave forms show the average of several hundred channel openings. In "flash" episodes, the flash is delivered if the channel remains open for 2 ms; the artifact is visible on the averaged trace. For the "control" episodes, the channel openings are recorded the same way, but there is no flash.

patch-clamp electrode. A window discrimiminator is activated when the signal is in the range corresponding to an open channel; the discriminator also has a "criterion" period—2 ms, in this case—to assure that the channel is neither a brief noise pulse nor a member of the brief class of channels described above. If these conditions are met, the flash lamp is triggered. While the lamp is recharging, the program gathers control channels according to the same conditions, but without flashing the lamp. The traces of Fig. 8 represent a retrospective artificial synchronization of the wave forms, so that all channels appear to open at the same time; in the experimental series, the flash occurs 2 ms later. On the average, the flash produces a more rapid closing, although the effect is less dramatic than with the whole-cell currents (Fig. 7) because of a lower flash intensity in the single-channel apparatus. This experiment suggests that it is possible to study single channels by "catching" them electrically and then manipulating them photochemically.

9.   THE PHOTOISOMERIZABLE OPEN-CHANNEL BLOCKER EW-1

Other azobenzene derivatives block acetylcholine receptors. The two major blocking mechanisms are (1) competitive antagonism, typified by *d*-tubocurarine, and (2)

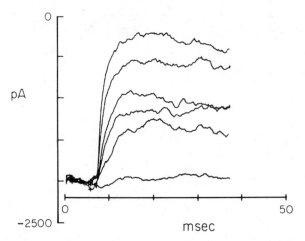

Figure 9. Open-channel blockade of agonist-induced current by the light-activated blocker, EW-1. Acetylcholine, 500 nM; EW-1, 2 μM; -100 mV; 15°C. Flashes of increasing intensity are delivered at 60-s intervals. For the control trace, with no decrease in the current, the flash is passed through a yellow filter (Schott OG-550) which transmits no light within the absorption spectrum of EW-1. There is no change in current. For the other traces, neutral-density filters are used (OD = 1.6, 1.2, 0.9, 0.5, and 0). Experiment by Daniel R. Kegel.

"open-channel blockade," typified by a large class of molecules such as procaine and other local anesthetics. The latter molecules appear to allow receptors to open as usual, then bind within or near the channel, blocking current flow. The photoisomerizable molecule, N-p-phenylazophenyl carbamylcholine (EW-1), is such a blocker (Lester et al., 1979). In an experiment with this drug, the cell is exposed both to a photostable agonist like ACh itself and to EW-1. The active *cis* isomer is produced by a flash. As a result, open channels are blocked, resulting in a decreased current (Fig. 9).

10. CONCLUSIONS

The gigaohm seal technique now allows the experimenter to study the results of photochemical perturbations at the level of single membrane channels and to correlate these recordings with the macroscopic signals. We look forward to applying this increased resolution to further studies on normal acetylcholine receptors, on reconstituted receptors in artificial membranes, and on modified receptors with specifically directed substitutions of individual amino acids. The results should lead to a clearer understanding of the relation between structure and the mechanisms of receptor activation and channel gating by this protein.

ACKNOWLEDGMENTS

This research was supported by the National Institutes of Health (grant NS-11756), by fellowships from the Muscular Dystrophy Association to L.D.C. and from the Del E. Webb Foundation to A.M.G. and L.D.C., and by a Fulbright–Hays travel grant to A.M.G.

REFERENCES

Bartels, E., N. H. Wassermann, and B. F. Erlanger (1971) Photochromic activators of the acetylcholine receptor. *Proc. Natl. Acad. Sci. USA*, **68**, 1820–1823.

Chabala, L. D., H. A. Lester, and R. E. Sheridan (1982) Single-channel currents from cholinergic receptors in cultured muscle. *Soc. Neurosci. Abstr.*, **8**, 498.

Chabala, L. D., A. M. Gurney, and H. A. Lester (1984) Patch-clamp studies of ACh channels activated by photoisomerizable agonists. *Biophys. J.*, **45**, 387a.

Chabala, L. D. and H. A. Lester (1984) Voltage dependence of acetylcholine receptor channel opening in rat myoballs. *J. Gen. Physiol.* **84**, 24a–25a.

Chabala, L. D. and H. A. Lester (1985) Kinetic and steady-state properties of acetylcholine receptor channels in voltage-clamped rat myoballs. *Biophys. J.* **47**, 257a.

Chabala, L. D., A. M. Gurney, and H. A. Lester (1985a) Photoactivation and dissociation of agonist molecules at the nicotinic acetylcholine receptor in voltage-clamped rat myoballs. *Biophys. J.*, **48**, 241–246.

Chabala, L. D., A. M. Gurney, and H. A. Lester (1985b) Dose–response of acetylcholine receptor channels opened by a flash-activated agonist in voltage-clamped rat myoballs. *J. Physiol.*, (in press).

Chabala, L. D., and H. A. Lester (1986) Activation of acetylcholine receptor channels by covalent bound agonists in cultured rat myoballs. *J. Physiol.*, (in press).

Colquhoun, D., and B. Sakmann (1981) Fluctuations in the microsecond time range of the current through single acetylcholine receptor channels. *Nature*, **294**, 464–466.

Hamill, O. P., A. Marty, E. Neher, B. Sakmann, and F. J. Sigworth (1981) Improved patch-clamp techniques for high-resolution current recording from cells and cell-free membrane patches. *Pflügers Arch.*, **391**, 85–100.

Jackson, M. B., B. S. Wong, C. E. Morris, H. Lecar, and C. N. Christian (1983) Successive openings of the same acetylcholine-receptor channel are correlated in their open times. *Biophys. J.*, **42**, 109–114.

Kao, P., A. J. Dwork, R.-R. Kaldany, M. L. Silver, J. Wideman, S. Stein, and A. Karlin (1984) Identification of the α subunit half-cystine specifically labeled by an affinity reagent for the acetylcholine receptor binding site. *J. Biol. Chem.* **259**, 11662–11665.

Kegel, D. R., B. D. Wolf, R. E. Sheridan, and H. A. Lester (1985) Software for electrophysiological experiments with a personal computer *J. Neurosci. Methods*, **12**, 317–330.

Labarca, P., J. Lindstrom, and M. Montal (1984) The acetylcholine receptor channel from *Torpedo californica* has two open states. *J. Neurosci.*, **4**, 502–507.

Lester, H. A., and L. D. Chabala (1984) Neither monoliganded nor desensitized receptors account for excess brief acetylcholine channels in cultured rat muscle. *Neurosci. Abstr.*, **10**, 12.

Lester, H. A., and J. M. Nerbonne (1982) Physiological and pharmacological manipulations with light flashes. *Annu. Rev. Biophys. Bioeng.*, **11**, 151–175.

Lester, H. A., M. E. Krouse, M. M. Nass, N. H. Wassermann, and B. F. Erlanger (1979) Light-activated drug confirms a mechanism of ion channel blockade. *Nature*, **280**, 509–510.

Lester, H. A., M. E. Krouse, M. M. Nass, N. H. Wassermann, and B. F. Erlanger (1980) A covalently bound photoisomerizable agonist. Comparison with reversibly bound agonists at *Electrophorus* electroplaques. *J. Gen. Physiol.*, **75**, 207–232.

Marty, A. and E. Neher (1982). Tight-seal whole-cell recording. In *Single-Channel Recording*, eds. Sakmann, B. and Neher, E., pp. 107–122. New York, Plenum.

Morris, C. E., B. S. Wong, M. B. Jackson, and H. Lecar (1983) Single-channel currents activated by curare in cultured embryonic rat muscle. *J. Neurosci.*, **3**, 2525–2531.

Nargeot, J., H. A. Lester, N. J. M. Birdsall, J. Stockton, N. H. Wassermann, and B. F. Erlanger (1982) A photoisomerizable muscarinic antagonist. Studies of binding and of conductance relaxations in frog heart. *J. Gen. Physiol.*, **79**, 657–678.

Nass, M. M., H. A. Lester, and M. E. Krouse (1978) Response of acetylcholine receptors to photoisomerizations of bound agonist molecules. *Biophys. J.*, **24**, 135–160.

Nerbonne, J. M., R. E. Sheridan, L. D. Chabala, and H. A. Lester (1983) *cis*-3,3'-Bis-[α-(trimethylammonium)methyl]azobenzene (*cis*-Bis-Q). Purification and properties at acetylcholine receptors of *Electrophorus* electroplaques. Mol. Pharmacol. **23**, 344–349.

Sakmann, B., and E. Neher, Eds. (1983) *Single-Channel Recording*. New York, Plenum.

Sheridan, R. E., and H. A. Lester (1982) Functional stoichiometry at the nicotinic receptor. The photon cross-section for phase 1 corresponds to two Bis-Q molecules per channel. *J. Gen. Physiol.*, **80**, 499–515.

Silman, I., and A. Karlin (1969) Acetylcholine receptor: Covalent attachment of depolarizing groups at the active site. *Science*, **164**, 1420–1421.

Sine, S., and J. H. Steinbach (1984) Activation of a nicotinic acetylcholine receptor. *Biophys. J.*, **45**, 175–184.

Walker, J. W., C. A. Richardson, and M. G. McNamee (1984) Effects of thio-group modifications of *Torpedo californica* acetylcholine receptor on ion-flux activation and inactivation kinetics. *Biochemistry*, **23**, 2329–2338.

Weinstock, M. M. (1983) Activation and desensitization of acetylcholine receptors in fish muscle with a photoisomerizable agonist. *J. Physiol. (Lond.)*, **338**, 423–433.

Index

Absorption, light, 72–77
  of Azo 1 dye, 263–267
  of calcium indicators, 330–331
  of chromophores, 407
  extrinsic, 135–136
  light scattering and, 187
  of M540 dye, 102–103, 107–108, 111–112
  of phenol red dye, 288
    dark adaptation of photoreceptor cells and,
      300–301, 305
    after intracellular injection into photo
      receptor cells, 292–293
    light-induced change in photoreceptor cells
      and, 298–300, 306
    in presence of rhabdoms, 290–291,
      297–298
  potential-dependent, 73–76
Acetoxy-methoxy dyes, 351–356
  AM-apda, 355–356
  AM-bapta, 353–355
  AM-quin, 351–352, 355
Acetoxymethyl esters, 335
Acetylcholine receptors:
  blocking of, 459–460
  photoisomerization at, 447–460
    of Bis-Q, 451–453, 458
    of bound agonist molecules, 457–459
    of open-channel blocker EW-1, 459–460
    optical system for studying, 448–451
    photochemistry of, 451
    of QBr, 451, 453–455, 458
    single-channel recordings of, 455–457
Acousto-optical deflectors in laser scanners, 215,
  222, 225
Actin, 61
Action potential, 137–148
  calcium, 146–148
  cAMP jumps and, in sinus venosus fibers,
    431–432
  depolarizing phase of, 143–148
  field stimulation and, 143–148
  ions and, 143–148, 151–155
  presynaptic, in real-time optical mapping, 172

  properties of, 140–143
  rising phase of, 143–148
  signal-to-noise ratio and, 213
  spread of, optical monitoring of, 211–225
    activation maps in, 216–219
    fluorescence and, 212–213
    laser scanner for, 215–225
    laser scanning of frog ventricle, 219
    laser scanning of mammalian heart,
      220–222
    temporal dispersion and, 140–141
Activation, endocardial and epicardial, 216–222
  of frog ventricle, 219
  of mammalian heart, 220–222
  maps of, 216–219
Actomyosin, 398, 411
Adenosine diphosphate (ADP):
  exchange of, with ATP, 390–393
  in muscle contraction, 398–400
Adenosine triphosphate (ATP):
  caged, see Caged ATP
  exchange of, with ADP, 390–393
  as leaving group, 390
  in muscle contraction, 397–405
    caged ATP and, 388, 399, 403–405
    laser pulse photolysis for, 399–403
    pH and, 404–405
  photoprotection and, 419
  photorelease and, 422–423
  red blood cells and, 207
Adenylic monophosphate, cyclic, see cAMP
  (cyclic adenylic acid)
ADP (adenosine diphosphate):
  exchange of, with ATP, 390–393
  in muscle contraction, 398–400
Adductor muscle, 399
Aequorea, 230
Aequorin, 230, 240–253, 336
  acetylated, 232–236
  arsenazo compared with, 249–251
  calcium-sensitive electrodes compared with,
    251–253
  in center of cell, 241–245

Aequorin (*Continued*)
    fluorescent dyes compared with, 331
    injection of, 245–249
        with phenol red dye, 247–249
    kinetics of, 234–235
    in voltage clamp experiments, 245–247
After-hyperpolarization of nerve terminal action
        potential, 141–143
Agonists, 451–459
    bound:
        photoisomerization of, 247–249
        reversibly, 451–453
    single-channel recordings of, 455–457
    tethered, 453–455
Allogromia, 9
Alpha-actinin, 53, 61
ALU-512 arithmetic processor, 18–22
AM-apda dye, 355–356
AM-bapta dye, 353–355
Amino acid/sodium cotransport, 205–206
Aminoglycoside antibiotics, 159
AMP (adenylic monophosphate), cyclic, *see*
        cAMP (cyclic adenylic acid)
Amphiuma red blood cells, 200
Amplifier noise, 92–93
Amplifiers:
    in optical monitoring of membrane potential,
        92–93
    in polychromators, 258–259
    in real-time optical mapping, 185
Amplitude calibration in real-time optical
        mapping, 190
AM-quin dye, 351–352, 355
Analogue image processing, 4, 6, 18–22
Anisotropy phosphorescence, 407, 411
Antibiotics, aminoglycoside, 159
Antiimmunoglobulin, 342
Antipyrylazo III dye, 233–234, 262
    in calcium measurement, 330–331
AP-512 analog processor, 18–22
Apple II microcomputer, 34–35, 39–40
Arc lamps, 84–86, 402–403
Arithmetic processor, ALU-512, 18–22
Arrhythmia, 222
Arsenazo III dye, 234, 262
    for calcium measurement, 330–331, 336
        in axons, 240, 249–251
        in cytoplasm, 255, 256
ATP, *see* Adenosine triphosphate (ATP)
Atrial fibers, 427–431
    nifedipine and, 433–436
Autofluorescence, 332, 335
Autosubtraction in video-enhanced microscopy, 8

AVEC methods of microscopy, 6, 9
Averaging for noise reduction, 7, 8, 55
Axons:
    of magnocellular neurons, 137
    squid, 7, 9–10
        axoplasm of, extruded, 7, 10
        fluorescence and absorption of dyes in,
            102
        in measurement of membrane potential,
            72–76, 79, 135
    unmyelinated, in hippocampus, 172–173
Axoplasm, extruded, 7, 10
Axotomy, 34
    zap, 43–50
        apparatus for, 44–48
        dyes in, 43–44, 46
        lasers in, 47, 49
        microirradiation for, 43–44
        in wind-sensitive interneurons, 48–50
Azo 1 dye, 261–282
    absorption in, 263–267
    birefringence in, 270–271, 278, 279
    calcium measurements from:
        advantages and disadvantages of, 279–282
        in fiber at rest, 271–276
        kinetic computations for, 270, 278
        stoichiometry of, 263, 267–268, 279–280
        in stretched fiber, 276–278
    magnesium and, 266–267
    proton binding to, 267–268
    signal-to-noise ratio of, 280
Azobenzene derivatives:
    acetylcholine receptor blockers, 459–460
    Bis-Q, 451–453, 458
    QBr, 451, 453–455, 458

Bacteria, halophilic, 353
Balanus, photoreceptors of, 286–287
Bapta dye, 349, 350
    loading of, 353–354
    physiological effects of, 357
    toxic effects of, 355–356
Barium current, 436–438
Bidirectional particle motion, 7, 9–10
Bilayer membranes, effects of dyes in,
        102
Binding of dye, 109–110
Birefringence, 72, 74–77
    in Azo 1 dye, 270–271, 278, 279
    extrinsic, 135–136
    potential-dependent, 74–76
Bis-Q, 451–453, 458
Bisulphite, 388–389

Bleaching of dyes, 33, 44, 80, 117–118
  in action potential studies, 141
  in molecular transport studies, 374–375. *See
    also* Fluorescence photobleaching
    recovery (FPR)
  in real-time optical mapping, 185, 188
Bleaching recovery, *see* Fluorescence
    photobleaching recovery (FPR)
Brain:
  frog optic tectum, 182–184
  mammalian cortex, 178–182
  slice preparation, 169–170
  spatiotemporal patterns of, 169–177
    fluorescent voltage-sensitive probes and,
      175–177
    light scattering and, 171–172, 187
    optical monitoring of, 177–178
    pre- and postsynaptic components of, 172
    spread of, 173–175
    unmyelinated axons and, 172–173
    voltage-sensitive dyes and, 169–170
Bromoxylenol blue dye, 287, 294–295
Bromphenol blue dye, 287, 294–295, 306
Buccal ganglion neurons from *Navanax*,
    121–129
  dye tests on, 126
  during feeding, 125–129
Buffers:
  calcium, 243–245, 308
    bapta dye as, 357
    quin dye as, 356–359
  pH, 295–297, 305–306
Bufo marinus, rod photoreceptors of, 352–353
Bullfrogs:
  atrial fibers of, 427–431
    nifedipine and, 433–436
  sinus venosus fibers of, 431–432

Caged ATP, 385–395
  calcium pump studies with, 393–394
  chemistry and photochemistry of, 386–390
  lasers in release of, 388, 403–405
  in muscle contraction, 399
    photolysis kinetics of, 388, 403–405
  photoprotection and, 419
  photorelease and, 422–423
  potential applications of, 394–395
  sodium pump studies of, 390–392
Caged calcium, 328, 343
Caged cAMP, 389
Calcium:
  buffering of, 243–245, 308, 356–357
  caged, 328, 343

exchange of, with sodium, 245
  membrane potential and, 250–251
  peripheral mitochondria and, 248–249
Calcium action potential, 146–148
Calcium antagonists, 432–439
  future development of, 439
  whole-cell recording techniques and, 436–438
Calcium buffers, 243–245, 308, 356–357
Calcium channel blockers, 432–439
  future development of, 439
  whole-cell recording techniques and, 436–438
Calcium current, inward, 143–148
  slow, 427–431, 433–436
Calcium-dependent potassium flux, 206–207
Calcium-independent luminescence, 233
Calcium indicators, 239–253
  absorbance indicators, 330–331
  aequorin, *see* Aequorin
  antipyrylazo III, 233–234, 262, 330–331
  arsenazo III, 234, 262, 330–331, 336
    in axons, 240, 249–251
    in cytoplasm, 255, 256
  bioluminescence and, 230
  calcium-sensitive electrodes, 233, 240,
    251–253, 328–330
    extracellular applications of, 330
  chemiluminescence as, 331–332
  cytoplasmic, 230, 255, 256
  fluorescent, 327–343, 347–362
    absorbance indicators compared with,
      330–331
    caged calcium and, 343
    calcium-sensitive electrodes
      compared with, 328–330
    chemiluminescence compared with, 331–332
    flow cytometry, 332, 335, 339–343
    fura-2, 332–338
      fluorescence microscopy and, 335–338
    indo-1, 332–335
    limitations of, 332
    loading of, 350–355
    nuclear magnetic resonance compared with,
      330
    photometric studies of, 359–362
    physiological effects of, 356–359
    toxic effects of, 355–356
  nuclear magnetic resonance as, 233, 330
  photoproteins as, 229–236
    introduction of, into cells, 235–236
    motion artifacts and, 230–232
    signal strength and, 232–233
    speed of response and, 233–235
  tetracarboxylate, *see* Tetracarboxylate dyes

Calcium influx, voltage-sensitive, 148
Calcium-mediated potassium conductance, 144
Calcium pump, 393–394
Calcium-regulated photoproteins:
    aequorin, *see* Aequorin
    definition of, 230
    obelin, 230, 232–236
        kinetics of, 235
Calcium-sensitive electrodes, 233, 240,
        251–253, 328–330
    extracellular applications of, 330
Cameras, video, 5
    in calcium measurement, 336–338, 361–362
    in real-time optical mapping, 186
    SIT (silicon intensifier target), 34–35, 43–44,
        54, 55, 361–362
cAMP (cyclic adenylic acid), 389
    jumps of, 423–425
        kinetic implications of, 430–432
        potassium current and, 431
        in sinus venosus fibers, 431–432
    slow inward calcium current and, 427–431
    tension development and, 429, 431
Carboxyfluorescein, as pH indicator, 312–324
    calibration of, 316–317
    compartment specificity of, 317–321
    drawbacks of, 324
    ionophore measurement by, 323–324
    loading of, 313–314
    mitochondrial pH and, 321–323
    monitoring of, 314–316
Cardiac glycosides, 222
Cation indicators, 323–324
CCD's (charge-coupled devices), 80
Cells:
    Ehrlich ascites tumor, *see* Ehrlich ascites
        tumor cells
    eukaryotic, fluorescence in, 202
    neurons, *see* Neurons
    pancreatic acinar, 350–352
    photoreceptor, *see* Photoreceptors, cells of,
        pH measurement of
    red blood, *see* Red blood cells
    smooth muscle, 52–53, 61
    spleen, 340–342
Cell surface proteins, 379–380
Cellular differentiation, 52
Central nervous system, neuronal activity
        monitoring in, 165–185, 189–193
    computers in, 168
    difficulties with, 189–190
    fluorescent voltage-sensitive probes and,
        175–177

light scattering and, 171–172, 187
pre- and postsynaptic components of, 172
principle of, 167–168
spatial resolution in, 191–192
spread of activity, visualization of, 173–175
unmyelinated axons and, 172–173
voltage-sensitive dyes and, 167–170,
        175–177
see also Brain
Centrifugation in intracellular pH monitoring,
        320
cGMP (cyclic guanine monophosphate), 389
    jumps of, 423–425, 429–431
Channel blockers, 432–439, 459–460
Charge-coupled devices (CCD's), 88
Chelators, 233
    fluorinated, 330
    tetracarboxylate, *see* Tetracarboxylate dyes
Chemical kinetics, *see* Kinetics
Chemical loading, 235–236
Chemiluminescence, 331–332
Chick dorsal root neurons, 9
Chromophores, 406–408
Cis isomers, 451
Compound optical signals, 154
Computer-aided light microscopy, 15–29
    contrast enhancement in, 23
    digital image processing by, 18, 53–55,
        60–61
    hardware for, 18–22
    images obtained by, 24–28
    pixel noise cancellation in, 26
    software for, 22–23
    spatial enhancement in, 23, 25–26
Computers:
    in digital imaging microscopy, 18, 53–55,
        60–61
    image enhancement by, 8
    in optical monitoring:
        of action potential, 138–139, 215–216
        of membrane potential, 93–95
    in real-time optical mapping, 168, 186
    video interfaces with, 39–42, 186
    in video microscopy of living neurons, 34–35,
        39–42
Concanavalin A, 342
Conduction velocity in real-time optical
        mapping, 173
Confocal microscopes, 90, 92
Contractile proteins, 53, 398, 411–412
Contrast enhancement, 6, 16, 23
Convective flow in molecular transport, 368,
        370–373

Cortex, mammalian, 178–182
Crayfish, rhabdoms of, 290–291, 297–298
Cross-bridge cycle in muscle contraction, 398–400
  photolysis and, 403–405, 410–412
Cyanine dyes, 112, 206–207
Cyclic AMP, *see* cAMP (cyclic adenylic acid)
Cyclic, *see* cGMP (cyclic guanine monophosphate)
Cytoplasm, pH in, 317–321, 324
Cytoplasmic calcium indicators, 230, 255, 256
Cytotoxicity of fluorescent calcium indicators, 355–356

Dark adaptation of photoreceptor cells, 300–301, 305
Dark noise, 82, 92–93, 259
Dendrites, photoinactivation of, 44–50
  apparatus for, 44–48
  in wind-sensitive interneurons, 48–50
Depolarizing phase of action potential, 143–148
Depth of focus of microscope lenses, 89–91
Deuterium oxide, 155–157
Dichlorophosphonazo III dye, 263
Dichroism, 274–277
Difference spectra, 105–107
Diffusion in molecular transport, 368–373
  in noninteracting systems, 379–380
  reversible bimolecular reaction and, 377–378
  systematic transport and, 376–377
  time course of, 370–373
Digital image processing, 4, 7–8, 16–18, 39–42
  computers in, 18
    hardware for, 53–55
    software for, 55, 60–61
  microscopes for, 51–62
    energy maxima in, 59–60
    filters in, 59–60
    image display in, 60–61
    image restoration in, 55–59
    image simplification in, 59–60
    noise reduction in, 55
    spatial inhomogeneity correction in, 55
Dimers, 102, 111–112
Diodes, silicon, 82–83
Directional sensitivity of interneurons, 48–50
Double-null point titration, 323–324
Drift in molecular transport, 368
  time course of, 370–373
Droop in real-time optical mapping, 185

Drug-receptor complexes, 448–460
  optical system for studying, 448–451
  photoisomerization and, 451–460
    Bis-Q and, 451–453, 458
    bound agonist molecules and, 457–459
    open-channel blocker EW-1 and, 459–460
    photochemistry of, 451
    QBr and, 451, 453–455, 458
    single-channel recordings of, 455–457
Dye(s), 32–34, 72–80
  absorption of, *see* Absorption, light
  acetoxy-methoxy, 351–356
    AM-apda, 355–356
    AM-bapta, 353–355
    AM quin, 351–352, 355
  antipyrylazo III, 233–234, 262
    in calcium measurement, 330–331
  arsenazo III, 234, 262
  Azo 1, *see* Azo 1 dye
  bapta, 349, 350
    loading of, 353–354
    physiological effects of, 357
    toxic effects of, 355–356
  binding of, 109–110
  birefringence of, *see* Birefringence
  bleaching by, 33, 44, 80, 117–118
    in action potential studies, 141
    in molecular transport studies, 374–375.
      See also Fluorescence photobleaching recovery (FPR)
    in real-time optical mapping, 185, 188
  bromoxylenol blue, 287, 294–295
  bromphenol blue, 287, 294–295, 306
  calcium indicators, *see* Calcium indicators
  carboxyfluorescein, *see* Carboxyfluorescein, as pH indicator
  cyanine, 112, 206–207
  dichlorophosphonazo III, 263
  dimers, 102, 111–112
  ejection of, from microelectrodes, 42–43
  fluorescence of, *see* Fluorescence
  fluorescent, *see* Fluorescent dyes
  fura-2, 332–338
    fluorescence microscopy and, 335–338
  hydrophobic, 112
  indo-1, 332–335
  intracellular injection of:
    iontophoretic, 175–176
    phenol red, 247–249, 287–288
    red-fluorescing, 65–68
  iontophoretic ejection of, 42–43
  ion transport processes and, 206–207
  lucifer yellow, 33, 49, 65–68

Dye(s) (*Continued*)
  membrane-bound, 102
  merocyanine, 135–136
    M540, *see* Merocyanine 540 (M540) in
      optical monitoring of membrane potential
    -oxazalone, 135, 154
    -rhodanine, 135, 138, 154
    signal-to-noise ratio of, 212
    WW375, 112
  metallochromic, 231
    antipyrylazo III, 233–234, 262
    arsenazo III, 234, 262
    dichlorophosphonazo III, 263
    kinetics of, 233–234
  monomers, 102, 111–112
  movement of, 40–42
  nitr-2, 343
  optical signal choice and, 77–78
  oxonol, 112
  pharmacological effects of, 79–80, 187
  phenol red, *see* Phenol red dye
  pH indicators, *see* pH indicators
  photodynamic damage by, 80, 117–118
    in real-time optical mapping, 185,
      188
  procion yellow, 32–33
  quin. *see* Quin dye
  red-fluorescing, 65–68
  redistribution (slow), 201
  rhodamines, 65–68
  saturation binding of, 109–110
  signal-to-noise ratio of, 73, 77–83
    action potential and, 213
    Azo 1 dye, 280
    ideal case of, 80–82
    noise types and, 82–83
    in optical monitoring of membrane
      potential, 140
  signal size and, 79, 107–111, 130
  6-carboxyfluorescein, 49
  slow, 201
  styryl, 135–136
  tetracarboxylate, *see* Tetracarboxylate dyes
  voltage-sensitive, 135–136
    fluorescent, 175–177, 188–189,
      222–225
    in optical monitoring of membrane
      potential, 212–213
    in real-time optical mapping, 167–170,
      175–177
  WW375, 112
  WW781, 212

EGTA, 361
Ehrlich ascites tumor cells, 312
  optical monitoring of, 199–207
    fluorescence/membrane potential calibration
      in, 201–204
    membrane resistance and, 204–205
    transport processes and, 205–207
Electrical monitoring, optical monitoring
      compared with, 134–137, 175–177, 189,
      192
Electrode calibration, 136
Electrodes, calcium-sensitive, 233, 240,
      251–253, 328–330
  extracellular applications of, 330
Electrogenic sodium/amino acid cotransport,
      205–206
Electrogenic sodium pump, 204–205
Electronic imaging, 4
  analogue, 4, 6, 18–22
  digital, *see* Digital image processing
Electrophorus electroplaques, 453–454, 458
Ellipsoidal mirror systems, 231–232
Endocardial activation, 219–220
Energy maxima in digital imaging microscopes,
      59–60
Epicardial activation, 219–220
Epifluorescence (epiillumination), in zap
      axotomy, 45–47
Epileptic activity, in mammalian cortex, 180–181
Eukaryotic cells, fluorescence in, 202
EW-1 open channel blocker, 459–460
Excess noise, 82–83
Excimer lasers, 402
Excitation, at nerve terminals, 133–160
  action potential in, 137–148
    depolarizing phase of, 143–148
    optical recording of, 137–140
    properties of, 140–143
  deuterium oxide and, 155–157
  extracellular calcium and, 141, 151–155
  hypertonic media and, 157
  secretion and, optical changes accompanying,
      148–151, 157–160
  temporal dispersion of, 154–155
Excitation-contraction coupling, 155–157
Excitation-secretion coupling:
  in neurohypophysis, 137–160
    action potential:
      and depolarizing phase of, 143–148
      and optical recording of, 137–140
      and properties of, 140–143
    deuterium oxide and, 155–157

extracellular calcium and, 141, 151–155
hypertonic media and, 157
optical changes accompanying, 148–151, 157–160
radioimmunoassays and, 148–151
Extracellular volume, 157, 158
Extraneous noise, 83
Extrasystole, 221–222

Fast optical signals, 73, 172
FB-512 frame buffer, 18–22
FCS, see Fluorescence correlation spectroscopy (FCS)
Fiber optics, 332
Field stimulation, action potential and, 143–148
Filters in digital imaging microscopes, 59–60
Flash lamps:
  in caged ATP release, 388
  xenon, 401, 424, 426–427, 433–435
    short-arc, 448–451
Flash photolysis experiments, 420–424
Flight muscle, 399
Flow cytometry, 332, 335, 339–343
  of lymphocytes, 339, 342–343
Flow in molecular transport, 368, 370–373
Fluorescein, as pH indicator, 312–324
  calibration of, 316–317
  compartment specificity of, 317–321
  drawbacks of, 324
  ionophore measurement by, 323–324
  loading of, 313–314
  mitochondrial pH and, 321–323
  monitoring of, 314–316
Fluorescence, 72–74, 77–78
  action potential spread and, 212–213
  anisotropy of, 407
  autofluorescence, 332, 335
  calibration of, into membrane potential, 201–204
  energy transfer of, 407–408
  extrinsic changes in, 135–136, 148–151
    stimulation rates of, 151
  intensity autocorrelation function of, 372–373
  intracellular pH monitoring by, 314, 316, 319–320
  light scattering and, 187
  of M540, 102, 111–112
  in membrane potential monitoring, 199–207
    fluorescence/membrane potential calibration in, 201–204
    membrane resistance and, 204–205
    transport processes and, 205–207

in molecular transport studies, 369, 372–373.
    See also Fluorescence correlation spectroscopy (FCS); Fluorescence photobleaching recovery (FPR)
  potential-dependent, 73–74
  prompt, 406
  quantitative analysis of, 39–42
  of quin dye, 356–357, 361
Fluorescence anisotropy, 411
Fluorescence correlation spectroscopy (FCS), 367–380
  applications of, 379–380
  experimental implementation of, 373–374
  FPR compared with, 369, 374–375
  multiple binding reactions and, 378–379
  reversible bimolecular reaction and, 377–378
  systematic transport and, 375–377
  theory of, 369–373
Fluorescence microscopy, 335–338
Fluorescence photobleaching recovery (FPR), 367–380
  applications of, 379–380
  experimental implementation of, 373–374
  FPR compared with, 369, 374–375
  multiple binding reactions and, 378–379
  reversible bimolecular reaction and, 377–378
  systematic transport and, 375–377
  theory of, 369–373
Fluorescent dyes, 32–33, 37–38, 44
  bapta, 349, 350
    loading of, 353–354
    physiological effects of, 357
    toxic effects of, 355–356
  as calcium indicators, 327–343, 347–362
    absorption indicators compared with, 330–331
    caged calcium and, 343
    calcium-sensitive electrodes compared with, 328–330
    chemiluminescence compared with, 331–332
    flow cytometry and, 339–343
    fura-2, 332–338
    indo-1, 332–335
    limitations of, 332
    loading of, 350–355
    nuclear magnetic resonance compared with, 330
    photometric studies of, 359–362
    physiological effects of, 356–359
    toxic effects of, 355–356
  fluorescein, see Fluorescein, as pH indicator

Fluorescent dyes (*Continued*)
  fura-2, 332–338
    fluorescence microscopy and, 335–338
    in hippocampal slices, 176–177
  indo-1, 332–335
  kinetics of, 233
  lipophilic, 201
  lucifer yellow, 33, 49, 65–68
  in membrane potential monitoring, 200–201, 207
  procion yellow, 32–33
  quin, 349–352
    Quin 2, 233, 329–330, 333–334
  red-fluorescing, 65–68
  signal-to-noise ratio of, 73, 77–83
    ideal case of, 80–82
    noise types and, 82–83
  voltage-sensitive, 175–177
    design and synthesis of, 188–189
    in laser scanning, 222–225
    in membrane potential monitoring, 212–213
  volume regulation and, 207
  yellow-fluorescing:
    lucifer yellow, 33, 49, 65–68
    procion yellow, 32–33
Fluorinated chelators, 330
Fluorophores, 331
Focus, depth of, of microscope lenses, 89–91
Force fields, 368
FPR, *see* Fluorescence photobleaching recovery (FPR)
Frame buffer, FB-512, 18–22
"Frame grabbing" of images, 40
Frequency-doubled ruby lasers, 388, 401–403
Frogs:
  atrial fibers of, 427–431
    nifedipine and, 433–436
  neurohypophysis of, 137–140
  optic tectum of, 182–184
  semitendinosus muscle of, 399
  sinus venosus fibers of, 431–432
  ventricles of, 219–220
Fura-2 dye, 332–338
  fluorescence microscopy and, 335–338

Ganglia neurons (invertebrates), 115–129
  buccal, from *Navanax*, 121–129
    dye tests on, 126
    during feeding, 125–129
  detectors for monitoring activity in, 118–119
  dyes for monitoring activity in, 117–118
  future of activity monitoring in, 130

  spatial resolution and activity monitoring in, 119–121
  supraesophageal, 121, 122
Gigaohm seal recording, 448–451, 460
Glutathione, 388–389
GMP, cyclic, *see* cGMP (cyclic guanine monophosphate)
Golgi technique, 32
Growth cone motility during microscopy, 36–38

Halophilic bacteria, 353
Heart:
  action potential spread in, 211–225
    activation maps for monitoring, 216–219
    frog ventricle, 219–220
    laser scanner for, 215–219, 222–225
    mammalian heart, 220–222
  atrial fibers of, 427–431
    nifedipine and, 433–436
  bullfrog, 427–432
  extrastimulus in, 221–222
  frog ventricle, 219–220
  rabbit, 220–222
  rat, 436–438
  sinus venosus fibers of, 431–432
  ventricular myocytes of, 436–438
Heavy water (deuterium oxide), 155–157
Helical xenon flash lamp, 401
Helisoma, 35, 36, 40–43
Hepatocytes, 312
HEPES, 295–297, 306
High extinction polarized light microscopy, 4
High-voltage electrical fields for photoprotein introduction into cells, 236
Hippocampus:
  fluorescence recording in, 176–177
  unmyelinated axons in, 172–173
Hormones:
  polypeptide, 368–369
  release of, 148
  response of single cells to, 52
Hydrogen, release of, from rhodopsin, 308
Hydrophobic dyes, 112
Hypertonic media, extracellular volume and, 157
Hypo-osmotic shock, 236
Hypothalamo-neurohypophyseal system, 137
Hypothalamus, 137, 138

Image enhancement techniques:
  computers in, 8
    hardware for, 18–22
    software for, 22–23
    see also Computer-aided light microscopy

contrast enhancement, 6, 16, 23
electronic, 4
    analogue, 4, 6, 18–22
    digital, *see* Digital image processing
    dyes for, *see* Dye(s)
    equipment for, 17. *See also specific
        instruments*
    microchannel plate, 47
    photographic, 4
    polarized light microscopy, 4
    spatial enhancement, 16
        in computer-aided light microscopy, 23,
        25–26
    three-dimensional, 53, 57–61
    see also Computer-aided light microscopy;
        Video-enhanced microscopy
Image processing circuit boards, IP-512, 18–22
Image recording:
    gigaohm seal, 448–451, 460
    intracellular electrical, 189
    limitations of, 186–188
    in membrane potential monitoring, 86–89
    multiple-site, 138–140. *See also* Multisite
        optical monitoring
    optical *vs.* electrical, 134–137, 168, 175–177,
        192
    in real-time optical mapping, 175–177
    single-channel, 455–457
    whole-cell, 436–438
    See also Optical monitoring
Indo-1 dye, 332–335
Infundibulum, 138
Interactive analysis of images, 60–61
Interneurons, wind-sensitive, 48–50
Intersystem crossing of electrons, 407
Intracellular electrical recordings, 189
Intracellular injection of dyes:
    aequorin, 245–249
    iontophoretic, 175–176
    phenol red, 247–249, 287–288
    red-fluorescing, 65–68
Intracellular messengers, *see* Photolabile
        intracellular messengers
Intracellular pH monitoring:
    absorption in, 314–317
        dark adaptation of photoreceptor cells and,
        300–301, 305
        after intracellular injection into
            photoreceptor cells, 292–293
        light-induced change in photoreceptor cells
            and, 298–300, 306
        in presence of rhabdoms, 290–291,
            297–298

centrifugation in, 320
electrophysiological responses of
    photoreceptor cells and, 293–295
fluorescence in, 314, 316, 319–320
ionophores and, 316–317, 323–324
light-induced change in photoreceptor cells
    and, 306–309
light scattering in, 316
optical recording from photoreceptor cells
    and, 288–290
pH buffers and, 295–297, 306
preparation for, with phenol red dye, 287–288
protein error and, 295–297, 305
real-time, 324
spatial properties of pH in cells and, 301–305
see also pH indicators
Intracellular population, analysis of, in real-time
    optical mapping, 190
Invertebrate ganglia neurons, 115–129
    buccal, from *Navanax,* 121–129
        dye tests on, 126
        during feeding, 125–129
    detectors for monitoring activity in, 118–119
    dyes for monitoring activity in, 117–118
    future of activity monitoring in, 130
    spatial resolution in, 119–121
    supraesophageal, 121, 122
Inward calcium current, 143–148
    slow, 427–431
        calcium antagonists and, 433–436
Inward sodium current, 146
Ionophores, intracellular pH and, 316–317,
    323–324
Ions:
    action potential and, 143–148, 151–155
    lipid-soluble synthetic, 201
    permeant, 200–201
    transport processes of, 205–207
Iontophoretic ejection of dyes, 42–43
Iontophoretic injection of dyes, 175–176
IP-512 image processing circuit boards, 18–22
Ischemia, 222
Isochrones, 219
Isomers, 451, 455
Iterative image restoration, 57–59

Kidney medulla, 392
Kinetics:
    of Azo 1 dye, 270, 278
    of caged ATP, 403–405
    of cAMP and cGMP jumps, 430–432
    of chelators, 233
    of fluorescent dyes, 233

Kinetics (*Continued*)
  of metallochromic dyes, 233–234
  of photoproteins, 234–235
Köhler illumination, 5

Lamps:
  arc, 402–403
  flash:
    in caged ATP release, 388
    xenon, 401, 424, 426–427, 433–435
    xenon short-arc, 448–451
  in muscle contraction studies, 401–403
  for optical monitoring of membrane potential,
    84–86
Lasers:
  in action potential monitoring, 215–225
    acousto-optical deflectors in, 215, 222, 225
    in frog ventricle, 219
    in mammalian heart, 220–222
  in ATP release, 397–405
    caged ATP and, 388, 403–405
    illumination sources for, 401–403
    photolysis and, 399–401
  excimer, 402
  frequency-doubled ruby, 388, 401–403
  in membrane potential monitoring, 86
  neodymium : YAG, 402
  nitrogen, 402
  in zap axotomy, 47, 49
Learning, neurons and, 116
Lethocerus, flight muscle of, 399
Ligands, external, 368–369
Light absorption, *see* Absorption, light
Light adaptation of photoreceptor cells,
    298–300, 306, 308–309
Light noise, 259
Light scattering, 90, 91
  absorption and, 187
  fluorescence and, 187
  in intracellular pH monitoring, 316
  neuron measurement and, 122, 124
  in photoreceptor cells, 304
  in real-time optical mapping, 171–172,
    187
  secretion and, 150–160
    deuterium oxide and, 155–157
    extracellular calcium and, 141, 151–155
    hypertonic media and, 157
    large intrinsic action potential and,
      157–160
  spatial resolution and, 191–192
Light sources for optical monitoring of
    membrane potential, 84–86

Limulus ventral photoreceptor cells, 350,
    353–355
  light scattering in, 304
  pH measurement of, 285–309
    dark adaptation and, 300–301, 305
    electrophysiological responses and,
      293–295
    intracellular injection and, 292–293
    light-induced change in, 298–300, 306,
      308–309
    optical recording from photoreceptors and,
      288–290
    pH buffers and, 295–297
    preparation for, with phenol red dye,
      287–288
    rhabdoms and, 290–291, 297–298
    spatial inhomogeneities in, 301–305
Lipid-soluble synthetic ions, 201
Lipophilic fluorescent dyes, 201
Liposomes, photoprotein introduction into cells,
    236
Low-light microscopy, 33, 36–38
Lucifer yellow dye, 33, 49, 65–68
Lymphocytes, flow cytometry of, 339, 342–343

Magnesium:
  Azo 1 dye and, 266–267
  preequilibration of photoproteins with,
    233–235
Magnocellular neurons, 137, 140
Membrane-bound dye, 102
Membrane potential:
  calcium entry and, 250–251
  electrical monitoring of, 134–137, 168,
    175–177, 192
  optical monitoring of, 71–96, 134–137,
    199–207
    accuracy of measurement in, 80–82
    amplifiers in, 92–93
    computers in, 93–95
    dyes for, 72–76, 79–80, 200–201, 297. *See
      also* Merocyanine 540 (M540) in optical
      monitoring of membrane potential
    electrogenic sodium pump in, 204–205
    extraneous noise in, 83
    fluorescence calibration and, 201–204
    future of, 95–96
    image-recording devices in, 86–89
    light sources for, 84–86
    M540 in, *see* Merocyanine 540 (M540) in
      optical monitoring of membrane potential
    microscopes for, 89–92
    noise in, 83

photodetectors in, 82–83
potential-dependent optical signals and, 73–77
real-time optical mapping, 175, 190
signal choice in, 77–78
spatial and temporal resolution and, 134, 135
transport processes and, 205–207
Membrane resistance in red blood cells and Ehrlich ascites tumor cells, 204–205
Memory, neurons and, 116
Mercury arc lamps, 84–86, 402–403
Mercury-xenon arc lamps, 84–86
Merocyanine 540 (M540) in optical monitoring of membrane potential, 101–112
absorption in, 102–103, 107–108, 111–112
difference spectra and, 105–107
fluorescence in, 102, 111–112
materials and methods for, 103–104, 111–112
signal size and, 107–111
Merocyanine dyes, 135–136
M540, see Merocyanine 540 (M540) in optical monitoring of membrane potential
merocyanine-oxazalone, 135, 154
merocyanine-rhodanine, 135, 138, 154
signal-to-noise ratio of, 212
WW375, 112
Messengers, intracellular, see Photolabile intracellular messengers
Metallochromic dyes, 231
antipyrylazo III, 233–234, 262
arsenazo III, 234, 262
dichlorophosphonazo III, 263
kinetics of, 233–234
Methane-diol-diacetate, 355–356
M540, see Merocyanine 540 (M540) in optical monitoring of membrane potential
Mice:
pancreatic acini of, 351–352
spleen cells of, 340–342
Microchannel plate image intensifier, 47
Microcomputers:
in digital imaging microscopy, 54, 55
in video microscopy of living neurons, 34–35, 39–40
Microelectrodes, 32
calcium-sensitive, 233, 328–330
extracellular applications of, 330
dye ejection from, 42–43
Microirradiation for zap axotomy, 43–44
Microscopes:
confocal, 90, 92
depth of focus of, 89–91

digital, 51–62
energy maxima in, 59–60
filters in, 59–60
image display in, 60–61
image restoration in, 55–59
image simplification in, 59–60
noise reduction in, 55
spatial inhomogeneity correction in, 55
flash lamps for, 426
light scattering in, 90, 91
for optical monitoring of membrane potential, 89–92
for photoisomerization studies, 450, 451
for real-time optical mapping, 185–186
spatial resolution in, 191
timers for, 41
for video-enhanced microscopy, 5
Microscopy:
AVEC methods of, 9
analogue image processing and, 6
computer-aided light, 15–29
contrast enhancement in, 23
digital image processing by, 18
hardware for, 18–22
images obtained by, 24–28
pixel noise cancellation in, 26
software for, 22–23
fluorescence, calcium indicators and, 335–338
growth cone motility during, 36–38
low-light, 33, 36–38
polarized light, 4
quantitative, 11
video-enhanced, see Video-enhanced microscopy
Zeiss Axiomat, 4, 5
Microsomal vesicles, 392
Microspectrophotometer, scanning, 289–290
Microtubules, 7, 9–10
Microviscosity, 332
Minicomputers:
in action potential monitoring, 215–216
in membrane potential monitoring, 93–94
Mirror systems, ellipsoidal, 231–232
Mitochondria:
calcium and, 248–249
pH in, 317–324
calibration of, 321–323
Monomers, 102, 111–112
Morphometry in action potential monitoring, 140
Mottle, digital subtraction of, 7

Movement in optical recording, 83
  calcium indicators and, 230–232
  in polychromators, 256
  in video-enhanced microscopy, 8–9
Movie film in optical monitoring of membrane
    potential, 87
MSORTV (Multiple site optical recording of
    transmembrane voltage), 138–140
Multisite optical monitoring, 71–73, 80–96
  accuracy of, 80–82
  amplifiers in, 92–93
  computers for, 93–95
  extraneous noise in, 83
  future of, 95–96
  image-recording devices for, 86–89
  light sources for, 84–86
  microscopes for, 89–92
  photodetectors for, 82–83
Muscle contraction, 397–412
  ADP in, 398–400
  applications of studies of, 408–411
  ATP release in, 397–405
    caged ATP and, 388, 399, 403–405
    laser pulse photolysis for, 399–403
    pH and, 404–405
  cross-bridge cycle in, 398–400
    photolysis and, 403–405, 410–412
  optical rulers and protractors for measuring,
    405–408
Myoballs, 451, 455
Myosin, 398, 411

NA (numerical aperture) of lenses, 89
Navanax inermis, buccal ganglion neurons from,
    121–129
  dye tests on, 126
  during feeding, 125–129
Neodymium : YAG laser, 402
Neomycin, 159
Nerve terminals, excitation and secretion at,
    133–160
  action potential:
    in depolarizing phase of, 143–148
    in optical recording of, 137–140
    in properties of, 140–143
  deuterium oxide and, 155–157
  extracellular calcium and, 141, 151–155
  hypertonic media and, 157
  optical changes accompanying, 148–151,
    157–160
  temporal dispersion of excitation and,
    154–155
Neural lobes, vasopressin release from, 159

Neurohypophyseal peptides, 137
Neurohypophysis:
  definition of, 138
  in excitation-secretion coupling, 137–160
    action potential:
      and depolarizing phase of, 143–148
      and optical recording of, 137–140
      and properties of, 140–143
    deuterium oxide and, 155–157
    extracellular calcium and, 141, 151–155
    hypertonic media and, 157
    optical changes accompanying, 148–151,
      157–160
  of *Xenopus,* 137–140
Neurons:
  chick dorsal root, 9
  ganglia, *see* Ganglia neurons (invertebrates)
  interneurons, wind-sensitive, 48–50
  learning and memory and, 116
  magnocellular, 137, 140
  regeneration of, 43
Neuropeptides, 135, 158–159
Neurophysins, 137
Neuroreceptors and neurotransmitters, response
    of single cells to, 52
Nicotinic acetylcholine receptors:
  blocking of, 459–460
  photoisomerization at, 447–460
    of Bis-Q, 451–453, 458
    of bound agonist molecules, 457–459
    of open-channel blocker EW-1, 459–460
    optical system for studying, 448–451
    photochemistry of, 451
    of QBr, 451, 453–455, 458
    single-channel recordings of, 455–457
Nifedipine, 432–438
  slow inward calcium current and, 433–436
  whole-cell recording techniques and, 436–438
Nitr-2 dye, 343
Nitroaromatics, 390
o-Nitrobenzyl derivatives, 419–424
Nitrogen lasers, 402
Noise:
  amplifier, 92–93
  averaging for reduction of, 7, 8, 55
  dark, 82, 92–93, 259
  excess, 82–83
  extraneous, 83
  light, 259
  in membrane potential measurements, 135
  movement, 83
  1/f, 82–83
  photon, 259

pixel, 7, 8, 26
shot, 80–83, 259
 limited (ideal) case of, 80–82
time-variant, 55
vibrational, 83
 see also Signal-to-noise ratio
Nuclear magnetic resonance, 233, 330
Null titration:
 double-null point, 323–324
 for fluorescence calibration, 201–204
 for intracellular pH calibration, 316
Numerical aperture (NA) of lenses, 89

Obelia, 230
Obelin, 230, 232–236
1/f noise, 82–83
Optical monitoring:
 multisite, 71–73, 80–96
  accuracy of, 80–82
  amplifiers in, 92–93
  computers for, 93–95
  extraneous noise in, 83
  future of, 95–96
  image-recording devices for, 86–89
  light sources for, 84–86
  microscopes for, 89–92
  photodetectors for, 82–83
 real-time, *see* Real-time monitoring
 simultaneous, of many neurons, 115–129
  in behaving animals, 124–129
  in buccal ganglion neurons, 121–124
  detector configuration in, 118–119
  dyes in, 117–118
  future of, 130
  spatial resolution in, 119–121
  in supraesophageal ganglion neurons, 121,
   122
 see also Image recording; *specific techniques*
Optical protractors and rulers in muscle
  contraction measurement, 405–408
Optical signals, 77–78
 absorption, *see* Absorption, light
 birefringence, *see* Birefringence
 calcium-independent luminescence, 233
 compound, 154
 dichroic, 274–277
 extrinsic (dye-dependent) changes in,
  135–136, 148–151
  stimulation rates of, 151
 fast, 73, 172
 fluorescence, *see* Fluorescence
 intrinsic changes in, 148–151
  excitation and secretion and, 157–160

potential-dependent, 73–77
size of, 79, 130
 calcium indicators and, 232–233
 in optical monitoring of membrane
  potential, 107–111
 real-time optical mapping and, 187
spike width of, 140–141, 144
time courses of, 74–75
undershoot, 141
wavelength dependence of, 150, 160
see also Signal-to-noise ratio
Optic tectum, 182–184
Oxidized cholesterol membranes, M540 in,
  102–112
 difference spectra and, 105–107
 dye structure and, 107–111
 optical studies of, materials and methods for,
  103–104, 111–112
 signal size and, 107–111
Oxonol dyes, 112
Oxytocin, 135

Pancreatic acinar cells, 350–352
Parallel recording in membrane potential
  monitoring, 87–89
Pars intermedia, 140
Pars nervosa, 138, 156–157
Parthenogenetic activation, 338
Patch-clamp experiments, 436–438, 450–451
Pattern analysis of single cells, 62
Peptide hormones, 137
pH:
 ATP release and, 404–405
 buffering of, 295–297, 305–306
 cytoplasmic, 317–321, 324
 jumps of, 422–423
 mitochondrial, 317–324
  calibration of, 321–323
 see also Intracellular pH monitoring; pH
  indicators
Pharmacological effects of dyes, 79–80,
  117–120, 187
Phase modulation video-enhanced microscopy,
  10–11
pH buffers, 295–297, 305–306
Phenol red dye, 247–249
 absorption of, 288
 as pH indicator in photoreceptor cells,
  285–309
 absorption of:
  after intracellular injection, 292–293
  rhabdoms and, 290–291, 297–298
 dark adaptations of cells and, 300–301, 305

Phenol red dye (*Continued*)
  electrophysiological responses of cells and,
    293–295
  light-induced change in cells and, 298–300,
    306, 308–309
  optical recording from, 288–290
  pH buffers and, 295–297, 305–306
  preparation for, 287–288
  spatial inhomogeneities of pH in cells and,
    301–305
pH indicators:
  intracellularly trapped, 311–324
    calibration of, 316–317
    compartment specificity of, 317–321
    drawbacks of, 324
    ionophore measurement by, 323–324
    loading of, 313–314
    mitochondrial pH and, 321–323
    monitoring of, 314–316
  phenol red dye, *see* Phenol red dye
pH jumps, 422–423
Phosphorescence, 407
Phosphorescence anisotropy, 407, 411
Photoaffinity labeling, 390
Photobleaching, 33, 44, 80, 117–118
  in molecular transport studies, 374–375. *See
    also* Fluorescence photobleaching
    recovery (FPR)
  in real-time optical mapping and, 185,
    188
Photobleaching recovery, *see* Fluorescence
    photobleaching recovery (FPR)
Photocathodes, 82, 88–89
Photocytes, 230
Photodetectors in membrane potential
    monitoring, 82–83
Photodiodes:
  matrix array of, 138–140
  silicon, 82–83
Photodynamic damage by dyes, 80, 117–118
  in real-time optical mapping, 185, 188
Photographic image enhancement, 4
Photoinactivation, 43–50
  dye in, 43
  of single identified dendrites in situ, 44–50
    apparatus for, 44–48
    in wind-sensitive interneurons, 48–50
Photoisomerization, 447–460
  of Bis-Q, 451–453, 458
  of bound agonist molecules, 457–459
  of open-channel blocker EW-1, 459–460
  optical system for studying, 448–451
  photochemistry of, 451
  of QBr, 451, 453–455, 458
  single-channel recordings of, 455–457

Photolabile intracellular messengers, 417–439
  calcium antagonists, 432–439
    future development of, 433–436
    whole-cell recording techniques and,
      436–438
  cAMP jumps and, 427–432
    kinetic implications of, 430–432
    in sinus venosus fibers, 431–432
    tension development and, 429, 431
  flash photolysis experiments on, 420–424
  future development of, 438–439
  o-nitrobenzyl derivatives, 419–424
  optical methods for physiological experiments
    with, 424–427
  reaction mechanisms of, 419–424
  slow inward calcium current and, 427–431
Photolysis:
  of caged ATP, 388, 403–405
  flash, 420–424
  in FPR, 371, 374–375
Photometers:
  single wavelength, 289
  spatial scanning, 290, 301–305
  spectrophotometers, 314–316
    scanning, 289–290
  in tetracarboxylate dye studies, 359–362
Photomultipliers, 231–232
Photon noise, 259
Photopigment, 350
Photoprotection, 419–420
Photoproteins, 230
  as biological calcium indicators, 229–236
    introduction of, into cells, 235–236
    kinetics of, 234–235
    motion artifacts and, 230–232
    signal strength and, 232–233
    speed of response and, 233–235
  calcium-regulated, 230
    aequorin, *see* Aequorin
    obelin, 230, 232–236
  fluorophores compared with, 331
Photoreceptors:
  cells of, pH measurement of, 285–309
    dark adaptation and, 300–301, 305
    electrophysiological responses and,
      293–295
    after intracellular injection with phenol red
      dye, 292–293
    light-induced change in, 298–300, 306,
      308–309
    light scattering in, 304
    optical recording in, 288–290
    pH buffers and, 295–297, 305–306
    preparation for, with phenol red dye,
      287–288

rhabdoms and, 290–291, 297–298
spatial inhomogeneities of pH in cells and, 301–305
rod, 350, 352–353
ventral eye, 350
Photorelease, 420–424
Photoremoval, 419–424
Physiological effects of bapta and quin dyes, 356–359
Pituitary, posterior, *see* Neurohypophysis
Pixel histogram stretching, 8
Pixel noise, 7, 8, 26
Point spread functions, three-dimensional, 57–59
Polarization rectifiers, 4
Polarized light microscopy, 4
Polychromators, 255–260
amplification in, 258–259
optical arrangement of, 256–258
resolution limits of, 259–260
signal-to-noise ratio in, 259–260
Polypeptide hormones, 368–369
Population activity, spatiotemporal patterns of, in mammalian brain slices, 169–177
fluorescent voltage-sensitive probes and, 175–177
light scattering and, 171–172
pre- and postsynaptic components of, 172
slice preparation and, 169–170
spread of activity, visualization of, 173–175
unmyelinated axons and, 172–173
voltage-sensitive dyes and, 167–170
Posterior pituitary, *see* Neurohypophysis
Postsynaptic activity in real-time optical mapping, 172
Potassium current:
calcium-dependent, 206–207
calcium-mediated, 144
cAMP jumps and, 431
Potential dependent optical signals, 73–77
Presynaptic action potential in real-time optical mapping, 172
Procambarus, rhabdoms of, 290–291, 297–298
Procion yellow dye, 32–33
Prompt fluorescence, 406
Propagation wave, 222
Protein error, intracellular pH and, 295–297, 305
Proteins:
actin, 61
alpha-actinin, 53, 61
calcium-regulated, 230
cell surface, 379–380
contractile, 53, 398, 411–412
distribution of, in single cells, 52

neurophysins, 137
photoproteins, *see* Photoproteins
Proton binding to Azo 1 dye, 267–268
Protractors, optical, 405–408
Pseudo-color conversion in video-enhanced microscopy, 8–9
Psoas muscle, 399
Pulsed arc lamps, 403

QBr, 451, 453–455, 458
Quantitative analysis of fluorescence, 39–42
Quantitative microscopy, 11
Quin dye, 349–352
as calcium buffer, 356–359
fluorescence of, 356–357, 361
loading of, 351–352, 354–355
photometric studies of, 359–362
physiological effects of, 356–359
Quin 2, 233, 329–330, 333–334
toxic effects of, 355–356

Rabbits:
heart of, 220–222
psoas muscle of, 399
Radechon photocathodes, 89
Radioimmunoassays, 148–151, 159
Rana temporaria muscle, calcium transients in, 268–270
Rapid scanning microspectrophotometer, 289–290
Rats:
cortex of, 178–182
ventricular myocytes of, 436–438
Real-time monitoring, 165–193
of brain:
apparatus for, 185–186
computers and, 168
difficulties with, 189–190
fluorescent voltage-sensitive probes and, 175–177
light scattering and, 171–172
limitations of, 186–188
slice preparation and, 169–170
spatial resolution in, 191–192
spread of activity and, 173–175
unmyelinated axons and, 172–173
in vitro, 177
in vivo, procedures for, 177–178
of *in vivo* frog optical tectum, 182–184
of *in vivo* mammalian cortex, 178–182
voltage-sensitive dyes and, 167–170, 188–189
of membrane potential, 175, 190
of pH in cells, 324

Recording, *see* Image recording
Red blood cells:
  Amphiuma, 200
  ATP and, 207
  flow cytometry of, 339, 342–343
  optical monitoring of, 199–207
    fluorescence/membrane potential calibration
      in, 201–204
    membrane resistance and, 204–205
    transport processes and, 205–207
Red-fluorescing dyes, 65–68
Redistribution dyes, 201
Reentrant impulse, 222
Refractory period in real-time optical mapping,
  173
Regeneration of neurons, 43
Reticulopodial movement, 9
Rhabdoms, 290–291, 297–298
Rhodamines, 65–68
Rhodopsin, 297–298, 308
Ringer's solution, 157
Rising phase of action potential, 143–148
Rod photoreceptors, 350, 352–353
Rotating display of images, 60–61
Rotenone, 318
Ruby lasers, frequency-doubled, 388, 401–403
Rulers, optical, 405–408

Sarcoplasmic reticulum (SR) membrane,
  393–394
Saturation binding of dyes, 109–110
Scallops, adductor muscle of, 399
Scanning spectrophotometer, 289–290
Scrape loading, 236
Screening, for signal size in dyes, 79
Sea urchin embryos, 336–338
Secretion, extracellular calcium and, 141,
  151–155
Secretion at nerve terminals, 133–160
  action potential in:
    depolarizing phase of, 143–148
    optical recording of, 137–140
    properties of, 140–143
  neuropeptide, 158–159
  optical changes accompanying, 148–151,
    157–160
  see also Excitation-secretion coupling
Selective axotomy, *see* Zap axotomy
Semitendinosus muscle, 399
Sensory interneurons, 48–50
Sequential subtraction in video-enhanced
  microscopy, 8–9
Serial recording in membrane potential
  monitoring, 87–89

Shot noise, 80–83, 259
Signal-to-noise ratio, 73, 77–79
  action potential and, 213
  of Azo 1 dye, 280
  ideal case of, 80–82
  of merocyanine dyes, 212
  noise types and, 82–83
  in optical monitoring of action potential,
    140
  in polychromators, 259–260
Silicon intensifier target cameras, *see* SIT
  (silicon intensifier target) cameras
Silicon photodiodes, 82–83
Simultaneous optical monitoring of many
  neurons, 115–129
  in behaving animals, 124–129
  in buccal ganglion neurons, 121–124
  detector configuration in, 118–119
  dyes in, 117–118
  future of, 130
  spatial resolution in, 119–121
  in supraesophageal ganglion neurons, 121,
    122
Single-channel recordings, 455–457
Single wavelength photometer, 289
Sinus venosus fibers, 431–432
SIT (silicon intensifier target) cameras, 34–35,
  43–44
  in digital imaging microscopes, 54, 55,
    361–362
6-carboxyfluorescein dye, 49
Skeletal muscle, 393
Slow dyes, 201
Slow inward calcium current, 427–431
  calcium antagonists and, 433–436
Smooth muscle cells, 52–53, 61
Sodium:
  cotransport of, with amino acids, 205–206
  exchange of, with calcium, 245
Sodium current, 146
Sodium efflux, uncoupled, 204–205
Sodium pump, 204–205, 390–392
Solid model image display, 61
Space-clamp experiments, 74
Sparrow criterion in video-enhanced microscopy,
  8
Spatial enhancement, 16, 23, 25–26
Spatial inhomogeneities, 55
  of pH, 301–305
Spatial resolution:
  light scattering and, 191–192
  membrane potential and, 134, 135
  in microscopes, 191
  neuronal activity and, 119–121, 191–192

in polychromators, 260
in real-time optical mapping, 191–192
Spatial scanning photometer, 290, 301–305
Spatiotemporal patterns of population activity in
brain slices, 169–177
fluorescent voltage-sensitive probes and,
175–177
light scattering and, 171–172
pre- and postsynaptic components of, 172
slice preparation and, 169–170
spread of activity, visualization of, 173–175
unmyelinated axons and, 172–173
voltage-sensitive dyes and, 167–170
Spectrofluorometers, 329
Spectrophotometers, 314–316
scanning, 289–290
Speed of response:
of Azo 1 dye, 279–280
of chelators, 233
of fluorescent dyes, 233
of metallochromic dyes, 233–234
of photoproteins, 234–235
Spleen cells, mouse, 340–342
Squid axons, 7, 9–10
axoplasm of, extruded, 7, 10
fluorescence and absorption of dyes in, 102
in measurement of membrane potential,
72–76, 79, 135
Squid synapses, in action potential
measurement, 137
SR vesicles, 393–394
Stains, see Dye(s)
Stereopairs in image display, 60–61
Stimulation rates of optical signal changes, 151
Stray light in video-enhanced microscopy, 7
Styryl dyes, 135–136
Sucrose, extracellular volume and, 158
Supraesophageal ganglion neurons, 121, 122
Synaptic transmission, action potential and, 137
Synaptic vesicles, 7

Television cameras:
in calcium measurement, 336–338
in real-time optical mapping, 186
for video-enhanced microscopy, 5
SIT (silicon intensifier target), 34–35,
43–44, 54, 55
Temporal dispersion, 140–141, 154–155
Temporal resolution, 134, 135, 260
Tethered agonists, 453–455
Tetracarboxylate dyes, 327–343, 347–362
absorbance indicators compared with,
330–331
Azo 1, see Azo 1 dye

caged calcium and, 343
calcium-sensitive electrodes compared with,
328–330
chemiluminescence compared with, 331–332
definition of, 262
flow cytometry and, 339–343
fura-2, 332–338
fluorescence microscopy and, 335–338
indo-1, 332–335
limitations of, 332
loading of, 350–355
nitr-2, 343
nuclear magnetic resonance compared with,
330
photometric studies of, 359–362
physiological effects of, 356–359
toxic effects of, 355–356
Tetraethylammonium in real-time optical
mapping, 173
Three-dimensional images, 52, 57–61
Thymocytes, 335–336
Time courses of optical signals, 74–75
Timers for microscopes, 41
Time-variant noise, 55
Tissue invasion, 154–155
Titration, null
double-null point, 323–324
for fluorescence calibration, 201–204
for intracellular pH calibration, 316
Toads, rod photoreceptors of, 350, 352–353
Toxicity of fluorescent calcium indicators,
355–356
Trans isomers, 451, 455
Transmembrane voltage, multiple site optical
recording of (MSORTV), 138–140
Transmitter release, aminoglycoside antibiotics
and, 159
Tungsten filament lamps, 84–86

Uncoupled sodium efflux, 204–205
Undershoot, after-hyperpolarization and, 141
Unmyelinated axons in hippocampus, 172–173

Vacuum photocathodes, 82, 88–89
Valinomycin, 202, 318
Vasopressin, 135, 159
Ventral eye photoreceptors, 350
Ventricular myocytes, 436–438
Vesicles:
effects of dyes in, 102
microsomal, 392
SR, 393–394
synaptic, 7
Vibrational noise, 83

Video-enhanced microscopy, 3–12
  applications of, 9–10
  in calcium measurements:
    extracellular, 141, 151–155
    intracellular, 336–338, 361–362
  cameras for, 5
    in real-time optical mapping, 186
    SIT (silicon intensifier target), 34–35,
      43–44, 54, 55, 361–362
  computer-video interfaces in, 39–42, 186
    in real-time optical mapping, 186
  future prospects for, 10–12
  Köhler illumination in, 5
  of living neurons, 34–43
    computers for, 34–35, 39–42
    dye ejection from microelectrodes and,
      42–43
    observation of morphological changes by,
      35–38
    quantitative analysis of fluorescent images
      by, 39–42
    video instrumentation for, 34–35
  microscopes for, 5
  motion detection and obliteration in, 8–9
  mottle, digital subtraction of, in, 7
  phase modulation, 10–11
  pixel histogram stretching in, 8
  pseudo-color conversion in, 8–9
  Sparrow criterion in, 8
  stray light in, 7
  subtraction in, 8–9
  television cameras for, 5
  in UV region of spectrum, 10
Vidicons, 5, 186
Viscosity, 332
Voltage-clamp experiments, 74–76, 135
  aequorin in, 245–247
  on atrial fibers, 433–435

Voltage-sensitive calcium influx, 148
Voltage-sensitive dyes, 135–136
  fluorescent, 175–177
    design and synthesis of, 188–189
    in laser scanning, 222–225
  in optical monitoring of membrane potential,
    212–213
  in real-time optical mapping, 167–170,
    175–177
Volume regulation, fluorescent dyes and, 207

Wavelength dependence, of optical signals, 150,
  160
Whisker barrels, in mammalian cortex, 180–181
Whole-cell recording techniques, 436–438
Wind-sensitive interneurons, 48–50
WW375 dye, 112
WW781 dye, 212

Xenon arc lamps, 403
Xenon flash lamps, 401, 424, 426–427,
  433–435
  short-arc, 448–451
Xenopus laevis, neurohypophysis of, 137–140

Yellow-fluorescing dyes:
  lucifer yellow, 33, 49, 65–68
  procion yellow, 32–33

Zap axotomy, 43–50
  apparatus for, 44–48
  dyes in, 43–44, 46
  epifluorescence in, 45–47
  lasers in, 47, 49
  microirradiation for, 43–44
  in wind-sensitive interneurons, 48–50
Zeiss Axiomat microscope, 4, 5